D0946991

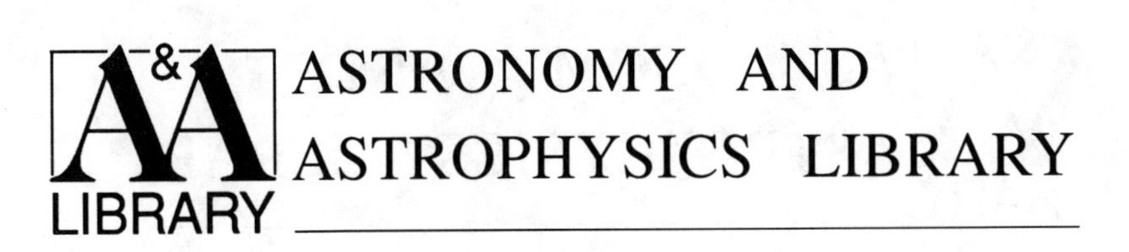
A&A
LIBRARY
ASTRONOMY AND
ASTROPHYSICS LIBRARY

ASTRONOMY AND ASTROPHYSICS LIBRARY

Tools of Radio Astronomy
By K. Rohlfs

Physics of the Galaxy and Interstellar Matter
By H. Scheffler and H. Elsässer

Galactic and Extragalactic Radio Astronomy 2nd Edition
Editors: G.L. Verschuur and K.I. Kellermann

Observational Astrophysics
By P. Léna

Astrophysical Concepts 2nd Edition
By M. Harwit

The Sun An Introduction
By M. Stix

Stellar Structure and Evolution
By R. Kippenhahn and A. Weigert

Relativity in Astrometry, Celestial Mechanics and Geodesy
By M.H. Soffel

The Solar System
By T. Encrenaz and J.-P. Bibring

Physics and Chemistry of Comets
Editor: W.F. Huebner

Supernovae
Editors: A.G. Petschek

Astrophysics of Neutron Stars
By V.M. Lipunov

Gravitational Lenses
By P. Schneider, J. Ehlers, and E.E. Falco

General Relativity, Astrophysics, and Cosmology
By A.K. Raychaudhuri, S. Banerji, and A. Banerjee

Stellar Interiors: Physical Principles, Structure, and Evolution
By C.J. Hansen and S.D. Kawaler

C.J. Hansen
S.D. Kawaler

Stellar Interiors

Physical Principles, Structure, and Evolution

With 84 Figures

Springer-Verlag
New York Berlin Heidelberg London Paris
Tokyo Hong Kong Barcelona Budapest

Carl J. Hansen
Joint Institute for Laboratory Astrophysics
University of Colorado
Boulder, CO 80309
USA

Steven D. Kawaler
Department of Physics and Astronomy
Iowa State University
Ames, IA 50011
USA

Library of Congress Cataloging-in-Publication Data
Hansen, Carl J.
Stellar interiors: physical principles, structure, and evolution / Carl J. Hansen, Steven D. Kawaler. -- 1st ed.
p. cm. -- (Astronomy and astrophysics library)
Includes bibliographical references and index.
ISBN 0-387-94138-X. -- ISBN 3-540-94138-X
1. Stars--Structure. 2. Stars--Evolution. 3. Astrophysics. I. Kawaler, Steven D. II. Title. III. Series.
QB808.H36 1994
523.8--dc20 91-27498
CIP

Cover art: Calligraphy by Sara Compton.

Printed on acid-free paper.

Production managed by Jim Harbison; manufacturing supervised by Vincent Scelta.
Camera-ready copy provided by the authors using Springer-Verlag's LATEX macro svsing.sty.
Printed and bound by Edwards Brothers, Inc., Ann Arbor, MI.
Printed in the United States of America.

9 8 7 6 5 4 3 2 1

ISBN 0-387-94138-X Springer-Verlag New York Berlin Heidelberg
ISBN 3-540-94138-X Springer-Verlag Berlin Heidelberg New York

Preface

That trees should have been cut
down to provide paper for this
book was an ecological affront.
From a book review.
— Anthony Blond (in the *Spectator*, 1983)

The first modern text on our subject, *Structure and Evolution of the Stars*, was published over thirty years ago. In it, Martin Schwarzschild described numerical experiments that successfully reproduced most of the observed properties of the majority of stars seen in the sky. He also set the standard for a lucid description of the physics of stellar interiors. Ten years later, in 1968, John P. Cox's two-volume monograph *Principles of Stellar Structure* appeared, as did the more specialized text *Principles of Stellar Evolution and Nucleosynthesis* by Donald D. Clayton—and what a difference ten years had made. The field had matured into the basic form that it remains today. The past twenty-plus years have seen this branch of astrophysics flourish and develop into a fundamental pillar of modern astrophysics that addresses an enormous variety of phenomena.

In view of this it might seem foolish to offer another text of finite length and expect it to cover any more than a fraction of what should be discussed to make it a thorough and self-contained reference.

Well, it doesn't. Our specific aim is to introduce only the fundamentals of stellar astrophysics. You will find little reference here to black holes, millisecond pulsars, and other "sexy" objects. Nor will you find a lot of sophisticated mathematics or physics. Those sorts of things are for the developing specialist. What you will find is a text designed for our target audience: the typical senior undergraduate or beginning graduate student in astronomy or astrophysics who wishes an overview of stellar structure and evolution with just enough detail to understand the general picture. She or he can go on from there to more specialized texts or directly to the research literature depending on talent and interests. To this end, this text presents the basic physical principles without chasing all the (interesting!) details.

On the other hand, like all authors, we have our own special interests and axes to grind, thus our last chapter is about asteroseismology, which we think is a really neat subject and one that is of growing importance. Also, Chapter 6 on stellar energy sources is rather long. This is because experience has taught us that the average beginning student knows practically nothing of nuclear physics from an astrophysical viewpoint. Our chapter on the use of the mixing length theory of convection (among other topics) might be longer than expected. The reason for this is that texts unfortunately seem to gloss over what assumptions are built into this make-shift but still surprisingly useful theory. We have attempted to correct this omission and discuss briefly some results of modern multidimensional numerical experiments in convection.

Our major bit of advice to the student is to have a decent background in the elements of astronomy and a firm grounding in physics. For the physics part, three years of undergraduate courses should suffice. That same student would also benefit from reading a first-year undergraduate nonscientist astronomy text. There are many good ones available. To help out here, our text includes a miniglossary in Appendix A.

We have no correspondingly firm advice for the instructor. From our combined experience of nearly thirty years, we understand that any instructor will do it her or his own way regardless of text. We do feel, however, that it is essential for the student in a course on stellar interiors, to very early on get an overview of the structure and evolution of a variety of objects very early on in a course about stellar interiors. Thus our text includes such an overview in Chapter 2 to provide a rationale for the student to go into further depth in the chapters that follow. (Too many texts get to the punch-line more than two thirds of the way into the book.) The first chapter is also required reading because it contains elementary material that will be used throughout the text. We also include a small number of problems that we think are interesting, relevant, and doable, at the ends of some chapters.

We have tried not to reserve separate chapters for what we consider to be specialized topics. Thus, for example, some elements of the effects of rotation are really only discussed in sections of the chapters on the sun (Chapter 8) and asteroseismology (Chapter 10). Magnetic fields make their brief appearance in Chapter 8 and in Chapter 9 on white dwarfs.

The tremendous advances in stellar astrophysics over the last thirty years have been made possible only through extensive computer modeling. Thus, to introduce the student to the computational techniques traditionally applied in this subject, we have included two computer programs on a PC diskette. These are written in `FORTRAN` and should require no special effort to run on any platform that supports this language. The first program, `ZAMS.FOR`, calculates decent zero-age hydrogen main sequence models. It has lots of output for the student to analyze. We have found it to be a useful practicum with which the student may explore the effects of, for example, composition changes, mass-luminosity relationships, and overall structural trends. The second program, `PULS.FOR`, uses some of the output from `ZAMS` to study the seismological properties of the zero-age main sequence models. Students can, for example, do quite a bit of helioseismology using these codes. Appendix C is an introduction to what may be found on the diskette and we advise the user to first read this appendix and to consult the `READ.ME` file on the diskette itself for further details.

Acknowledgments: We wish to thank our many senior colleagues and graduate students for numerous reprints, suggestions, and comments—some of which we had the good sense to follow. The bad parts are solely our fault. In particular we thank: W.D. Arnett, M.C. Begelman, P.A. Bradley, P. Brassard, J.I. Castor (for much material in Chapter 5), M. Fardel, G. Fontaine, T.P. Hartsell, R.E. Nather, M.S. O'Brien, W.D. Pesnell, M. Rast, K. Serizawa, V. Trimble, H.M. Van Horn, and D.E. Winget. And many kudos to our editors at Springer-Verlag. The senior author (C.J.H.) also sends astral thanks and appreciation to J.P. Cox, who spent many of his valuable hours discussing stars and such with me before his untimely death. His spirit hovers over the better parts of this text. Finally, both authors wish to thank their families and especially their wives: may they not become computer widows again.

The text was set in LaTeX by the authors.

Carl J. Hansen
University of Colorado / Boulder

Steven D. Kawaler
Iowa State University / Ames

November 1993

Contents

1
Preliminaries

Are you sitting comfortably? Then I'll begin.
From a B.B.C. radio program.
— J.S. Lang (1950)

If you want to gain an important insight into what makes stars work, go out and look at them for a few nights. You will find that they appear to do nothing much at all except shine steadily. This is certainly true from a personal or historical perspective and, from fossil evidence, we can extend this to a roughly three-billion-year span for the sun. The reason for this relative tranquility is that stars are, on the whole, very stable objects in which self-gravitational forces are delicately balanced by steep internal pressure gradients. The latter require high temperatures. In the deep interior of a star these temperatures are measured in (at least) millions of degrees Kelvin and, in most instances, are sufficiently high to initiate the thermonuclear fusion of light nuclei. The power so produced then gradually leaks out through the remaining bulk of the star and finally gives rise to the radiation we see streaming off the surface. The vast majority of stars spend most of their active lives in such an equilibrium state, and it is only the gradual transmutation of elements by the fusion process that eventually causes their structure to change in some marked way.

This chapter will introduce some concepts and physical processes that, when tied together, will enable us to paint a preliminary picture of the stellar interior and to make estimates of the magnitudes of various quantities such as pressure, temperature, and lifetimes. Later chapters will expand on these concepts and processes and bring us up to date on some modern developments in stellar structure and evolution. If, in reading this chapter, you begin to get lost, we suggest you review Appendix A on some properties of stars and nomenclature. For those of you who have no background in the subject at all, a good first-year undergraduate text on astronomy for the nonscientist may be in order. There are several excellent texts available. Some portions of the material we present will also make considerable demands on your understanding of physical processes. We assume that you have a decent background in undergraduate physics. If not, you will have to catch up and review that material.

1.1 Hydrostatic Equilibrium

We first consider the theoretician's dream: a spherically symmetric, nonrotating, nonmagnetic, single, etc., star on which there are no net forces acting and, hence, no accelerations. There may be internal motions, such as those associated with convection, but these are assumed to average out overall. We wish to find a relation that expresses this equilibrium. For this, assume that the stellar material is so constituted that internal stresses are isotropic and thus reduce to ordinary pressures, and define the following quantities, which will be used throughout this text:

radius: r is the radial distance measured from the stellar center (cm)

total stellar radius: $\mathcal{R}$

density: $\rho(r)$ is the mass density at r (g cm^{-3})

temperature: $T(r)$ is the temperature at r (deg K)

pressure: $P(r)$ is the pressure at r (dyne cm^{-2} = erg cm^{-3})

mass: $\mathcal{M}_r$ is the mass contained within a sphere of radius r (g)

total stellar mass: $\mathcal{M} = \mathcal{M}_\mathcal{R}$

luminosity: $\mathcal{L}_r$, the rate of energy flow through a sphere at r (erg s^{-1})

total stellar luminosity: $\mathcal{L} = \mathcal{L}_\mathcal{R}$

local gravity: $g(r)$ local acceleration due to gravity (cm s^{-2})

gravitational constant $G = 6.6726 \times 10^{-8}$ g^{-1} cm^3 s^{-2}

solar mass: $\mathcal{M}_\odot = 1.989 \times 10^{33}$ g

solar luminosity: $\mathcal{L}_\odot = 3.847 \times 10^{33}$ erg s^{-1}

solar radius: $\mathcal{R}_\odot = 6.96 \times 10^{10}$ cm

Note that all the above are expressed in cgs units. There is really no good reason for this, but cgs seem to be the units of choice for most of those researchers dealing with stars. MKS could just as well be used instead. We will, of course, use solar units (e.g., $\mathcal{R}/\mathcal{R}_\odot$) when appropriate. Other physical constants may be found in Appendix B.

What we shall do now is investigate the balance of forces within a star in equilibrium. From elementary considerations, the local gravity at radius r is $g(r) = G\mathcal{M}_r/r^2$ and

$$\mathcal{M}_{r+dr} - \mathcal{M}_r = d\mathcal{M}_r = 4\pi r^2 \rho(r)\, dr \tag{1.1}$$

is the mass contained within a spherical shell of infinitesimal thickness dr at r. The integral of (1.1) then yields the mass within r:

$$\mathcal{M}_r = \int_0^r 4\pi r^2 \rho\, dr. \tag{1.2}$$

Either (1.1) or (1.2) will be referred to as the *mass equation*, the *equation of mass conservation*, or the *continuity equation*. If we consider a 1-cm^2 element of area on the surface of the shell at r, there is then an inwardly directed gravitational force on a volume 1 cm^2 $\times$ dr of

$$\rho g\, dr = \rho \frac{G\mathcal{M}_r}{r^2}\, dr. \tag{1.3}$$

To counterbalance this force we must rely on an imbalance of pressure forces; that is, the pressure $P(r)$ pushing outwards against the inner side of the shell must be greater than the pressure acting inward on the outer face. The net force outwards is $P(r) - P(r + dr) = -(dP/dr)\, dr$. Adding the gravitational and differential pressure forces then yields

$$\rho \ddot{r} = -\frac{dP}{dr} - \frac{G\mathcal{M}_r}{r^2}\rho \tag{1.4}$$

as the equation of motion.

By hypothesis, all net forces are zero, with $\ddot{r} = 0$, and we obtain the *equation of hydrostatic* (or mechanical) *equilibrium*

$$\frac{dP}{dr} = -\frac{G\mathcal{M}_r}{r^2}\rho = -g\rho. \tag{1.5}$$

Since $g, \rho \geq 0$, then $dP/dr \leq 0$ and the pressure must decrease outwards everywhere. If this condition is violated anywhere within the star, then hydrostatic equilibrium is impossible and local accelerations must occur.

We can obtain the hydrostatic equation in yet another way and, at the same time, introduce some new concepts.

1.2 An Energy Principle

The preceding was a local approach to mechanical equilibrium because only local quantities at r were involved (although a gradient did appear). What we shall do now is take a global view wherein equilibrium is posed as an integral constraint on the structure of the entire star. You are to imagine that the equilibrium star is only one of an infinity of possible configurations and the trick is to find the right one. Each configuration will be specified by an integral function so constructed that the equilibrium star is represented by a stationary point in the series of possible functions. This begins to sound like a problem in classical mechanics and the calculus of variations—and it is. The function in question is the total stellar energy, and we shall now construct it.

The total gravitational potential energy, Ω, of a self-gravitating body is defined as the *negative* of the total amount of energy required to disperse all mass elements of the body to infinity. The zero point of the potential is taken as the final state after dispersal. (In other words, it is the energy required to assemble the star, in its current configuration, by collecting material from the outside universe.) We can get to the dispersed state by successively peeling off spherical shells from our spherical star. Suppose we have already done so down to an interior mass of $\mathcal{M}_r + d\mathcal{M}_r$ and we are just about to remove the next shell, which has a mass $d\mathcal{M}_r$. To move this shell outwards from some radius r' to $r' + dr'$ requires $(G\,\mathcal{M}_r/r'^2)\,d\mathcal{M}_r\,dr'$ units of work. To go from r to infinity then gives a contribution to Ω of (remembering the minus sign for Ω)

$$d\Omega = -\int_r^\infty \frac{G\,\mathcal{M}_r}{r'^2}\,d\mathcal{M}_r\,dr' = -\frac{G\,\mathcal{M}_r}{r}\,d\mathcal{M}_r.$$

To disperse the whole star requires that we do this for all $d\mathcal{M}_r$ or,

$$\Omega = -\int_0^{\mathcal{M}} \frac{G\,\mathcal{M}_r}{r}\,d\mathcal{M}_r. \tag{1.6}$$

The potential energy thus has the units of $G\,\mathcal{M}^2/\mathcal{R}$ and we shall often write it in the form

$$\Omega = -q\,\frac{G\,\mathcal{M}^2}{\mathcal{R}}. \tag{1.7}$$

For a uniform density sphere, with ρ constant, it is easy to show that the pure number q is equal to 3/5. (This should be familiar from electrostatics, where the energy required to disperse a uniformly charged sphere to infinity is $-3e^2\mathcal{R}/5$.) Because density almost always decreases outwards for equilibrium stars, the value of 3/5 is, for all practical purposes, a lower limit with $q \geq 3/5$.

For the sun, $G\mathcal{M}_\odot{}^2/\mathcal{R}_\odot \approx 3.8 \times 10^{48}$ ergs. If we divide this figure by the present solar luminosity we find a characteristic time (the Kelvin-Helmholtz time scale) of about 3×10^7 years. More will be said about this time scale later on.

If we neglect gross mass motions or phenomena such as turbulence, then the total energy of the star is Ω plus the total internal energy arising from microscopic processes. Let E be the local specific internal energy in units of ergs per gram of material. It is to be multiplied by ρ if energy per unit volume is desired. (Thus E will sometimes have the units of erg cm^{-3} but you will either be forewarned by a statement or the appearance of those units.) The total energy, W, is then the sum of Ω and the mass integral of E

$$W = \int_{\mathcal{M}} E \, d\mathcal{M}_r + \Omega = U + \Omega, \tag{1.8}$$

which also defines the total internal energy

$$U = \int_{\mathcal{M}} E \, d\mathcal{M}_r. \tag{1.9}$$

The statement now is that the equilibrium state of the star corresponds to a stationary point with respect to W. This means that W for the star in hydrostatic equilibrium is an extremum relative to all other possible configurations the star could have (with the possible exception of other extrema). What we are going to do to test this idea is to perturb the star away from its original state in an *adiabatic* but otherwise arbitrary and infinitesimal fashion. The adiabatic part can be satisfied if the perturbation is performed sufficiently rapidly that heat transfer between mass elements cannot take place (as in an adiabatic sound wave). We shall show later that energy redistribution in normal stars takes place on time scales longer than mechanical response times. On the other hand, we also require that the perturbation be sufficiently slow that kinetic energies of mass motions can be ignored.

If δ represents either a local or global perturbation operator (depending on the obvious context), then the stellar hydrostatic equilibrium state is that for which

$$(\delta W)_{\mathrm{ad}} = 0$$

where the "ad" subscript denotes "adiabatic." Thus if arbitrary, but small, adiabatic changes result in no change in W, then the initial stellar state is in hydrostatic equilibrium. To show this, we must examine how U and Ω change when ρ, T, etc., are varied adiabatically. We thus have to look at the pieces of

$$(\delta W)_{\mathrm{ad}} = (\delta U)_{\mathrm{ad}} + (\delta \Omega)_{\mathrm{ad}}.$$

A perturbation δ causes U to change by δU with

$$U \longrightarrow U + \delta U = U + \delta \int_{\mathcal{M}} E \, d\mathcal{M}_r = U + \int_{\mathcal{M}} \delta E \, d\mathcal{M}_r.$$

The last step follows because we choose to consider the change in specific internal energy of a particular mass element $d\mathcal{M}_r$. (This is a *Lagrangian* description of the perturbation about which more will be said in Chapter 10.) Now consider δE. What we do is to label each mass element of $d\mathcal{M}_r$ worth of matter and see what happens to it (and E) when its position r, and ρ, and T are changed.

For an infinitesimal and reversible change (it would be nice to be able to put the star back together again), the combined first and second laws of thermodynamics state that

$$dQ = dE + P\,dV_\rho = T\,dS. \tag{1.10}$$

Here dQ is the heat added to the system, dE is the increase in internal specific energy, and $P\,dV_\rho$ is the work done by the system on its surroundings if the "volume" changes by dV_ρ. This volume is the *specific volume*, with $V_\rho = 1/\rho$ and is that associated with a given gram of material. It has the units of cm^3 g^{-1}. (The symbol V will be reserved for ordinary volume with units of cm^3.) The entropy S, and Q, are also mass-specific quantities. If we replace the differentials in the preceding by δs, then the requirement of adiabaticity ($\delta S = 0$) immediately yields $(\delta E)_{\rm ad} = -P\,\delta V_\rho$. Thus, $(\delta U)_{\rm ad} = -\int_\mathcal{M} P\,\delta V_\rho\,d\mathcal{M}_r$.

What is δV_ρ? From the definition of the specific volume and the mass equation (1.1),

$$V_\rho = \frac{1}{\rho} = \frac{4\pi r^2\,dr}{d\mathcal{M}_r} = \frac{d(4\pi r^3/3)}{d\mathcal{M}_r}. \tag{1.11}$$

To make life easy, we restrict all perturbations to those that maintain spherical symmetry. Thus if the mass parcel $d\mathcal{M}_r$ moves at all, it moves only in the radial direction to a new position $r+\delta r$. Perturbing V_ρ in (1.11) is then equivalent to perturbing r or

$$V_\rho \longrightarrow V_\rho + \delta V_\rho = \frac{d[4\pi(r+\delta r)^3/3]}{d\mathcal{M}_r} = V_\rho + \frac{d(4\pi r^2\delta r)}{d\mathcal{M}_r} \tag{1.12}$$

to first order in δr where we assume that $|\delta r/r| \ll 1$. (Later we will call this sort of thing "linearization.") The variation in total internal energy is then given by

$$(\delta U)_{\rm ad} = -\int_\mathcal{M} P\,\frac{d(4\pi r^2\delta r)}{d\mathcal{M}_r}\,d\mathcal{M}_r. \tag{1.13}$$

We now introduce two boundary conditions. The first is obvious: we do not allow the center of our spherically symmetric star to move. This amounts to requiring that $\delta r(\mathcal{M}_r = 0) = 0$. The second is called the "zero boundary condition on pressure" and it requires that the pressure at the surface vanish. Thus, $P_S = P(\mathcal{M}_r = \mathcal{M}) = 0$. This last is perfectly reasonable in this context because, in our idealized star, the surface is presumably where the mass runs out and we implicitly assume that no external pressures have been applied. (Later in Chapter 7, we will have to worry quite a bit more about this "surface." It is more subtle than it may appear.) Now integrate (1.13) by parts, apply the boundary conditions to the resulting constant term, and find

$$(\delta U)_{\rm ad} = \int_{\mathcal{M}} \frac{dP}{d\mathcal{M}_r} 4\pi r^2 \delta r \, d\mathcal{M}_r .$$

The corresponding analysis for $(\delta\Omega)_{\rm ad}$ yields

$$\Omega \longrightarrow \Omega + \delta\Omega = -\int_{\mathcal{M}} \frac{G\mathcal{M}_r}{r+\delta r} \, d\mathcal{M}_r = \Omega + \int_{\mathcal{M}} \frac{G\mathcal{M}_r}{r^2} \delta r \, d\mathcal{M}_r$$

to first order in δr after expansion of the denominator in the first integral.

Putting it all together we find

$$(\delta W)_{\rm ad} = \int_{\mathcal{M}} \left[\frac{dP}{d\mathcal{M}_r} 4\pi r^2 + \frac{G\mathcal{M}_r}{r^2} \right] \delta r \, d\mathcal{M}_r .$$

The aim is now to see what happens when this expression is set to zero. Is hydrostatic equilibrium regained? This is an exercise from the calculus of variations (as in Goldstein 1981). If δr is indeed arbitrary (subject to restrictions of symmetry), then the only way $(\delta W)_{\rm ad}$ can vanish is for the integrand to vanish identically; that is, we must have

$$\frac{dP}{d\mathcal{M}_r} = -\frac{G\mathcal{M}_r}{4\pi r^4} . \tag{1.14}$$

The equation of hydrostatic equilibrium follows immediately after the mass equation (1.1) is used to convert the differential from $d\mathcal{M}_r$ to dr. The version (1.14) is Lagrangian in nature (the independent variable is $d\mathcal{M}_r$) and, after introducing acceleration in the appropriate place, is often used in one-dimensional hydrodynamical studies of stars.

You may profit from considering the significance of $(\delta^2 W)_{\rm ad}$, which is the second variation of W. As discussed by Chiu (1968, §2.12), the sign of the second variation determines whether the equilibrium configuration is mechanically stable or unstable to small perturbations. This is like asking whether a pencil balanced on its point is "stable."

1.3 The Virial Theorem and Its Applications

We shall now derive the virial theorem and, from it, obtain some interesting and useful relations between various global stellar quantities such as W and Ω. This will be primarily an exercise in classical mechanics at first, but the utility of the virial theorem in making simple estimates of temperature, density, and the like, will soon be apparent. In addition, the theorem will be applied to yield estimates for some important stellar time scales. Most texts on stellar interiors contain some discussion of this topic. We shall follow Clayton (1968, chap. 2). A specialized reference is Collins (1978).

Consider the scalar product $\sum_i \mathbf{p}_i \bullet \mathbf{r}_i$ where $\mathbf{p}_i$ is the vector momentum of a free particle of mass m_i located at position $\mathbf{r}_i$ and the sum is over all particles comprising the star. If the mechanics are nonrelativistic, then recognize that

$$\frac{d}{dt}\sum_i \mathbf{p}_i \bullet \mathbf{r}_i = \frac{d}{dt}\sum_i m_i \dot{\mathbf{r}}_i \bullet \mathbf{r}_i = \frac{1}{2}\frac{d}{dt}\sum_i \frac{d}{dt}(m_i r_i^2) = \frac{1}{2}\frac{d^2 I}{dt^2},$$

where I is the moment of inertia, $I = \sum_i m_i r_i^2$. On the other hand, the derivative of the original sum yields

$$\frac{d}{dt}\sum_i \mathbf{p}_i \bullet \mathbf{r}_i = \sum_i \frac{d\mathbf{p}_i}{dt} \bullet \mathbf{r}_i + \sum_i \mathbf{p}_i \bullet \frac{d\mathbf{r}_i}{dt}.$$

The last term is just $\sum_i m_i v_i^2$ (v_i is the velocity of particle i) and is equal to twice the total kinetic energy, K, of all the free particles in the star. Furthermore, take note of Newton's law

$$\frac{d\mathbf{p}_i}{dt} = \mathbf{F}_i$$

where $\mathbf{F}_i$ is the force applied to particle i. Putting this together we then have

$$\frac{1}{2}\frac{d^2 I}{dt^2} = 2K + \sum_i \mathbf{F}_i \bullet \mathbf{r}_i. \tag{1.15}$$

The last term is the virial of Clausius, but to make any use of it all of the $\mathbf{F}_i \bullet \mathbf{r}_i$ must be specified.

The most important factor in the virial is the mutual gravitational interaction of the particles in the star. (And, remember, we are still ignoring magnetic fields, etc. These make their own kinds of contributions.) To treat gravity, let $\mathbf{F}_{ij}$ be the gravitational force on particle i due to the presence of particle j. Because such forces are equal and opposite, $\mathbf{F}_{ij} = -\mathbf{F}_{ji}$. You may verify by direct construction (with, say, three particles) that

$$\sum_i \mathbf{F}_i \bullet \mathbf{r}_i = \sum_{\substack{i,j \\ i<j}} (\mathbf{F}_{ij} \bullet \mathbf{r}_i + \mathbf{F}_{ji} \bullet \mathbf{r}_j)$$

where the sum is to be taken over all i and j provided that $i < j$. Hereafter in this section, this convention will be assumed and the limits on the sum will not be given.

From elementary physics, the Newtonian gravitational force is

$$\mathbf{F}_{ij} = -\frac{Gm_i m_j}{r_{ij}^3}(\mathbf{r}_i - \mathbf{r}_j)$$

where r_{ij} is the interparticle distance $r_{ij} = |\mathbf{r}_i - \mathbf{r}_j|$. The gravitational contribution to the virial thus becomes

$$\sum \mathbf{F}_{ij} \bullet (\mathbf{r}_i - \mathbf{r}_j) = -\sum \frac{Gm_i m_j}{r_{ij}} = \text{Virial}$$

using the equal and opposite expression to obtain the first term. It should be apparent that the last sum (with minus sign) is just the negative of the work required for dispersal to infinity; that is, we have recovered Ω. Thus we have

$$\text{Virial} = \Omega.$$

Combining this with (1.15) we obtain

$$\tfrac{1}{2}\frac{d^2 I}{dt^2} = 2K + \Omega \tag{1.16}$$

as the "virial theorem" which we will often refer to as just "virial."

Note that this expression refers to quantities derived from sums (or integrals) over the whole star. If we had chosen instead to consider only a portion of the star—as, say, defined by a sphere of radius $r_S \leq \mathcal{R}$ and volume V—then I, K, and Ω would refer only to that portion. However, the spherical shell containing material within radii $r_s < r \leq \mathcal{R}$ would contribute an additional term to the right-hand side of (1.16) given by $-3P_S V$ where P_S is the pressure at r_S. If $r_S \rightarrow \mathcal{R}$ and $P_S \rightarrow 0$ (as in a zero boundary condition on pressure), then (1.16) is unchanged because we have just encompassed the whole star and no external pressures act at $\mathcal{R}$. (For a derivation of this additional term see, for example, Cox 1968 §17.2 or Clayton 1968, pp.134–135.) We will not have occasion to use this term, but its possible presence should be kept in mind.

We now interpret what the energy K represents. For example, is it U or, if not, how does it differ? We had

$$2K = \sum_i m_i v_i^2 = \sum_i \mathbf{p}_i \bullet \mathbf{v}_i. \tag{1.17}$$

The scalar product of $\mathbf{p}$ and $\mathbf{v}$ measures the rate of momentum transfer and, hence, from the kinetic theory of gases, must be related to the pressure. In the continuum limit of an isotropic gas, pressure is given by

$$P = \tfrac{1}{3}\int_p n(\mathbf{p})\, \mathbf{p} \bullet \mathbf{v}\, d^3\mathbf{p} \tag{1.18}$$

where $n(\mathbf{p})$ is the number density of particles with momentum $\mathbf{p}$ and the integration is over all momenta. The units of $n(\mathbf{p})$ are number $\text{cm}^{-3}\ p^{-3}$. Since the sum in (1.17) includes all particles, it should be clear that (1.18) need only be integrated over total volume V (in cm^3) to obtain an expression for K; namely,

$$2K = 3\int_V P\, dV. \tag{1.19}$$

Furthermore, since $d\mathcal{M}_r = \rho\, d(\frac{4}{3}\pi r^3) = \rho\, dV$, we find

$$2K = 3\int_{\mathcal{M}} \frac{P}{\rho}\, d\mathcal{M}_r \tag{1.20}$$

and the virial theorem becomes

$$\frac{1}{2}\frac{d^2I}{dt^2} = \int_{\mathcal{M}} \frac{3P}{\rho}\, d\mathcal{M}_r + \Omega. \tag{1.21}$$

We now apply this to stars by looking into some possible choices for the equation of state.

1.3.1 Application: Global Energetics

Consider a simple, but useful, relation between pressure and internal energy of the form

$$P = (\gamma - 1)\rho E \tag{1.22}$$

where γ is a constant and E is still in erg g^{-1}. This is usually called a "γ–law equation of state" and is not just of academic interest. For example (and as will be shown later), for a monatomic ideal gas $\gamma = c_P/c_V = 5/3$ where c_P and c_V are, respectively, the specific heats at constant pressure and volume. In this instance $P = \frac{2}{3}\rho E$. For radiation or a completely relativistic Fermi gas $\gamma = 4/3$. Since $2K = 3(\gamma - 1)\int E\, d\mathcal{M}_r$—from combining (1.20) and (1.22)—then $K = \frac{3}{2}(\gamma - 1)U$. Thus $K = U$ only if $\gamma = 5/3$. Note that a γ of 5/3 does not necessarily mean the gas is ideal and monatomic.

The virial theorem is now

$$\frac{1}{2}\frac{d^2I}{dt^2} = 3(\gamma - 1)U + \Omega. \tag{1.23}$$

If we let $W = U + \Omega$, as in (1.8), then the theorem becomes

$$\frac{1}{2}\frac{d^2I}{dt^2} = 3(\gamma - 1)W - (3\gamma - 4)\Omega. \tag{1.24}$$

For hydrostatic equilibrium d^2I/dt^2 must be zero and W is related to Ω by

$$W = \frac{3\gamma - 4}{3(\gamma - 1)}\Omega, \tag{1.25}$$

which shows explicitly the relation between W and Ω for hydrostatic stars with the γ-law equation of state.

Since the energy W is that which is available to do useful work, a dynamically stable star should have $W < 0$. Otherwise, the star would have enough energy, at least in principle, to completely disperse all or part of itself. Equation (1.25) then implies that a star in hydrostatic equilibrium should have a γ that exceeds 4/3. However, and as we shall find later on, even this condition does not always guarantee safety. The star could contain a potentially explosive fuel which, if ignited, could also cause W to exceed zero for a time. In addition, we do not necessarily expect the total energy to remain absolutely and forever constant. After all, stars do shine and lose energy in doing so.

We shall now explore some consequences of energy losses due to radiation where the energy source is gravitational energy released by contraction.

1.3.2 Application: The Kelvin-Helmholtz Time Scale

Barring bizarre circumstances, a star derives its energy to shine from three sources: internal energy, thermonuclear fuel, and gravitational contraction. Some, or all, are used at one time or another. Here we briefly examine the last source. A more complete treatment will be deferred to Chapter 6, where stellar energy sources are discussed in more depth.

Suppose a star contracts very gradually while maintaining sphericity and hydrostatic equilibrium at all times. (Realize that contraction cannot occur without some acceleration unless all mass elements are just coasting. What we mean here is that hydrostatic equilibrium is to be maintained *almost* exactly.) As the star contracts, Ω and, possibly, W change. Denote these changes by $\Delta\Omega$ and ΔW. If γ remains constant as contraction proceeds, then (1.25) implies

$$\Delta W = \frac{3\gamma - 4}{3(\gamma - 1)} \Delta\Omega. \tag{1.26}$$

Because we cannot follow the star's progress exactly (at least at this stage in this text) we use some dimensional arguments to estimate what Ω and W do upon contraction.

Let $\mathcal{R}$ be the total stellar radius (or some other representative radius) and $\Delta\mathcal{R}$ be its change through some stage in the contraction. We assume $\gamma > 4/3$. From (1.7), $\Omega \propto -G\mathcal{M}^2/\mathcal{R}$, which implies that $\Delta\Omega \propto (G\mathcal{M}^2/\mathcal{R}^2)\Delta\mathcal{R}$ for constant q. Since $\Delta\mathcal{R} < 0$ for contraction, then $\Delta\Omega$ is also negative in these circumstances and the star sinks deeper into its own potential well. This means that energy has been liberated in some form. The virial result (1.26) also implies that $\Delta W < 0$ and thus the system as a whole has lost energy. What exactly is the energy budget here? Well, part of what has been made available goes into internal energy. This may be seen from (1.23) (with d^2I/dt^2 set to zero for equilibrium), which becomes

$$\Delta U = -\frac{1}{3(\gamma - 1)} \Delta\Omega \tag{1.27}$$

and yields $\Delta U > 0$ for contraction. Of the $|\Delta\Omega|$ units of energy made available, ΔU is used to "heat" up the star. The rest is lost from the system. At this stage in our discourse, it is simplest to assume that this energy has been radiated from the stellar surface during the contraction;

that is, power has been expended in the form of luminosity. Note that if $\gamma = 5/3$ (as for an ideal monatomic gas), then $\Delta U = -\Delta W = -\Delta\Omega/2$ and the split between internal energy and time-integrated luminosity is equal. Note also that if an increase in temperature is associated with the increase in U, then the star has an overall specific heat that is negative: a loss of total energy means an increase in temperature. This phenomenon is an important self-regulating mechanism for normal stars. Finally, if $\gamma = 4/3$, then $\Delta W = 0$ and all the energy goes into increasing U and the star need not radiate at all.

Suppose we now extend the above analysis and hypothesize that contraction is solely responsible for maintaining stellar luminosities. For an ideal gas star with $\gamma = 5/3$, $\Delta W = \Delta\Omega/2 = (q/2)\left(G\mathcal{M}^2/\mathcal{R}^2\right)\Delta\mathcal{R}$. If we equate $-dW/dt$ to the luminosity $\mathcal{L}$ (as a power output) then

$$\mathcal{L} = -\frac{dW}{dt} = -\frac{q}{2}\frac{G\mathcal{M}^2}{\mathcal{R}}\left(\frac{d\mathcal{R}/dt}{\mathcal{R}}\right). \tag{1.28}$$

It is clear that if $\mathcal{L}$ is kept constant then this equation defines a characteristic e-folding time for radius decrease of

$$t_{\rm KH} \approx \frac{q}{2}\frac{G\mathcal{M}^2}{\mathcal{L}\mathcal{R}} \tag{1.29}$$

where the "KH" subscript is for the originators of the idea, Baron W.T. Kelvin and H.L.F. Helmholtz. Choosing a representative value of q of 3/2 (which is about right for the sun),

$$t_{\rm KH} \approx 2 \times 10^7 \left(\frac{\mathcal{M}}{\mathcal{M}_\odot}\right)^2 \left(\frac{\mathcal{L}}{\mathcal{L}_\odot}\right)^{-1} \left(\frac{\mathcal{R}}{\mathcal{R}_\odot}\right)^{-1} \text{ years.} \tag{1.30}$$

We know that a figure of 2×10^7 years for radius changes for the sun cannot be correct from fossil evidence: terrestrial life is the same now (except for relatively inconsequential developments) as it was many millions of years ago. Any major structural change in the sun would have had profound consequences for life and they are not seen. However, we will find that most stars do depend on (or, more accurately, are forced into) gravitational contraction at some stage of evolution, and the corresponding time scales can be comparatively very short.

1.3.3 Application: A Dynamic Time Scale

Consider a star in hydrostatic equilibrium and composed purely of an ideal gas so that $W = \Omega/2 \approx -G\mathcal{M}^2/\mathcal{R}$. If, by some magic, an internal process were to instantaneously take place whereby $\gamma \to 4/3$ but W did not change significantly, then $d^2I/dt^2 \approx -G\mathcal{M}^2/\mathcal{R}$ from (1.24). By dimensional arguments, $I \approx \mathcal{M}\mathcal{R}^2$, so we may define a time scale $t_{\rm dyn}$ by

$d^2I/dt^2 \approx I/t_{\rm dyn}^2 \approx \mathcal{M}\mathcal{R}^2/t_{\rm dyn}^2$. Equating the two expressions for d^2I/dt^2 yields $t_{\rm dyn}^2 \approx \mathcal{R}^3/G\mathcal{M}$ or

$$t_{\rm dyn} \approx \frac{1}{[G\langle\rho\rangle]^{1/2}} \tag{1.31}$$

where $\langle\rho\rangle \approx \mathcal{M}/\mathcal{R}^3$ is approximately the average density. The dynamic time scale $t_{\rm dyn}$ is then a measure of the e-folding time for changes in radius as the star makes readjustments in structure. (In this example, d^2I/dt^2 is negative and the star collapses.) For the sun $t_{\rm dyn}$ is about an hour, which is many orders of magnitude shorter a time than $t_{\rm KH}$.

Expression (1.31) is the "period–mean density relation" and it will come up again when we discuss variable stars.

1.3.4 Application: Estimates of Stellar Temperatures

We can squeeze even more out of the virial theorem. Consider a star of uniform density and temperature composed of a monatomic ideal gas. The internal energy density

$$E = \tfrac{3}{2}nkT = \tfrac{3}{2}\rho\frac{N_{\rm A}kT}{\mu}\ \text{erg cm}^{-3}. \tag{1.32}$$

Here n is the number density of free particles (in number cm^{-3}), k and $N_{\rm A}$, respectively, are Boltzmann's and Avogadro's constants, and μ is the mean molecular weight (usually in amu) per ion or atom of the stellar mixture. The quantity μ will be discussed in more detail shortly but, for now, regard it as that thing which makes $n = \rho N_{\rm A}/\mu$. For a typical stellar mixture of elements it is of order unity.

Multiplying E by the stellar volume V yields U and, since $\rho V = \mathcal{M}$, we find $U = \frac{3}{2}\mathcal{M}N_{\rm A}kT/\mu$. On the other hand, $U = -\Omega/2$ from the virial theorem (1.23) for the $\gamma = 5/3$ gas, and $\Omega = -\frac{3}{5}G\mathcal{M}^2/\mathcal{R}$ for the constant-density sphere. Equate the two forms for U, solve for T in terms of ρ, $\mathcal{M}$, and μ, eliminate $\mathcal{R}$ by way of the density, and find

$$T = 4.09 \times 10^6\, \mu \left(\frac{\mathcal{M}}{\mathcal{M}_\odot}\right)^{2/3} \rho^{1/3}\ \text{K}. \tag{1.33}$$

Before discussing the numerical results obtainable from this expression, it is worthwhile deriving the main components of it from another perspective.

The Lagrangian expression for the equation of hydrostatic equilibrium (1.14) is useful in this regard. In dimensional form it states that P is proportional to $G\mathcal{M}^2/\mathcal{R}^4$. But P also varies as $\mathcal{M}T/\mathcal{R}^3\mu$ after density has been eliminated in the ideal gas law, $P = nkT$. After equating the two versions of P we find (1.33) (but not the constant). The point is that if $\mathcal{R}$ is made smaller, for example, then ρ increases as $1/\mathcal{R}^3$ and, consequently, so would the ideal gas pressure were T to stay constant. This dependence of P on $\mathcal{R}$ is not strong enough, however, because P must also increase as $1/\mathcal{R}^4$ for hydrostatic equilibrium independent of the temperature. Thus the ideal gas equation of state and hydrostatic equilibrium demand that T must increase as $1/\mathcal{R} \propto \rho^{1/3}$.

Figure 1.1 shows (1.33) plotted as $\log T$ versus $\log \rho$ for $\mu = 1$ with $\mathcal{M}$ ranging between 0.3 and 100 $\mathcal{M}_\odot$. As a typical star, consider the present-day sun, which has an average density of $\langle \rho \rangle \approx 1.4$ g cm^{-3} and a central density of approximately 80 g cm^{-3}. If "average" may be identified with the quantities in (1.33), then an average temperature for the sun is a few million degrees. Even though it does not make a lot of sense to talk about an average temperature for a star, we note that the central temperature for the present-day sun is $T_c \approx 15 \times 10^6$ K, which is not too different from the number just found. As we shall see later, a temperature greater than about 10^6 K is just what is needed to initiate hydrogenic nuclear fusion in stars. A star thus produces energy by nuclear fusion because hydrostatic equilibrium requires high temperatures.

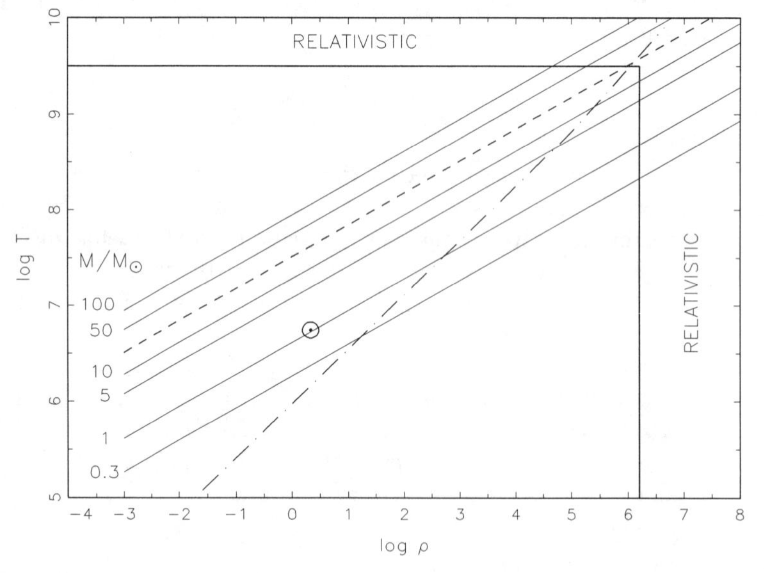

FIGURE 1.1. The ideal gas virial result for temperature versus density for various masses (in solar units) from equation (1.33). Radiation pressure dominates above the dashed line and degenerate electrons must be considered below the dotted-dashed line. Regions where relativistic effects are important are indicated. The location of a constant-density "sun" is shown by the $\odot$.

Figure 1.1 has other lines on it that partition the $\log \rho$–$\log T$ plane into regions where equations of state other than the ideal gas law hold sway. The "degeneracy" boundary defines that region where Fermi-Dirac degenerate electrons begin to play a major role. Above a line corresponding to about

$25\ \mathcal{M}_\odot$, radiation pressure (with a γ of 4/3 and $P = \frac{1}{3}aT^4$) becomes important. The areas beginning at $\rho \approx 10^6$ g cm^{-3} and $T \approx m_e c^2/k \approx 5 \times 10^9$ K ($m_e c^2$ is the electron rest mass energy) are regions where relativistic effects come in. All of these domains have their own peculiarities, which can greatly modify the simple picture built up thus far. But we shall have to wait for Chapter 3 to see what they are.

1.3.5 Application: Another Dynamic Time Scale

We have already found one dynamic time scale associated with global readjustments of the moment of inertia when hydrostatic equilibrium is seriously thrown out of kilter. We now consider perturbations in structure induced by small-amplitude sound waves and, specifically, compute how long it takes an adiabatic sound wave to travel from, say, the center of a star to the surface and back to the center again. (A discussion of mechanisms that might make such waves, or cause them to be reflected, we postpone until Chapter 10.) If the stellar sound speed is v_s, taken as constant for now, and Π is the "period" for one complete traversal, then

$$\Pi = \frac{2\mathcal{R}}{v_s}. \tag{1.34}$$

From elementary physics, the square of the local adiabatic sound speed is given by

$$v_s^2 = \left(\frac{dP}{d\rho}\right)_{\rm ad} = \Gamma_1 \frac{P}{\rho} \tag{1.35}$$

where Γ_1, our first "adiabatic exponent," is

$$\Gamma_1 = \left(\frac{d\ \ln P}{d\ \ln \rho}\right)_{\rm ad} = \frac{\rho}{P}\left(\frac{dP}{d\rho}\right)_{\rm ad}. \tag{1.36}$$

We shall find later that Γ_1, which measures how pressure changes in response to changes in density under adiabatic conditions, is of order unity.

If hydrostatic equilibrium is closely maintained while the weak sound wave passes through the star, then version (1.21) of the virial theorem yields

$$-\Omega = 3\int_{\mathcal{M}} \frac{P}{\rho}\, d\mathcal{M}_r = 3\int_{\mathcal{M}} \frac{v_s^2}{\Gamma_1}\, d\mathcal{M}_r \approx \frac{3v_s^2}{\Gamma_1}\mathcal{M}$$

where the sound velocity and Γ_1 appearing on the right-hand side represent suitable averages of those quantities. Since $|\Omega| \approx G\mathcal{M}^2/\mathcal{R}$ we then obtain an estimate for the period of $\Pi \approx \left(\mathcal{R}^3/G\mathcal{M}\right)^{-1/2}$. Constants of order unity (such as Γ_1) have been set to unity. After eliminating mass and radius in favor of density we find

$$\Pi \approx \frac{1}{\left[G\langle\rho\rangle\right]^{1/2}} \approx \frac{.04}{\left[\langle\rho\rangle/\langle\rho_\odot\rangle\right]^{1/2}} \text{ days.} \tag{1.37}$$

The final factor of 0.04 days comes from taking care with some of those quantities of order unity and inserting information which will be dealt with in Chapter 10.

Expression (1.37) is the same as $t_{\rm dyn}$ of (1.31) and rightly so because they both describe mechanical phenomena involving the whole star. A more careful analysis of how standing sound waves behave, however, introduces an additional factor of $(3\Gamma_1 - 4)^{1/2}$ in the denominator of (1.37). Again, a "gamma" of 4/3 will do curious things—as is obvious if $\Gamma_1 < 4/3$. We will postpone this discussion until it is time to examine variable stars.

1.4 The Constant-Density Model

We are now going to construct a model of a star by insisting that density be everywhere constant. Of course, in real life, we cannot do this—the run of density is determined by many factors—but the model does have some utility. The constant density "model" of the previous section was somewhat of a fudge. There we claimed that the star was in hydrostatic equilibrium, at constant temperature, and the ideal gas law was responsible for the pressure. A little thought, but not very much, should convince you that those conditions are contradictory. They imply that the pressure must be constant and yet hydrostatic equilibrium is still satisfied. We will make amends now.

If we set $\rho = \rho_c =$ constant, with "c" meaning center, then the mass equation (1.1) yields $\mathcal{M}_r = \frac{4}{3}\pi r^3 \rho_c$. This last expression is true up to the surface where $r = \mathcal{R}$ and $\mathcal{M}_r = \mathcal{M}$. Thus, after some trivial algebra,

$$\mathcal{M}_r = \frac{r^3}{\mathcal{R}^3}\mathcal{M}.$$

This is now used in the Lagrangian form of the hydrostatic equilibrium equation (1.14) to rid ourselves of r. The pressure gradient is then

$$\frac{dP}{d\mathcal{M}_r} = -\frac{G\mathcal{M}}{4\pi\mathcal{R}^4}\left(\frac{\mathcal{M}_r}{\mathcal{M}}\right)^{-1/3}.$$

Integrate this using the zero pressure boundary condition at $\mathcal{R}$ to find

$$P = P_c\left[1 - \left(\frac{\mathcal{M}_r}{\mathcal{M}}\right)^{2/3}\right] = P_c\left[1 - \left(\frac{r}{\mathcal{R}}\right)^2\right] \tag{1.38}$$

where P_c is the central pressure (at $\mathcal{M}_r = 0$) with

$$P_c = \frac{3}{8\pi}\frac{G\mathcal{M}^2}{\mathcal{R}^4} = 1.34 \times 10^{15}\left(\frac{\mathcal{M}}{\mathcal{M}_\odot}\right)^2\left(\frac{\mathcal{R}}{\mathcal{R}_\odot}\right)^{-4}\ \text{dyne cm}^{-2}. \tag{1.39}$$

The numerical value for P_c can be shown to be a *lower limit* for central pressures in hydrostatic objects if it is assumed that ρ always decreases outwards. This assumption is correct except for some very unusual circumstances (which may, in any case, signal an incipient instability in structure). That P_c is a lower limit seems reasonable because stronger concentrations of mass toward the center than that of constant density imply stronger gravitational fields which, in turn, require higher pressures to maintain equilibrium (as in, for example, the "linear star model" of Stein 1966 who uses a linear run of density with radius to illustrate properties of simple stellar models).

A simple exercise for the reader is to verify that the above expressions for pressure and mass distribution satisfy the equilibrium version of the virial theorem (1.21) with $d^2I/dt^2 = 0$ and $\Omega = -\frac{3}{5}G\mathcal{M}^2/\mathcal{R}$.

To find a temperature distribution we have to specify an equation of state and we again choose the monatomic ideal gas as a useful example with $P = nkT$. But, before we reach our objective, we shall first find out how to compute n or, equivalently, the mean molecular weight μ.

1.4.1 Calculation of Molecular Weights

We assume that the gas is composed of some mixture of neutral atoms, ions (in various stages of ionization), and electrons but, overall, the gas is electrically neutral. These are the free particles comprising n. First collect the ions and neutral atoms together into nuclear isotopic species, calling all of them "ions" for now, and denote a specific species by an index i. Thus, for example, assign some particular index to all the ions of ^{4}He. Each nucleus of index i has an integer nuclear charge Z_i and a nuclear mass number, in amu (atomic mass units), of A_i. For ^{4}He, $Z_i = 2$ and $A_i = 4$. (The atomic mass of ^{4}He is not exactly four, but this is close enough.) Furthermore, let X_i be the *fraction by mass* of species i in the mixture such that $\sum_i X_i = 1$. Thus, for example, if 70% of the mass of a sample of matter were composed of species i, then $X_i = 0.7$. The ion number density, in units of cm^{-3}, of a given species i is then

$$n_{\mathrm{I},i} = \frac{(\text{mass/unit volume) of } i}{(\text{mass of 1 ion) of } i} = \frac{\rho X_i N_\mathrm{A}}{A_i} \tag{1.40}$$

where $N_\mathrm{A} = 6.022137 \times 10^{23}$ mole^{-1}. The total for all ions is

$$n_\mathrm{I} = \sum_i n_{\mathrm{I},i} = \rho N_\mathrm{A} \sum_i \frac{X_i}{A_i}. \tag{1.41}$$

Now define μ_I as the "total mean molecular weight of ions" such that

$$n_\mathrm{I} = \frac{\rho N_\mathrm{A}}{\mu_\mathrm{I}} \tag{1.42}$$

or

$$\mu_{\rm I} = \left[\sum_i \frac{X_i}{A_i} \right]^{-1} . \tag{1.43}$$

The ion mean molecular weight is then a sort of mean mass of an "average" ion in the mixture and it contains all the information needed to find the number density of ions.

The electrons are a bit more difficult to treat. To find out how many free electrons there are we must have prior knowledge of what are the states of ionization for all species. This information is difficult to come by and we will defer until later just how to do it. We skirt the issue for now and assume that some good soul has done the work for us and has supplied us with the quantities y_i that contain what we want. These y_i are defined such that the number density of free electrons associated with nuclear species i is given by

$$n_{{\rm e},i} = y_i \, Z_i \, n_{{\rm I},i} = \rho N_{\rm A} \left(\frac{X_i}{A_i} \right) y_i Z_i . \tag{1.44}$$

Thus, out of the Z_i electrons that a particular ion of species i could possibly contribute to the free electron sea, only the fraction y_i are, on average, actually free. We shall call y_i the "ionization fraction." A value $y_i = 1$ then means that the species is completely ionized whereas $y_i = 0$ implies complete neutrality. The total electron number density is therefore

$$n_{\rm e} = \sum_i n_{{\rm e},i} = \rho N_{\rm A} \sum_i \left(\frac{X_i}{A_i} \right) y_i Z_i = \frac{\rho N_{\rm A}}{\mu_{\rm e}} , \tag{1.45}$$

which also defines $\mu_{\rm e}$, the "mean molecular weight per free electron." (Note that in no way are we assigning a "weight" to the electron.) Thus

$$\mu_{\rm e} = \left[\sum_i \frac{Z_i X_i y_i}{A_i} \right]^{-1} . \tag{1.46}$$

If you look carefully at the way $\mu_{\rm e}$ is constructed you will realize that it is the ratio of the total number of nucleons (protons plus neutrons) contained in all nuclei to the total number of free electrons in any sample of the material.

Finally, from the definition of n as the sum of $n_{\rm I}$ and $n_{\rm e}$, we easily find that the total mean molecular weight is

$$\mu = \left[\frac{1}{\mu_{\rm I}} + \frac{1}{\mu_{\rm e}} \right]^{-1} \tag{1.47}$$

with

$$n = n_{\rm I} + n_{\rm e} = \frac{\rho N_{\rm A}}{\mu} . \tag{1.48}$$

For relatively unevolved stars, in which nuclear transformations have not progressed to any great extent, the major nuclear constituents are hydrogen (^{1}H) and helium (^{4}He). We shall refer to their mass fractions (X_i) as, respectively, X and Y. All else shall collectively be called "metals" (or, sometimes, "heavies") and their mass fraction is denoted by Z (not to be confused with ion charge). A typical value of Z might be, at most, a few percent. Obviously

$$X + Y + Z = 1. \tag{1.49}$$

A catalogue of the relative abundances of metals seen on the surfaces of most stars, including the sun, reveals that the dominant heavy elements are carbon, nitrogen, oxygen, and neon. Elements heavier than those, up to nickel, contribute a little, and past there we find only traces. For the most part, the isotopes of the major heavy elements fall along the "valley of beta-stability" for which $Z_i/A_i \approx 1/2$. The same value of charge to mass number also applies to ^{4}He.

In the deep stellar interior, hydrogen, helium, and most of the metals are completely ionized ($y_i = 1$). If, in addition, metals comprise only a minor fraction of the total, with $Z \ll 1$, then you may use the results of the above analysis to find the following convenient approximation for μ_e:

$$\mu_e \approx \frac{2}{1+X} \, . \tag{1.50}$$

Note, however, that in detailed modeling of stars, this is to be used with great caution; ionization might not be complete (or elements may even be completely neutral) and abundances may be quite strange.

The ion mean molecular weight can be similarly approximated under the same conditions as above with the additional observation that Z is small compared to an average A ($A = \langle A_i \rangle \approx 14$ or so). The result is

$$\mu_I \approx \frac{4}{1+3X} \, . \tag{1.51}$$

Using (1.50) and (1.51), an approximation for the total mean molecular weight is then

$$\mu \approx \frac{4}{3+5X} \, . \tag{1.52}$$

For a star just beginning the longest active period of its natural life—a "zero-age main sequence" star (ZAMS)—typical abundances are $X \approx 0.7$, $Y \approx 0.3$, and $Z \approx 0.03$ (or somewhat less for stars formed early on in Galactic history). These correspond to $\mu_I \approx 1.3$, $\mu_e \approx 1.2$, and $\mu \approx 0.6$.

Now that we have reasonable approximations for the molecular weights and are assured that typical values are near unity, we return to the constant-density model.

1.4.2 The Temperature Distribution

Taking $P = \rho N_A kT/\mu$, and using the central pressure of (1.39), the central temperature for the constant-density model becomes

$$T_c = \frac{1}{2}\frac{G\mathcal{M}}{\mathcal{R}}\frac{\mu}{N_A k} = 1.15 \times 10^7 \,\mu\left(\frac{\mathcal{M}}{\mathcal{M}_\odot}\right)\left(\frac{\mathcal{R}}{\mathcal{R}_\odot}\right)^{-1} \text{K}. \tag{1.53}$$

The temperature distribution with respect to r and $\mathcal{M}_r$ is of the same form as that of the pressure in (1.38) with P_c replaced by T_c.

For solar values of mass and radius, this central temperature is remarkably close to that of the present-day sun found from sophisticated solar models and is an improvement over the virial "average" estimate of (1.33). The constant-density model result for T_c is higher than that from the virial because we have found the detailed run of pressure in the model and not just some average pressure.

As noted earlier, however, we cannot just assume a density distribution and expect such a stellar model to satisfy all the equations of stellar structure. We now discuss some of these additional constraints and equations.

1.5 Energy Generation and Transport

One goal of the effort in fusion energy research is to heat up a plasma containing potential thermonuclear fuel to temperatures exceeding about a million degrees and then physically contain it for a sufficiently long period of time. Most stars do that as a matter of course. They have the temperatures, containment mechanism, fuel, and time, and can fuse together light elements into heavier ones and, by doing so, release energy. We shall not discuss here precisely what kinds of thermonuclear burning take place in stars (see Chapter 6) but we shall extend our notion of equilibrium to include energy generation and how it is balanced by the leakage of energy through the star. In particular, suppose some sort of nuclear burning is taking place within a given localized gram of material. If the energy generated in that gram is not transferred elsewhere, then a nonequilibrium condition holds and the material heats up. If, on the other hand, we succeed in somehow removing energy as fast as it is liberated, and no faster, then we say the material is in "thermal balance." (Note that this term is not universally used by all authors in this context.) The sample of material is, of course, not strictly in equilibrium because, in the case of fusion, the composition is changing with time as more massive nuclear species are produced—but usually very slowly. We shall return to that problem later.

To express thermal balance quantitatively, consider a spherically symmetric shell of mass $d\mathcal{M}_r$ and thickness dr. Within that shell denote the power generated per gram as ε (erg g^{-1} s^{-1}). We shall refer to it as the "energy generation rate." The total power generated in the shell is $4\pi r^2 \rho \varepsilon \, dr = \varepsilon \, d\mathcal{M}_r$.

To balance the power generated, we must have a net flux of energy leaving the shell. If $\mathcal{F}(r)$ is the flux (in units of erg cm^{-2} s^{-1}), with positive values implying a radially directed outwards flow, then $\mathcal{L}_r = 4\pi r^2 \mathcal{F}(r)$ is the total power, or luminosity, in erg s^{-1}, entering (or leaving) the shell's inner face, and $\mathcal{L}_{r+dr} = 4\pi r^2 \mathcal{F}(r + dr)$ is the luminosity leaving through the outer face at $r + dr$. The difference of these two terms is the net loss or gain of power for the shell. For thermal balance that difference must equal the total power generated within the shell. That is,

$$\mathcal{L}_{r+dr} - \mathcal{L}_r = d\mathcal{L}_r = 4\pi r^2 \rho \varepsilon \, dr,$$

which yields the differential equation

$$\frac{d\mathcal{L}_r}{dr} = 4\pi r^2 \rho \varepsilon. \tag{1.54}$$

We will refer to this (or a more general version of it) as the "energy equation." Its Lagrangian form is

$$\frac{d\mathcal{L}_r}{d\mathcal{M}_r} = \varepsilon \tag{1.55}$$

by way of the mass equation. Note that we have used total differentials here. If the "equilibrium state" were also a function of time, then partials would appear instead. Note also that other energy sources, such as gravitational contraction, are being completely ignored at the moment.

Since, for now, we are only considering $\varepsilon \geq 0$, then $\mathcal{L}_r \geq 0$ also. Thus $\mathcal{L}_r$ must either be constant (in regions where $\varepsilon = 0$) or increase monotonically with r or $\mathcal{M}_r$. We will demonstrate later that ε is usually a strong function of temperature and, because temperature is expected to decrease outwards in a star, ε should be largest in the inner stellar regions provided that fuel is present. Thus $\mathcal{L}_r$ should increase rapidly from the center, starting from zero, and then level out to its surface value of $\mathcal{L}$. There are exceptions to these statements for highly evolved stars, but they will suffice for now.

Future discussions will make extensive use of a power law expression for ε of the form

$$\varepsilon = \varepsilon_0 \rho^\lambda T^\nu \tag{1.56}$$

where ε_0, λ, and ν are constants over some sufficiently restricted range of T, ρ, and composition. As important examples, consider briefly the two ways that stars burn hydrogen into helium (^{4}He). These are the proton-proton (pp) chains, and the carbon-nitrogen-oxygen (CNO) cycles. The first is, for the most part, a simple sequence of nuclear reactions, starting with one involving two protons, that gradually add protons to intermediate reaction products to eventually produce helium. The second cycle uses C, N, and O as catalysts to achieve the same end. For typical hydrogen-burning temperatures and densities ($T \gtrsim$ a few million degrees, ρ of order one to one hundred g cm^{-3}), the temperature and density exponents ν and λ are given in Table 1.1. We also give the exponents for the "triple-alpha" reaction, which effectively combines three ^{4}He nuclei to make one nucleus of ^{12}C at temperatures exceeding 10^8 K. The constant term ε_0 shall not concern us for the present, and the derivation of all these numbers will be given in Chapter 6.

TABLE 1.1. Temperature and density exponents

Energy Generation Mode for ε	λ	ν
pp-chains	1	≈ 4
CNO-cycles	1	≈ 15
Triple-α	2	≈ 40

On the hydrogen-burning main sequence, the pp chains dominate for stars of mass less than about one solar mass, but the CNO cycles take over for more massive stars. This sensitivity to mass just reflects the combined factors of the general tendency of temperatures to increase with mass (see Fig. 1.1) and the relative values of the temperature exponents, ν, for the two modes of energy generation.

The total energy released in the conversion of hydrogen to helium is approximately 6×10^{18} ergs for every gram of hydrogen consumed. To get an idea of what this might represent, a simple calculation will easily convince you that the sun, with its present-day hydrogen content of roughly 70% by mass, could continue to shine for almost 10^{11} years at its present luminosity just by burning all its available hydrogen.

What about the other factor in thermal balance? What determines $\mathcal{L}_r$? As we shall see, there are three major modes of energy transport: radiation (photon) transfer, convection of hotter and cooler mass elements, and heat conduction, with the first two being most important for most stars. (White dwarfs depend heavily on the last mode, but those stars are in a class by themselves.)

For those of us concerned primarily with the interiors of stars, it is fortunate that the transfer of energy by means of radiation is easily described. Except for the very outermost stellar layer, the energy flux carried by radiation obeys a Fick's law of diffusion; that is, the flow is driven by a gradient of a quantity having something to do with the radiation field. The form is $\mathcal{F}(r) = -\mathcal{D}\, d(aT^4)/dr$ where aT^4 is the radiation energy density and $\mathcal{D}$ is a diffusion coefficient. We shall show in Chapter 4 that the important part of $\mathcal{D}$ is the "opacity," κ, which, by its name alone, lets you know how the flow of radiation is hindered by the medium through which it passes. We suspect that $\mathcal{D}$ should be inversely proportional to κ. Without further ado, multiply $\mathcal{F}(r)$ by $4\pi r^2$ to obtain a luminosity, put in the relevant factors in $\mathcal{D}$ to be derived later, and find

$$\mathcal{L}_r = -\frac{4\pi r^2 c}{3\kappa\rho} \frac{d\,aT^4}{dr}. \tag{1.57}$$

The alternative Lagrangian form is

$$\mathcal{L}_r = -\frac{(4\pi r^2)^2 c}{3\kappa} \frac{d\,aT^4}{d\mathcal{M}_r}. \tag{1.58}$$

We shall have ample opportunity to use both of these forms.

The calculation of opacities is no easy matter and there is a whole industry set up for just that purpose. Later on we shall outline briefly what goes into them but, for now, we write a generic opacity in the power law form

$$\kappa = \kappa_0 \rho^n T^{-s} \;\; \text{cm}^2\ \text{g}^{-1}. \tag{1.59}$$

As in the case of ε, the coefficients and powers κ_0, n, and s are constants. Important examples are electron Thomson scattering opacity ($n = s = 0$), which is important for completely ionized stellar regions, and Kramers' opacity ($n = 1$, $s = 3.5$), which is characteristic of radiative processes involving atoms.

The luminosity carried by the transport of hot or colder material, which we call convection, is a good deal more difficult to treat. We shall give a simple prescription in the section below along with simple ideas that tie together what has been discussed thus far.

1.6 Stellar Dimensional Analysis

Some texts on stellar evolution (and see especially Cox 1968, chap. 22) discuss the topics of "homology" and "homologous stars." These terms describe sequences of simple spherical stellar models in complete equilibrium where one model is related to any of the others by a simple change in scale. More specifically, we assume that the models all have the same constituent physics (equation of state, opacity, etc., as given by power laws), the same uniform composition, and that $\mathcal{M}_r$ and r are related as follows. If one of the stars in the homologous collection is chosen as a reference star—call it star 0 and refer to it by a zero subscript—then these relations must apply in order that the stars be homologous to one another:

$$r = \frac{\mathcal{R}}{\mathcal{R}_0} r_0 \tag{1.60}$$

and

$$\mathcal{M}_r = \frac{\mathcal{M}}{\mathcal{M}_0} \mathcal{M}_{r,0} \tag{1.61}$$

where those quantities not subscripted with a zero refer to any another star in the collection. These relations mean that the stars have the same relative mass distribution such that radius r and the mass interior to that radius are related by simple ratios to the corresponding quantities in the reference star. We may also replace these equations by their derivatives keeping $\mathcal{R}$, $\mathcal{R}_0$, $\mathcal{M}$, and $\mathcal{M}_0$ all constant.

A consequence of the above is that the mass equations (1.1) for two stars may be divided one by the other to give a relation between the densities at equivalent mass points:

$$\rho = \rho_0 \frac{d\mathcal{M}_r}{d\mathcal{M}_{r,0}} \frac{1}{dr/dr_0} \left(\frac{r}{r_0}\right)^{-2} = \rho_0 \left(\frac{\mathcal{M}}{\mathcal{M}_0}\right) \left(\frac{\mathcal{R}}{\mathcal{R}_0}\right)^{-3} . \tag{1.62}$$

For stars of constant, but differing, densities this is obvious. It would also be an obvious result in a comparison of *average* densities between *any* two stars. However, (1.62) is true in general only for homologous stars.

What follows is a simplified treatment of homologous stars using a form of dimensional analysis. We shall follow the scheme of Carson (1986), and the results obtained will turn out to be identical to those obtained from standard homology arguments. They will also be very useful for estimating how various stellar quantities such as mass, radius, etc., are related. Again, however, the results are not to be used blindly.

We start by writing the Lagrangian version of the equation of hydrostatic equilibrium (1.14) in a form that emphasizes the dependence of pressure on mass and radius. Fundamental constants, such as G, could be retained but, at the end, it would be apparent that they were not needed. We then have

$$P \propto \frac{\mathcal{M}^2}{\mathcal{R}^4} \tag{1.63}$$

where $\mathcal{M}$ and $\mathcal{R}$ are chosen to represent mass and radius variables as in the spirit of (1.62). The pressure is specified in power law form in the same way as was done for the energy generation rate and opacity. Thus write

$$P = P_0 \rho^{\chi_\rho} T^{\chi_T} . \tag{1.64}$$

The constants P_0 (which will not be needed), χ_ρ, and χ_T, are assumed to be the same for all stars in the set under consideration. Note that (1.64) may also be written in differential form as

$$d \ln P = \chi_\rho \, d \ln \rho + \chi_T \, d \ln T . \tag{1.65}$$

If (1.63) and (1.64) are equated, we then arrive at a relation between $\mathcal{R}$, ρ, T, and $\mathcal{M}$, which is also written in logarithmic derivative form; namely,

$$4 \, d \ln \mathcal{R} + \chi_\rho \, d \ln \rho + \chi_T \, d \ln T = 2 \, d \ln \mathcal{M} . \tag{1.66}$$

The plan is now to treat the energy equation (1.55), the power law form of the energy generation rate (1.56), the diffusive radiative transfer equation

(1.58), and the power law opacity (1.59) in the same way as we did earlier. The aim will be to construct separate $\mathcal{R}$, ρ, T, and $\mathcal{L}$ versus $\mathcal{M}$ relations as

$$\mathcal{R} \propto \mathcal{M}^{\alpha_{\mathcal{R}}} \tag{1.67}$$

$$\rho \propto \mathcal{M}^{\alpha_\rho} \tag{1.68}$$

$$T \propto \mathcal{M}^{\alpha_T} \tag{1.69}$$

$$\mathcal{L} \propto \mathcal{M}^{\alpha_{\mathcal{L}}} \tag{1.70}$$

where the exponents α are to be determined. We have the requisite number of equations to do this. For example, equations (1.67–1.70) may be inserted into (1.66) to yield one relation between the αs:

$$4\alpha_{\mathcal{R}} + \chi_\rho \alpha_\rho + \chi_T \alpha_T = 2$$

where a common factor of $d \ln \mathcal{M}$ has been divided out. If this sort of thing is done for, in order, the mass equation, the equation of hydrostatic equilibrium (just done), the energy equation, and, finally, the transfer equation, we then obtain the matrix equation

$$\begin{pmatrix} 3 & 1 & 0 & 0 \\ 4 & \chi_\rho & 0 & \chi_T \\ 0 & \lambda & -1 & \nu \\ 4 & -n & -1 & 4+s \end{pmatrix} \begin{pmatrix} \alpha_{\mathcal{R}} \\ \alpha_\rho \\ \alpha_{\mathcal{L}} \\ \alpha_T \end{pmatrix} = \begin{pmatrix} 1 \\ 2 \\ -1 \\ 1 \end{pmatrix}. \tag{1.71}$$

The determinant of the matrix on the left-hand side of (1.71) is

$$\mathrm{D_{rad}} = (3\chi_\rho - 4)(\nu - s - 4) - \chi_T(3\lambda + 3n + 4) \tag{1.72}$$

where the "rad" subscript reminds us that energy transfer is by radiation in this case. We assume here that $\mathrm{D_{rad}}$ is not zero but it could be for some particular combination of temperature and density exponents. The latter circumstance leads to some strange situations, which we will discuss later.

The solutions to (1.71) are then (adapted from Carson 1986, with the correction of a minor typographical error in $\alpha_{\mathcal{R}}$):

$$\alpha_{\mathcal{R}} = \tfrac{1}{3}\left[1 - 2(\chi_T + \nu - s - 4)/\mathrm{D_{rad}}\right] \tag{1.73}$$

$$\alpha_\rho = 2(\chi_T + \nu - s - 4)/\mathrm{D_{rad}} \tag{1.74}$$

$$\alpha_{\mathcal{L}} = 1 + \left[2\lambda(\chi_T + \nu - s - 4) - 2\nu(\chi_\rho + \lambda + n)\right]/\mathrm{D_{rad}} \tag{1.75}$$

$$\alpha_T = -2(\chi_\rho + \lambda + n)/\mathrm{D_{rad}} \tag{1.76}$$

where these are to be used in (1.67–1.70) only in the situation where radiation is assumed to carry all the luminosity (or where radiation transfer seems to dominate).

If energy transport is primarily by means of convection, then the above analysis has to be modified and we include that analysis for completeness (although, as we shall see, the results are of limited use). We shall have to wait until Chapter 5 to explore convection in detail, but it will have to suffice for now to state that vigorous and efficient convection implies that the dependence of temperature on density as a function of radius is adiabatic. Specifically this means that

$$T(r) \propto \rho(r)^{\Gamma_3-1} \tag{1.77}$$

where $(\Gamma_3 - 1)$ is the adiabatic thermodynamic derivative

$$\Gamma_3 - 1 = \left(\frac{d \ln T}{d \ln \rho}\right)_{\text{ad}} \tag{1.78}$$

similar to Γ_1 of (1.36). Γ_3 is also of order unity and we shall see much more of these Γs later. This relation replaces the radiative transfer equation of the preceding analysis and means that the last row in the matrix of (1.71) is replaced by $(0,\ \Gamma_3 - 1,\ 0,\ -1)$ and the last element of the right-hand side constant column vector is now zero. A simple calculation yields the determinant for the new system

$$\mathrm{D}_{\text{conv}} = (3\chi_\rho - 4) + 3\chi_T(\Gamma_3 - 1) \tag{1.79}$$

and the new exponents α are

$$\alpha_{\mathcal{R}} = (1 - 2/\mathrm{D}_{\text{conv}})/3 \tag{1.80}$$

$$\alpha_\rho = 2/\mathrm{D}_{\text{conv}} \tag{1.81}$$

$$\alpha_{\mathcal{L}} = 1 + 2[\nu(\Gamma_3 - 1) + \lambda]/\mathrm{D}_{\text{conv}} \tag{1.82}$$

$$\alpha_T = 2(\Gamma_3 - 1)/\mathrm{D}_{\text{conv}}. \tag{1.83}$$

How well does this analysis work? The stars that we think we know the most about are located on the hydrogen main sequence. For the most part these stars are nearly homogeneous in composition and their masses, luminosities, and radii are relatively well-determined. Figure 1.2, constructed primarily from data given in Allen (1973, §100, and see Table 3–6 in Mihalas and Binney 1981) illustrates the observed relation between these three quantities.

From our previous discussion we expect that stars on the upper (more massive part of the) main sequence should have higher central temperatures just because they are more massive. The appropriate opacity law to use in this case is electron scattering for which $n = s = 0$. Similarly, the energy is generated primarily by the CNO cycles and thus, from Table 1.1, $\lambda = 1$ and $\nu \approx 15$. Although, as we shall show, the inner regions of these stars are convective, radiative transport of energy still dominates in the outer regions from which the power finally escapes. Finally, although radiation pressure is important, the pressure is mostly determined by the ideal gas

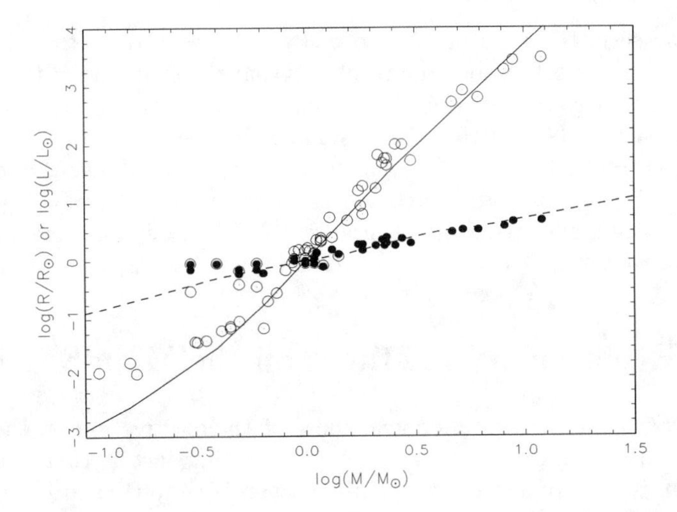

FIGURE 1.2. Luminosity and radius versus mass for main sequence stars. All quantities are in solar units. The solid (dashed) line is that for luminosity (radius) adapted from material in Allen (1973). Open (filled) dots are luminosity (radius) for components of binaries, from Harris *et al.* (1963) and Böhm (1989).

law for which $\chi_\rho = \chi_T = 1$. If these stars represent, roughly, a homologous sequence, then the preceding analysis should give values of $\alpha_\mathcal{R}$ and $\alpha_\mathcal{L}$ that reproduce the slopes in Figure 1.2. Using equations (1.72) through (1.76) and the exponents just quoted, we find that $\alpha_\mathcal{R} = 0.78$, and $\alpha_\mathcal{L} = 3.0$. A fit to the slopes in Figure 1.2 for stars with masses greater than about one solar mass yields

$$\frac{\mathcal{R}}{\mathcal{R}_\odot} \approx \left(\frac{\mathcal{M}}{\mathcal{M}_\odot}\right)^{0.75} \tag{1.84}$$

and

$$\frac{\mathcal{L}}{\mathcal{L}_\odot} \approx \left(\frac{\mathcal{M}}{\mathcal{M}_\odot}\right)^{3.5} \tag{1.85}$$

where the sun is not only used for normalization of the various quantities, but it appears as the reference star in the homologous set of stars. Obviously the homology relations have done very well. In addition, $\alpha_T = 0.22$ and $\alpha_\rho = -1.33$ so that temperature should increase with mass on the upper main sequence whereas density should decrease. We state now, without further proof, that this is indeed what happens. Stellar models show that central (as a homologous point) temperatures and densities do just this and the exponents are just about what we find.

The lower (less massive) main sequence is more difficult to treat. The pp chains dominate the energy generation rate and Kramers' opacity (with $n = 1$, $s = 3.5$) operates through much of the star but, and especially for very low mass stars, convection carries the luminosity almost everywhere.

This may seem to be no problem because we have derived the homology relations for convection, but the trouble is that the structure of these stars may be almost solely determined by what happens at the very outermost radiative surface (see Chapter 7). Assuming that radiative transfer is most important results in better agreement with the observations. (Try putting in an ideal gas, etc., into the convective analysis with $\Gamma_3 \approx 5/3$ and see if the resulting exponents match Figure 1.2. Then do the same with the radiative relations.) We shall have to do a lot more to get these stars right.

1.7 Evolutionary Lifetimes on the Main Sequence

It is a fact of life that stars spend most of their active life on the main sequence converting hydrogen to helium. Another fact is that when approximately 10% of a star's original hydrogen is converted to helium, the star undergoes rapid structural transformations that cause its luminosity and/or effective temperature to change enough so that it can no longer be called a main sequence star. (Why this happens is a later subject.) Thus the main sequence lifetime is geared to the rate at which fusion reactions take place. To estimate that time, t_{nuc}, all we have to do is calculate how much energy is released by burning 10% of the star's available hydrogen and compare it to the main sequence luminosity. From the figures quoted before for the energy release per gram in hydrogen burning, it is evident that

$$t_{\text{nuc}} \approx \frac{0.1 \times 0.7 \times \mathcal{M} \times 6 \times 10^{18}}{\mathcal{L}} \text{ s} \tag{1.86}$$

or, after converting to years and solar units,

$$t_{\text{nuc}} \approx 10^{10} \left(\frac{\mathcal{M}}{\mathcal{M}_\odot}\right) \left(\frac{\mathcal{L}}{\mathcal{L}_\odot}\right)^{-1} \text{ years.} \tag{1.87}$$

Note that a factor of 0.7 appears in (1.86). This is the typical value of the hydrogen mass fraction X given previously.

To eliminate the luminosity in (1.87), use the mass-luminosity relation (1.85) and find, for upper main sequence stars,

$$t_{\text{nuc}} \approx 10^{10} \left(\frac{\mathcal{M}}{\mathcal{M}_\odot}\right)^{-2.5} \text{ years.} \tag{1.88}$$

The main sequence lifetime of the sun is thus expected to be around 10^{10} years. More massive stars have shorter lifetimes because they are so profligate in using up their fuel to maintain their high luminosities. Stars on the lower main sequence with masses not much less than the sun have lifetimes that exceed present estimates for the age of the galaxy and universe.

This simple theoretical result explains why the main sequence for clusters (all of whose stars are assumed to have been formed at nearly the same time) terminates at the "turn-off point" leaving only the lower mass stars, and why rough estimates may be made for the ages of those clusters (although more is involved than what we have implied).

Now that we have completed a brief summary of some basic material, the next chapter will contain an overview of stellar evolution and some of the kinds of stars evolution produces. After that we shall get down to business, deriving the equations governing stellar structure and evolution, and developing the necessary input physics required to find their solutions.

1.8 Exercises

1. This is a little exercise in some items that this book does *not* cover but which are essential to an understanding of stars. It has to do with spectral classification of stars in the UBV photometric system and some other matters. We recommend that you browse through the second and third chapters of Mihalas and Binney (1981). Appendix A also contains some information. Most of what you need for this exercise may also be found in Allen (1973, chap. 10). Note that you will have to look up numbers in tables and these tables are not always entirely consistent: it's still not an exact science. Some of the answers you get for the following questions will therefore be estimates but they will be good ones. In any case, you are told that a star has been observed with a UBV color index of $B - V = 1.6$ and that interstellar reddening is negligible. In addition, the parallax of the star is $\pi = 0.25$ seconds of arc, and its apparent visual magnitude is $m_V = 9.8$. Detailed spectroscopy also reveals that the star has all the characteristics of a main sequence star (luminosity class V).
   (a) What is the spectral class of the star?
   (b) What is the distance to the star (in parsecs), its distance modulus, its absolute magnitude (M_V), bolometric correction ($B.C.$), bolometric magnitude ($M_{\rm bol}$), and luminosity (in $\mathcal{L}_\odot$)?
   (c) What is its effective temperature ($T_{\rm eff}$) and radius (in $\mathcal{R}_\odot$)?
   (d) Estimate the mass of the star (in $\mathcal{M}_\odot$).

2. A useful (albeit not terribly realistic) model for a homogeneous composition star may be obtained by assuming that the density is a linear function of radius. (See Stein 1966.) Thus assume that

$$\rho(r) = \rho_c \left[1 - r/\mathcal{R}\right]$$

where ρ_c is the central density and $\mathcal{R}$ is the total radius where zero boundary conditions, $P(\mathcal{R}) = T(\mathcal{R}) = 0$, apply.

(a) Find an expression for the central density in terms of $\mathcal{R}$ and $\mathcal{M}$. (You will have to use the mass equation.)

(b) Use the equation of hydrostatic equilibrium and zero boundary conditions to find pressure as a function of radius. Your answer will be of the form $P(r) = P_c \times$(polynomial in $r/\mathcal{R}$). What is P_c in terms of $\mathcal{M}$ and $\mathcal{R}$? (It should be proportional to $G\mathcal{M}^2/\mathcal{R}^4$.) Express P_c numerically with $\mathcal{M}$ and $\mathcal{R}$ in solar units.

(c) In this model, what is the central temperature, T_c? (Assume an ideal gas.) Compare this result to that obtained for the constant-density model. Why is the central pressure higher for the linear model whereas the central temperature is lower?

(d) Verify that the virial theorem is satisfied and write down an explicit expression for Ω (i.e., what is q of Eq. 1.7?).

3. We shall discuss completely degenerate electron equations of state in Chapter 3, but we can use them now without explaining what they are. If the electrons are nonrelativistic, then the power law exponents for pressure of equation (1.64) can be shown to be $\chi_\rho = 5/3$ and $\chi_T = 0$. Use this information to find the exponent $\alpha_\mathcal{R}$ in $\mathcal{R} \propto \mathcal{M}^{\alpha_\mathcal{R}}$ of (1.67). You will find that it does not matter whether the star is fully convective or fully radiative; you get the same answer from the homology relations. (The answer may be found in §3.5.2.)

1.9 References and Suggested Readings

The following format for references will be used throughout this text. In addition to listing the sources, we will occasionally make editorial comments leading the reader to where we believe especially good discussions of some material can be found. General references are usually listed first. These are then followed by those keyed to sections within a chapter.

General References

Many of the quotes found at the beginning of the chapters are from

▷ *The Oxford Dictionary of Quotations*, 3d ed. 1980 (Oxford: Oxford University Press).

▷ Metcalf, F. 1986, *The Penguin Dictionary of Modern Humorous Quotations* (London: Penguin Books Ltd.).

The monograph by

▷ Cox, J.P. 1968, *Principles of Stellar Structure*, in two volumes (New York: Gordon and Breach),

which was written with the aid of R.T. Giuli may still be *the* text on stellar structure. You can sometimes find it in used bookstores, but even then its price is beyond the means of the average student. What you will *not* find in this work are modern discussions of some topics such as evolution in close binary systems, supernova models, magnetic fields, rotation, etc. Do not let this discourage you. The care paid to detail and accuracy, and the clarity of style, are worth it. You will note, incidently, that we have attempted to conform to Cox's nomenclature for various quantities but there is no true standard. You may have to do some translation if you consult other texts.

We must also guide you to the excellent text by

▷ Kippenhahn, R., and Weigert, A. 1990, *Stellar Structure and Evolution* (Berlin: Springer-Verlag).

This work was released relatively late during the completion of our own text and so, except for just in a few places, we have not referenced it as much as we might have wished. The authors pioneered much of the work in stellar structure and evolution over the last twenty-five years, and their text contains a wealth of detail regarding the results of stellar modeling. Although much of their philosophy and nomenclature differ from what you will find here, both texts supplement each other in many respects.

You can now purchase a paperback version (1983) of the text by

▷ Clayton, D.D. 1968, *Principles of Stellar Evolution and Nucleosynthesis* (New York: McGraw-Hill).

It, like Cox (1968), is a bit outdated, but the last four chapters on nuclear reactions and nucleosynthesis are still the clearest and most complete. There are also excellent sections on the calculation of opacities and other quantities discussed from a nice physical viewpoint.

Two other texts worthy of mention are

▷ Böhm-Vitense, E. 1992, *Introduction to Stellar Astrophysics: Stellar Structure and Evolution* (Cambridge: Cambridge University Press),

which is the third volume in a three-volume series, and

▷ DeLoore, C.W., and Doom, C. 1992, *Structure and Evolution of Single and Binary Stars* (Hingham, Mass.: Kluwer).

This last text may prove especially useful for its treatment of binary system, which is a topic we only touch upon in Chapter 2.

The text by

▷ Mihalas, D., and Binney, J. 1981, 2d ed. *Galactic Astronomy* (San Francisco: Freeman)

has a wealth of material on stars and other matters astronomical and astrophysical. We recommend it strongly as a general reference for all students. Yet another is the monograph by

▷ Jaschek, C., and Jaschek, M. 1987, *The Classification of Stars* (Cambridge: Cambridge University Press).

As the title implies, this work describes how and why stars are classified observationally. Most sciences start off with observation and classification so the importance of such work should not be underestimated.

The relatively new text (in two volumes)

▷ Shu, F.H. 1991, 1992, *The Physics of Astrophysics*, Vols. 1–2 (Mill Valley, CA: University Science Books)

offers an interesting alternative to gathering together many texts to fill in the physics you need for astrophysics. The two volumes are at the graduate level but Shu gives enough introductory material for an undergraduate to follow the presentation. Not all topics are covered but this work may fit many of your needs. The total cost, however, is not insubstantial.

▷ Allen, C.W. 1973, *Astrophysical Quantities* 3d ed. (London: Athlone)

is the most popular compendium of astrophysical lore, tables, etc., to be found in a single volume. It should be on your shelf (if you can afford it). Another reference to look into is

▷ Lang, K.R. 1991, *Astrophysical Data: Planets and Stars* (Berlin: Springer-Verlag).

§**1.2**: An Energy Principle

▷ Goldstein, H. 1981, *Classical Mechanics* 2d ed. (Reading: Addison-Wesley)

is a standard text on classical mechanics. Many of us were raised on it.

If you can find the text by

▷ Chiu, H.-Y. 1968, *Stellar Physics*, Vol. 1 (Waltham, MA: Blaisdell),

we suggest you browse through its chapters. It is unfortunate that the second volume has never appeared. The first volume covers topics you cannot find in other standard texts in stellar astrophysics. It is now out of print.

§**1.3**: The Virial Theorem and Its Applications

The short monograph by

▷ Collins, G.W., II 1978, *The Virial Theorem in Stellar Astrophysics* (Tucson: Pachart)

contains many applications and variations on the virial theorem plus detailed derivations. Our discussion of the theorem does not include important topics such as magnetic fields, rotation, and relativistic effects. You will find them in Collins. The references to Cox (1968) and Clayton (1968) are listed above.

§**1.4**: The Constant-Density Model

▷ Stein, R.F. 1966, in *Stellar Evolution* (New York: Plenum Press), Eds. Stein and Cameron, pp. 3–82.

If you can find this symposium volume, Stein's article is worth the effort. In it he uses simple models to bring out important points in stellar structure and evolution.

§**1.6**: Stellar Dimensional Analysis

The *Observatory* often publishes useful short articles that deserve more exposure. Among these are often amusing commentaries on astronomical subjects and historical articles. The reference to

▷ Carson, T.R. 1986, *Observatory*, **106**, 71

may be found there.

The observational data for the mass-luminosity and mass-radius relations of Figure 1.2 are from Allen (1973) (see above) and

▷ Harris, D.L., III, Strand, K.Aa., and Worley, C.E. 1963, in *Basic Astronomical Data*, Ed. K.Aa. Strand (Chicago: University of Chicago Press), p. 273

▷ Böhm, C. 1989, *Ap. Space Sci.*, **155**, 241.

The first reference is in one in a series of books which, though somewhat outdated, still contain much useful material.

2

An Overview of Stellar Evolution

I never know how much of what I say is true.
— Bette Midler (1980)

The preceding chapter introduced some of the ingredients necessary for understanding simple stellar models. We now follow an entirely different path and assume that all the theoretical tools are in hand and evolved stellar models have already been constructed for our pleasure and enlightenment. All we have to do is study the results. The details of how some of these results come about are left to the remainder of the book.

It is impossible to present all that has been learned about the structure and evolution of stars in one self-contained package. In part, this is because there are serious gaps in our understanding but, moreover, extensive and internally consistent calculations are often difficult to find in the literature. Each investigator's views are also somewhat unique, and the trend is toward specialization. The sources for the overview given here are several and we make no pretense of suggesting that what you read here is the last word. In fact, we shall often have to state that there is no common agreement on the causes of some phenomenon or, at worst, concede that what we observe is still shrouded in mystery. What is encouraging, however, is the progress that *has* been made in the last few decades.

The scheme of this chapter is to outline the life history of different kinds of stars from their birth to some appropriate end. Again, you will want to consult Appendix A at the back of this text to find the meaning of some terms or concepts. Note also that we shall sometimes make statements of a cause-and-effect kind without explaining how such things come about.

All we can do is to suggest patience until a later chapter arrives with an explanation.

But, before we proceed, it is essential that we introduce the *Hertzsprung–Russell diagram*—abbreviated as the H–R diagram—which we shall use extensively. This two-dimensional diagram is the astronomer's way of characterizing important observational properties of stars. The vertical axis is a measure of the power output of a star while the abscissa tells us the color or, equivalently, the temperature of the visible surface. The units used for the axes depend on context and who is presenting them. An observer will usually express power in magnitudes of one sort or the other. A theoretician usually prefers luminosity (and the conversion from magnitude to luminosity is sometimes no easy matter). Similarly, the observer will indicate color as a difference in magnitudes between two spectral bands but the theoretician uses *effective temperature*, $T_{\rm eff}$, which is a theoretical construct. The relation between luminosity, total stellar radius, and $T_{\rm eff}$ is

$$\mathcal{L} = 4\pi\sigma\mathcal{R}^2T_{\rm eff}^4 \tag{2.1}$$

where Stefan-Boltzmann's constant $\sigma = 5.6705 \times 10^{-5}$ erg cm^{-2} K^{-4} s^{-1}. There are some subtleties to what is meant by radius and effective temperature but, in the simplest definition, $\mathcal{R}$ is the radius of the visible surface (photosphere) and $T_{\rm eff}$ is the temperature on that surface. Thus (2.1) is the blackbody radiant luminosity emitted from the surface of a sphere of radius $\mathcal{R}$ whose surface temperature is $T_{\rm eff}$. The effective temperature of the sun is $T_{\rm eff}(\odot) = 5778$ K. In solar units for $\mathcal{L}$ and $\mathcal{R}$, (2.1) becomes

$$\frac{\mathcal{L}}{\mathcal{L}_\odot} = 8.973 \times 10^{-16} \left(\frac{\mathcal{R}}{\mathcal{R}_\odot}\right)^2 T_{\rm eff}^4. \tag{2.2}$$

We shall usually use the $\mathcal{L}$–$T_{\rm eff}$ version of the H–R diagram. One major convenience in doing so is that it is very easy to place lines of constant radius on such a diagram. Note, however, that the effective temperature scale runs from right to left with the highest temperatures appearing on the left (for historical reasons). Note also, the H–R diagram gives no further information than $\mathcal{L}$, $T_{\rm eff}$, and $\mathcal{R}$. It says nothing (at least directly) about stellar mass, composition, or state of evolution.

An example of an H–R diagram is shown in Figure 2.1 from the review article by Iben (1991). It shows typical ranges of stellar luminosities and effective temperatures and three lines of constant radius that can be deduced from (2.1) or (2.2). Nearby and bright stars are also indicated (from data listed in Allen 1973). It is clear that most of these stars lie along a relatively well-defined locus, which is called the "main sequence." Others are collectively called "giants" (because of their large size) while a small number have radii of about $10^{-2}\,\mathcal{R}_\odot$ and these are the "white dwarfs." There are other kinds of stars than those shown in the figure and part of the task of this chapter is to explore possible evolutionary relationships between these diverse objects.

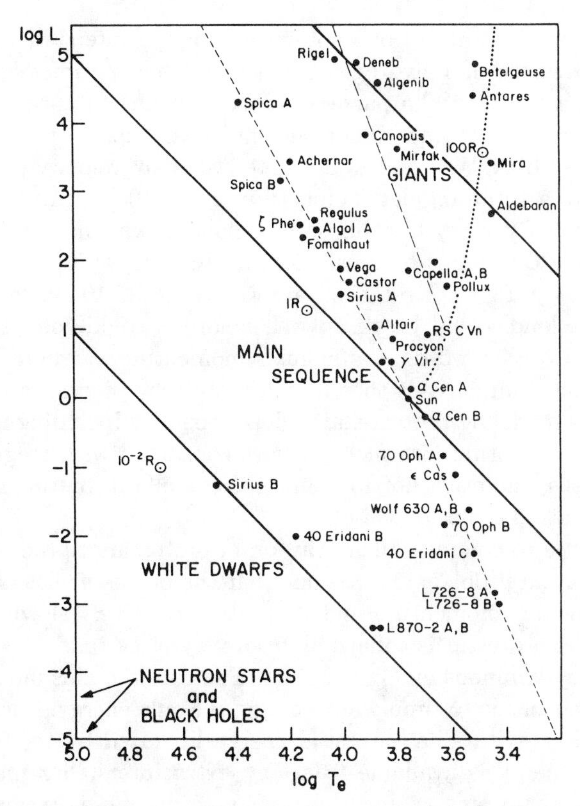

FIGURE 2.1. An illustrative Hertzsprung-Russell diagram showing nearby and bright stars as seen from the earth. Reproduced with permission from Iben (1991).

2.1 Pre–Main Sequence Evolution

The first step in a study of stellar structure and evolution is to form stars. That is no easy task. The brief pre–main sequence phase is the link between interstellar clouds and young stars, yet the theoretical study of star formation is a subject still in its infancy. This is despite a wealth of observational data on very young stellar objects (see the review by Shu, Adams, and Lizano 1987 for a flavor of the observations). This area remains a frontier because the physical conditions (i.e., densities, temperatures, and sources and sinks of energy) prevailing at the earliest time in a star's infancy are very different from those found in fully cooked stars. Stellar structure and evolution codes (which are designed to compute the hydrostatic evolution of stars) cannot accurately handle the important early phases such as dynamical collapse of a protostar.

As an illustration of the kinds of difficulties one encounters in the game of star formation, consider the "angular momentum problem." Assume that we want to make a star out of a spherical clump of interstellar stuff. With an average density of 1 hydrogen atom per cubic centimeter, we need a sphere with a radius of 2.5 parsecs (7.8×10^{18} cm) to encompass 1 $\mathcal{M}_\odot$ of interstellar matter. Since this cloud must participate in the overall rotation of the Milky Way, it has a built-in angular velocity of 10^{-15} s^{-1}, and therefore a total angular momentum of 5×10^{55} g cm^2 s^{-1}. One can easily see that if you try to collapse this cloud down to stellar dimensions (by decreasing the radius by 8 orders of magnitude) and conserve angular momentum as you go, the rotation rate would exceed 10 cycles per second! Clearly, the cloud would break up well before it could collapse to stellar dimensions. In other words, the angular momentum per unit mass of the interstellar medium exceeds that found in stars by several orders of magnitude. Thus modeling protostellar collapse requires including mechanisms for angular momentum loss such as those connected with magnetic fields, accretion disks, and other horrible things that stellar evolution calculations try to avoid at all costs.

All attempts to compute the evolution of protostars tiptoe past the angular momentum problem and assume that the protostar has already shed its excess angular momentum and is rotating relatively slowly (or not at all). Since the protostar is collapsing from very large radii, it will first appear as a very luminous and cool object. The core remains far too cool for thermonuclear fusion to supply any energy; the only energy source available to the star is gravitational potential energy liberated by the contraction. (Additional energy is available from any extra interstellar material that accretes onto the protostar, but this is also a form of gravitational energy.) At this stage, the time scale for radius change is the Kelvin–Helmholtz time scale (see §1.3.2) and, in this respect, the evolution of the protostar is described very well by equations (1.28)–(1.30).

Because of the high total luminosity, the interior of a nonaccreting hydrostatic protostellar model is convective. For fully convective stars, it is not possible to build a hydrostatic model if the temperature falls below a certain value; thus there is a "forbidden region" in the H–R diagram (and we shall see more of this in §7.3.3). The boundary of this forbidden region is almost vertical for a given mass star, so that as a star contracts it evolves at roughly constant $T_{\rm eff}$ along the boundary of the forbidden region (the so-called "Hayashi track," after C. Hayashi). Eventually, as the luminosity of the contracting protostar drops and the central temperature rises, the interior can become radiative. Further evolution of the star leads to an increase in the effective temperature, and a gradual lengthening of the time scale for evolution. Finally, if the mass of the star exceeds about 0.1 $\mathcal{M}_\odot$, the central temperature and density reach the point where thermonuclear fusion of hydrogen into helium becomes an important energy source, and the star settles onto the main sequence. Examples of such pre–main sequence evolutionary tracks are shown in Figure 2.2, along with the observed locations of T Tauri stars, which are those still in the process of

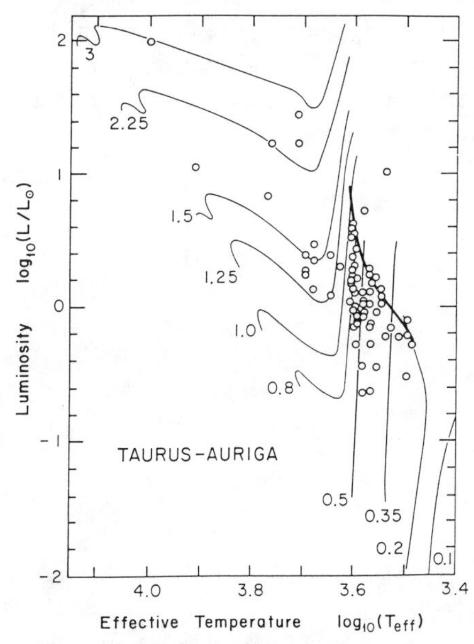

FIGURE 2.2. Shown are pre–main sequence evolutionary tracks adopted by Stahler (1988) from various sources. Masses are in solar units. Also shown are the observed locations of a number of T Tauri stars. Reproduced with permission.

contracting to the main sequence. The sense of evolution is from the upper right (cool temperatures) to the main sequence, where stellar masses are indicated.

2.2 Single Stars On and Near the Main Sequence

Upon arrival on the main sequence, a star is just beginning the longest active portion of its life. This is evident from studies of stellar populations that show, at least for stars in the solar neighborhood, that stars with the right combinations of luminosity and effective temperature to qualify them as main sequence or near–main sequence stars comprise approximately 80% of *all* stars. All other stars (except for most white dwarfs) are in a more transitory stage of evolution.

The time of arrival on the main sequence will be called "time zero" and a crucial step in modeling stars is to start there. A "zero age main sequence," or ZAMS, refers to that locus of points on the H–R diagram which describes a collection of stars satisfying the following requirements: they are chemically homogeneous with identical or very similar compositions, they are in hydrostatic and thermal balance, and energy is derived solely from nuclear burning. This is an operational definition used by the theoretician for constructing *ab initio* ZAMS stellar models (and, hopefully, those models will correspond to real main sequence stars). It says nothing about how the star got to the main sequence nor, in the usual case, does it contain information about the effects of magnetic fields, rotation, etc., which may differ from star to star on a main sequence.

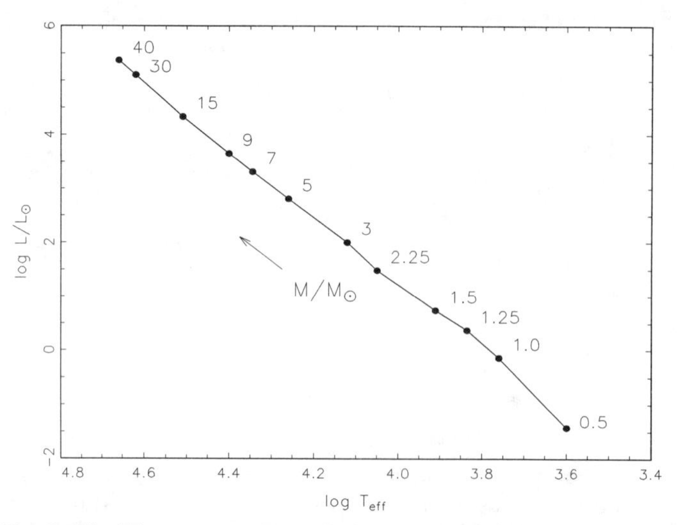

FIGURE 2.3. The Hertzsprung–Russell diagram for ZAMS stars compiled from Iben (1965) and Brunish and Truran (1982). Masses are in $\mathcal{M}_\odot$ and the composition is $X = 0.70$ and $Z = 0.02$.

To construct a ZAMS star model, only mass and composition need be specified. Specification of composition requires that the isotopic abundances of all elements be settled upon in advance. This constitutes the composition "mix." In addition, the equation of state, opacity, and energy generation rate must all be known for that mix at any temperature or density that could conceivably arise as the model is constructed. (How to construct or "compute" a model is an art to be elaborated upon in Chapter 7.) We also assume that heat transfer by radiation and conduction is well understood and that, if necessary, convective heat transfer is handled properly. That is actually a lot to assume!

But, given any arbitrary mass and mix, are we guaranteed that a stellar model exists; that is, can we find a solution to the problem as posed? With respect to composition the answer is clearly "no." (We do not even care to contemplate a star made of lead!) The same is true for mass. As an example, objects of near-Jovian mass exist right next door but they are not (usually) called stars. This is because they are too lightweight and cool to use thermonuclear fusion as an energy source at any stage in their careers. We thus suspect that there may be a minimum mass (for a given mix) below which there are no ZAMS stars. The lower mass limit for ZAMS stars with hydrogen-rich compositions is presently thought to be around 0.1 $\mathcal{M}_\odot$. Stars with very large masses are also penalized but for different reasons: truly massive hydrogen-rich stars are plagued by loss of mass from their surfaces or by dynamical instabilities to such an extent that their formation on a classical ZAMS seems to be precluded. The upper mass limit is uncertain but a value near 100 $\mathcal{M}_\odot$ is indicated.

A more esoteric question is whether more than one ZAMS solution may exist for a given mass and mix. Surprisingly, the answer is "yes." ZAMS models may be constructed that have the same mass and mix but whose luminosities, effective temperatures, and overall structures differ. (The reason for this is that equations of state, opacities, and nuclear generation rates are so complex that they may take on very different qualities depending on temperature and density.) This nonuniqueness is a violation of the so-called "Vogt–Russell theorem," which states that, if a solution exists, it must be unique. Unfortunately, the "theorem" is overly restrictive and does not apply in practice. Multiple solutions to the ZAMS problem (and others) that may exist, however, seem to involve structures that cannot be produced in nature as far as we know, although those solutions may have their own theoretical interest. The following discussion will consider only standard solutions that mimic nature in important respects. Furthermore, only hydrogen-rich ZAMS are of immediate interest, although, again, a theoretician may deal with more general objects.

Shown in Figure 2.3 is the ZAMS for a Pop I (relatively high metal content) composition and, in Figure 2.4, low-mass main sequences for the three compositions $(Y, Z) = (0.1, 10^{-4})$, $(0.3, 10^{-4})$, and $(0.3, 0.04)$. The first composition is unrealistic because such low helium concentrations are not seen in the atmospheres of main sequence stars. The second and third could represent extreme Pop II (low content of metals) and Pop I stars although the large value of Y may be somewhat too large for Pop II objects. Figures 2.5 and 2.6 show the corresponding luminosity versus mass relations.

The limits on mass used for these figures, of 0.50 $\mathcal{M}_\odot$ to 40 $\mathcal{M}_\odot$, were chosen for the following reasons. Above something like 40 $\mathcal{M}_\odot$, mass loss from the surface is so rapid that it is not clear how to construct completely realistic models since mass loss in stages just prior to the ZAMS strongly influences the final structure. The reason for terminating the ZAMS at 0.50 $\mathcal{M}_\odot$ is that stars at and below this mass have main sequence lifetimes that far exceed the probable age of the universe and are thus not of immediate interest for stellar evolution. (They are important for other reasons.) We have discussed main sequence lifetimes in the first chapter—as in (1.88)—but a very useful elaboration is due to Iben and Laughlin (1989). They offer the following expression as an approximation for the total elapsed evolutionary time, $t_{\rm evol}$, from the main sequence to the planetary nebula stage for stars with $0.6 \leq \mathcal{M}/\mathcal{M}_\odot \leq 10$:

$$\log t_{\rm evol} = 9.921 - 3.6648 \log \left(\frac{\mathcal{M}}{\mathcal{M}_\odot} \right)$$

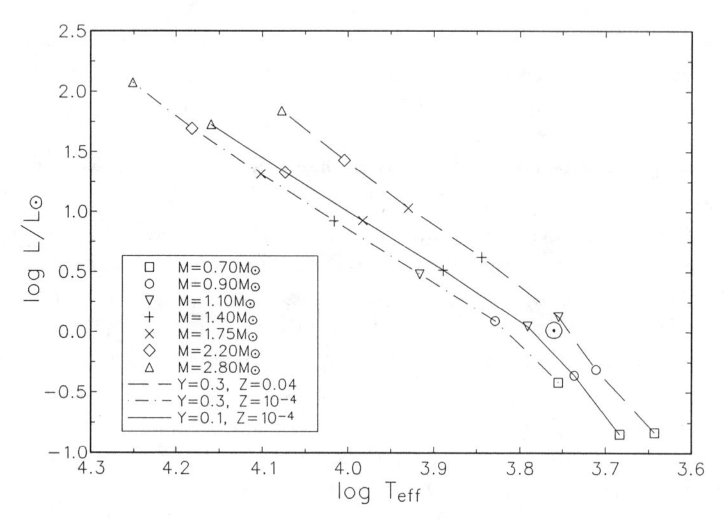

FIGURE 2.4. The theoretical ZAMS for low-mass stars from the calculations of Mengel et al. (1979). The results for three compositions are shown, as is the location of the present-day sun ($\odot$).

$$+ 1.9697 \left[\log\left(\frac{\mathcal{M}}{\mathcal{M}_\odot}\right)\right]^2 - 0.9369 \left[\log\left(\frac{\mathcal{M}}{\mathcal{M}_\odot}\right)\right]^3 \qquad (2.3)$$

where t_{evol} is in the units of years. This analytic approximation to evolutionary lifetimes should include the effects of composition but is adequate for many purposes. A major portion of t_{evol} is spent near the main sequence. Thus a star with mass 0.70 $\mathcal{M}_\odot$ has $t_{\text{evol}} \sim 35$ Gyr (1 Gyr is 10^9 years) and this age is very long indeed. We shall return shortly to this issue of the importance of main sequence lifetimes and the age of our galaxy or the universe.

Also noted in Figure 2.4 for the lower ZAMS is the location of the present-day sun. The unevolved sun, with (Y, Z) of near $(0.28, 0.02)$, must also lie among the curves in those figures and, from interpolation by eye, you should be able to place it close to the $\odot$. This means that the sun has not changed appreciably in either luminosity or effective temperature during the 4.5×10^9 years of its total life; the sun is still a main sequence star, albeit not a ZAMS star.

Other features of Figures 2.4 and 2.6 are the shifts of the various curves with composition. The qualitative behavior of these shifts may be reproduced using modifications of the homology relations discussed in §1.6—as follows.

Major opacity sources for the lower main sequence are of the Kramers' type with $n = 1$ and $s = 3.5$ as discussed in the previous chapter (and see Eq. 1.59). For Z greater than about 10^{-4} and X normal, an important part of the opacity is due to bound-free transitions (of which more later),

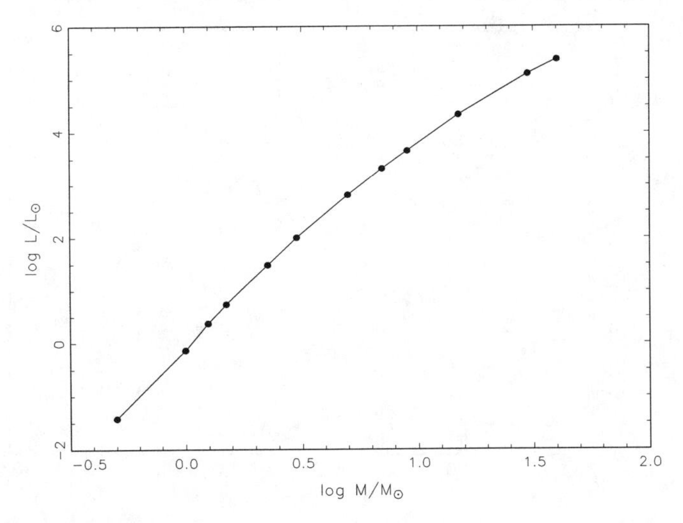

FIGURE 2.5. Luminosity versus mass for the main sequence compiled from the same sources as Figure 2.3.

which have the rough functional form

$$\kappa \propto Z(1+X)\rho T^{-3.5}. \tag{2.4}$$

You may easily verify that for X between 0.7 and 0.9 (as in the figures), the composition-dependent part of κ varies as $ZX^{0.44}$ where the exponent for X is a good approximation. Similarly, the mean molecular weight used in the ideal gas law appropriate for the lower main sequence may be recast as $\mu \propto X^{-0.57}$ (from Eq. 1.52) if we assume complete ionization. The pp chains supply the energy for low-mass hydrogen-burning stars and the rate for those chains depends on the square of the hydrogen concentration (because two protons are involved in starting off the chains). Hence, from (1.56) and Table 1.1,

$$\varepsilon_{\rm pp} \propto X^2 \rho T^4. \tag{2.5}$$

The rest is a simple exercise in manipulating the equilibrium equations of stellar structure, as was done for homologous stars in §1.6, but retaining the composition dependence in κ (assuming radiative transfer), ε, and pressure in the ideal gas law. The following results are obtained and their derivation will be left to the reader as an exercise:

$$\mathcal{L} \propto Z^{0.35} X^{1.55}\, T_{\rm eff}^{4.12}; \tag{2.6}$$

$$\mathcal{R} \propto Z^{0.15} X^{0.68} \mathcal{M}^{0.077}; \tag{2.7}$$

$$\mathcal{L} \propto Z^{-1.1} X^{-5.0} \mathcal{M}^{5.46}; \tag{2.8}$$

$$T_{\rm eff} \propto Z^{-0.35} X^{-1.6} \mathcal{M}^{1.33}. \tag{2.9}$$

Apart from some minor differences in the numerical values of the exponents,

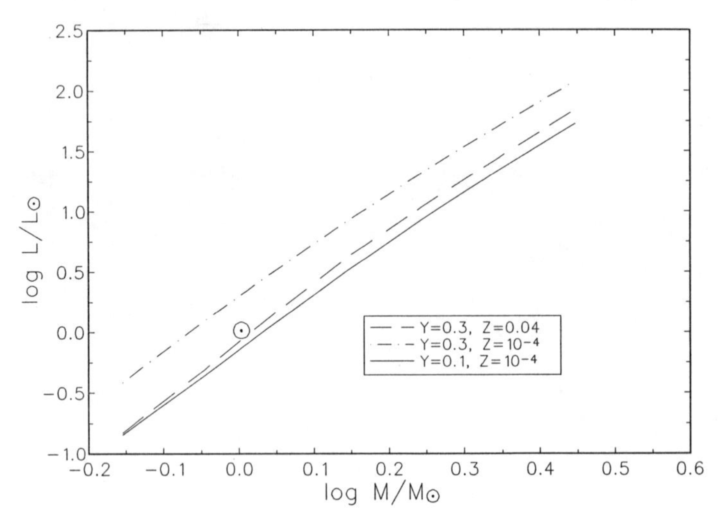

FIGURE 2.6. Luminosity versus mass for the lower ZAMS from the same source as Figure 2.4

these results are the same as those quoted in the informative review article by Sandage (1986) on the concept and history of population types for stars.

These relations, and specifically (2.6) and (2.8), do not do too badly in reproducing the various slopes deducible from Figures 2.4 and 2.6, as you may check by carefully picking numbers off the figures or, better yet, by consulting the original tables in Mengel et al. (1979). Only the luminosity dependence on Z for (2.8) is amiss because opacities for cool stars are considerably more complicated than the one we have chosen (but the exponent does have the correct sign).

An important consequence of the Z dependence in (2.6) is that stars with low concentrations of metals should form a ZAMS that lies below one for Pop I stars on the H–R diagram. This is best seen in Figure 2.4, where models with $Z = 10^{-4}$ (– · –) have lower luminosities at constant effective temperature than those with $Z = 0.04$ (– –). Or, the lower Z models are hotter (bluer) at the same luminosity. Such lower luminosity stars are observed among the field stars in the solar neighborhood, for example, and they constitute the "subdwarfs." These stars have high space velocities and low metal content—as evidenced by ultraviolet excesses in their colors—and are characteristic of Pop II objects. Thus the solar neighborhood contains a mixture of population types with differing initial ZAMS positions. This situation is the cause of considerable confusion because it may mask the effects of evolution within a population.

Tables 2.1 and 2.2 list some properties of representative ZAMS models from various sources. Along with the model mass and composition, each model is keyed by a model number (the first column) to help bridge across the tables. The fourth and fifth columns of Table 2.1 list the model luminos-

TABLE 2.1. Zero age main sequence models

No.	$\mathcal{M}/\mathcal{M}_\odot$	(X, Y)	$\log \mathcal{L}/\mathcal{L}_\odot$	$\log T_{\rm eff}$	$\mathcal{R}_{10}$	ref.
1	60	(0.74, 0.24)	5.701	4.683	70.96	(1)
2	40	(0.74, 0.24)	5.345	4.642	56.89	(1)
3	30	(0.74, 0.24)	5.066	4.606	48.53	(1)
4	20	(0.74, 0.24)	4.631	4.547	38.73	(1)
5	15	(0.74, 0.24)	4.292	4.498	32.89	(1)
6	10	(0.74, 0.24)	3.772	4.419	25.94	(1)
7	7	(0.74, 0.24)	3.275	4.341	20.99	(1)
8	5	(0.74, 0.24)	2.773	4.259	17.18	(1)
9	3	(0.74, 0.24)	1.951	4.118	12.76	(1)
10	2	(0.74, 0.24)	1.262	3.992	10.30	(1)
11	1.75	(0.74, 0.24)	1.031	3.948	9.683	(1)
12	1.50	(0.74, 0.24)	0.759	3.892	9.141	(1)
13	1.30	(0.74, 0.24)	0.496	3.834	8.831	(1)
14	1.20	(0.74, 0.24)	0.340	3.800	8.650	(1)
15	1.10	(0.74, 0.24)	0.160	3.771	8.035	(1)
16	1.00	(0.74, 0.24)	-0.042	3.752	6.934	(1)
17	0.90	(0.74, 0.24)	-0.262	3.732	5.902	(1)
18	0.75	(0.73, 0.25)	-0.728	3.659	4.834	(2)
19	0.60	(0.73, 0.25)	-1.172	3.594	3.908	(2)
20	0.50	(0.70, 0.28)	-1.419	3.553	3.553	(3)
21	0.40	(0.70, 0.28)	-1.723	3.542	2.640	(3)
22	0.30	(0.70, 0.28)	-1.957	3.538	2.054	(3)
23	0.20	(0.70, 0.28)	-2.238	3.533	1.519	(3)
24	0.10	(0.70, 0.28)	-3.023	3.475	0.805	(3)
25	0.08	(0.70, 0.28)	-3.803	3.327	0.650	(3)

ity and effective temperature and the sixth column gives the model radius in units of 10^{10} cm. (We shall occasionally use the subscript notation S_n to denote the value of a quantity S in units of 10^n.) $\mathcal{R}$ is, of course, derivable from the relation $\mathcal{L} = 4\pi\sigma\mathcal{R}^2 T_{\rm eff}^4$. The references in the last column are as follows:

(1) Models with this reference number have been generated using the computer code **ZAMS** discussed in Appendix C. This code uses simple physics and analytic fits to opacities and energy generation rates. It should not be used for ZAMS models with masses much less than $1\ \mathcal{M}_\odot$ and the results for the most massive stars are suspect because mass loss is not taken into account. Aside from these caveats, the models are respectable. You should be able to reproduce the table entries with this reference number (and much more besides) using this code.

(2) These models are from VandenBerg, D.A., Hartwick, F.D.A., Dawson, P., and Alexander, D.R. 1983, *Ap.J.*, **266**, 747. As with the models

TABLE 2.2. ZAMS models (continued)

No	$\mathcal{M}/\mathcal{M}_\odot$	$T_{c,6}$	ρ_c	$\log P_c$	q_c	q_{env}
1	60	39.28	1.93	16.22	0.73	0
2	40	37.59	2.49	16.26	0.64	0
3	30	36.28	3.05	16.29	0.56	0
4	20	34.27	4.21	16.37	0.46	0
5	15	32.75	5.48	16.44	0.40	0
6	10	30.48	8.33	16.57	0.33	0
7	7	28.41	12.6	16.71	0.27	0
8	5	26.43	19.0	16.84	0.23	0
9	3	23.47	35.8	17.06	0.18	0
10	2	21.09	47.0	17.21	0.13	neg.
11	1.75	20.22	66.5	17.25	0.11	neg.
12	1.50	19.05	76.7	17.28	0.07	neg.
13	1.30	17.66	84.1	17.28	0.03	neg.
14	1.20	16.67	85.7	17.26	0.01	10^{-7}
15	1.10	15.57	84.9	17.22	0	5×10^{-5}
16	1.00	14.42	82.2	17.17	0	0.0035
17	0.90	13.29	78.5	17.11	0	0.020
18	0.75	10.74	81.5	–	0	–
19	0.60	9.31	79.1	–	0	–
20	0.50	9.04	100	17.10	0	–
21	0.40	8.15	104	17.04	0	–
22	0.30	7.59	107	17.05	*	1
23	0.20	6.53	180	17.24	*	1
24	0.10	4.51	545	17.68	*	1
25	0.08	3.30	775	17.83	*	1

of reference (3), they contain much more sophisticated physics than does ZAMS.

(3) These very low mass models are from the "MM EOS" sequence of Dorman, B., Nelson, L.A., and Chau, W.Y. 1989, *Ap.J.*, **342**, 1003; and see Burrows, A., Hubbard, W.B., and Lunine, J.I. 1989, *Ap.J.*, **345**, 939.

The central temperature (in units of 10^6 K) is $T_{c,6}$, and ρ_c and P_c are, respectively, the central density and pressure in cgs units. These are listed in Table 2.2. Finally, the last two columns in that table list q_c and q_{env}. The quantity q_c is the fractional mass of a possible convective core in a model (see Chapter 5). For example, in a model of $\mathcal{M} = 60\ \mathcal{M}_\odot$ the inner 73% of the mass is convective starting from model center. The corresponding quantity q_{env} is the fractional mass contained in a fully or partially convective envelope. (The term "envelope" is often used in stellar astrophysics to refer to the outer layers of a star, although the exact meaning will depend on context.) For our purpose here, if q_{env} is not zero, then it is the fractional mass measured from the model surface inward to a level where convection ceases. Thus, for example, the outer 0.35% of the mass of the model numbered 16 (1 $\mathcal{M}_\odot$) is entirely or partially convective. This is a ZAMS model of the sun and, it turns out, it is completely convective from just under the photosphere inward to that mass level. This corresponds, however, to the outer 17% of the radius. A listing of "neg." for q_{env} means that a negligible fraction of the envelope is convective (say, less than 10^{-8} in mass). A "0" in that column means that there is no convection. Finally, a "–" implies that the information was not available to us. Now for what may be learned from the models.

For the higher mass stars, and keeping composition fixed, radius is seen to increase with mass as expected from the homology relation (1.84) where $\mathcal{R} \propto \mathcal{M}^{0.75}$. Since strict hydrostatic equilibrium holds for these models, equation (1.63) plus (1.84) implies that $P \propto \mathcal{M}^{-1}$. If this pressure is taken as the central pressure, then P_c should decrease with mass. It does, although not as fast as homology would imply. The relation of density to mass and radius of (1.62) combined with (1.84) yields $\rho \propto \mathcal{M}^{-5/4}$, and this general behavior is shown in Table 2.2 where ρ_c decreases with mass. We already know that luminosity increases with mass (from 1.85), and it is an easy matter to show that T_{eff} and T_c do so also. In summary, ZAMS stars of high mass get bigger, brighter, and less dense as mass increases.

Two of the ingredients of the above homology analysis were that the ideal gas law was appropriate for the pressure and that energy was carried exclusively by radiation. Neither of these assumptions is strictly true. If you carefully examine the central properties of the 60 $\mathcal{M}_\odot$ model, for example, you will find that the combination of T_c, ρ_c, the given composition (with the results of §1.4.1), and the ideal gas law do not add up to P_c; the value for P_c in the table is too large by a factor of about 1.6. The reason for this is that for massive ZAMS stars, radiation pressure plays a significant role in supporting the deep stellar interior. This is reviewed in §7.2.7 in terms of a simple but elegant model.

Radiation plays another important role in massive main sequence stars. We have seen from the tables that massive stars are very luminous. As this power emerges from the very outermost layers of the star, it must eventually appear in the form of radiation to be passed out to space. This radiation, however, exerts pressure and it is conceivable that this pressure, pushing outward on the stellar material, may be strong enough to overcome the inwardly directed force of gravity. We address this problem more fully in §7.3.2, where an estimate is made of what luminosity is required to just balance gravity. This "Eddington Limit" is $\mathcal{L}_{Edd}/\mathcal{L}_\odot \approx 3.5 \times 10^4\ \mathcal{M}/\mathcal{M}_\odot$ (see Eq. 7.127). If the luminosity exceeds this limit for a given mass, then hydrostatic equilibrium is in jeopardy. For the 60 $\mathcal{M}_\odot$ of Table 2.1, the luminosity is $5 \times 10^5\ \mathcal{L}_\odot$, whereas its Eddington luminosity is $\mathcal{L}_{Edd} \approx 2 \times 10^6\ \mathcal{L}_\odot$. This is perilously close.

It is also obvious that convection is important in these massive stars. The sixth column of Table 2.2 reveals that a major fraction of the deep interior mass is convective. The reason for this is as follows. Recall that massive ZAMS stars produce energy by burning hydrogen via the CNO cycles (see §1.5). This mode of energy generation has a very large temperature exponent ($\nu \approx 15$), which means that the energy generation rate must increase rapidly near the stellar center where temperatures are higher. This presents a problem for the star because all the power generated must pass through the surface of a relatively small sphere. If this power is to be transported by radiation, then equations (1.57–1.58) imply that the local temperature gradient must be large and negative. As we shall show in Chapter 5 (with an example for massive ZAMS stars), this is the kind of situation that causes the stellar material to become convective. Convection in this situation can transport a lot of power because hot dense material is transported to cooler regions where heat is then deposited. The net result for the star is that the temperature gradient (and other physical parameters of the structure) are adjusted so that a state of thermal balance is produced wherein the local rate of energy production is balanced by losses (as discussed in §1.5). If you run the **ZAMS** code discussed in Appendix C and for these massive models, you will find that convection transports practically all of the power through the stellar core.

As we go "down" the main sequence to less massive stars, the size of the convective core shrinks until at about 1.2 $\mathcal{M}_\odot$ it disappears entirely. This is mainly due to the decreasing importance of the CNO cycles in favor of the pp chains, which have a much lower temperature sensitivity ($\nu \approx 4$). On the other hand, and at about the same mass, Table 2.2 reveals that the very outermost layers of the star start to becomes convective. The reason for this differs from that given above. Here the effect is primarily due to the transition from Thomson scattering opacity to one of Kramers' form associated with the ionization and recombination of hydrogen and helium in the outer layers. The cooler outer layers of the cooler stars favor Kramers' opacity. Graphical examples of stellar opacities are given in Chapter 4 (Figs. 4.2 and 4.3) and these opacities are *large* when ionization and recombination processes are taking place. Since power transported by radiation is adversely affected by large opacities—as is evident from equation (1.57) where luminosity is inversely proportional to opacity—the temperature gradient must again be large and negative for this power to be driven through the outer layers. The result is that the stellar material becomes convectively unstable. Thus there is a dividing line for a ZAMS at a little over a solar mass where radiation and convection play their respective roles in transporting energy through various regions of the star. The result is that upper and lower main sequence stars are different in important respects.

The tendency of ever-increasing depth of envelope convection reaches a limit in the stars of the lower ZAMS at about 0.3 $\mathcal{M}_\odot$, where the entire star becomes convective. (The "$*$" in the q_c column of Table 2.2 indicates that this entry no longer means anything in this context.) Detailed modeling of these dim, cool, and dense stars is extraordinarily difficult. Their equations of state are so strongly influenced by the close proximity of neighboring atoms that the gas (if it can be called that) is no longer ideal or even perfect. The opacities in the outer layers are dominated by radiative transitions among molecules of many kinds and a good number of these have not yet been adequately investigated in the laboratory, nor have that many been computed theoretically.

Oddly enough, we can make approximate models of these stars just because they are completely convective. To see how this comes about, recall the comment made about convection and homology in §1.6 where we stated without proof (see Chapter 5 for more details) that vigorous convection implies a simple relation between the temperature and density as given by (1.77). The key quantity in that equation is the adiabatic exponent Γ_3, which is 5/3 for an ideal, monatomic, and nonionizing gas (and see §3.7). For completely convective and homogeneous stars with that ideal equation of state, the homology analysis of §1.6 may be extended to produce "polytropic models," or "polytropes," which essentially give the entire structure of the star. Polytropes are discussed more fully in §7.2 but the idea is simple enough. It is *assumed* that pressure is everywhere related to density by $P(r) = K\rho^{1+1/n}(r)$ where K is some constant and n is the constant "polytropic index." In the case we are considering here, it can be shown that $n = 3/2$. If the pressure versus density relation is applied to the equations of hydrostatic equilibrium and mass conservation and the proper boundary conditions are set, it is then relatively easy to calculate the ratio of central to mean density for the resulting polytropic model as $\rho_c/\langle\rho\rangle = 5.991$ (see Table 7.1). The mean density of a star using the units of Table 2.2 is

$$\langle\rho\rangle = 475\,\mathcal{R}_{10}^{-3}\,\mathcal{M}/\mathcal{M}_\odot \tag{2.10}$$

and you may indeed verify that $\rho_c/\langle\rho\rangle \approx 6$ for models with masses of 0.30 $\mathcal{M}_\odot$ or less. Note that we have not said it is easy find the total radius for these stars: it isn't.[1]

Another result from polytrope theory is that the central temperature for index 3/2 polytropes consisting of an ideal gas is given by

$$T_c = 1.235 \times 10^7\,\mu \left(\frac{\mathcal{M}}{\mathcal{M}_\odot}\right)\left(\frac{\mathcal{R}}{\mathcal{R}_\odot}\right)^{-1} . \tag{2.11}$$

If the mean molecular weight is $\mu \approx 0.6$, then this result predicts central temperatures of only a few million degrees Kelvin for the least massive

---

[1]ZAMS stars of greater mass have considerably larger values for the ratio of central to mean density and correspond more to polytropes of index $n = 3$ as discussed in §7.2.

ZAMS models in the tables and herein lies part of the reason the tables end at 0.08 $\mathcal{M}_\odot$.

In constructing ZAMS models we have insisted that they be in complete equilibrium. But this may not be possible for some masses. In particular, any attempt to construct equilibrium ZAMS models of normal composition with masses less than about 0.08 $\mathcal{M}_\odot$ are bound to fail. The problem, at one level, is that central temperatures for such low mass objects are too low for hydrogen burning to maintain thermal balance: the star leaks energy faster than it can be supplied by thermonuclear fusion. You might think that contraction and the associated rise in temperature would increase the rate of thermonuclear energy production so as to eventually bring about thermal balance but this is not the case: the object just cannot do it. But the problem is more subtle than this. We have to ask how superlow mass objects form in the first place. It is supposed that they form from the interstellar medium in much the same way as do normal stars and then gradually contract. But, contrary to the evolution of normal stars, these stars keep contracting and the gravitational energy released by the contraction is always a significant (or the only) source of energy. In this sense they never reach what we call the ZAMS but evolve to yet lower luminosities and effective temperature over very long time scales.

The term "brown dwarfs" is now commonly applied to objects with masses less than approximately 0.08 $\mathcal{M}_\odot$ although, at the time of this writing, none has been unambiguously identified from optical or infrared observations. Of course it may be argued with some justification that there is one in our immediate neighborhood; namely, we have Jupiter with a mass of 318 $\mathcal{M}_\oplus \approx 0.001\mathcal{M}_\odot$! For a review of very low mass stars—including possible brown dwarf candidates—see Liebert and Probst (1987) and other references cited at the end of this chapter.

Low mass main sequence stars, although not very luminous, are an important component of the general stellar population because there are so many of them. One way to demonstrate this is to find the rate at which stars of different mass are created. E.E. Salpeter, in a pioneering study in 1955, estimated this "birth rate function," $\psi_s \, d\mathcal{M}$, for stars in the local galactic disk. The following expression, quoted from Shapiro and Teukolsky (1983) (and see Mihalas and Binney 1981), gives the Salpeter rate of star formation in a sample pc^3 per year as

$$\psi_s \, d\left(\frac{\mathcal{M}}{\mathcal{M}_\odot}\right) = 2 \times 10^{-12} \left(\frac{\mathcal{M}}{\mathcal{M}_\odot}\right)^{-2.35} d\left(\frac{\mathcal{M}}{\mathcal{M}_\odot}\right) \ \mathrm{pc}^{-3}\,\mathrm{yr}^{-1} \qquad (2.12)$$

which is supposed to hold over the mass range $0.4 \lesssim \mathcal{M}/\mathcal{M}_\odot \lesssim 10$. Despite the uncertainties associated with the result—especially for the least massive stars—it gives the flavor of more modern and complicated analyses in that lower mass stars are more likely to be formed than those of greater mass. Combine this with the observation that lower mass stars live longer on the main sequence and we conclude that they must be very common indeed.

We now turn our attention to how stars evolve off of the main sequence, with the warning that the following simple discussion leaves out many of the important details. Some of those details will be picked up later but, for now, the reader is advised to read a section and then go to sample papers in the literature to see how complex the subject really is.

2.3 Evolution of Single Stars Off the Main Sequence

The evolution of stars following the ZAMS phase results in a rich variety of intermediate and final configurations. The principal parameter that determines the future course of evolution for a single star is its main sequence mass. We have loosely referred to upper and lower main sequences but, to be more specific about how stars with different ranges of initial mass on the ZAMS evolve, we shall refer to *upper*, *intermediate*, *lower*, and *unevolved lower* ZAMS. The last is the easiest to explain. Present estimates of the Hubble time range from 10 to 20 Gyr. If that time is roughly the age of the universe, then we certainly do not expect to find stars older than that. Using either (1.88) or (2.3) as a guide to main sequence lifetimes, stars whose masses are less than about $0.8\mathcal{M}_\odot$ cannot have exhausted their central hydrogen fuel supply and evolved off the main sequence even had they been formed at the birth of the Milky Way. For this reason, we shall leave the evolution of single stars with masses less than about $0.8\ \mathcal{M}_\odot$ to astronomers of a few eons hence. Note, however, if stars with low masses are members of binary or multiple systems, then their evolution may be of considerable interest, and we shall return to them later.

In the following we shall discuss the evolution of the different mass ranges of stars only up to a certain point. After we finish, later sections will carry their evolution forward in time to later phases.

Following Iben and Renzini (1983) we define the *lower* ZAMS to be that consisting of those stars whose masses lie between about 0.8–1.0 $\mathcal{M}_\odot$ and 2–2.3 $\mathcal{M}_\odot$ where the precise limits depend on composition. From the previous discussion, the least massive of these are just capable of evolving off the mass sequence in a Hubble time. So why are lower ZAMS stars different from those of mass exceeding about 2.3 $\mathcal{M}_\odot$? Figure 2.7 from Iben (1985), shows the evolutionary development of central temperature and density for single stars of mass 1, 2, 7, and 15 $\mathcal{M}_\odot$. The sense of time is indicated by the arrows on the "tracks" and note that temperature and density usually increase with time. For the moment, concentrate on the 1 and 2 $\mathcal{M}_\odot$ tracks. Just before the line labeled "helium ignition" those tracks merge and, at this point, the core of helium built up by the burning of hydrogen looks very much the same independent of mass. By this time, the stars are well off the main sequence and are red giants. The main point here is that shortly after the merger of the tracks, the central regions become electron degenerate and the physics of the Pauli exclusion principle for electrons provide the main pressure support for the core. This is indicated on the figure by the dashed line mysteriously labeled "$\epsilon_F/kT = 10$." (We shall explain in detail what this means in Chapter 3.) To the right of the line the stellar material is degenerate and to the left, nondegenerate wherein the usual ideal gas provides pressure support. It is these features of approach to degeneracy and convergence of tracks to a common electron degenerate helium core that define the lower mass sequence and so strongly effect the subsequent evolution (of which more later).

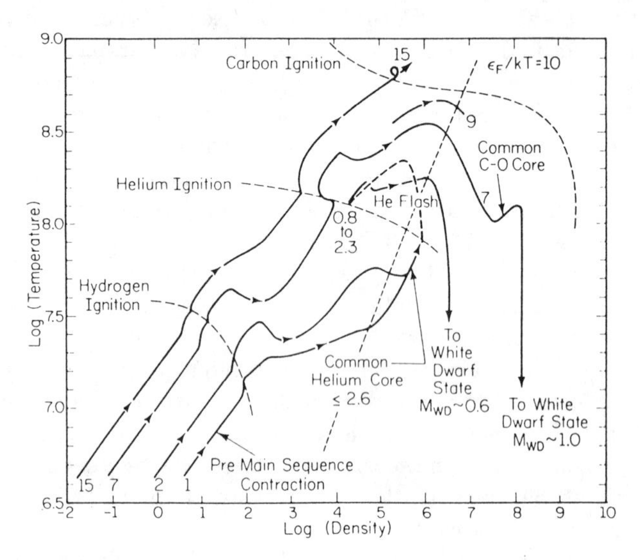

FIGURE 2.7. Central density versus central temperature for evolving stellar models. Reproduced, with permission, from I. Iben Jr. 1985, "The Life and Times of an Intermediate Mass Star," in *Quarterly Journal of the Royal Astronomical Society*, Volume 26, published by Blackwell Scientific Publications.

The *intermediate* ZAMS stars evolve off the main sequence and never form a degenerate helium core. The upper mass limit for these stars is about 8 to 9 $\mathcal{M}_\odot$. The lower limit is not well determined but must somehow merge with the previous class of stars and is thus around 3 $\mathcal{M}_\odot$.[2] Figure 2.7 indicates that "helium ignition" takes place in the nondegenerate helium core. This means that helium begins to burn and is gradually converted to carbon and oxygen in an environment where an ideal gas con-

[2]You should realize that we are being deliberately fuzzy about what constitutes the endpoints of these various mass ranges of stars. The later evolution is so different for them that minor changes in composition, constituent physics, methods of calculation, and so on, can drastically change what happens to models with masses around the endpoints.

trols the equation of state. Following the buildup of a carbon-oxygen core, the density increases sufficiently so that the core becomes degenerate and, in a fashion similar to the lower ZAMS stars, the tracks merge (as denoted by "Common C-O Core" in the figure).

The *upper* ZAMS stars follow a different course of evolution and, as we shall see, some of them end their lives as supernovae. As mentioned before, the maximum mass observed approaches 100 $\mathcal{M}_\odot$ and we shall take that as the upper limit in mass in subsequent discussions. If we disregard mass loss for now, what distinguishes these stars from those of intermediate mass is that, after ignition of helium, a carbon-oxygen core is built up that stays nondegenerate all the way through the point where carbon begins to burn. If and when electron degeneracy takes hold is what distinguishes the lower, intermediate, and upper main sequences.

In all cases, the main feature of evolution near the main sequence is the conversion of hydrogen to helium as an energy source. However, a price is paid for this conversion. According to (1.52) the mean molecular weight of a standard mix containing 70% hydrogen is $\mu \approx 0.6$ assuming complete ionization. However, as hydrogen burning proceeds in the hot central regions, the central value of μ increases towards 4/3 corresponding to pure ionized helium. This increase in μ means that the number of free particles has decreased by a factor of almost one half and so would the ideal gas pressure if temperature and density were to remain fixed. To make up the deficit, central temperatures and densities must increase through contraction of the central regions to maintain pressure support of the overlying mass. This picture may be complicated by core convection, which tends to mix processed material with overlying hydrogen-rich material, but the outcome is clear: a helium core is gradually built up and that core contracts and heats up.

Following exhaustion of hydrogen in the core, the growing helium remnant is surrounded by an active shell of burning hydrogen, which supplies the power for the star. The core itself, however, has no energy source of its own (except some input from contraction) and hence it tends to be isothermal since no temperature gradients are required. It is at this point—and we shall elaborate more on this shortly—that the structure of the star begins to change radically and the main sequence phase ends. The departure of model Pop I stars from the main sequence is shown in Figure 2.8 from Iben (1967). Note that the "zig-zags" in the H–R diagram are mass-dependent. Lower main sequence stars (those whose primary energy source is the pp chain) depart almost vertically from the main sequence, while more massive stars move nearly horizontally when they leave the main sequence. For the moment, we are interested primarily in those portions of the tracks numbered from (1) to (5): by point (5) the star certainly cannot be considered a main sequence object.

The time scales associated in going from one point to the next for some of the tracks of Figure 2.8 are listed in Table 2.3 (adapted from Iben 1967) and, from these, it seems safe to define the main sequence for a given mass as that long-lived phase (1)–(2). The relatively rapid phase (2)–(3) signals contraction of the central regions as hydrogen is rapidly depleted. The subsequent evolution to point (6) (for masses less than 15 $\mathcal{M}_\odot$) of the *red giant branch* (RGB) is characterized by a continual expansion and reddening of the star to lower effective temperatures. We now partially address the issue of why this happens.

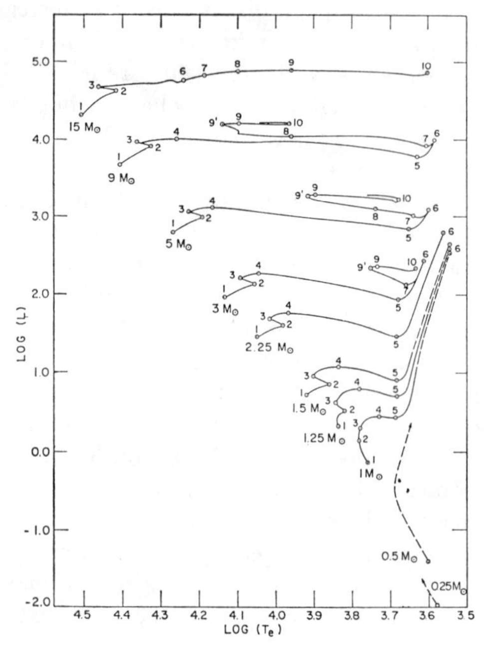

FIGURE 2.8. Representative theoretical evolutionary tracks for stars of different masses. Reproduced, with permisssion, from Iben (1967), *Annual Review of Astronomy and Astrophysics*, Volume 5, ©1967 by Annual Reviews, Inc.

TABLE 2.3. Stellar lifetimes where (i)–(i+1) is interval (in yrs). From Iben (1967).

$\mathcal{M}_\odot$	(1)–(2)	(2)–(3)	(3)–(4)	(4)–(5)
9.00	2.14×10^7	6.05×10^5	9.11×10^4	1.48×10^5
5.00	6.55×10^7	2.17×10^6	1.37×10^6	7.53×10^5
3.00	2.21×10^8	1.04×10^7	1.03×10^7	4.51×10^6
2.25	4.80×10^8	1.65×10^7	3.70×10^7	1.31×10^7
1.50	1.55×10^9	8.10×10^7	3.49×10^8	1.05×10^8
1.25	2.80×10^9	1.82×10^8	1.05×10^9	1.46×10^8

In an early 1942 study Schönberg and Chandrasekhar demonstrated that when an isothermal helium core is built up to a mass corresponding to about 10% of the initial hydrogen mass of the star, it is no longer possible to maintain quasi-hydrostatic equilibrium for the core of the model star if pressure support is due to an ideal gas. Other studies, including evolutionary calculations, support this by showing that the core contracts and heats rapidly and the increase in temperature causes a corresponding rise in energy generation rate and luminosity. The envelope (that region outside of the hydrogen-burning shell), however, responds by expanding rapidly and the star soon leaves for other climes on the H–R diagram. This signals the end of the main sequence phase of evolution for the star, but it will have spent some 70% to 80% of its active fuel-burning life in that phase.

The enormous increase in radius that accompanies hydrogen exhaustion moves the star into the red giant region of the H-R diagram. While the transition to red giant dimensions is a fundamental result of all evolutionary calculations, a convincing yet intuitively satisfactory explanation of this dramatic transformation has not been formulated. Our discussion of this phenomenon follows that of Iben and Renzini (1984) although we must state that it is not the whole story.[3]

Actual evolutionary calculations suggest that the envelope expansion for upper main sequence stars is roughly homologous. This means that if a given mass element is followed in time, then the local gradient of luminosity scales as

$$l' \equiv \frac{d \ln \mathcal{L}_r}{d \ln r} = 3n - s \tag{2.13}$$

where n and s are the opacity power law exponents for ρ and T respectively. (This follows easily from the dimensional arguments of the previous chapter using an ideal gas and diffusive radiative energy transport). Since r in this expression is at a fixed mass level, then dr may be regarded as δr, which is the change in r over some evolutionary time period. The corresponding change in $\mathcal{L}_r$, $\delta \mathcal{L}_r$, can be related to δr by the above equation.

The dominant opacity source in the envelope of the more massive stars is electron scattering. In this case, n and s are nearly zero in the bulk of the envelope and only approach their Kramers' values of 1 and 3.5 in the very outermost layers. From the evolutionary calculations it is found that l' is small, but positive, so that the radius expands with increasing core luminosity to maintain constant $\mathcal{L}_r$ in the envelope. As expansion proceeds and the envelope material cools further, both n and s increase through the

[3]Other attempts include: Eggleton and Faulkner (1981); Weiss (1983); Yahil and Van den Horn (1985); Applegate 1988; Whitworth (1989); Renzini et al. (1992). Bhaskar and Nigam (1991) use an interesting set of dimensional arguments plus notions from polytrope theory. We suspect the answers may lie in their paper but someone has yet to come along and translate the mathematics into an easily comprehensible physical picture.

interior toward their Kramers' values, and l' becomes smaller. This means that an increase in luminosity at the given mass level implies a greater change in r. Thus the expansion accelerates. Finally, l' goes to zero or turns negative (as it must because $l' = -0.5$ for pure Kramers') at some depth in the envelope. A value of l' approaching zero means that $1/l' \rightarrow \infty$ and *any* increase in $\mathcal{L}_r$ results in dramatic expansion. The result is a *rapid* expansion of the envelope with a corresponding decrease in T_{eff} and the star heads to the cool side of the H–R diagram. In this case, additional expansion of the envelope decreases the amount of luminosity it can transport, and some of the core luminosity becomes trapped in the envelope. This trapped energy forces further expansion, and further trapping of energy. Catastrophe is ultimately avoided because efficient convection finally takes over the chore of energy transport. The star is now well on its way to the red giant stage. At this stage we are roughly at point (5) on the tracks of Figure 2.8 and are ready to ascend a Hayashi track up to the red giants.

This picture has to be amended somewhat because it is not true for all stars. Those with masses less than about 1.2 $\mathcal{M}_\odot$ already have convective envelopes of some depth on the main sequence so that they can cope, though not entirely, with the rising core luminosity. For those lower main sequence stars, we have already found that the helium core becomes dense and degenerate so that core gravities are high. High gravities at the core edge mean high temperatures and, in turn, an acceleration of hydrogen shell burning and luminosity. This ultimately drives the star upward in luminosity in the H–R diagram and along the Hayashi track to the red giant region.

2.4 Late Stages of Evolution

The post–main sequence evolution of single stars will be discussed in order of increasing ZAMS mass: lower, intermediate, and upper.

2.4.1 Lower Main Sequence Stars

A most exhaustive survey of the evolution of lower mass stars has been undertaken by Mengel et al. (1979), who computed 247 evolutionary sequences for stars with masses between approximately 0.5 and 4 $\mathcal{M}_\odot$ and of various compositions. This extensive computation was done for the purpose of studying ensembles of stars that were formed at nearly the same time such as galactic and globular clusters. For our discussion we also assume that these coevally formed stars had the same composition independent of mass. The composite H–R diagram for stars in such a cluster changes in a systematic way as time progresses. From the previous discussion, we know that more massive stars leave the main sequence first and the less massive have to wait their turn. Thus the main sequence will "peel down" toward

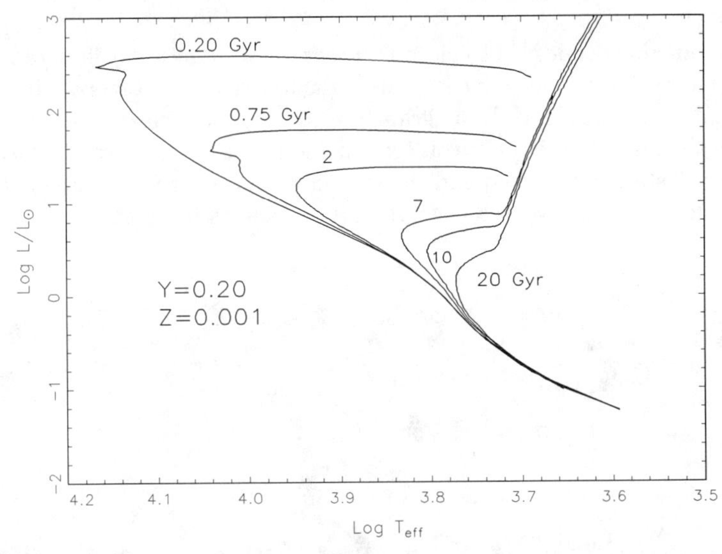

FIGURE 2.9. Representative theoretical isochrones for clusters of the indicated ages. Adapted from the *Revised Yale Isochrones*, Green et al. (1987). Note that times are in Gyr (10^9 year).

the giant branch with time. Using grids of models like those of Mengel et al. (1979), we can compute the locus of points corresponding to the positions of stars of different masses at a given time. These lines, known in the business as "isochrones," can then, in principle, be compared with observed cluster H–R diagrams to find the age of a cluster.

In practice, this is a bit more difficult than it sounds because, for example, the initial composition of the stars must somehow be determined either by direct observation or by means of the isochrone-cluster comparison itself. Neither of these is easy to accomplish for many clusters. In particular, the surface helium abundance for globular cluster stars is virtually impossible to determine directly at this time. A sample set of isochrones is shown in Figure 2.9 and they do look like the H–R diagrams of moderately old clusters after the conversion of $\mathcal{L}$ and T_{eff} to broadband magnitudes and colors has been made. It is easy to see the time development from such diagrams and, from them, an age for the cluster may be derived if the cluster falls into the time domain covered by the isochrones. The first item of information to look for in an observed cluster H–R diagram is where the main sequence ends (i.e., where stars start peeling off the main sequence). This is the "turnoff" point and stars at that point are getting ready to leave the main sequence. Hence, if we can deduce the main sequence lifetime for those stars, we have the age of the cluster, since all stars in the cluster are assumed to have been born at roughly the same time. This age is directly keyed to the mass and composition of those stars at the turnoff point. One of the major observational difficulties in doing this (aside from converting observed magnitudes and colors to theoretical quantities) is determining just where the turnoff point is. This is demonstrated in Figure 2.10, where the color-magnitude H–R diagram for the globular cluster M3 is shown (from the informative review of Renzini and Pecci 1988). Here, the locations of 10,637 stars are plotted and the estimated turnoff point is designated by the label "TO." As you can see, it isn't as easy as it sounds.

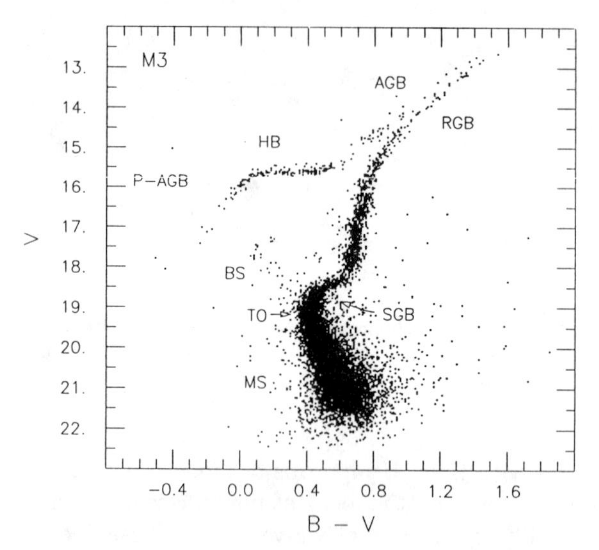

FIGURE 2.10. The observational Hertzsprung–Russell diagram for the globular cluster M3. The turnoff point is labeled TO. Reproduced, with permission, from Renzini and Pecci (1988), *Annual Review of Astronomy and Astrophysics*, Volume 26, ©1988 by Annual Reviews Inc.

Having established the luminosity of the turnoff point and using a spectroscopic determination of the metallicity of the stars in the cluster, the age can then be found (making reasonable assumptions regarding the helium abundance). As quoted by Iben and Renzini (1984), the relationship between cluster age, luminosity at turnoff ($\mathcal{L}_{\rm TO}$), and composition is

$$\begin{aligned}\log \mathcal{L}_{\rm TO} \approx\ & \left[0.019\,(\log Z)^2 + 0.065 \log Z + 0.41Y - 1.179\right] \log t_9 \\ & + 1.246 - 0.028\,(\log Z)^2 - 0.272 \log Z - 1.073Y \qquad (2.14)\end{aligned}$$

where t_9 is the cluster age in units of 10^9 years and $\mathcal{L}_{\rm TO}$ is in $\mathcal{L}_\odot$. This expression adequately reflects the results of evolutionary calculations for $-4 \le \log Z \le -1.4$, $0.2 \le Y \le 0.3$, and $0.2 \le t_9 \le 25$, which are ranges of general interest. Note that increasing either Y or Z, for fixed $\mathcal{L}_{\rm TO}$, decreases the age. If this expression is too daunting, a simpler (but more restrictive) one may be found in Sandage *et al.* (1981) as their equation (17).

The evolution onto the red giant branch (RGB) is indicated in Figures 2.9–2.11 by the sharp upward swing at effective temperatures $\log T_{\rm eff} \approx 3.7$

($(B-V) \approx 0.8$). The path taken on the RGB is dictated by the presence of deep convection in the envelope and thus the evolution up the giant branch approaches the Hayashi track described in the sections on pre–main sequence evolution (§§2.1 and 7.3). Evolution along the giant branch is much more rapid than on the main sequence. Typically, it takes five times longer to reach the turnoff than to evolve from the turnoff to the tip of the giant branch. Hence the range of masses represented by stars from turnoff and up the giant branch is extremely narrow. This means that in a given cluster, all stars from the tip of the giant branch down to the turnoff have almost the same mass. For this reason, isochrones for the giant branch (and subgiant branch) are almost indistinguishable from evolutionary tracks for a single mass.

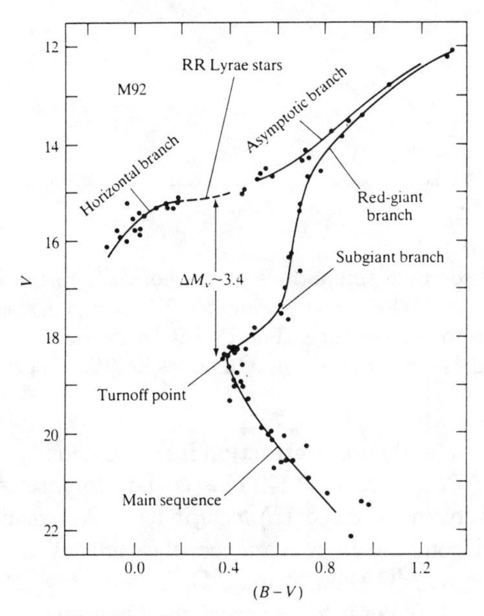

FIGURE 2.11. Schematic H-R diagram for the globular cluster M92. The solid lines correspond to the indicated evolutionary stages. From *Galactic Astronomy*, 2d ed., by D. Mihalas and J. Binney (1981). Copyright ©1991 by W.H. Freeman and Company. Reprinted with permission.

The lower mass stars, by definition, soon develop a common electron degenerate helium core. An important characteristic of the equation of state for degenerate electrons is that pressure is relatively insensitive to temperature; that is, increasing the temperature (but not by too much) has little effect on pressure. The core of the evolving RGB star reaches this condition when the central density and temperature (ρ_c and T_c) approach 4×10^5 g cm^{-3} and $\sim 5 \times 10^7$ K as shown in Figure 2.7. At this stage contraction is aided by the emission of neutrinos from the stellar core. Because stellar material (at least under the conditions considered here) is virtually transparent to neutrinos, they are an additional energy sink for the core (see Chapter 6). This not only accelerates the contraction, it may also induce a mild temperature inversion in the core so that the maximum core temperature, $T_{\rm max}$, need not be at stellar center.

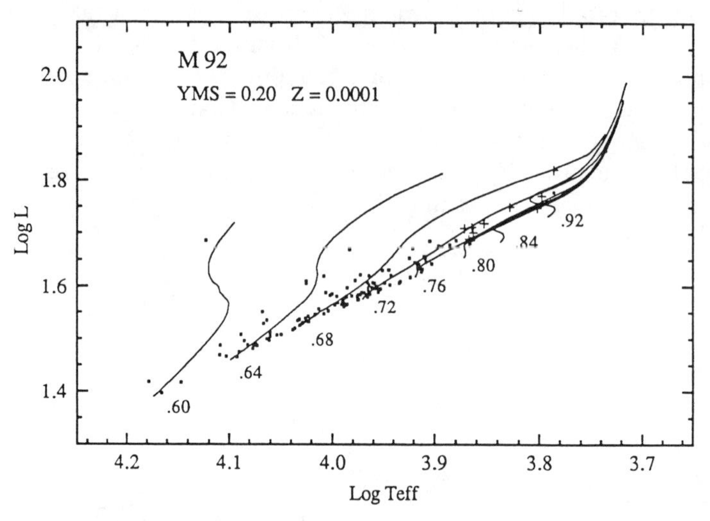

FIGURE 2.12. Theoretical simulations of the horizontal branch of M92 by Lee, Demarque, and Zinn (1987). The parameter YMS is the mass fraction of helium while on the main sequence. The evolutionary tracks and points are for post–He-flash models, with the indicated masses in $\mathcal{M}_\odot$. Reprinted with permission.

Further evolution and core contraction increase density and temperature still further until $T_{\rm max}$ nears 10^8 K. This is the characteristic temperature at which helium burns via the *triple alpha* (or 3α) reaction(s) whereby three ^{4}He nuclei combine in a sequence of reactions to produce ^{12}C. A further α capture on ^{12}C then forms ^{16}O. The energy released by the 3α reaction is about 6×10^{17} ergs for every gram of helium converted to carbon. Note that this is down by a factor of ten from the corresponding figure for hydrogen-burning: the star has to burn more material for each erg gained.

The ignition of helium under electron degenerate conditions is almost an explosive affair because of the low-temperature sensitivity of pressure versus the high-temperature sensitivity of the 3α reaction ($\nu \approx 40$). Because of the neutrino-induced temperature inversion, helium is most likely ignited off-center in the degenerate core. When reactions begin, the liberated energy increases the temperature locally but, because of the degeneracy, the rising temperature does not cause an increase in the pressure. The latter means that the core cannot relieve and cool itself by expansion and the increasing temperatures just cause the 3α to run all the faster, and so on. The result is an explosive runaway or *helium flash*. Just how explosive the helium flash can be depends on how (and by whom) the evolution is calculated.

The major evidence that leads us to believe that the helium flash is not explosive enough to disrupt the star in a serious fashion is that stars in the post-helium flash stage are observed and they constitute the horizontal branch of globular clusters, for example.

The mass of the helium core at the onset of the helium flash is (as quoted by Iben and Renzini 1984 with a minor simplification by us)

$$\mathcal{M}_{\text{core}} \approx 0.476 - 0.221(Y - 0.3) - 0.009(3 + \log Z) - 0.023(\mathcal{M} - 0.8) \quad (2.15)$$

where masses are in solar units. For the range of masses in this section (0.8 to about $2.3\mathcal{M}_\odot$) core masses all lie close to $0.5\mathcal{M}_\odot$. The presumed reason why that core survives the helium flash is that, at some stage in the flash, convective mass transport enables the excess heat generated by burning to be transported away from the site of the flash. How this might be accomplished has been examined by many researchers but is still not well understood (and see §§4.3.2–4.3.4 of Iben and Renzini 1984).

Having survived the helium flash, the star now quietly burns helium in its core while a thin hydrogen-burning shell continues to add helium to that core. For stars with low Z Population II compositions, this phase corresponds to the "horizontal branch" or HB in globular clusters. The location of the HB for low Z globular cluster stars is shown schematically in Figure 2.11. Luminosities are typically 30 to 100 $\mathcal{L}_\odot$. The luminosity of the HB at an effective temperature of $\log T_{\text{eff}} = 3.85$ is (from Iben and Renzini 1984)

$$\log \mathcal{L}_{3.85} \approx 1.73 + 1.40(Y_{\text{HB}} - 0.3) + 0.073(3 + \log Z) \quad (2.16)$$

where $\mathcal{L}_{3.85}$ is in solar units and Y_{HB} is the surface abundance of helium on the HB star.[4] This last may not be the same as that on the ZAMS, and its presence in (2.16) allows for the possibility of mixing at some prior stage. Note that these luminosities are considerably lower than those reached at the "tip" of the RGB where the helium flash was first initiated. The determination of masses on the HB is more difficult and must be approached in an indirect, but very illuminating, fashion.

We shall have more to say about intrinsically variable stars later in §2.6.1 but the following has direct bearing on HB evolution. Population II stars on the HB having effective temperatures within about 500 K of $\log T_{\text{eff}} \approx 3.85$ are RR Lyrae variable stars. These have periods, Π, of regular light variability ranging from a couple of hours to a day with an average period of half a day. The theoretical dependence of their periods on mass, luminosity, and color has been thoroughly investigated. A sample theoretical result from Iben (1971) for the "fundamental mode" of variability (see later), is

---

[4]The difference between the luminosity of the HB and that of the turnoff point for a globular cluster is also an ingredient in determining the age of the cluster. See §5 of Iben (1991).

that the period is given by

$$\log \Pi \approx -0.340 + 0.825(\log \mathcal{L} - 1.7)$$
$$-3.34(\log T_{\rm eff} - 3.85) - 0.63(\log \mathcal{M} + 0.19) \qquad (2.17)$$

where Π is in days, and mass and luminosity are in solar units. Thus if $\mathcal{L}$ and $T_{\rm eff}$ are determined and Π is observed, then $\mathcal{M}$ follows. Using this relation, it is found that typical RR Lyrae masses are around 0.7 $\mathcal{M}_\odot$ with a spread of about 0.2 $\mathcal{M}_\odot$. We can turn this around by using the period-mean density relation of (1.37) (as a first approximation) and find that a period of 0.5 days implies a mean density of about 10^{-2} g cm^{-3}. Combining this figure with $0.6\mathcal{M}_\odot$, for example, yields a radius of $\sim 4\mathcal{R}_\odot$. You may easily check that $\log T_{\rm eff}$ of 3.85 and a $\log(\mathcal{L}/\mathcal{L}_\odot) \approx 1.6$ are consistent and this last figure and places the object on the HB at the location of the RR Lyrae variables.

The critical point about the masses quoted above for RR Lyrae stars, and therefore HB stars, is that they are less than the masses of their progenitor main sequence stars. If these results are correct, then mass must have been lost during the time elapsed between the main sequence and the HB stages. Such mass loss from red giants is observed, but the physical mechanism is not well understood. The article by Dupree (1986) reviews what is presently known about mass loss from cool stars of various kinds. For spectral class M supergiants typical mass loss rates are $\dot{\mathcal{M}} \sim 10^{-6}$ $\mathcal{M}_\odot$ per year with flow velocities at large distances from the star of $\sim$ 20 km s^{-1}. (These stars are also characterized by atmospheres that extend far beyond the visible photosphere.) It does not take a long time to lose a tenth of a solar mass from a star at this rate. It may also be that the helium flash induces rapid mass loss but, since the duration of the flash is expected to be short, observers have not yet captured that event.

Other evidence indicates that mass loss results in a range of remnant mass after the helium flash. To reproduce accurately the observed shapes of horizontal branches in globular clusters with a fixed core mass, one must allow for different envelope (i.e., outer hydrogen layer) masses. The more material outside the core that remains following helium ignition, the redder the HB star is. Thus the observed color distribution of HB stars can be explained if those stars are burning helium in their cores following the helium flash, and that they differ only in the amount of envelope material remaining. Recent calculations (Lee, Demarque, and Zinn 1987) suggest that for a core mass of $0.5\mathcal{M}_\odot$, the mass of the HB stars ranges from $0.6\mathcal{M}_\odot$ to $0.9\mathcal{M}_\odot$.

An estimate of the lifetime of the core helium-burning stage on the HB, from Iben (1974), is given by

$$\log t_{\rm HB} \approx 7.74 - 2.2(\mathcal{M}_{\rm core} - 0.5) \;\; \text{years} \qquad (2.18)$$

where $\mathcal{M}_{\rm core}$ is given by (2.15). (The details of this time scale depend on how convection is treated and on what the precise mix of heavy elements is.) A simple calculation, similar to that done for main sequence lifetimes in the previous chapter but using helium-burning energy yields and HB luminosities, yields an HB "main sequence" lifetime of $\sim 10^7$ years; that is, the HB star evolves off the HB in much the same fashion as happens on the ZAMS.

The subsequent evolution parallels departure from the ZAMS. Helium and hydrogen shell sources provide the luminosity (although perhaps not both at all times). The star moves back to the red giant branch but at higher luminosities than before. This is indicated in Figure 2.11 by the track labeled "Asymptotic branch." A more modern name is "asymptotic giant branch" or "AGB." (References from the older literature may refer to this as the "second giant branch." We suggest Iben and Renzini 1983 and 1984 and Iben 1991 for those wishing to review AGB evolution.) This branch skims the original RGB (hence "asymptotic") just to the blue in temperature. As the stars ascend this branch things become increasingly complicated. Mass loss is detected in these stars with rates as high as $\sim 10^{-4}\ \mathcal{M}_\odot\ \mathrm{yr}^{-1}$.

In addition, theoretical calculations indicate that a thermal instability, due partially to interactions between hydrogen- and helium-burning shells, may occur later in AGB evolution. The hydrogen shell usually provides most of the luminosity in this phase. However, material processed through the hydrogen shell is converted to helium and "dumped" down onto the helium layer. Eventually, the base of the helium shell is compressed to the point where helium-burning occurs in a mildly degenerate environment. The helium shell then undergoes a "flash" similar to, but much less dramatic than, the core helium flash. At the onset of this flash, the stellar luminosity increases rapidly and accompanying expansion pushes the hydrogen-burning shell outward, and to low temperatures, where it is effectively extinguished. As the helium shell drops in luminosity following the flash, the hydrogen shell eventually recovers and the star again evolves, quiescently burning hydrogen in a shell until the next thermal pulse. Unlike the core flash, these shell flashes can repeat periodically and the entire cycle takes about 10^5 years, whereas the initial pulse lasts only a few years. Were we able to observe an AGB star undergoing a pulse—which is unlikely because of the short time scales involved—we would see sharp fluctuations in luminosity followed by a rapid decline, and then a gradual recovery to when the next pulse appears.

A by-product of flashes in the helium-burning convective shell is the production of both carbon- and neutron-rich isotopes of heavier elements. Because of the deep convection zones in these stars, these nuclei, and those processed from prior CNO hydrogen-burning stages, may be "dredged up" to where they may finally be seen at the photosphere. This is an exciting area of research, which combines both "standard" stellar evolution calculations and stellar nucleosynthesis of some elements.

The AGB is also the domain of the *long-period* variable stars, which include the *Mira* variables. These all have periods of order a year (see what the period-mean density relation has to say about this) and most are not entirely regular in their variations.

A guide to the evolution in AGB phase follows from the early results of Paczyński (1970) and Uus (1970). The core of the late AGB star is composed of carbon and oxygen and (see Fig. 2.7) this core is electron degenerate. Overlying the core are the helium- and hydrogen-burning shells, which provide most of the stellar luminosity, yet it lies well within the inner one percent of the stellar radius! Paczyński and Uus demonstrated that the evolution of stars of intermediate mass is essentially governed by the mass of the core and, in particular, there is a direct relation between stellar luminosity and mass.[5] A rough form of this relationship is

$$\mathcal{L}/\mathcal{L}_\odot \approx 6 \times 10^4 (\mathcal{M}_c/\mathcal{M}_\odot - 0.5) \tag{2.19}$$

and we offer this in the spirit that it gives the flavor of what happens for the lower main sequence stars. Thus if the mass lost by the AGB star leaves some remnant of the hydrogen shell intact, then the immediate effect is that mass loss on the AGB does not affect the star's luminosity.

With all this going on, the next stage of evolution is associated with extensive mass loss from the star. How this occurs is not fully determined, but some combination of various types of variability, pulses, "superwinds," and ordinary mass loss must eventually conspire to rid the star of a substantial fraction of its original mass and prepare it for the next stage.[6] The subsequent evolution is to the Planetary Nebula Nucleus (PNN) stage where the fires of nuclear burning gradually shut off. (For a review of planetary nebulae and their hot central stars see Kaler 1985.) On the H–R diagram the track to the PNNs is to the hotter effective temperatures at nearly constant luminosity and then down to the white dwarfs on a track of nearly constant radius. A sample track from the HB to the white dwarfs for a 0.6 $\mathcal{M}_\odot$ HB model that does not lose mass on the AGB (!) is shown in Figure 2.13. Keep in mind, however, that the true track of a remnant object in the H-R diagram during the shedding of its envelope may be complicated by late thermal pulses, reignition of nuclear-burning shells, and episodic mass loss events.

The final products of this evolution, the white dwarfs, are common but intrinsically faint objects. They have electron degenerate cores comprising almost all of the mass of the star. Because a gas of degenerate electrons is highly resistant to compression, this core cools down at nearly constant

[5]Other references to this intriguing problem are: Iben 1977; Havazelet and Barkat 1979; Wood and Zarro 1981; Tuchman *et al.* 1983.

[6]Illustrative hydrodynamic calculations of superwinds associated with Mira-like variables are discussed by Bowen and Willson (1991). This is a promising area of study that impinges on several aspects of later stellar evolution.

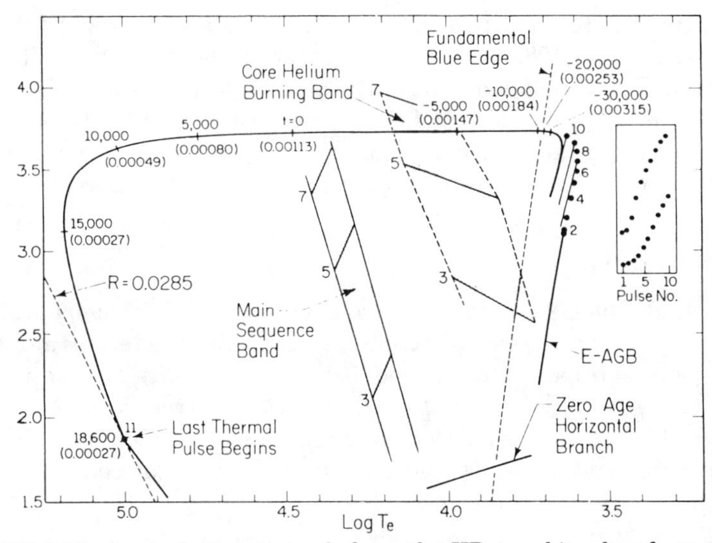

FIGURE 2.13. An evolutionary track from the HB to white dwarf stage. The original mass on the HB is 0.6 $\mathcal{M}_\odot$. Times (positive and negative) are measured in years from a point in the track at $\log T_{\rm eff} = 4.5$. From Iben and Renzini (1983). Reprinted, with permission, from *Annual Review of Astronomy and Astrophysics*, Volume 21, ©1983 by Annual Reviews Inc.

radius, transporting its remnant thermal energy by conduction (as in a metal) to the outer layers. These outer layers are nondegenerate and act as an insulating blanket, resisting and slowing down the release of energy to space. Masses for single white dwarfs seem to be clustered tightly around 0.6 $\mathcal{M}_\odot$ indicating, again, the role of mass loss in prior evolutionary stages.

An important feature of white dwarfs is that there is a maximum mass for these objects beyond which no equilibrium structures that are supported by electron degeneracy can exist. For a white dwarf composed of helium or other elements arising from normal thermonuclear processes, this maximum mass, known as the *Chandrasekhar limit*, is about 1.4 $\mathcal{M}_\odot$. It should be evident that this limit places severe constraints both on some early evolutionary stages, and on the evolution of white dwarfs themselves. Not indicated in Figure 2.13 is that there is a cutoff at $\mathcal{L} \approx 10^{-4.5}\mathcal{L}_\odot$ below which little (or no) single white dwarfs are found in the solar neighborhood. As will be discussed in Chapter 9, this cutoff reflects the finite age of the disk of our galaxy.

Before going on to discuss the evolution of more massive stars, it should be pointed out that, despite the uncertainties in details, the evolution of lower-mass stars seems to be well mapped out. Much effort has gone into comparing the theoretical tracks with observed H–R diagrams for globular clusters. A major difficulty in this program has been that helium in cool stars is not observable because temperatures are too cool to ionize it and, hence, give rise to atomic transitions. Thus Y is not determined directly. Iben and Renzini (1984) and Iben (1991) tell the detective tale of how the combination of turnoff points, statistics of stars in the RGB, HB, and AGB phases, and the properties of RR Lyrae variables yield $Y \approx 0.3$ (or a trifle less) and a mean age of globular clusters of 15.5 ± 3 Gyr. The final results are not yet in, but the interim report, complicated as it is, is worth reading.

2.4.2 Intermediate-Mass Stars

Single stars of mass between about 3 and 8–9 $\mathcal{M}_\odot$ begin helium burning on the RGB under nondegenerate conditions. Central temperatures at ignition are, as in lower-mass stars, around 10^8 K but densities are considerably lower with $\rho_c \sim 10^4$ g cm^{-3}. In this case the material is not degenerate and pressure may easily respond to the increased energy output from the burning. Helium ignition is nonexplosive and the stellar core expands to regulate the burning to reasonable levels.

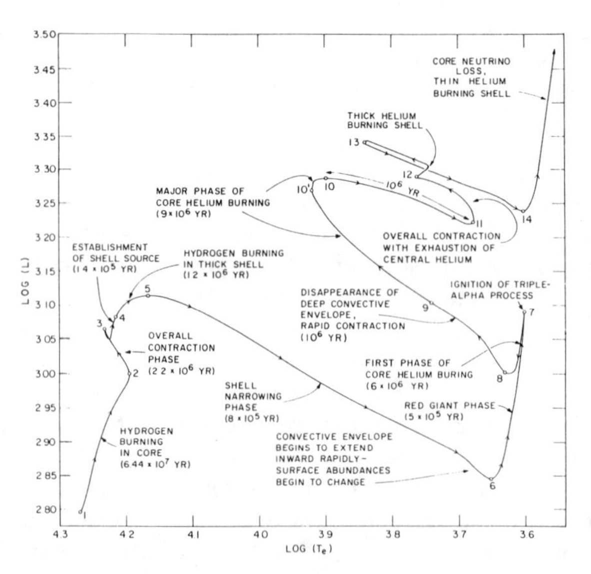

FIGURE 2.14. The evolutionary track of a 5 $\mathcal{M}_\odot$ star from the ZAMS to the AGB taken from Iben (1967). Reprinted, with permisssion, from *Annual Review of Astronomy and Astrophysics*, Volume 5, ©1967 by Annual Reviews Inc.

Quiescent core helium-burning takes place analogous to the HB and evolution proceeds to the blue at roughly constant luminosity as helium is consumed. This phase is illustrated in Figure 2.8 for a variety of masses as points 7–9. (Time scales for the evolution from one point to the next in the figure may be found in Iben 1967). During this relatively long-lived excursion the star may cross the Cepheid strip and there become a Pop I

Classical Cepheid variable—these will be discussed in §2.6.1. Unfortunately, the precise path of the star's evolution in the H–R diagram is dependent on the details of the calculation and differs from author to author but all agree on general trends.

Exhaustion of helium in the core, and its replacement by carbon and oxygen signal the end of this phase and the star heads to the AGB. A complete track to this stage is shown in Figure 2.14 for a 5 $\mathcal{M}_\odot$ model star. Not shown are the effects of thermal pulses which, as in lower-mass stars, also take place.

The structure of a 7 $\mathcal{M}_\odot$ star at one point in the AGB phase is shown in Figures 2.15 and 2.16 from Iben (1985). The central carbon-oxygen core has a radius of a few $\times\ 10^{-3}\ \mathcal{R}_\odot$ and a mass of 0.95 $\mathcal{M}_\odot$. This corresponds to an average density of about 10^7 g cm^{-3}. At the same time, the core temperature has risen to 4×10^8 K at a point midway through the core. The central temperature has been depressed by energy losses in the form of neutrinos to $T_c \approx 2 \times 10^8$ K. This places the model at the stage indicated by "Common C-O Core" in Figure 2.7 and the core is highly electron degenerate. The structure of the remainder of the star is in striking contrast to the core with an overall radius of 480 $\mathcal{R}_\odot$ and average density of $\sim 10^{-7}$g cm^{-3}. It is not clear how this fluffy, low-gravity, high-luminosity envelope responds to the urging of thermal pulses, stellar winds, and other possible phenomena, but some mass loss must occur. Just how much is still an important question. There is evidence that some young open clusters have single white dwarfs in them and, from the age of these clusters, it appears that their progenitors had an initial ZAMS mass of at least 5–8 $\mathcal{M}_\odot$. If true, then this means that at least some stars in the intermediate-mass range lose some, if not all, of their envelopes (a few $\mathcal{M}_\odot$). On the other hand, if the 7 $\mathcal{M}_\odot$ star considered above were to be left with its 0.95 $\mathcal{M}_\odot$ core intact as a proto-white dwarf, then the final object would have a mass that is well outside the normal white dwarf mass range of 0.6 ± 0.1 $\mathcal{M}_\odot$. However, these high-mass white dwarfs would be comparatively rare and their unusually large masses could be accommodated in the tail of the mass distribution.

An interesting question arises in the event that substantial portions of the envelopes of these intermediate-mass stars are not lost. In that case active evolution must proceed. If the carbon-oxygen core continues to contract and heat up to some degree, then we have a potentially explosive situation. This is indicated in Figure 2.7 by the line labeled "carbon ignition." The heavy ion reaction involving two carbon nuclei is almost as temperature sensitive as the 3α reaction and, in this instance, carbon burning may take place in an even more degenerate environment. Might we then expect a "carbon flash" with intensities rivaling, or perhaps greatly exceeding, the helium flash? The preponderance of opinion is that these intermediate-mass stars escape this fate by mass loss. But this is not to say that all of these stars are aware of this opinion.

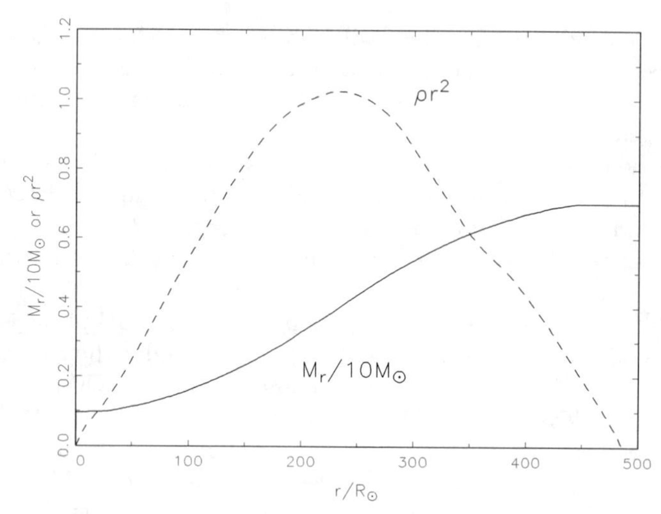

FIGURE 2.15. The internal structure of a $7\mathcal{M}_\odot$ AGB model from the center to the stellar surface. The quantity ρr^2 is in the units of 10^{-2} g $\mathcal{R}_\odot^2$ cm^2. Adapted from Iben (1985).

2.4.3 Upper Main Sequence Stars

Stars of mass greater than about 9 $\mathcal{M}_\odot$ avoid the helium flash and do not end up with a degenerate carbon-oxygen core after quiescent helium burning. However, to make things interesting, their high luminosities induce mass loss on a grand scale on the main sequence and beyond.

Observations of almost all bands in the electromagnetic spectrum, from the radio to UV and perhaps beyond, have established that luminous stars of all kinds lose mass. For massive blue stars, the generally accepted cause of the mass loss is radiation pressure acting on strong spectral lines. Other processes may play a role, such as intrinsic stellar variability in these stars, but little has been done to explore their importance.

Figure 2.17, from Chiosi and Maeder (1986), shows the observational correlation between mass loss rates ($\dot{\mathcal{M}}$ in $\mathcal{M}_\odot$ per year) and luminosity. You will note that there is a spread in the mass loss rate and certain classes of stars stand out. Among these are the *Wolf-Rayet* (WR) objects, which are characterized by strong emission lines that dominate the optical spectrum. These stars lie near the main sequence but their spectra indicate a strong deficiency or absence of hydrogen. There is also evidence that the products of hydrogen or helium-burning have been brought to their surfaces. They are thus highly evolved but, in some sense, young. (See Abbott and Conti 1987 for a review of these stars.) The properties of other stars (except PNN) in the hot and luminous region of the H–R diagram have been reviewed by Conti (1978) and Chiosi and Maeder (1986).

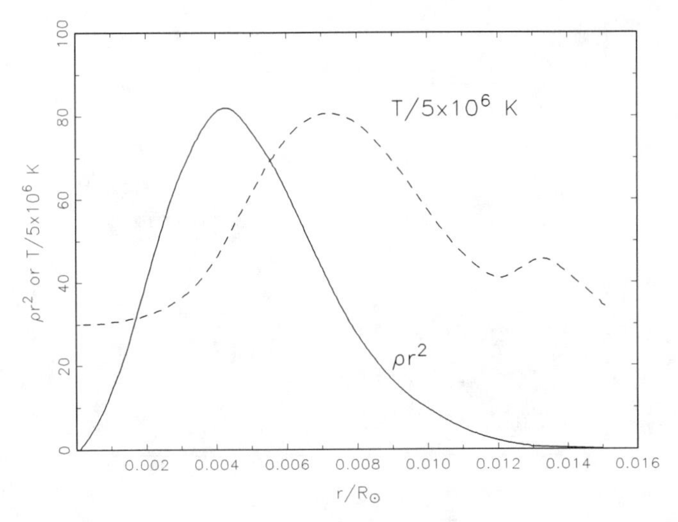

FIGURE 2.16. Structural properties of the C/O core of the $7\mathcal{M}_\odot$ AGB model of Figure 2.15. The units of ρr^2 are now g $\mathcal{R}_\odot^2$ cm^2 and temperature is scaled by 5×10^6 K.

The data in Figure 2.17 are well represented by the relation

$$\left(\frac{\dot{\mathcal{M}}}{\mathcal{M}_\odot}\right) = 10^{-14.97}\left(\frac{\mathcal{L}}{\mathcal{L}_\odot}\right)^{1.62} \quad \text{yr}^{-1} \tag{2.20}$$

from Garmany and Conti (1984). More general fits to the data have included effective temperature or radius but this will do for us. If (2.20) holds on the ZAMS (and we ignore pre–main sequence evolution still) then the mass-luminosity relation (1.85) may be used to eliminate luminosity in the above. Further, if $\dot{\mathcal{M}}/\mathcal{M}_\odot$ is then divided into $\mathcal{M}/\mathcal{M}_\odot$, we obtain the ZAMS mass loss time scale

$$t_{\text{ml}} \approx 10^{14.97}\left(\frac{\mathcal{M}}{\mathcal{M}_\odot}\right)^{-4.67} \quad \text{years.} \tag{2.21}$$

Compare this to the main sequence lifetime of (1.88) and find

$$\frac{t_{\text{ml}}}{t_{\text{nuc}}} = 10^{5.0}\left(\frac{\mathcal{M}}{\mathcal{M}_\odot}\right)^{-2.17} \quad \text{years.} \tag{2.22}$$

Taking this at face value, but remembering that approximate exponents may be dangerous, a 60 $\mathcal{M}_\odot$ ZAMS star would have a mass loss time scale 14 times longer than its ZAMS evolution time. That would seem to imply that the effect of mass loss is small. However, we must remember that a decrease in mass means a decrease in central temperature and, hence, for the CNO cycles, a more dramatic decrease in energy generation, and so

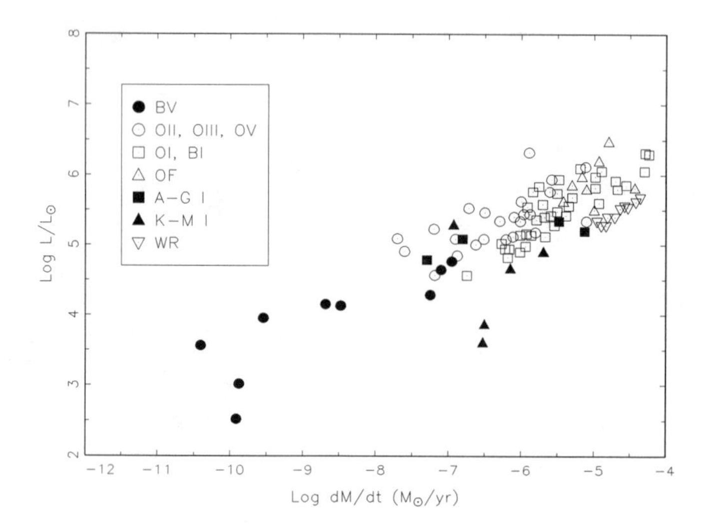

FIGURE 2.17. Empirical relation between mass loss rate $d\mathcal{M}/dt$ and luminosity. Adapted from Chiosi and Maeder (1986).

on. When done properly, the evolution of an upper main sequence star on the H–R diagram depends sensitively on the rate of mass loss. This is illustrated in Figure 2.18 from Brunish and Truran (1982) using one prescription for mass loss rates. (More recent and extensive compilations of massive star evolution off the main are due to Maeder 1990 and Chin and Stothers 1990.) The general effects of mass loss compared to "standard" constant mass evolution near the main sequence is that post–main sequence luminosities are lower (although the stars tend to be overluminous for their mass), main sequence lifetimes are somewhat longer, and the helium core at the end of hydrogen burning is smaller.

The behavior of more advanced evolutionary phases depends on details that are still a bit fuzzy. The general trends are summarized in Chiosi and Maeder (1986). Computed H–R diagram locations and inferred statistics in the more long-lived stages are, however, in reasonable accord with observations of massive stars. The main features of the most advanced, and short-lived, stages are thought to be relatively well understood and have their own special fascination.

Nuclear burning in the later stages progresses through the production of heavy elements up to those in the mass range around iron (although the precise paths through the nuclear-burning stages depend on the initial mass). One has to imagine the structure of the star at that time: a core composed of "iron peak" material surrounded by successive shells of decreasing molecular weight material with each shell having been produced by a prior burning stage. This picture has been given the flavorful title of "the onion-skin model." The outer appearance of the star probably de-

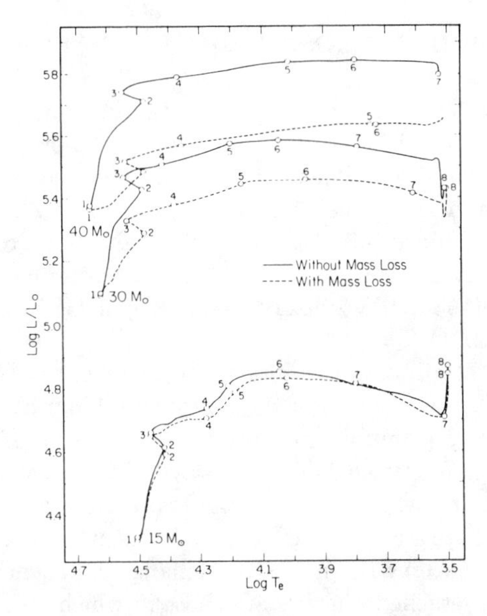

FIGURE 2.18. Evolutionary tracks for massive stars with and without mass loss. Reproduced, with permission, from Brunish and Truran (1982).

pends on initial total mass. For stars with masses greater than perhaps 20 to 40 $\mathcal{M}_\odot$, the helium core is thought to have been exposed by extensive mass loss and thus should resemble a WR star (see Maeder 1990). Objects with lower masses retain at least part of their original envelope and are most likely in a red giant stage. What is of most interest here are the inner portions of the star.

As thermonuclear burning produces successively heavier elements the energy yield per gram progressively becomes smaller. Far more energy is released in the conversion of hydrogen to helium than in adding on the last few nucleons to make an element with 50 or 60 nucleons. A basic fact of nuclear physics, which is taken advantage of by those in the nuclear weapons business, is that both fusion of light elements and fission of very heavy elements is exoergic. The middle mass regime around iron is useless; those materials do not constitute a nuclear fuel. Stars that were first formed on the upper main sequence eventually end up with a core of iron peak matter of total mass perhaps exceeding 1 $\mathcal{M}_\odot$. To continue the discussion past this point would infringe on the subject matter of §2.6.3—which is supernovae—and thus we stop here.

2.5 Evolution in Close Binary Systems

We define a close binary system as one containing two stars in which the evolution of either star is strongly influenced by the presence of the other. In this instance, the discussion of the previous sections must be modified to include stellar interactions. We shall not emphasize this subject but important text references are Tassoul (1978), and Shapiro and Teukolsky (1983), and you should consult articles such as those by Shu and Lubow (1981), Trimble (1983), Paczyński (1971), Kopal (1959), Pringle (1985), and §3 of Iben (1991).

Well over half of all stars in the sky are members of binary (or multiple) systems and probably half of these are close binaries. The mechanism through which they primarily interact is their mutual gravitational field and this may manifest itself is several ways. The most extreme example is the contact binary system wherein the two stars are in a state of semicoalescence and share a common envelope. A less extreme example is where the stars are just touching but may still exchange mass, energy, and angular momentum. Semidetached binaries are those in which one member may be losing mass (some to the partner star) because portions of its surface are on a gravitational and centrifugal equipotential surface in common with the companion star; that is, it is not clear who owns the surface. The last are the detached systems. Even here, however, stellar winds may be exchanged or tidal distortions or thermal heating effects may influence evolution.

Of prime importance for gauging the interactions of close binaries is the nature of the common gravitational and centrifugal potential field formed by the partners. The immediately relevant factors are the stellar masses, binary separation, shapes of the orbits, and the orbital period. These are, of course, not all independent of one another. In addition, contributions due to nonsphericity of the stars induced by rotation or tidal effects must be included.

This is probably impossible for the general situation. The simplest case is that of two detached point stars in circular orbits about one another. This situation is adequately treated in classical mechanics texts and requires only an application of Kepler's laws.

The next step of any practicality is to assume that one of the two stars is a rigid sphere while the other is tidally distorted and that both stars are rotating with the same frequency as the circular orbital period. Thus, viewed from a corotating frame, the system is fixed in appearance. To achieve this circularization of orbit and synchronization of spin requires dissipation of orbital kinetic energy and the application of torques at a prior stage in the system. Whether this can be done in most situations is a matter of controversy (Shu and Lubow 1981). However, for detached binaries this rigid body orbit can be shown to have the least mechanical energy and this corotation model seems to be applicable to many real systems.

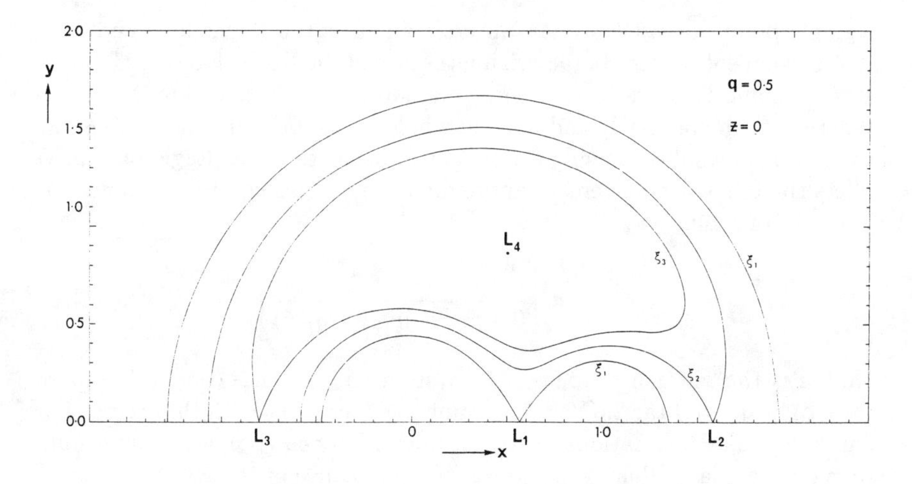

FIGURE 2.19. A sample set of equipotential surfaces in the Roche model. The picture is taken in the plane of the common orbit (the x–y plane) and the point mass stars are located at the positions $x = 0$ and $x = 1$ (in units we need not worry about). The mass ratio is two to one. By symmetry, the system is to be reflected about the x–axis. The ξ_i represent the equipotential surfaces. Note the locations of the three Lagrangian points L_1, L_2, and L_3. From Kitamura (1970). Reprinted by permission of Kluwer Academic Publishers.

The figure of the tidally distorted star and the resultant equipotentials in the corotation model is extremely difficult to compute because, among other things, we have to know something about the internal constitution of the stars—and thus we shall back off at this point. What is tractable is the computation of equipotentials for two point masses in orbit about each other: this involves pure classical mechanics. A particularly useful model is due to Roche wherein the two point mass stars are in circular orbits about each other. This "Roche model" has several interesting features and a sample set of equipotentials (gravitational plus centrifugal) are shown in Figure 2.19. The figure is drawn in a frame corotating with the two stars and is thus static in time. Close to each star the equipotentials are nearly circular and are dominated by the $1/r$ gravitational terms. Moving outwards, however, centrifugal effects and gravitational contributions from the partner star become increasingly important and the equipotentials take on a teardrop shape. The tips of the teardrops finally merge at the point labeled L_1 (the inner Lagrangian point) and its associated equipotential surface, the *Roche lobe.* (The figure of this surface is like a three-dimensional dumbbell with an infinitesimal connecting point.) At far distances the equipotentials again become circular because the two stars begin to behave gravitationally like a single point mass. Of primary interest for us is the Roche lobe and L_1.

A convenient way of characterizing the size of the Roche lobe is as follows. If we assume that the Roche surface associated with one star is not too different from spherical (and, on the whole, it is not), then define $\mathcal{R}_{\rm RL}$ as the radius of a sphere having the same volume as the lobe. Eggleton (1983) offers the following convenient approximation for this equivalent radius for "Star 1" of a binary:

$$\mathcal{R}_{\rm RL,1} \approx a \left[\frac{0.49\, q^{2/3}}{0.6\, q^{2/3} + \ln\left(1 + q^{1/3}\right)} \right] . \tag{2.23}$$

Here a is the semimajor axis of the system and $q = \mathcal{M}_1/\mathcal{M}_2$ is the mass ratio of Star 1 to Star 2. (You may want to experiment with this expression using various combinations of its parameters. Then compare your results to typical stellar radii.) Now, on to the significance of the Roche lobe.

First consider a star whose radius is less than $\mathcal{R}_{\rm RL}$ of its own Roche lobe. The surface of that star must be coincident with an equipotential within its Roche lobe else the surface would have to readjust to unbalanced gravitational and centrifugal forces. Suppose now that the star expands through evolutionary processes or that its $\mathcal{R}_{\rm RL}$ becomes smaller as a result of changes in the orbital characteristics of the system. If this keeps up, then what happens when the surface coincides with, or "fills," the Roche lobe? In that case a parcel of matter at the surface of the star may as well belong to either star. The first, and inner, Lagrangian point, L_1 plays a crucial role at this stage. If a particle at L_1 is given any small initial outward velocity, it will fall inward toward the companion star. (This is like having two connected potential wells filled with water. If we add just a little more water to one, it will spill over into the other. See Fig. 1.5 of Pringle 1985.) It is in this way that mass may be transferred from the lobe filling star to its companion and thus form a semidetached system. The Lagrange points labeled L_2 and L_3 have similar properties and, with particular reference to L_2, it still remains a serious question of how much mass and angular momentum may be lost from the entire system by mass leaking out of that outer Lagrangian point in the event one or both stars are so large as to encroach upon it.

A simple evolutionary example of how mass transfer may come about is the following. Suppose that a binary system is formed containing two detached ZAMS stars, one somewhat more massive than the other. The more massive star will evolve faster than its companion and eventually grow in size as it leaves the main sequence. If the masses and orbital parameters are such that the more massive star expands sufficiently so as to fill its Roche lobe, then mass transfer will take place in the (now semidetached) system and the less massive star will be the beneficiary. Just how this takes place depends on several as yet not clearly determined factors. However, to consider one possibility, if the size of the mass-receiving star (Star 2)

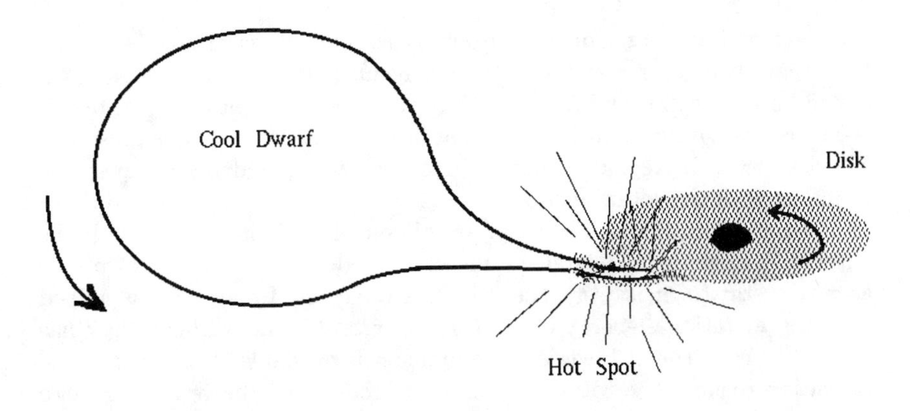

FIGURE 2.20. A schematic picture of an accretion disk. The cool star, in this case, overflows its Roche lobe and matter spills out of the L_1 point in a stream finally forming a disk. Where the stream meets the already formed disk, we expect a hot spot.

is less than a critical size (depending on the mass ratio of the system), then the inflowing mass from the L_1 point will spiral in toward it. The spiraling is a natural consequence of the rotation of the system and Coriolis forces. If the mass transfer rate is large enough, then an *accretion disk* will form around Star 2. Such structures have been under intensive theoretical investigation during recent years and deserve some discussion at this time. More extensive treatments may be found in the review article by Pringle (1981) and in Shapiro and Teukolsky (1983, Chapter 14).

If the mass spiraling toward Star 2 is dense enough, then collisions within the gas may well convert some of its kinetic energy to internal heat with little loss of angular momentum. The result is that the flow will gradually establish itself in circular Keplerian orbits about Star 2. Those orbits, from the symmetry of the entire system, should be nearly coplanar with the binary orbital plane and form a disk-like structure. A generic picture of the system in shown in Figure 2.20. As shown, the disk extends from a considerable fraction of the radius of the Roche lobe of Star 2 inward to nearly the star itself. This comes about because internal viscous mechanisms, of which there may be a few, continually cause kinetic energy to be converted to heat and hence cause the disk to radiate. In response, material gradually settles deeper into the gravitational well of Star 2. But to do so it must lose angular momentum. This is accomplished by viscous shear in the disk—Keplerian orbits do not constitute solid body rotation—which transports some gas and angular momentum outwards. This outward flow is smaller than that headed for Star 2 and the net result is that matter is finally accreted onto Star 2. How the matter is accreted is a complex subject but at some distance away from the stellar surface a transition region is presumed to exist in which Keplerian motion gives way to allow deposited material to merge with the rotating star.

The energetics of disk accretion are of considerable interest here. If the outer disk edge is at a large radius compared to the radius ($\mathcal{R}$) of the accreting star (of mass $\mathcal{M}$), then the finally accreted material has gained a total gravitational energy of $G\mathcal{M}/\mathcal{R}$ per gram. Some of this energy has gone to heating the disk while the remainder is available to the star or the transition region. The rate at which this occurs must be proportional to the rate of mass accretion, which we denote by $\dot{\mathcal{M}}$. If the system is in a steady state, then we may equate the rate of heating in the disk to a disk luminosity that manifests itself in various forms of emitted radiation. In the simple case of no magnetic fields or other special effects, disk luminosities for two classes of mass-receiving star are (see Shapiro and Teukolsky 1983, §14.5)

$$\mathcal{L}_{\text{disk}} \sim \frac{1}{2}\frac{G\dot{\mathcal{M}}\mathcal{M}}{\mathcal{R}} \sim 10^{34}\,\dot{\mathcal{M}}_9 \quad \text{erg s}^{-1} \tag{2.24}$$

for white dwarfs, and

$$\mathcal{L}_{\text{disk}} \sim \frac{1}{2}\frac{G\dot{\mathcal{M}}\mathcal{M}}{\mathcal{R}} \sim 10^{37}\,\dot{\mathcal{M}}_9 \quad \text{erg s}^{-1} \tag{2.25}$$

for neutron stars or black holes. Here $\dot{\mathcal{M}}_9$ is in units of $10^{-9}\ \mathcal{M}_\odot$ per year, which is a rate associated with mass transfer in real systems (to within an order of magnitude or so). Note that one unit of $\dot{\mathcal{M}}_9$ implies greater, or much greater, than solar luminosities for $\mathcal{L}_{\text{disk}}$. Note further the factor of 1/2 multiplying $G\dot{\mathcal{M}}\mathcal{M}/\mathcal{R}$ in these expressions: the remainder of the power goes to the star or boundary layer of the disk.

Mass transfer may also occur without complete filling of the Roche lobe. If, for example, vigorous mass loss is taking place from the surface of one of the stars in a close binary system, then some of that mass may be transferred to the companion even if the mass-losing star does not fill its own Roche lobe. Such might be the case for a system containing an upper main sequence star or a very luminous star on the RGB or AGB. If the stars are close enough, some of the mass streaming out of the Roche surface may well end up being captured in the companion's potential well.

There are many possibilities for close binary mass transfer. All that is required is stellar expansion at some stage or a means of decreasing separation distance. Prime evolutionary candidates include, but are not restricted to: evolution off the main sequence, rapid expansion up the RGB, and AGB evolution. These three common possibilities are called Cases A,B, and C, respectively (Paczyński 1971). As an example of how complicated things can get, consider Case A.

As the more massive star in an initially detached system begins its expansion off the main sequence and eventually fills its Roche lobe, some mass is transferred to the companion. The transfer rate may vary in time, giving rise to systems of different observational character (e.g., β Lyrae or Algol systems). Some mass is probably lost from the system, as is angular momentum (neglecting these last two constitutes the "conservative case"). Most studies indicate that the mass-losing star maintains a size roughly equal to the size of its Roche lobe and continues evolving in that fashion until whatever is left is no longer capable of expansion. (Note that in following this evolution the binary parameters period, separation, mass ratio, etc., change so that the size of the Roche radius $\mathcal{R}_{\rm RL}$ also changes.) The remnant at this stage may consist of merely a bare helium core whose hydrogen-rich envelope has been lost. In the meanwhile, the companion may not have been idle: it has been gaining mass and evolving in response to its own needs. If its initial mass was sufficiently small compared to its partner, then it may evolve off the main sequence well after the time that the more massive star has ceased losing mass. If enough time has passed, then the latter could have cooled down to become a white dwarf. The active star may now evolve off the main sequence and transfer mass back to its now white dwarf partner. It does not take too much effort to conceive of scenarios where this back and forth mass tossing may take place and each scenario can correspond to some observed stellar phenomenon. Some of these will be brought up shortly; a more complete list may be found in Trimble (1983).

2.6 Special Kinds of Stars

We shall now take a brief tour of those kinds of stars that bear special mention. Some of these have already been alluded to or discussed briefly in previous sections.

2.6.1 Intrinsically Variable Stars

By intrinsically variable stars we mean those that vary in their light output in a periodic or semiperiodic fashion and where the cause of variability is due solely to internal processes within the star. They are uniquely valuable in stellar astronomy because, as in terrestrial seismology, temporal variations in the behavior of their surface layers allow us to probe deep into their interiors and thus enrich our understanding of stellar structure.

We have already referred to some variable stars in our discussion of post–main sequence evolution but the list is much longer. Table 2.4 lists most known kinds of variable stars along with some of their properties. Be forewarned, however, that the names given to some classes of variable stars differ from author to author and, worse yet, there are occasional disputes

TABLE 2.4. Intrinsically variable stars

Kind of Variable	Period	Pop	Spec	M_V	R or NR
RR Lyrae	1.5–24 h	II	A2–F2	$\langle M_V \rangle \approx 0.6$	radial
Classical Cepheids	1–50 d	I	F6–K2	-6 to -0.5	radial
W Virginis	2–45 d	II	F2–G6	-3 to 0	radial
RV Tauri	20–150 d	II	G–K	~ -3	radial
Red Semiregular	100–200 d	I and II	M,N,R,S	-3 to -1	radial
Long period, Miras	100–700 d	I and II	M,N,R,S	-2 to $+1$	radial
The sun	5–10 min	I	G2	$+4.83$	nonradial
β-Cepheids (β CMa)	3–6 h	I	B1–B2	-4.5 to -3.5	both(?)
53 Per variables	0.5–2 d	I	O9–B5	-5 to -4	nonradial
ζ Oph	hours	I	O	-6 to -5	nonradial
Rapid Ap	5–15 min	I	$\sim$A5	$+2$ to $+2.5$	nonradial
δ-Sct (Dwarf Cepheid)	0.5–5 h	I	A5–F5	$+2$ to $+3$	nonradial
DO, DB, DA WDV's	100–1000 s	I(?),II	O,B2,A0	$+2,+7,+8$	nonradial

R or NR refers to Radial or Nonradial, Spec is Spectral type.

over the actual existence of some types. In addition, it is probably true (or it is our prejudice) that practically *all* stars will turn out to be variable when looked at closely enough. Some of the points we have just made are illustrated in Figures 2.21 and 2.22, which show the locations of variable stars in the H–R diagram. Almost everywhere you look there are variable stars except in those places where you do not expect stars in the first place. Our compilation in Table 2.4 is thus a bit conservative and minimalist and derives from standard lists (primarily from Cox 1980 and Unno et al. 1989). And now we take a short tour of the variable star H–R plane.

The RR Lyrae, Classical Cepheid (Type I Cepheid), and W Virginis (Type II Cepheid) variables all lie along a well-defined "Cepheid instability strip" on the H–R diagram. (See Fig. 2.21 and note that the width of the strip is between only some 600 K and 1100 K wide.) The first two are probably the best known and most completely investigated of all variable stars and their inner workings are, for the most part, well understood. All three of these classes share a common mode of variability: they pulsate (or oscillate) with simple motions whereby all elements of the star move periodically in and out while maintaining spherical symmetry. Such motion is referred to as "radial" (to distinguish it from something much more complicated). If the motions are such that the elements move in unison (no nodes in displacements) then the mode of motion is called the "fundamental" (and most of the above variables do just that). We may estimate the period, Π, of variability by using the period-mean density relation of (1.37), which described the time taken for a sound wave to traverse the star from center to surface and back again. This works remarkably well and means

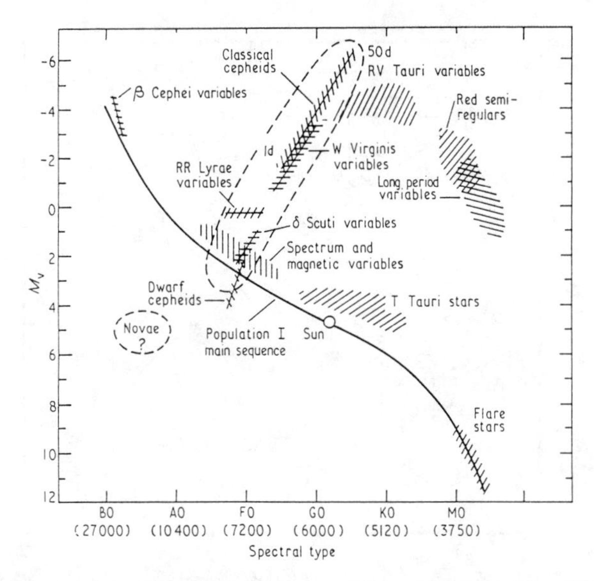

FIGURE 2.21. The location of some variable stars on the H–R diagram. Reprinted, with permission, from J.P. Cox, "Pulsating Stars," in *Reports on Progress in Physics*, Volume 37, 1974, Published by IOP Publishing Limited.

that the underlying motion is the result of a well-organized pressure wave coursing back and forth through the star in a purely radial manner. But what starts the wave in the first place? The answer, which was unraveled by Cox and Whitney (1958) and Zhevakin (1953), involves a classic application of the principles of stellar structure combined with hydrodynamics and radiative transfer (and we ignore convection for now).

A nonvariable star in hydrostatic and thermal equilibrium when perturbed in some way so as to induce mass motions will gradually return to its original state. The reason for this is that a parcel of mass that has been compressed, for example, is hotter than its surroundings and thus (normally) radiates to its surroundings. Thus the $P\,dV$ work that went into compressing the parcel is effectively dissipated and the parcel loses some of the "bounce" that might cause it to reexpand—as a necessary prerequisite for a sound wave. If this behavior is mostly the rule in the star, then the energy that originally went into perturbing the star will gradually leak out through the star and eventually be radiated from the surface. In a variable star—of those we understand—the situation is reversed: there are regions where compression results in heat effectively leaking *into* the compressed parcel and thus giving the parcel *more* bounce than usual. (We are simplifying a very complex process here and will make partial amends in Chapter 10.) If these regions dominate, then a perturbation will be driven and grow in time and not be damped out. The regions in a variable star that are responsible for this driving are those in which atomic ionization

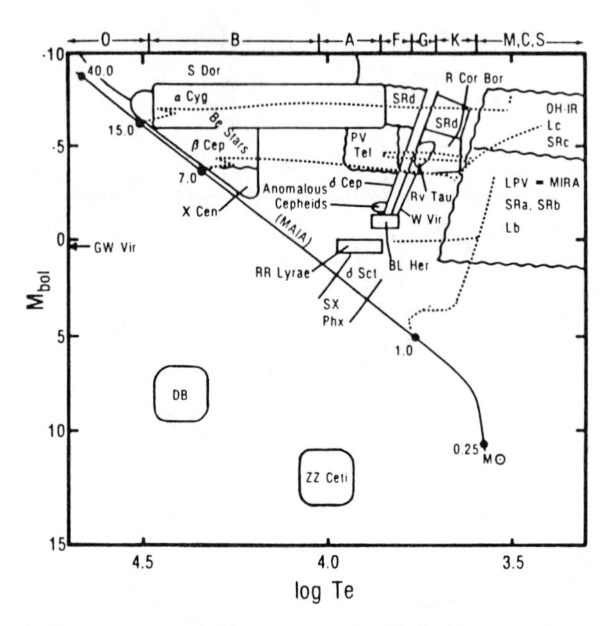

FIGURE 2.22. Even more variable stars on the H–R diagram. Reproduced, with permission, from Becker (1987).

and recombination of abundant elements are actively taking place. Such "ionization zones" are important for more reasons than just stellar variability. It is the effect of ionization on equations of state and opacity that eventually makes variable stars expand and contract in radius and wax and wane in light output. The ionization of most importance for variables in the Cepheid instability strip is that associated with the removal of the second electron from helium ("second helium ionization"). The reason why these variables are so tightly constrained in the strip is that at the relatively constant effective temperature of the strip ($8500 \lesssim T_{\rm eff} \lesssim 4000$ K) that ionization zone lies at a depth in the star that is especially propitious for driving of pulsations.

Since the processes of driving and damping of pulsations depend on mechanical effects (sound wave perturbations) plus the leakage of radiation or heat, there should be two time scales of relevance. The mechanical part is taken care of by the period–mean density relation. A rough estimate of the time required for thermal readjustment in a star is the Kelvin–Helmholtz time of §1.3.2. Taking the ratio of these two times we find

$$\frac{t_{\rm dyn}}{t_{\rm KH}} \approx 5 \times 10^{-12} \left(\frac{\mathcal{M}}{\mathcal{M}_\odot}\right)^{-5/2} \left(\frac{\mathcal{R}}{\mathcal{R}_\odot}\right)^{5/2} \left(\frac{\mathcal{L}}{\mathcal{L}_\odot}\right). \tag{2.26}$$

For a typical Classical Cepheid variable with $\mathcal{M} \sim 5\mathcal{M}_\odot$, $\mathcal{L} \sim 10^3\mathcal{L}_\odot$, and $\mathcal{R} \sim 30\mathcal{R}_\odot$, this yields $t_{\rm dyn}/t_{\rm KH} \sim 10^{-7}$. If we associate $t_{\rm KH}$ with the time it takes thermal effects to drive the star to pulsate at observable amplitude, then it requires about 10^7 periods of oscillation (since $\Pi \sim t_{\rm dyn}$) before the star really feels the effects of driving. But pulsate it does. Looked at another way, the dominant process in this Cepheid variable is the mechanical motion: the thermal (or "nonadiabatic") effects are comparatively weak. Thus the internal "heat engine" of the variable has, in mechanical terms, a high "Q." This is true for almost all variable stars except for those of very high luminosity and large radius, such as red supergiants.

The RR Lyrae and Classical Cepheid variables play another important role in astronomy besides being seismological benchmarks for stellar structure. They are both primary standard candles that provide the intermediate steps in determining distances on a cosmic scale (and, hence, the Hubble constant). It would be instructive to review the crucial role stars play in this enterprise but that would take us far afield. We suggest you read Rowan-Robinson (1985) for that story. However, if you consult Table 2.4, you will find that the RR Lyrae variables have a mean absolute visual magnitude of $\langle M_V \rangle \approx 0.6$: the spread around the mean is only about one-half magnitude. Thus (aside from annoying things like reddening and bolometric corrections), an RR Lyrae star seen in a distant globular cluster as a horizontal branch star stands out as a beacon whose distance modulus is easily found.

Cepheid variables are a bit more complicated standard candles. As seen in Figure 2.21 they occupy a good portion of the Cepheid instability strip with a wide range in luminosities and, hence, radii. In evolutionary terms they are in the stage of helium core burning and, from our previous discussion, the more massive the star the higher its luminosity. A relation due to Iben and Tuggle (1972) (and see §6 of Iben 1991) describing the relation between the two for massive Pop I stars ($X = 0.7$, $Z = 0.02$) in this stage is

$$\frac{\mathcal{L}}{\mathcal{L}_\odot} \approx 2.3 \left(\frac{\mathcal{M}}{\mathcal{M}_\odot} \right)^{3.99} . \tag{2.27}$$

Accepting this, we then use the approximate constancy of effective temperature in the strip, equate $\mathcal{L} = 4\pi\sigma\mathcal{R}^2 T_{\rm eff}^4$ to $\mathcal{L} \propto \mathcal{M}^4$ of the above, and find $\mathcal{R} \propto \mathcal{M}^2/T_{\rm eff}^2$. It is then an easy matter to apply the period–mean density relation and obtain an approximate "period-luminosity-color" (PLC) relation for the Classical Cepheids of $\mathcal{L} \propto \Pi^{8/5} T_{\rm eff}^{24/5}$. In other words, measure the period and effective temperature (color) of a distant Cepheid, apply this relation (with constants included), and find the time-averaged luminosity and eventually the distance.

This result is of course suspect because it relies on theory. A "true" PLC relation must follow from calibration of observations of Cepheids. For this, we refer you to the extensive review of Feast and Walker (1987). Since this text is primarily theoretical, our choice for a PLC relation is from Iben and Tuggle (1972), which derives from a combination of stellar evolution and pulsation studies:

$$\log\left(\frac{\mathcal{L}}{\mathcal{L}_\odot}\right) = -17.1 + 1.49 \log \Pi + 5.15 \log T_{\text{eff}} \tag{2.28}$$

where the period, Π, is in days. (Compare the exponents to those found above.) There should be some composition-dependent terms in this expression but, as Feast and Walker (1987) point out in any case, this gives quite decent results (with the theory tending to give slightly higher luminosities than the observations).

Moving off the Cepheid strip we find the β Cepheid (or β Canis Majoris) variables, which lie on a narrow strip in the H–R diagram that veers off the upper main sequence at around spectral class $B1$ ($\mathcal{M} \approx 10\mathcal{M}_\odot$). These variables, which have been observed for many years, have proven to be a thorn in the side of theoreticians, and many attempts have been made to explain why they vary. The most promising, and very recent, work makes use of newly computed opacities. As an example, you may wish to read Moskalik and Dziembowski (1992).

It is now believed that nonradial variable stars are the most common of all periodically variable stars. Thus, for example, the sun is listed in Table 2.4. Even though its variability is detectable only by use of sensitive instruments, we now know that it quivers and oscillates in some 10^7 independent ways. We shall have much more to say about this in Chapter 10, but recent studies are well on their way to resolving the interior of our nearest star by using these oscillations as seismological probes—and, if the sun is variable, then other sun-like stars should also be.

Among the very hot stars in the upper left-hand portion of the H–R diagram we find those whose variability is revealed through periodic changes in the profiles of (primarily) absorption lines. Most, if not all, are thought to be multiperiodic nonradial oscillators. They constitute a major challenge for the theorist because the cause of variability is not known. Part of the difficulty is that their atmospheres are dynamic and show clear evidence for mass loss and we have no idea how this interacts with the underlying mechanism that drives the star to pulsate.

The "rapidly oscillating Ap stars" are characterized by low amplitude, short-period photometric variations, strong magnetic fields, and enhanced surface abundances of exotic elements such as strontium and europium (among others less exotic). The observed short-period light variations are modulated in amplitude by the rotation of the star and it is thought that the pulsations are carried around by an off-axis magnetic field as the star rotates. This is the "oblique pulsator model" due to D. Kurtz who is responsible for discovering and defining this class of variable star. (A recent review with pertinent references is Kurtz 1990.) The oscillations most certainly represent nonradial acoustic motions such as those seen in the sun, but at much larger amplitudes. It is not yet clear what causes the oscillations or whether the magnetic field plays a role in driving them. Clues may be forthcoming from the δ Scuti variables (which are nearby in the H–R diagram) but these stars have, for the most part, no chemical peculiarities.

Finally, and with apologies to all the other kinds of variables, we have the variable white dwarfs (WDVs). These were discovered only within the last 20 years but the cause of their variability is already relatively well understood. As in the Cepheid variables, ionization zones are responsible. Since white dwarfs are so common (they comprise some 25% of all stars in the solar neighborhood), it is believed that they constitute the majority of all variable stars. Three classes are now recognized (although, as in most classification schemes, there will probably be more to come) and these are the GW Vir (variable DO white dwarfs), DBV (variable DB white dwarfs, and ZZ Ceti (variable DA white dwarfs) stars. The spectroscopic designations for white dwarfs (in general) refer primarily to differences in surface composition. We have given them more ordinary spectral types in Table 2.4 but we have done so only to emphasize that each class appears to be restricted to a rather narrow temperature range (as indicated in Fig. 2.22). They are interesting for a number of reasons. First of all, their pulsations do not represent acoustic waves as is the case for almost all other variables. Instead, imbalanced buoyancy forces result in gravity waves very much like those seen in the earth's oceans. As in the oceans, there may be waves with differing length scales and periods and, indeed, all of these variables are multiperiodic. There is also the following. The hottest of them (the GW Vir stars) are stars that have just left the post–AGB PNN phase. Since we still do not understand in detail how mass is ejected from red supergiants to make planetary nebulae, it is hoped that seismological investigations of their remnant cores will provide some of the missing links in our arguments. The coolest WDVs (the ZZ Ceti stars), on the other hand, are among the oldest stars in the disk of our galaxy and, if we can date them by seismological means, we then have a way of peering back to the time of earliest star formation in the disk (Winget et al. 1988, Iben and Laughlin 1989). These variables are considered in more detail in Chapter 10.

2.6.2 Cataclysmic Variables

Cataclysmic variables are eruptive close binary systems consisting of a white dwarf and a late type main sequence star, a red giant star, or, in some rare circumstances, another white dwarf. They are characterized by abrupt increases in light output with time scales typically of a day followed by a decline in light lasting from weeks to perhaps a year. The time interval between eruptions varies from days to only having been seen once in recorded history. During quiescence they are relatively inconspicuous. Orbital periods range from a little more than one hour to 15 hours for most systems although those containing a red giant may have periods in the hundreds of days. The rare examples with two white dwarfs may have periods of only 20 minutes but there exists some ambiguity in the interpretation of the observations. There are three major subclasses of these objects: the dwarf novae, classical novae, and the magnetic cataclysmic stars. (We shall only mention is passing the "novalike" and "recurrent novae": it isn't clear at this time just where they belong in the overall scheme of things.)

Recommended readings include: Robinson (1976), Bath (1985), Wade and Ward (1985), Starrfield and Snijders (1987), Szkody and Cropper (1988), and Starrfield (1988). A very readable review of evolutionary scenarios for cataclysmic variables can be found in the article by Webbink (1989). Several conference proceedings have been devoted to the subject (and some of the recommended readings may be found there). Older texts that are still worth reading are Payne-Gaposchkin (1957), and Glasby (1970). An extensive catalogue of these systems (and others) is due to Ritter (1987).

Dwarf Novae

Dwarf novae erupt repetitively (but not with a regular period) with intervals between outbursts of tens to hundreds of days. The duration of the outburst may vary but, for the majority of systems, there is a correlation between the duration of outburst and the interval of time before the next one takes place: the longer the duration, the longer the interval. At outburst peak, the optical luminosity is typically around $\mathcal{L} \approx 10^{34}$ erg s^{-1} with a (usually) rapid rise to the peak and a slower decline. The total energy released for an outburst is estimated to be of the order 10^{38}–10^{39} ergs. With this combination of readily identifiable characteristics, it is not surprising that some 300 dwarf novae are now known.

Because the dwarf novae are binary systems we can view them from different aspects as the two stars orbit and perhaps eclipse each other. The following is a generic description of what information has been derived from observations of the binary light curves of many systems. (We shall note later some variations on the following themes.) There is clear evidence for a stream of matter headed in a trajectory from the cool star to the vicinity of the white dwarf that reveals itself by a "hump" in the light curve. The increase in luminosity at the hump is thought to be due to collision of the infalling material from the secondary with an accretion disk surrounding the white dwarf thus causing a "hot spot." In most systems, the emergence of the hot spot also coincides with a noisy "flickering" in the light output, which most likely reflects the violence of the collision process. In strong support of the presence of a disk is the spectrum of the light emitted by the system: it is consistent with that expected from a thin but (often) optically thick bright disk and this light from the disk, in most cases, outshines both white dwarf and secondary stars. In rare cases Doppler-shifted atomic lines are observed that directly indicate rotation of material around the white dwarf. It is thus clear that dwarf novae systems represent mass overflow through a Roche lobe resulting in the formation of an accretion disk.

What is not so clear, and is still controversial, is what causes the eruptions. Two perfectly reasonable models are prime contenders. If the mass-losing secondary star is subject to instabilities within its Roche surface or at the inner Lagrangian point that cause variations in the amount of mass fed to the disk, then a higher than normal transfer rate will cause the disk to gain more energy and brighten and supplement the amount of material crashing down onto the white dwarf. A lull in to the transfer rate, on the other hand, will result in a quiescent state. The rhythm of outbursts is then set by the secondary. But if the secondary is a well-behaved star we are led to the competing model. As mass is steadily fed to the disk in this model, the disk grows in size and gradually brightens as it stores mass. Theoretical calculations have shown that this is a potentially unstable situation. If the accretion continues unabated, conditions in the disk may reach a point where the physics of ionization of hydrogen and helium cause what may best be described as a phase transition in the properties of the disk. The end result is a change in the mass-storage capabilities of the disk from one where additional mass may be easily accommodated and the disk is cool to one in which the disk rapidly heats up, glows more brightly, and dumps material down onto the white dwarf. Neither theory nor observations are yet up to discriminating between these models. The combination of observation and modeling of disk structures do lead to estimates of how much the total mass flow through the disk is modulated between outbursts and quiescent states. During quiescence the mass transfer rate estimate is $\dot{\mathcal{M}} \sim 5 \times 10^{-11}$ $\mathcal{M}_\odot$ yr^{-1}, which is boosted to $\dot{\mathcal{M}} \sim 5 \times 10^{-9}$ $\mathcal{M}_\odot$ yr^{-1} during outburst. The factor of 100 between these numbers is roughly consistent with the difference of power output between the two states. An order of magnitude (or so) estimate for the mass of the disk is found by multiplying the quiescent mass transfer rate by a typical time between outbursts (say a month) and yields $\mathcal{M}_{\text{disk}} \sim 10 \times 10^{-11}$ $\mathcal{M}_\odot$. Compared to the mass of the stars involved, this is a remarkably small number considering what the disk is able to do.

The dwarf novae are not a completely homogeneous class of objects and there are well-recognized subclasses named after their prototypes. The "Z Camelopardalis" systems, for example, have normal outbursts but, occasionally, instead of returning to the usual quiescent state, the light output from the system stays roughly constant at a level intermediate between quiescence and outburst peak for a few days to months. (It is somewhat difficult to reconcile these "standstills" with a long-lasting disk instability.) "SU Ursae Majoris" dwarf novae sometimes undergo "super-outbursts" during which the light output far exceeds that of a normal outburst. Finally, and as far as we shall go, the "U Geminorum" variables are those that fit into neither of the above subclasses and which may be thought of as more the prototype dwarf novae. Sample visual light curves for the three major classes are shown in Figure 2.23.

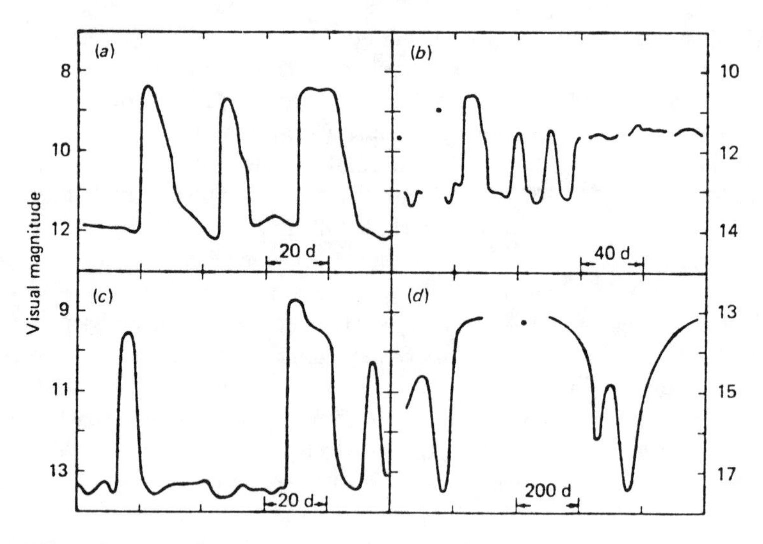

FIGURE 2.23. Visual light curves for three classes of dwarf novae: (a) a U Gem DN; (b) Z Camelopardalis; (c) an SU UMa DN. (Panel d is for a weird beast.) From Wade and Ward (1985), ©Cambridge University Press, reproduced with permission.

Lastly, we mention the presence of periodic optical oscillations that have been seen in some dwarf novae at outburst. These have periods ranging from 15 to 30 seconds and their cause in unknown. If they originate from the white dwarf, there is the possibility of using them to probe the interiors of these objects as is done for the variable white dwarfs.

Magnetic Cataclysmic Variables

If the white dwarf in a cataclysmic system has a very strong magnetic field, then the preceding scenario of Roche lobe overflow, hot spot, and disk has to be modified. Material flowing out of the inner Lagrangian point senses the magnetic field and, rather than be captured into a Keplerian disk, is funneled down the field lines onto the magnetic pole of the white dwarf. These "AM Herculis" systems radiate circularly polarized light and have "high" and "low" states of light output, each of which may last from months to years. In the rare cases where measurements have been successful, surface magnetic field strengths of around 10^7 gauss have been deduced for the white dwarf surface. (Such fields have also been observed in a few single white dwarfs. See Chapter 9.) Finally, some AM Her systems emit very strongly in the X-ray and this is perhaps not too surprising: material is funneled directly down an accretion column to the white dwarf without having to be stored in a disk as in the dwarf novae.

(Classical) Novae

The most energetic of the cataclysmics are the (classical) novae, which may yield a total of 10^{44}–10^{45} ergs upon eruption with some 10^{38} ergs being radiated in the optical. Unlike the dwarf novae, the cause of the outburst is the thermonuclear detonation of hydrogen-rich material that has gradually accumulated on the surface of the white dwarf by way of an accretion disk. The two types of objects also differ because there is firm evidence of mass ejection from the erupting novae to the extent of 10^{-5}–10^{-4} $\mathcal{M}_\odot$ with ejection velocities of order 1000 km s^{-1}. The time between eruptions may be estimated by dividing the mass lost by the mass transfer rate through the disk during quiescence. With $\dot{\mathcal{M}} \sim 10^{-8}$–$10^{-9}$ $\mathcal{M}_\odot$ yr^{-1} (similar to that for dwarf novae during outburst) this yields times in the thousands to millions of years. Thus if we find a well-documented account of a novae having erupted within historical times it is unlikely that we shall see it go off again within our lifetime. This is consistent with the observational definition of classical novae: we have seen a particular object erupt only once. But just wait long enough and some future astronomer will have the pleasure of seeing it erupt once more. The "recurrent novae" erupt frequently (and sometimes regularly) but it is not clear whether these less violent events are scaled-down classical novae or more energetic dwarf novae. We leave them to some future text.

There are two major subclasses of novae, which are classified by how rapidly their light output declines after maximum light and by expansion velocity. These distinctions characterize the speed classes "fast" and "slow." And slow may be slow indeed: the nova RR Tel has been in outburst for nearly forty years. What is surprising about all classes of classical novae is that the post-nova object (after everything has quieted down) appears identical to the prenova object (in the few cases where we have information prior to the eruptive event).

The physics of the eruption is reasonably clear. As material rich in hydrogen gradually accumulates on the surface of the white dwarf it is compressed and heated in the dwarf's strong gravitational field ($\sim 10^6$ earth gravities). When a critical mass of 10^{-5}–10^{-4} $\mathcal{M}_\odot$ has finally been deposited, the temperature and density at the base of this material have risen to $\sim 10^8$ K and 10^3–10^4 g cm^{-3}; a combination sufficient to initiate a runaway thermonuclear explosion using hydrogen as fuel. It has been established that CNO nuclei play a major role in controlling how the explosion proceeds. In normal CNO hydrogen-burning in main sequence stars the nuclei carbon, nitrogen, and oxygen act as catalysts in the conversion of hydrogen to helium. This is also true for a classical nova but because the explosion is far more rapid than slow burning on the main sequence, many of the intermediate nuclei produced do not have enough time to decay by positron emission until the explosion is well under way and material is already being ejected. The energy released by decays at later times helps power the expansion of the ejecta. It also appears that a successful fast nova requires that the material accreted from the secondary be overabundant in CNO nuclei as compared to the sun and the atmospheres of most other normal stars. What causes this overabundance in the outer layers of the secondary is not known but observations of the ejected matter confirm that C, N, and O are indeed overabundant along with other nuclei such as neon and magnesium (although the presence of the latter may reflect the composition of the underlying white dwarf).

Classical novae show a wide range of observable properties and these must reflect the influence of several parameters. Among these are the white dwarf mass and luminosity, the total mass accreted before the explosion is triggered, the rate of mass accretion, and the composition of the accreted material. Exploration of this large parameter space is one of the most challenging areas of theoretical stellar astrophysics.

X–Ray Bursters

From our brief discussion of close binary systems we know it is possible for evolution to have proceeded along different paths to have produced another kind of compact object other than a white dwarf, namely, a neutron star or black hole. Because of the deeper gravitational wells for these objects, we expect energetics on a grander scale than if only a white dwarf were involved and, indeed, they are. One class of such objects are the X-ray bursters (of Type I), which are mainly associated with globular cluster and Galactic bulge Pop II systems. Here, short-duration (1 to 10 seconds) X-ray bursts are produced by thermonuclear events on the surfaces of weakly magnetized neutron stars in direct analogy to what we have seen for classical novae. In both cases potential fuel is accreted, and compressed and heated, in a medium that has a low specific heat and one in which pressure is relatively insensitive to changes in temperature. The difference is that these conditions are fulfilled more strongly in the neutron star case because the gravitational field at the surface of the neutron star is so much stronger than that of the white dwarf. The theory was first discussed by Hansen and Van Horn (1975) who posed a very simplistic, but generally successful, model that preceded the actual discovery of the bursters. Since a complete discussion of these fascinating objects (and their relatives) is inappropriate here, we refer you to Shapiro and Teukolsky (1983), Taam (1985), and Hartman and Woosley (1988).

2.6.3 Supernovae

As far as we know, the most spectacular thing that can happen to a star is that it becomes a supernova. There are two major types of supernovae (SN) distinguished primarily by their spectra during and after eruption, the time evolution of their light output (i.e., their "light curves"), and, to some extent, by their population class. Both types are rare and it is estimated that our galaxy experiences only one such event for each type every 40 years or so (Tammann 1982). Rarer yet are those actually observed in our galaxy, where the last was described by Kepler in 1604 C.E.[7] The remaining twenty or so that should have erupted in the 400-year interval since that time were presumably obscured from view by dust in our galactic plane. We do, however, see the gaseous remains of past supernovae sprinkled around our galaxy and we believe that pulsars (or black holes) are the solid remnants for one class of SN. Obscuration poses less of a problem for observing supernovae in external galaxies where the count is now up to about 700—a number that includes the recent and very important supernova SN1987A in the Large Magellanic Cloud (LMC) at a relatively close distance of 50 kpc. (The designation refers to the year of discovery and the alphabetical ordering of discovery in that year.) With one possible exception (SN1961V), SN1987A is the *only* supernova whose presupernova progenitor was observed and spectroscopically classified before the explosion. What we say about the progenitors of all other supernovae is based purely on inference.

The various types of supernova share two things in common:

- The total energy released is of the order 10^{51} ergs in the form of photons, neutrinos, and the kinetic energy of mass motion. Which of these three elements dominates depends on supernova class but, in any case, the energy released in the form of photons appears to be relatively minor. Expansion velocities of ejected material usually exceed 10^4 km s^{-1} during the early stages of explosion.
- Thermonuclear burning of stellar material during the explosion produces heavy elements and the integrated effect of such processing is the major source of the abundance patterns of heavy elements seen in nature.

What follows is a brief description of what is seen from these objects and how we think they work. Primary references in the review literature are Wheeler (1981), Trimble (1982, 1983), Woosley and Weaver (1986), Weiler and Sramek (1988), Arnett et al. (1989), and Bethe (1990). Collections of papers include those edited by Brown (1988) and Petschek (1990).

Type I Supernovae

Type I supernovae (SN I) are characterized by an absence of hydrogen lines in their spectra either at early times during the explosion or later on. Their light curves are remarkably uniform with one SN I looking very much

[7] We use the nondenominational designation C.E., for "common era," instead of the more usual A.D.

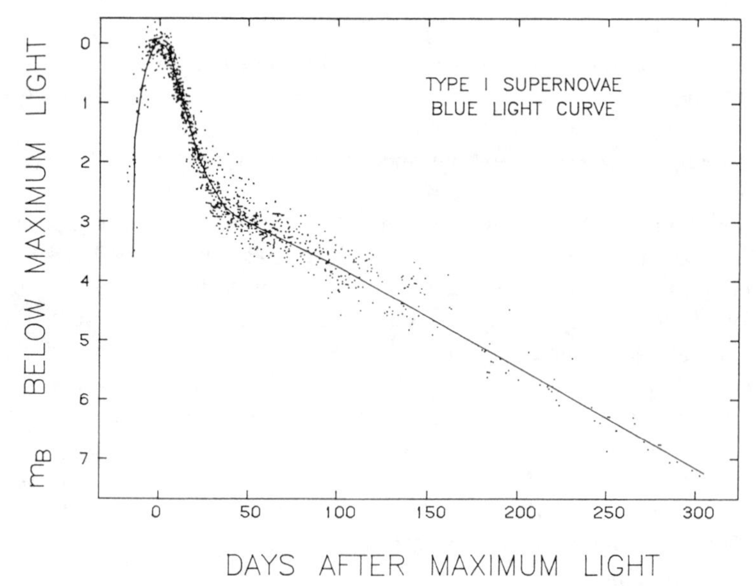

FIGURE 2.24. A composite of the light curves of 38 Type I supernovae in the blue portion of the visible. The ordinate measures the decrease in light output from that of maximum light as measured in apparent magnitude. The solid line represents the mean light curve for the class. The figure is reproduced, with permission, from Doggett and Branch (1985).

like another, as may be seen in the composite set of light curves shown in Figure 2.24 (from Doggett and Branch 1985). (We pass by, for now, any of the "peculiar" Type I supernovae.) We see a fast rise (of a few days) to maximum light followed by a steep falloff (about a month's worth), and then a long decline with an e–folding time for magnitude decrease of approximately 70 days. The spectra of well-studied SN I events reveal the presence of elements, with atomic masses ranging from helium through iron (and cobalt, which will turn out to be very significant) although not all these are present in all examples nor at all times through the observed light curve. Hydrogen is not seen (by definition) and this is a clue to what are the progenitors of SN I; they must be highly evolved stars that have lost all, or essentially all, of their hydrogen during evolutionary phases prior to the event. Additional evidence is available from details in their spectra, in what kinds of galaxies they are found, and where they are found in those galaxies.

Table 2.5, adapted from Wheeler (1990), lists elements seen in the spectra from three subclasses of Type I SN both near time of maximum light and at about six months past maximum. The presence, or lack, of these elements defines the subclasses. Since the subclass SN Ia is the most commonly observed (and we do have a good model), we shall concentrate our attention on them.

TABLE 2.5. Elements observed in the spectra of Type I SN.

Subclass	Near maximum	About 6 months
SN Ia	O, Mg, Si, S, Ca, Fe	Fe, Co
SN Ib	O, Ca, Fe	O, Ca, Mg
SN Ic	He, Fe, Ca	O, Mg

Type Ia supernovae are seen in all types of galaxies and, in spiral galaxies, they tend to lie between spiral arms or in the halo where we find the older stars (Weiler and Sramek 1988). This points to their being Pop II objects. (Classification in astronomy is very conservative and the juxtaposition of Type I and Pop II is just an unfortunate consequence.) The combination of Pop II and no detectable hydrogen in the spectrum of the supernovae immediately (with some hindsight) suggests that white dwarfs are involved. The standard model for the event is a close binary system containing a white dwarf and a companion mass-losing star (Whelan and Iben 1973). This sounds like a nova system but the crucial difference is that it is supposed that the white dwarf is very close to its Chandrasekhar limit either because that is the way it was made or by accretion from the companion. If matter is accreted onto such a white dwarf, the inevitable consequence is a reduction in size and compressional heating of the interior. If the added heat cannot be transported away sufficiently rapidly, then potential thermonuclear fuels—such as the carbon and oxygen thought to make up most of a white dwarf—may be ignited. Because of the extremely high densities in the interior, any ignition of fuel initiates a runaway explosion and a supernova is born.

To demonstrate the possible energetics of such an explosion, consider the thermonuclear burning of pure carbon in the form of ^{12}C under these conditions. If the burning is not somehow controlled then a sequence of reactions rapidly processes the carbon to elements in the mass range of iron, which have the highest binding energy per nucleon. It has been known for a long time (e.g., Fowler and Hoyle 1964, Truran et al. 1966) that the most abundant of these elements produced is the terrestrially radioactive nucleus ^{56}Ni, which has a half-life against electron decay of 6.1 days with a mean electron energy of 1.72 MeV. The energy released by the formation of ^{56}Ni (from 4+2/3 ^{12}C nuclei) is 8.25×10^{-5} erg. Were we to convert 1 $\mathcal{M}_\odot$ in this way, the total energy release would be about 10^{52} erg with plenty to spare for a supernova.

Details of the explosion process have been explored by many investigators (see, for example, the review by Woosley and Weaver 1986) using state-of-the-art hydrodynamic computer codes but some uncertainties remain. Two crucial parameters are the rate of mass accretion onto the white dwarf and the mass of the white dwarf. Different choices of combinations of these parameters lead to quite different events. It is possible, for example, to completely disrupt the white dwarf and leave no stellar remnant or to have a partial explosion that leaves behind a white dwarf of lesser mass. Other possibilities include the detonation of helium in a white dwarf that has not converted all of that element to carbon and oxygen. In any case, it is not thought that a neutron star—and, perhaps, pulsar—would be left behind as a remnant.

One very attractive feature of this model is that ^{56}Ni is one of the major nuclear by-products of the explosion. It decays to ^{56}Co which, in turn, decays to ^{56}Fe by electron emission with a half-life of 77.12 days. ^{56}Fe is, by far, the most abundant isotope of iron in nature and it is presumably what we see in the short- and long-term spectra of SN Ia (see Table 2.5). In addition, the electron emitted in the decay of ^{56}Co carries an energy of 3.58 MeV, which can be used to heat the expanding material and—to top it off—the decay time scale closely matches the typical SN Ia falloff time in magnitude. This pretty picture was anticipated as far back as 1962 and we recommend your reading Arnett et al. (1989) for a full recounting. In that review you will also learn (and we shall come back to this shortly) that gamma-ray lines from ^{56}Co have been detected from SN1987A. Except for nuisance details, the picture is now nearly complete.

The situation is not nearly as complete for either SN Ib or SN Ic. In those objects, iron is seen early on but not at late times. It is suspected that a thermonuclear event involving massive stars is the cause, but realistic modeling has not yet been done. We shall leave them for the future.

Type II Supernovae

Figure 2.25 shows the composite light curves for SN I and two subclasses of Type II supernova (from Doggett and Branch 1985). These two are designated SN II-P and SN II-L where the "P" stands for "plateau" and the "L" for "linear," which are designations that reflect the appearance of the light curves. The elements seen in the spectra of these include hydrogen and iron at early times and the presence of hydrogen is what distinguishes this class from Type I supernovae. However, these are not Pop II objects: they are not seen in elliptical galaxies (yet) but are associated with the young and massive stars that define the spiral arms of spiral galaxies or with the same kinds of stars in irregular galaxies.

Rather than give a general discussion of Type II supernovae, we shall concentrate on SN1987A in the LMC, which was first observed visually and photographically on February 24, 1987. Despite the fact that this is not a typical SN II object—its light curve and spectrum are almost unique, and it is intrinsically dimmer than what is typical—the basic physical processes driving the explosion are most certainly those of other SN II. (There are a small number of other "peculiar" supernovae, which we neglect here.) In

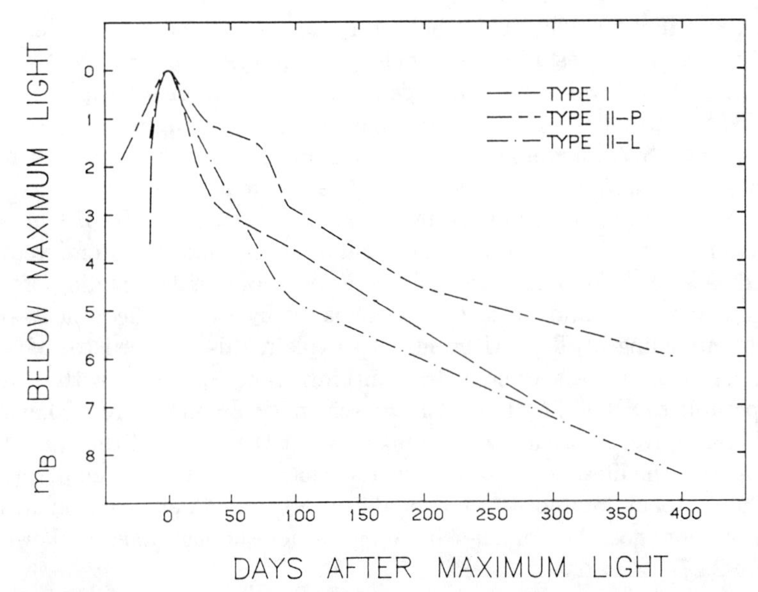

FIGURE 2.25. A composite of the light curves in blue light of supernovae of Type I and the two subclasses of Type II called SN II-P and SN II-L. Note the plateau-like feature at about 50 days for SN II-P, which contrasts with the linear decline for SN II-L. Reproduced, with permission, from Doggett and Branch (1985).

addition, we have a wealth of information concerning this object because of its relative proximity. We shall rely on the review by Arnett et al. (1989) in what follows and we strongly suggest you read that article. The review by Bethe (1990) gives an excellent overview of the physics involved.

First of all, we know which star exploded. It was Sanduleak–69 202, which was a B3 I blue supergiant with $\mathcal{L} \approx 1.1 \times 10^5\ \mathcal{L}_\odot$ and $T_{\rm eff} \approx 16000$ K. From various lines of evidence, it is estimated that the main sequence mass of Sanduleak–69 202 was in the range 16 to 22 $\mathcal{M}_\odot$ and that during its presupernova evolutionary stages it lost perhaps a few solar masses of its hydrogen-rich envelope. Although the star was certainly a Pop I object, its original composition was metal-poor compared to objects of similar mass in our galaxy: low-metallicity stars are characteristic of stars in the neighborhood of SN1987A and for the LMC in general.[8] A low metallicity must affect the evolution of massive stars because opacities are reduced in the envelope and the lowered abundance of carbon, nitrogen, and oxygen means that CNO burning cannot proceed as rapidly as it does in stars of

---

[8]This may not be the whole story because Podsiadlowski (1992) made convincing arguments that the precursor was a member of a binary system. Its evolution may thus have been affected in ways difficult to determine now.

higher metal abundance. The whole story has not yet been unraveled, but SN1987A ended up as a blue star before it exploded. Normal SN II events are thought to involve *red supergiants* and this difference explains why SN1987A is peculiar. (However, comparatively low-luminosity supernovae such as SN1987A may be much more common than we think: we just have a harder time finding them than we do the "normal" brighter objects.)

Perhaps the two most important observations made of SN1987A are the detection of neutrinos *prior to* the optical discovery and the later detection of radioactive ^{56}Co. These are the two keys to our understanding of how the star exploded and both were anticipated by earlier theoretical work on the modeling of Type II events. To explain this we need to delve in more detail into the thermonuclear burning stages of a star with a mass comparable to SN1987A. This is shown schematically in Figure 2.26, where each box represents an active burning stage at the center of the star. Also indicated is the lifetime of each stage, the central density and temperature, the total stellar luminosity and, finally, the total power given off in the form of neutrinos. An "onion-skin" diagram for the last stage is shown in Figure 2.27.

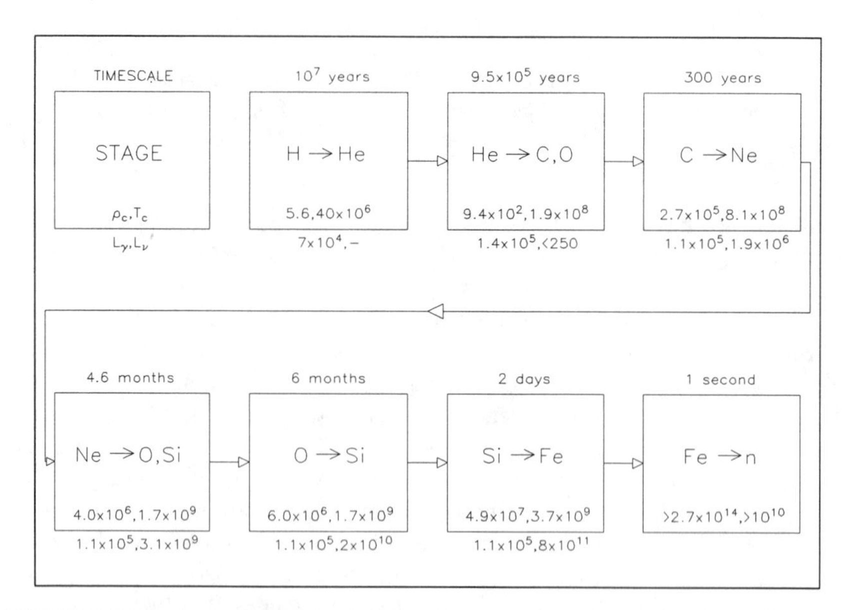

FIGURE 2.26. A representation of the thermonuclear burning stages of a star similar to SN1987A. The first box is a key to the notations in the boxes following—each of which represents a stage. Central density is in the units of g cm^{-3}, T_c in K, and photon and neutrino luminosities are in the units of $\mathcal{L}_\odot$. The progression of arrows indicates the arrow of time. The figure is adapted from Table 1 of Arnett et al. (1989).

We shall discuss these burning stages more fully in Chapter 6 but note the following: successively heavier nuclei are produced in the stellar center (as expected); the burning time scales decrease as time goes on; energy losses from neutrino emission soon outstrip other losses by the time carbon burning (to neon) commences; the optical luminosity stays practically constant past carbon burning; central temperatures and densities continually increase: the next to final stage results in the manufacture of Fe (generic iron peak material). This iron-producing stage, which takes only a couple of days, is the crucial time for the presupernova because iron peak nuclei can yield no further energy release from nuclear burning: that material is the ultimate stellar cinder. At the same time, however, energetic neutrinos are formed in the hot and dense central regions, pass unhindered through the remainder of the star, and are lost to space carrying off with them nearly as much energy in just two days as the sun has emitted in photons since it was born! This is a stellar disaster. But the rest of the star does not know about it—yet. Viewed from the outside, the star has not changed appearance for the last 300 years or so but this is because the normal processes involving the diffusion of photons from the interior take a long time.

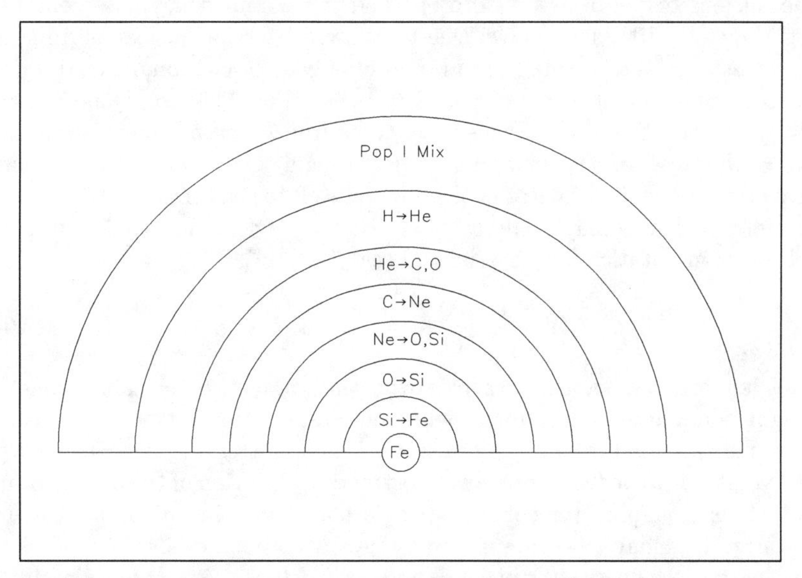

FIGURE 2.27. An onion-skin diagram for the last stage of Type II presupernovae. The thickness of the layers is not to scale.

The amount of iron core that is formed by the burning of silicon is approximately 1 $\mathcal{M}_\odot$ and has a radius near that of a white dwarf. This core (or perhaps the innermost part of it), having no further source of nuclear energy production and losing energy from neutrinos, now collapses on a time scale of seconds or less. As it does so, the temperature rises rapidly until the radiation bath of high-energy photons (in the form of gamma rays) interacts with the iron peak nuclei and effectively boils off their constituent nucleons. What speeds the process along is that at the very high densities encountered during the collapse, electrons gain enough energy that they may be captured on nuclei, thus converting protons into neutrons plus neutrinos. These processes continue until the stellar plasma of the core is reduced to a sea of mostly neutrons at a density comparable to or exceeding that of nuclear matter (2.7×10^{14}g cm^{-3}) confined in a radius measured in only tens of kilometers. Further collapse is effectively halted by the very stiff equation of state of nuclear matter.

What has happened is that all the nuclear burning stages the core has experienced—from hydrogen-burning through to the production of iron peak material—have been unraveled. This means that all the energy produced in those stages must be repaid back along with the energy lost through the emission of neutrinos. To estimate what is owed we consider first the nuclear energetics. The binding energy per nucleon of nuclei in the iron peak is about 9 MeV/nucleon. (This is the energy required to disperse the constituent nucleons to infinity. See Chapter 6.) If an average nucleus in this peak has 56 nucleons, it then requires about 8×10^{-4} ergs to reduce the nucleus to neutrons and protons and, for a unit solar mass containing about 2×10^{55} such nuclei, the total owed back for nuclear burning is about 2×10^{52} ergs. Finally (and as probably a lower limit), two days of neutrino emission at a rate of 8×10^{11} $\mathcal{L}_\odot$ (see Fig. 2.26) totals more than 10^{50} ergs. Where does all this energy come from? Actually, this is easy to answer because we have caused an object roughly the size of a white dwarf to collapse down its gravitational potential well to something the size of 10 km in radius. You may easily calculate that if 1 $\mathcal{M}_\odot$ is involved, the total release of gravitational energy is, and see §1.2,

$$|\Delta\Omega| \approx \frac{G(\mathcal{M}_\odot)^2}{10^6} \approx 3 \times 10^{53} \quad \text{erg.} \tag{2.29}$$

This is more than enough to repay debts and, in fact, is the same order of magnitude number as we found when discussing Type I supernovae. How this energy is used to blow up the star is still a matter of some controversy but what is involved is some way of abstracting a portion of the energy from out of the collapsed (or collapsing) core and depositing it further out in the star. This may be accomplished by having core material "bounce" as it reaches nuclear density (or beyond) and, as it bounces, collide with infalling material thus forming a shock that propagates outward lifting off most or all of the remainder of the star. The alternative is to produce enough high-energy neutrinos (by various processes ultimately relying on the energy released by collapse) so that some fraction of them might interact with overlying material to the extent that they effectively push off the outer layers. The calculations are very difficult and depend on many physical parameters (plus how the numerical work is done). Either way of doing things, however, and in the best of worlds, yields supernova energies in the proper range and, more to the point, predicts that high fluxes of neutrinos will pass out of the star before the event is seen optically, and that the violence of the event will cause nuclear processing of ejected material to iron peak nuclei—including large amounts of ^{56}Ni.

Both of these conditions are met for SN1987A. Neutrinos were detected about a quarter of a day before optical discovery, with energies within the proper range and over a time scale (5 to 10 seconds) that seems reasonable with the time scale estimated for their production. Furthermore, gamma-ray lines of ^{56}Co were detected well after the event, consistent with the early production of ^{56}Ni. Modeling of a later light curve powered by radioactive decay of these isotopes (plus ^{57}Co, ^{44}Ti, and ^{22}Na) also appears to be successful.[9] One further question remains, and this is whether a compact neutron-rich or black hole remnant lurks within the exploding debris. All models point to a remnant neutron star but none has been observed as yet. In its simplest manifestation it would appear as a pulsar but this requires that the neutron star have a strong magnetic field and that it be rotating (as the least of the requirements). We're all waiting.

2.7 Concluding Remarks

There are several kinds of stars we have not really discussed here. For example, pulsars are observed and they are very important for our understanding of supernova evolution and dynamics. The same is true for neutron stars which, through rotation, power the pulsar mechanism. We have also made little attempt to describe the entire zoo of known stars—normal or not. A good observational overview of stars is to be found in Jaschek and Jaschek (1987) who, generally speaking, use spectral type classification as a platform to guide us through the zoo.

The remainder of this text will deal primarily with the application of relatively simple physics to help us understand what goes on inside the kinds of stars discussed in this chapter.

2.8 Exercises

1. By some combination of means, a binary system has been observed and the following parameters determined for it:

   - The system has zero eccentricity; i.e., the orbits are circular.
   - The mass of the primary (the brighter star) is $\mathcal{M}_1 = 5\,\mathcal{M}_\odot$.

[9] The evolution of the the multiwavelength light curves of SN1987A is reviewed in McCray (1994).

- The inclination of the system is $i = 30°$.
- The period is $P = 31.86$ days.
- The maximum velocity of the primary along the line of sight to us is $V_r = 10.17$ km s^{-1}.

We assume that both primary and secondary stars were formed at the same time on the ZAMS and that further evolution has been such that neither mass nor angular momentum loss from either star or the system has occurred since that time. This means that the orbital parameters have not changed since the system was formed. The following questions require that you read up on Kepler's laws as applied to binary systems: see, for example, Mihalas and Binney (1981), pp. 79–86.

(a) What is the numerical value of the semimajor axis, a, of the system? Compare this figure to the distance of the planet Mercury from the sun.

(b) What is the mass, $\mathcal{M}_2$, of the secondary?

(c) After what period of time following ZAMS formation will the primary expand to fill its Roche lobe as a result of normal evolution? You will need radius versus time information to answer this. This information can be found by reading $\mathcal{L}$ and T_{eff} from Figure 2.8 and computing $\mathcal{R}$ from equation 2.1. For a larger version of Figure 2.8, see Figure 6-16 of Clayton (1968). You may also "cheat" and use the original source for these figures: Iben, I. Jr. 1966, *Ap.J.*, **143**, 483.

2. A Classical Cepheid variable with a period of 10 days is seen in a distant galaxy. Its observed color and apparent visual magnitudes are, respectively, $(B - V)_0 = 0.7$ and $m_V = 14$. If we assume there is no dust or gas between us and the star, estimate the distance to the galaxy using material from, say, Chapter 10 of Allen (1973, 3d ed). Also check how well the PLC relation given earlier in this text works given all this information.

3. As an exercise of your skills in homology or dimensional analysis try the following:

(a) Verify relations (2.6)–(2.9) for the lower main sequence,

$$\mathcal{L} \propto Z^{0.35} X^{1.55} T_{\text{eff}}^{4.12}; \tag{2.30}$$

$$\mathcal{R} \propto Z^{0.15} X^{0.68} \mathcal{M}^{1/13}; \tag{2.31}$$

$$\mathcal{L} \propto Z^{-1.1} X^{-5.0} \mathcal{M}^{5.46}; \tag{2.32}$$

$$T_{\text{eff}} \propto Z^{-0.35} X^{-1.6} \mathcal{M}^{1.33}. \tag{2.33}$$

(b) Now do the same thing for the upper main sequence where electron scattering and the CNO cycles are important. Still assume diffusive radiative transfer and the ideal gas equation of state. Since the rate for the CNO is proportional to the abundance of CNO nuclei times that of protons (see Chapter 6), take $\varepsilon_{\rm CNO} \propto XZ\rho T^{15}$.

4. We now have the tools to investigate a curious class of stars called "blue stragglers" that continues to baffle astronomers. One model for these stars is that they mix up their insides somehow so that their composition is always homogeneous. Construct a family of homologous stars in which the mean molecular weight μ is kept as an independent variable (i.e., $\mathcal{L} \propto \mu^{\beta_{\mathcal{L}}} \mathcal{M}^{\alpha_{\mathcal{L}}}$, etc.). Assume CNO burning ($\varepsilon \propto \rho T^{15}$) and Kramers opacity ($\kappa \propto \rho T^{-3.5}$). For simplicity, neglect the weak composition dependence of ε and κ, and assume a nondegenerate ideal gas equation of state.

   (a) How does the main sequence, $\mathcal{L}$ vs. $T_{\rm eff}$, vary with μ? Explain what is happening in physical terms.

   (b) What is the functional form of an evolutionary track for a homogeneous star in the H-R diagram? Draw an H-R diagram showing the evolutionary track in relation to the main sequence.

   (c) How does the luminosity of such a homogeneously evolving star change with μ? Such a star burns about 10 times as much fuel as a normally evolving star before depleting hydrogen. Qualitatively, will the homogeneous star live 10 times as long as a normal star? Why or why not?

   (d) Draw a schematic H-R diagram of a moderately old cluster as it would look if some small fraction of all stars underwent homogeneous evolution. Qualitatively, how would this diagram be modified if the "homogeneous" stars actually retained a small outer portion of their envelopes as unmixed hydrogen?

   (e) How much energy is required to mix the interior of such a star from gravitational considerations? Is such mixing then feasible? Can you suggest some ways in which intermediate-mass main sequence stars could mix themselves up?

You may wish to check out detailed computations of such mixed models. See, for example, Saio, H. and Wheeler, J.C. (1980), *Ap.J.*, **242**, 1176.

2.9 References and Suggested Readings

Introductory Remarks

We shall refer to

▷ Mihalas, D., and Binney, J. 1981, *Galactic Astronomy*, 2d ed. (San Francisco: Freeman)

several times in this text. Their discussion of stellar evolution is not complete but what they have is worth reading.

▷ Iben, I. Jr. 1991, *Ap.J. Suppl.*, **76**, 55

contains a personal account of Iben's work and, as is usual in his papers, the reference list is exhaustive. The compendium by

▷ Allen, C.W. 1973, *Astrophysical Quantities* (London: Athlone)

should be available to you as a general reference for facts.

▷ Lang, K.R. 1991, *Astrophysical Data: Planets and Stars* (Berlin: Springer-Verlag)

is also useful in this regard.

▷ Kippenhahn, R., and Weigert, A. 1990, *Stellar Structure and Evolution* (Berlin: Springer-Verlag)

devote several chapters to reviewing the different phases of evolution and special kinds of stars. We suggest you read our single chapter first and then go on to their extended discussion. They give many more detailed results from model calculations than do we.

§**2.1**: Pre–Main Sequence Evolution

The pre–main sequence tracks of Figure 2.2 are from the review article by

▷ Stahler, S.W. 1988, *Pub. Astron. Soc. Pacific*, **100**, 1474,

which we recommend highly for an "easy" introduction to the subject. A more advanced discussion may be found in

▷ Shu, F.H., Adams, F.C., and Lizano, S. 1987, *Ann. Rev. Astron. Ap.*, **25**, 23

and, for T Tauri stars,

▷ Bertout, C. 1989, *Ann. Rev. Astron. Ap.*, **27**, 351.

§**2.2**: Single Stars On and Near the Main Sequence

The ZAMS results shown in Figure 2.3 are taken from

▷ Iben, I. Jr. 1965, *Ap.J.*, **141**, 993

▷ Brunish, W.M., and Truran, J.W. 1982, *Ap.J. Suppl.*, **49**, 447.

The composition-dependent lower ZAMS figures use material from

▷ Mengel, J.G., Sweigart, A.V., Demarque, P., and Gross, P.G. 1979, *Ap.J. Suppl.*, **40**, 733.

For similar results see the references given in Tables 2.1 and 2.2 in addition to

▷ VandenBerg, D.A., and Bell, R.A. 1985, *Ap.J. Suppl.*, **58** 561

▷ D'Antona, F. 1987, *Ap.J.*, **329**, 653.

A good review of consequences of uncertainties in constructing models for low-mass stars may be found in

▷ Renzini, A., and Pecci, F.F. 1988, *Ann. Rev. Astron. Ap.*, **26**, 199.

The estimate of evolutionary time from the main sequence to the planetary nebula stage for low-mass stars is from

▹ Iben, I. Jr., and Laughlin, G. 1989, *Ap.J.*, **341**, 312.

The sources they use for the fit we quote include some of the above references.

The homology relations for the lower main sequence are discussed by

▹ Sandage, A., 1986, *Ann. Rev. Astron. Ap.*, **24**, 421.

This is a nice introductory paper.

The number of literature citations pertaining to very low mass stars and brown dwarfs has increased during the past few years partially because of the intense interest in how to "hide" mass in galaxies (as part of the "missing mass problem"). Among the many useful references are

▹ VandenBerg, D.A., Hartwick, F.D.A., Dawson, P., and Alexander, D.R. 1983, *Ap.J.*, **266**, 747

▹ Nelson, L.A., Rappaport, S.A., and Joss, P.C. 1986, *Ap.J.*, **311**, 226

▹ Liebert, J., and Probst, R.G. 1987, *Ann. Rev. Astro. Ap.*, **25**, 473

▹ Burrows, A., Hubbard, W.B., and Lunine, J.I. 1989, *Ap.J.*, **345**, 939

▹ Dorman, B., Nelson, L.A., and Chau, W.Y. 1989, *Ap.J.*, **342**, 1003

▹ Stevenson, D.J. 1991, *Ann. Rev. Astron. Ap.*, **29**, 163

▹ Burrows, A., Hubbard, W.B., Saumon, D., and Lunine, J.I. 1993, *Ap.J.*, **406**, 158.

Estimates for probable luminosity functions for brown dwarfs are discussed in

▹ Laughlen, G., and Bodenheimer, P. 1993, *Ap.J.*, **403**, 303.

The birth rate function, $\psi_s\, d\mathcal{M}$, quoted earlier is due to

▹ Salpeter, E.E. 1955, *Ap.J.*, **121**, 161.

For more complete discussions of this function and associated statistics of stars see

▹ Mihalas, D., and Binney, J. 1981, *Galactic Astronomy*, 2d ed. (Freeman: San Francisco), chap. 4

▹ Shapiro, S.L., and Teukolsky, S.A. 1983, *Black Holes, White Dwarfs, and Neutron Stars* (New York: Wiley & Sons), §1.3.

§**2.3**: Evolution of Single Stars Off the Main Sequence

▹ Iben, I. Jr., and Renzini, A. 1983, *Ann. Rev. Astron. Ap.*, **21**, 271

is only one of a series of reviews by one or both of these authors. Others include

▹ Iben, I. Jr. 1967, *Ann. Rev. Astron. Ap.*, **5**, 571

▹ Iben, I. Jr. 1974, *Ann. Rev. Astron. Ap.*, **12**, 215

▹ Iben, I. Jr. 1985, *Quart. Jour. Royal Astron. Soc.*, **26**, 1.

The original paper concerning the hydrostatic stabilty of isothermal cores near the main sequence phase is

- ▹ Schönberg, M., and Chandrasekhar, S. 1942, *Ap.J.*, **96**, 161.

Attempts to clarify why main sequence stars eventually evolve to be red giants are discussed in

- ▹ Eggleton, P. P., and Faulkner, J. 1981, in *Physical Processes in Red Giants*, Eds. I. Iben Jr. and A. Renzini (Dordrecht: Reidel), p. 179
- ▹ Weiss, A. 1983, *Astron. Ap.*, **127**, 411
- ▹ Iben, I. Jr., and Renzini, A. 1984, *Phys. Reports*, **105**, 329, §5.3
- ▹ Yahil, A., and Van den Horn, L. 1985, *Ap.J.*, **296**, 554
- ▹ Applegate, J.H. 1988, *Ap.J.*, **329**, 803
- ▹ Whitworth, A.P. 1989, *M.N.R.A.S.*, **236**, 505
- ▹ Bhaskar, R., and Nigam, A. 1991, *Ap.J.*, **372**, 592
- ▹ Renzini, A., Greggio, L., Ritossa, C., and Ferrario, L. 1992, *Ap.J.*, **400**, 280.

You might also look into the commentary by

- ▹ Tayler, R.J. 1988, *Nature*, **335**, 14.

§**2.4**: Late Stages of Evolution

There is a rich literature concerning the later stages of evolution of stars. For reviews we suggest:

- ▹ Iben, I. Jr. 1991, *Ap.J. Suppl.*, **76**, 55 (and all the references therein)
- ▹ Iben, I. Jr., and Renzini, A. 1984, *Phys. Reports*, **105**
- ▹ Chiosi, C., Bertelli, G., and Bressan, A. 1992, *Ann. Rev. Astron. Ap.*, **30**, 235.

The Yale Isochrones of Figure 2.9 are taken from data of

- ▹ Green, E. M., Demarque, P., and King, C. R. 1987, *The Revised Yale Isochrones and Luminosity Functions*, Yale University Observatory report (New Haven, Conn.).

You may also make up your own isochrones using the model results of, for example,

- ▹ Mengel, J.G., Sweigart, A.V., Demarque, P., and Gross, P.G. 1979, *Ap.J. Suppl.*, **40**, 733.

With further effort you could try to reproduce, using theoretical results, the observational H–R diagram of Figure 2.10 from

- ▹ Renzini, A., and Pecci, F.F. 1988, *Ann. Rev. Astron. Ap.*, **26**, 199.

To match turnoff points, equation (2.14) may be used from

- ▹ Iben, I. Jr., and Renzini, A. 1984, *Phys. Reports*, **105**

or the simpler form from

- ▹ Sandage, A., Katem, B., and Sandage, M. 1981, *Ap.J. Suppl.*, **46**, 41.

The periods of RR Lyrae stars, as a function of mass, $T_{\rm eff}$, and luminosity are adequately represented by (2.17) from

▷ Iben, I. Jr. 1971, *Pub. Astron. Soc. Pacific.*, **83**, 697;
see also
▷ Iben, I. Jr. 1974, *Ann. Rev. Astron. Ap.*,**12**, 215
for more information about HB stars. Core masses of helium-burning HB stars are discussed by
▷ Lee, Y.-W., Demarque, P., and Zinn, R. 1987, in *I.A.U. Colloqium #95, the Second Conference on Faint Blue Stars*, Eds. A.G.D. Philip, D. Hayes, and J. Liebert (L. Davis: Schnectady), p. 137.
How such stars have already lost mass on the RG branch is reviewed by
▷ Dupree, A. K. 1986, *Ann. Rev. Astron. Ap.*, **24**, 377.
See also the paper by
▷ Bowen, G.H., and Willson, L.A. 1991, *Ap.J. Letters*, **375**, L53
for a discussion of hydrodynamic effects on mass loss among cool and luminous stars.

An older but still worthwhile paper to peruse about the relation between AGB core mass and luminosity is
▷ Paczyński, B. 1970, *Acta Astron.*, **6**, 426.
Further refinements on this topic are due to
▷ Uus, U. 1970, *Nauch. Inform. Akad. Nauk. USSR*, **17**, 30
▷ Iben, I. Jr. 1977, *Ap.J.*, **217**, 788
▷ Havazelet, D., and Barkat, Z. 1979, *Ap.J.*, **233**, 589
▷ Wood, P.R., and Zarro, D.M. 1981, *Ap.J.*, **247**, 247
▷ Tuchman, Y., Glasner, A., and Barkat, Z. 1983, *Ap.J.*, **268**, 356.

Subsequent evolution of low-mass stars takes us to the land of the planetary nebulae. These are reviewed by
▷ Kaler, J.B. 1985, *Ann. Rev. Astron. Ap.*, **23**, 89.
The sample evolutionary track to the white dwarfs shown in Figure 2.13 is from
▷ Iben, I. Jr., and Renzini, A. 1983, *Ann. Rev. Astron. Ap.*, **21**.
More detailed tracks and results are discussed in Chapter 9. A review of white dwarf masses may be found in
▷ Weidemann, V. 1990, *Ann. Rev. Astron. Ap.*, **28**, 103.

▷ Chiosi, C., and Maeder, A. 1986, *Ann. Rev. Astron. Ap.*, **24**, 329
review the status of mass loss for luminous blue stars and
▷ Abbott, D.C., and Conti, P.S. 1987, *Ann. Rev. Astron. Ap.*, **25**, 113
discuss WR objects; see also
▷ Conti, P.S. 1978, *Ann. Rev. Astron. Ap.*, **16**, 371.
Equation (2.20) for a simple mass loss rate is from
▷ Garmany, C.D., and Conti, P.S. 1984, *Ap.J.*, **284**, 705.
Theoretical evolutionary tracks for mass-losing stars are given by
▷ Brunish, W.M., and Truran, J.W. 1982, *Ap.J.*, **256**, 247
▷ Maeder, A. 1990, *Astron. Ap. Suppl.*, **84**, 139

▹ Chin, C., and Stothers, R.B. 1990, *Ap.J. Suppl.*, **73**, 821.

§**2.5**: Evolution in Close Binary Systems

Chapters 2 and 16 of

▹ Tassoul, J.-L. 1978, *Theory of Rotating Stars* (Princeton: Princeton University Press)

contain useful information about stars in close binary systems.

▹ Shapiro, S.L., and Teukolsky, S.A. 1983, *Black Holes, White Dwarfs, and Neutron Stars* (New York: John Wiley & Sons)

emphasize those systems in which one of the members is a compact object. The excellent review by

▹ Trimble, V. 1983, *Nature*, **303**, 137

gives (in her usual inimitable style) an overview of the evolution of stars in binary systems. Another fine account can be found in

▹ Webbink, R. 1989, *American Scientist*, **77**, 248.

An important early review for the student is by

▹ Paczyński, B. 1971, *Ann. Rev. Astron. Ap.*, **9**, 183.

The short review by

▹ Shu, F.H., and Lubow, S.H. 1981, *Ann. Rev. Astron. Ap.*, **19**, 277

should also be consulted for technical details on mass, energy, and angular momentum transfer. Several articles in the 1985 work

▹ *Interacting Binary Stars*, Eds. J.E. Pringle and R.A. Wade (Cambridge: Cambridge University Press)

are also of interest. The reference to Pringle (1985) may be found there. We also suggest your looking into the now classic work of

▹ Kopal, Z. (1959) *Close Binary Systems* (London: Chapman and Hall)

It is as good an introduction to the mechanical theory of close binary systems as we know.

The approximation to the equivalent Roche lobe radius is due to

▹ Eggleton, P.P. 1983, *Ap.J.*, **268**, 368.

See also the older work by Paczyński (1971). Figure 2.18 is taken from

▹ Kitamura, M. 1970, *Ap. Space Sci.*, *7*, 272.

This reference contains a wealth of numeric and graphical information on the "Roche geometry."

We have not gone into the physics of accretion disks which, in fairness, would deserve a whole chapter. A good introduction is the review by

▹ Pringle, J.E. 1981, *Ann. Rev. Astron. Ap.*, **19**, 137.

§**2.6**: Special Kinds of Stars

The two main textbook references for the theory of variable stars are

▹ Cox, J.P. 1980, *Theory of Stellar Pulsation* (Princeton: Princeton University Press)

▹ Unno, W., Osaki, Y., Ando, H., Saio, H., and Shibahashi, H. 1989, *Nonradial Oscillations of Stars*, 2d ed. (Tokyo: University of Tokyo Press).

Our Chapter 10 discusses more general considerations in asteroseismology.

Figure 2.21 is from

▹ Becker, S.A. 1987, in *Stellar Pulsation*, Eds. A.N. Cox, W.M. Sparks, and S.G. Starrfield (Berlin: Springer-Verlag), p. 16.

This symposium proceedings, given in memory of J.P. Cox, is an excellent compendium of modern research in variable stars. We also recommend the 1988

▹ *Advances in Helio– and Asteroseismology*, *IAU Symposium 123*, Eds. J. Christensen-Dalsgaard, S. Frandsen (Dordrecht: Reidel).

▹ Jaschek, C., and Jaschek, M. 1987, *Classification of Stars* (Cambridge: Cambridge University Press)

is particularly useful in placing variable stars in the context of nonvariables. Some of you might also wish to peruse

▹ Hoffmeister, C., Richter, G., and Wenzel, W. 1985, *Variable Stars* (Berlin: Springer-Verlag).

Key papers from the early literature on how variable stars work are

▹ Cox, J.P., and Whitney, C.A. 1958, *Ap.J.*, **127**, 561

▹ Zhevakin, S.A. 1953, *Russian A.J.*, **30**, 161.

A very useful reference that describes how cosmological distances and time scales are derived (usually virtually all the tools of astronomy) is

▹ Rowan-Robinson, M. 1985, *The Cosmological Distance Ladder* (New York: Freeman).

The mass luminosity and PLC relations for massive core helium-burning stars is from

▹ Iben, I. Jr., and Tuggle, R.S. 1972, *Ap.J.*, **173**, 135

and see

▹ Iben, I. Jr. 1991, *Ap.J. Suppl.*, **76**, 55.

Observational constraints on PLC relations are discussed in

▹ Feast, M.W., and Walker, A.R. 1987, *Ann. Rev. Astron. Ap.*, **25**, 345.

Information on the rapidly oscillating Ap stars may be found in

▹ Unno, W., Osaki, Y., Ando, H., Saio, H., and Shibahashi, H. 1989, *Nonradial Oscillations of Stars*, 2d ed. (Tokyo: University of Tokyo Press)

▹ Kurtz, D.W. 1990, *Ann. Rev. Astron. Ap.*, **28**, 607.

A sample of what is going on with variable stars using new opacities is

▹ Moskalik, P., and Dziembowski, W.A. 1992, *Astron. Ap.*, **256**, L5.

There will be more to come.

Three papers pertaining to the age of the Galactic disk and the evolution of white dwarfs are

▹ Winget, D.E., Hansen, C.J., Liebert, J., Van Horn, H.M., Fontaine, G., Nather, R.E., Kepler, S.O., and Lamb, D.Q. 1988, *Ap.J.*, **315**, L77

▹ Iben, I. Jr., and Laughlin, G. 1989, *Ap.J.*, **341**, 312

▹ Wood, M.A. 1992, *Ap.J.*, **386**, 539.

A recent review is due to

▹ D'Antona, F., and Mazzitelli, I. 1990, *Ann. Rev. Astron. Ap.*, **28**, 139.

For another perspective using radioactivity-based chronologies see

▹ Cowan, J.J., Thielemann, K.-R., and Truran, J.W. 1991, *Ann. Rev. Astron. Ap.*, **29**, 447.

Our main references for cataclysmic variables are the following reviews:

▹ Robinson, E.L. 1976, *Ann. Rev. Astron. Ap.*, **14**, 119

▹ Wade, R.A., and Ward, M.J. 1985, in *Interacting Binary Stars*, Eds. J.E. Pringle and R.A. Wade (Cambridge: Cambridge University Press)

▹ Bath, G.T. 1985, *Reports Prog. Phys.*, **48**, 483

▹ Starrfield, S., and Snijders, M.A. J. 1987, in *Exploring the Universe with the IUE Satellite*, Ed. Y. Kondo (Dordrecht: Reidel), p. 377

▹ Szkody, P., and Cropper, M. 1988, and Starrfield, S. 1988, in *Multiwavelength Astrophysics*, Ed. F.A. Córdova (Cambridge: Cambridge University Press), p. 109

▹ Starrfield, S. 1988, in *Multiwavelength Astrophysics*, Ed. F.A. Córdova (Cambridge: Cambridge University Press), p. 159.

The cited texts are

▹ Payne-Gaposchkin, C. 1957, *The Galactic Novae* (Amsterdam: North-Holland)

▹ Glasby, J.S. 1970, *The Dwarf Novae* (London: Constable and Co.).

A catalogue of cataclysmic systems is due to

▹ Ritter, H. 1987, *Astron. Ap. Suppl.*, **70**, 335;

see also

▹ Duerbeck, H.W. 1987, *Space Sci. Rev.*, **45**, 1

▹ Downes, R.A., and Shara, M.M. 1993, *Pub. Astron. Soc. Pacific*, **105**, 127.

X-ray bursters are discussed in some detail in

▹ Shapiro, S., and Teukolsky, S. A. 1983, *Black Holes, White Dwarfs, and Neutron Stars* (New York: Wiley & Sons)

▹ Taam, R.E. 1985, *Ann. Rev. Nucl. Part. Sci.*, **35**, 1

▹ Hartman, D., and Woosley, S.E. 1988, in *Multiwave Astrophysics*, Ed. F.A. Córdova (Cambridge: Cambridge University Press), p. 189.

The first exploratory calculations were reported by

▹ Hansen, C.J., and Van Horn, H.M. 1975, *Ap.J.*, **195**, 735.

A collection of references concerning supernovae (and SN1987A) include

▹ Wheeler, J.C. 1981, *Reports Prog. Phys*, **44**, 85

▹ Trimble, V. 1982, *Rev. Mod. Phys.*, **54**, 1183

▹ Trimble, V. 1983, *Rev. Mod. Phys.*, **55**, 511

▹ Doggett, J.B., and Branch, D. 1985, *Astron. J.*, **90**, 2303

▹ Woosley, S.E. and Weaver, T.A. 1986, *Ann. Rev. Astron. Ap.*, **24**, 205

▹ Weiler, K.W., and Sramek, R.A. 1988, *Ann. Rev. Astron. Ap.*, **26**, 295

▹ Brown, G.E. 1988, Editor *Theory of Supernovae* in *Phys. Reports*, **163**

▹ Arnett, W.D., Bahcall, J.N., Kirshner, R.P., and Woosley, S.E. 1989, *Ann. Rev. Astron. Ap.*, **27**, 629

▹ Petschek, A. 1990, Editor *Supernovae* (New York: Springer)

▹ Bethe, H.A. 1990, *Rev. Mod. Phys.*, **62**, 801.

We also recommend

▹ Bethe, H.A. 1993, *Ap.J.*, **412**, 192

which contains a nice analytic way of looking at SN1987A. A complete review of the dynamics and composition of the ejecta from SN1987A will appear in

▹ McCray, R. 1994, *Ann. Rev. Astron. Ap.*, **31**, in press.

Other references cited are

▹ Tammann, G. 1982, in *Supernovae: A Study of Current Research*, Eds. M.J. Rees and R.J. Stoneham (Reidel: Dordrecht), p. 371

who discuss supernovae rates. An updated review of this question is due to

▹ van den Bergh, S., and Tammann, G.A. 1991, *Ann. Rev. Astron. Ap.*, **29**, 263.

▹ Fowler, W.A., and Hoyle, F. 1964, *Ap.J. Suppl.*, **9**, 201

▹ Truran, J.W., Cameron, A.G.W., and Gilbert, A.A. 1966, *Can. J. Phys.*, **44**, 576

▹ Whelan, J., and Iben, I. Jr. 1973, *Ap.J.*, **186**, 1007

▹ Wheeler, J. C. 1990, in Brown (1990)

▹ Podsiadlowski, P. 1992, *Pub. Astron. Soc. Pacific.*, **104**, 717.

§**2.7**: Concluding Remarks

▹ Jaschek, C., and Jaschek, M. 1987, *Classification of Stars* (Cambridge: Cambridge University Press).

§**2.8**: Exercises

▹ Clayton, D.D. 1968, *Principles of Stellar Evolution and Nucleosynthesis* (New York: McGraw-Hill)

- ▹ Allen, C.W. 1973, *Astrophysical Quantities* 3d ed. (London: Athlone)
- ▹ Mihalas, D., and Binney, J. 1981, *Galactic Astronomy*, 2d ed. (San Francisco: Freeman)
- ▹ Saio, H. and Wheeler, J.C. 1980, *Ap.J.*, **242**, 1176.

3

Equations of State

The worth of a State, in the long run,
is the worth of the individuals composing it.
— John Stuart Mill (1806–1873)

What is Matter?—Never mind.
What is Mind?—no matter.
— from *Punch* (1855)

The equations of state appropriate to the interiors of most stars are simple in one major respect: they may be derived using the assumption that the radiation, gas, fluid, or even solid, is in a state of *local thermodynamic equilibrium* or LTE. By this we mean that at a particular position in the star complete thermodynamic equilibrium is as very nearly true as we could wish. It is only near the stellar surface or in highly dynamic events, such as in supernovae, where this assumption may no longer be valid.

The reasons why LTE works so well are straightforward: particle–particle and photon–particle mean free paths are short and collision rates are rapid compared to other stellar length or time scales. (A major exception to this rule involves nuclear reactions, which are usually slow.) Thus, to a high degree of approximation, two widely separated regions in the star are effectively isolated from one another as far as the thermodynamics are concerned and, for any one region, the Boltzmann populations of ion energy levels are consistent with the local electron kinetic temperature.[1]

---

[1]For further discussions of the conditions for LTE see Cox (1968, §7) and

One typical scale length in a star is the *pressure scale height*, λ_P, given by

$$\lambda_P = -\left(\frac{d\ln P}{dr}\right)^{-1} = \frac{P}{g\rho} \tag{3.1}$$

where the equation of hydrostatic equilibrium has been used to eliminate dP/dr. The constant-density star discussed in the first chapter easily yields an estimate for this quantity of

$$\lambda_P\,(\rho = \text{constant}) = \frac{\mathcal{R}^2}{2r}\left[1 - \left(\frac{r}{\mathcal{R}}\right)^2\right]$$

using the run of pressure given by equation (1.38). The central value of λ_P is infinite but through most of the constant-density model it is of order $\mathcal{R}$. Near the surface it decreases rapidly to zero. We compare these lengths to photon mean free paths, λ_{phot}, which we construct from the opacity by

$$\lambda_{\text{phot}} = (\kappa\rho)^{-1} \ \text{cm}. \tag{3.2}$$

This quantity is a measure of how far a photon travels before it is either absorbed or scattered into a new direction (see Chapter 4). Note that opacity has the units of $\text{cm}^2\ \text{g}^{-1}$.

For electron scattering, which is the smallest opacity in most stellar interiors, later work will show that $\kappa \approx 1\ \text{cm}^2\ \text{g}^{-1}$. If we consider the sun to be a typical star and set $\mathcal{R} = \mathcal{R}_\odot$ and $\rho = \langle\rho_\odot\rangle \approx 1\ \text{g cm}^{-3}$ in the above, we then find λ_{phot} is at most a centimeter and $\lambda_P \sim 10^{11}$ cm through the bulk of the interior. Thus λ_{phot} is smaller than λ_P by many orders of magnitude. We could also have compared λ_{phot} with a temperature scale height and found the same sort of thing because, for the sun, the temperature decreases by only $10^{-4}\ \text{K cm}^{-1}$ on average from center to surface.

Another simple calculation yields an estimate of how much of a star is *not* in LTE. If the photon mean free path is still of order 1 cm, then the relative radius at which the pressure scale height is equal to the photon mean free path is $(r/\mathcal{R}) \approx 1 - 10^{-11}$ using the constant-density model. This means, as a crude estimate, that it is within only the last one part in 10^{11} of the radius that the assumption of LTE fails. In more realistic models, the assumption of LTE breaks down within the region of the stellar photosphere which is usually the only part of a star we actually can see.

In the following sections we shall quote some results from statistical mechanics, which will eventually be used to derive equations of state for stellar material consisting of gases (including photons) in thermodynamic equilibrium. Because several excellent texts on statistical mechanics are available

---

Mihalas (1978, Chapter 5). Note that different regions cannot be entirely isolated from one another because, otherwise, energy could not flow between them. Chapter 4 will consider this issue.

for reference, many results will be stated without proof. One particular text we recommend is Landau and Lifshitz (1958, or later editions) for its clean style and inclusion of many fundamental physical applications. Additional material may be found in the monograph by Pippard (1957) and in Cox (1968) and Kippenhahn and Weigert (1990).

3.1 Distribution Functions

The "distribution function" for a species of particle measures the number density of that species in the combined 6–dimensional space of coordinates plus momenta. If that function is known for a particular gas composed of a combination of species, then all other thermodynamic variables may be derived given the temperature, density, and composition. For the next few sections we shall assume that the gas, including electrons and photons, is a perfect (sometimes called ideal) gas in that particles comprising the gas interact so weakly that they may be regarded as noninteracting as far as their thermodynamics is concerned. They may, however, still exchange energy and other conserved properties. Before writing down the distribution function for a perfect gas we first introduce what may be an unfamiliar thermodynamic quantity.

The variables of thermodynamic consequence we have encountered thus far are P, T, ρ (or $V_\rho = 1/\rho$), S, E, Q, and various isotopic number densities, n_i (see §1.4.1). The latter have been, and will be, given in the units of number cm^{-3}. We also introduce N_i, which is the number density of an ith species in the units of number per gram of material with $N_i = n_i/\rho$. It is the Lagrangian version of n_i and it will prove useful because it remains constant even if volume changes.

Another useful thermodynamic quantity is the *chemical potential*, μ_i, defined by

$$\mu_i = \left(\frac{\partial E}{\partial N_i}\right)_{S,V}$$

as associated with an ith species in the material (not to be confused with μ_I, the ion molecular weight). If there exist "chemical" reactions in the stellar mixture involving some subset of species (ions, electrons, photons, molecules, etc.) whose concentrations could, in principle, change by dN_i as a result of those reactions, then thermodynamic (and chemical) equilibrium requires that

$$\sum_i \mu_i \, dN_i = 0. \tag{3.3}$$

Changing N_i by dN_i in a real mixture usually means that other components in the mixture must change by an amount related to dN_i so that not all the dN_i are independent.

As an example, consider the ionization-recombination reaction

$$\mathrm{H}^+ + \mathrm{e}^- \longleftrightarrow \mathrm{H}^0 + \gamma \tag{3.4}$$

where H^0 is neutral hydrogen—assumed to have only one bound state in the following discussion—H^+ is the hydrogen ion (a proton), and e^- is an electron. We shall neglect the photon that appears on the right-hand side of (3.4) in the following because, as we shall show, its chemical potential is zero and will not enter into the application of (3.3). The double-headed arrow is to remind us that the reaction proceeds equally rapidly in both directions in thermodynamic equilibrium. Now write (3.4) in the algebraic form

$$1\,\mathrm{H}^+ + 1\,\mathrm{e}^- - 1\,\mathrm{H}^0 = 0$$

where the coefficients count how many individual constituents are destroyed or created in a single reaction. A more general form for this equation is

$$\sum_i \nu_i \, \mathrm{C}_i = 0. \tag{3.5}$$

The C_i represent H^+, H^0, and e^- in the example and the ν_i, or *stoichiometric coefficients*, are the numerical coefficients. Obviously the concentrations, N_i, are constrained in the same way as the C_i. Thus if N_1 changes by some arbitrary amount dN_1, then the ith concentration changes according to

$$\frac{dN_i}{\nu_i} = \frac{dN_1}{\nu_1}.$$

Equation (3.3) then becomes

$$\sum_i \mu_i \frac{dN_1}{\nu_1} \nu_i = \frac{dN_1}{\nu_1} \sum_i \mu_i \nu_i = 0$$

or, since dN_1 is arbitrary,

$$\sum_i \mu_i \nu_i = 0. \tag{3.6}$$

This is the equation for *chemical equilibrium*, which must be part of thermodynamic equilibrium when reactions are taking place.[2]

As another simple, and useful, example consider a classical blackbody cavity filled with radiation in thermodynamic equilibrium with the walls of the cavity. Equilibrium is maintained by the interaction of the photons with material comprising the walls but the number of photons, N_γ, fluctuates about some mean value; that is, photon number is not strictly conserved.

[2]We exclude thermonuclear reactions from this discussion for the present because they may proceed very slowly and, usually, only in one direction during stellar nuclear burning.

Therefore dN_γ need not be zero. Nevertheless, reactions in the cavity must satisfy a symbolic relation of the form $\sum \mu_i \, dN_i + \mu_\gamma \, dN_\gamma = 0$ with $dN_i = 0$. The last two statements can only be reconciled if

$$\mu_\gamma = 0 \quad \text{for photons.} \tag{3.7}$$

It is for this reason that photons were not included in the ionization and recombination reaction of (3.4): the vanishing of μ_γ makes its presence superfluous in the chemical equilibrium equation (3.6).

It is reasonable, and correct, to expect that given T, ρ, and a catalogue of what reactions are possible, we should be able to find all the N_i for a gas in thermodynamic equilibrium. In other words, information about N_i is contained in μ_i for the given T and ρ. In a real gas this connection is difficult to establish because it requires a detailed knowledge of how the particles in the system interact. For a perfect gas things are easier. Any text on statistical mechanics may be consulted for what follows.

The relation between the number density of some species of elementary nature (ions, photons, etc.) in coordinate-momentum space and its chemical potential in thermodynamic equilibrium is found from statistical mechanics to be

$$n(p) = \frac{1}{h^3} \sum_j \frac{g_j}{\exp\left\{\left[-\mu + \mathcal{E}_j + \mathcal{E}(p)\right]/kT\right\} \pm 1} \,. \tag{3.8}$$

We call $n(p)$ the distribution function for the species. The various quantities are:

μ is the chemical potential of the species

j refers to the possible energy states of the species (e.g., energy levels of an ion)

$\mathcal{E}_j$ is the energy of state j referred to some reference level

g_j is the degeneracy of state j (i.e., the number of states having the same energy $\mathcal{E}_j$)

$\mathcal{E}(p)$ is the kinetic energy as a function of momentum p

a "+" in the denominator is used for Fermi-Dirac particles (fermions of half-integer spin) and a "−" for Bose-Einstein particles (bosons of zero or whole integer spin)

h is Planck's constant

$n(p)$ is in the units of number per $(\text{cm} - \text{unit momentum})^3$ where the differential element in coordinate-momentum space is $d^3\mathbf{r}\, d^3\mathbf{p}$.

To retrieve the physical space number density, n (cm^{-3}), for the species from (3.8) we need only integrate over all momentum space which, from standard arguments, is assumed to be spherically symmetric; that is,

$$n = \int_p n(p)\, 4\pi p^2\, dp \ \ \mathrm{cm}^{-3}. \tag{3.9}$$

Because we shall want to eventually consider relativistic particles, the correct form of the kinetic energy for a particle of rest mass m is given by

$$\mathcal{E}(p) = \left(p^2c^2 + m^2c^4\right)^{1/2} - mc^2 \tag{3.10}$$

which reduces to $\mathcal{E}(p) = p^2/2m$ for $pc \ll mc^2$ in the nonrelativistic limit, and $\mathcal{E}(p) = pc$ for extremely relativistic particles or those with zero rest mass.

We shall also need an expression for the velocity which, from Hamilton's equations, is

$$v = \frac{\partial \mathcal{E}}{\partial p}\,. \tag{3.11}$$

This is the velocity to use in the following kinetic theory expression for isotropic pressure (as in eq. 1.18)

$$P = \tfrac{1}{3} \int_p n(p)\, pv\, 4\pi p^2\, dp\,. \tag{3.12}$$

Finally, the internal energy is simply

$$E = \int_p n(p)\, \mathcal{E}(p)\, 4\pi p^2\, dp\,. \tag{3.13}$$

That completes all that we shall need to construct practical equations of state in the following applications.

3.2 Blackbody Radiation

Photons are massless bosons of unit spin. Since they travel at c, they only have two states (two spin orientations or polarizations) for a given energy and thus the degeneracy factor in (3.8) is $g = 2$. From before, $\mu_\gamma = 0$ and $\mathcal{E} = pc$. Because there is only one energy level (no excited states), $\mathcal{E}_j$ may be taken as zero. Putting this together, we find that the photon number density is given by

$$n_\gamma = \frac{8\pi}{h^3} \int_0^\infty \frac{p^2\, dp}{\exp\left(pc/kT\right) - 1} \quad \mathrm{cm}^{-3}. \tag{3.14}$$

Let $x = pc/kT$ and then use the integral

$$\int_0^\infty \frac{x^2\,dx}{e^x - 1} = 2\,\zeta(3) = 2(1.202\cdots)$$

where $\zeta(3)$ is a Riemann Zeta function, to find

$$n_\gamma = 2\pi\,\zeta(3)\left(\frac{2kT}{ch}\right)^3 \approx 20.28\,T^3 \quad \text{cm}^{-3}. \tag{3.15}$$

Find, in similar fashion, that the radiation pressure is

$$P_\text{rad} = \left(\frac{k^4}{c^3h^3}\frac{8\pi^5}{15}\right)\frac{T^4}{3} = \frac{aT^4}{3} \quad \text{dyne cm}^{-2}\,, \tag{3.16}$$

and that the energy density is

$$E_\text{rad} = aT^4 = 3P_\text{rad} \quad \text{erg cm}^{-3} \tag{3.17}$$

where a is the radiation constant $a = 7.566 \times 10^{-15}$ erg cm^{-3} K^{-4}. Note that (3.17) is a γ-law equation of state $P = (\gamma - 1)E$ (as in eq. 1.22 after E in that equation is converted to energy per unit volume) with $\gamma = 4/3$. Thus a star whose equation of state is dominated by radiation is in danger of approaching the $\gamma = 4/3$ limit discussed in Chapter 1.

It will be convenient for later purposes to define the energy density per unit frequency or wavelength in the radiation field. These energy densities are usually designated by u (with an appropriate subscript). Recall that frequency is given by $\nu = \mathcal{E}/h = pc/h$ and wavelength by $\lambda = c/\nu$. If u_p is the energy density per unit momentum (that is, the integrand of 3.13 with $E_\text{rad} = \int_0^\infty u_p\,dp$) and u_ν and u_λ are the corresponding densities per unit frequency and wavelength, then you may easily show

$$u_\nu\,d\nu = \frac{8\pi h\nu^3}{c^3}\frac{1}{e^{h\nu/kT} - 1}\,d\nu \quad \text{erg cm}^{-3}\ \text{Hz}^{-1}\ \text{Hz} \tag{3.18}$$

and

$$u_\lambda\,d\lambda = \frac{8\pi hc}{\lambda^5}\frac{1}{e^{hc/\lambda kT} - 1}\,d\lambda \quad \text{erg cm}^{-3}\ \text{cm}^{-1}\ \text{cm}\,. \tag{3.19}$$

Associated quantities are the *frequency-dependent Planck function*

$$B_\nu(T) = \frac{c}{4\pi}u_\nu \tag{3.20}$$

and the *integrated Planck function*

$$B(T) = \int_0^\infty B_\nu(T)\,d\nu = \frac{ca}{4\pi}T^4 = \frac{\sigma}{\pi}T^4 \tag{3.21}$$

where σ is the Stefan-Boltzmann constant. We shall make use of these functions when we discuss radiative transfer in the next chapter.

3.3 Ideal Monatomic Gas

As we shall soon show, the Boltzmann distribution for an ideal gas is characterized by $(\mu/kT) \ll -1$. We start off by asserting that this inequality holds for a sample of gas.

Assume, for simplicity, that the gas particles are nonrelativistic with $\mathcal{E} = p^2/2m$, $v = p/m$, and that they have only one energy state $\mathcal{E} = \mathcal{E}_0$. These could be, as examples, elementary particles, or a collection of one species of ion in a given state. If $(\mu/kT) \ll -1$, then the term ± 1 in the denominator of (3.8) may be neglected compared to the exponential and the gas becomes purely classical in character with no reference to quantum statistics. The expression for the number density is then

$$n = \frac{4\pi}{h^3} g \int_0^\infty p^2 e^{\mu/kT} e^{-\mathcal{E}_0/kT} e^{-p^2/2mkT}\, dp. \tag{3.22}$$

The integral is elementary and yields μ in terms of number density,

$$e^{\mu/kT} = \frac{nh^3}{g(2\pi mkT)^{3/2}} e^{\mathcal{E}_0/kT}. \tag{3.23}$$

Because we require $\exp(\mu/kT) \ll 1$ (since $\mu/kT \ll -1$), the right-hand side of (3.23) must be small. Thus $nT^{-3/2}$ cannot be too large. If this is not true, then other measures must be taken. For example, if μ/kT is negative but not terribly less than -1, it is possible to expand the original integrand for n (with the ± 1 statistics term retained) in a power series and then integrate. The additional terms obtained, assuming convergence of the series, represent Fermi-Dir ac or Bose-Einstein corrections to the ideal gas. This is done for fermions in Chiu (1968, Chapter 3) and in Chandrasekhar (1939, Chapter X), for example. In any event, μ may be computed once n and T are given. We assume here that (3.23) is by far the largest contribution to any expansion leading to an expression for μ for given n and T.

It is easy to take logarithmic differentials of n that yield the following expressions, and you may easily verify from the literature that they are the distribution functions for a Maxwell-Boltzmann ideal gas:

$$\frac{dn(p)}{n} = \frac{4\pi}{(2\pi mkT)^{3/2}} e^{-p^2/2mkT} p^2\, dp \tag{3.24}$$

and, in energy space,

$$\frac{dn(\mathcal{E})}{n} = \frac{2}{\pi^{1/2}} \frac{1}{(kT)^{3/2}} e^{-\mathcal{E}/kT} \mathcal{E}^{1/2}\, d\mathcal{E}. \tag{3.25}$$

A similar procedure involving the neglect of the ± 1 statistical factor equivalent to what was done for (3.23) yields the pressure

$$P = g\frac{4\pi}{h^3}\frac{\pi^{1/2}}{8m}(2mkT)^{5/2} e^{\mu/kT} e^{-\mathcal{E}_0/kT} \tag{3.26}$$

or, after substituting for $e^{\mu/kT}$ of (3.23),

$$P = nkT \quad \text{dyne cm}^{-2} \tag{3.27}$$

which comes as no surprise. This last result is true even if the particles are relativistic (as an easy calculation will show). The internal energy is

$$E = \tfrac{3}{2}nkT \quad \text{erg cm}^{-3} \tag{3.28}$$

using the same procedures. (Note that if reactions are present that change the relative concentrations of particles, then E must contain information about the energetics of such reactions; see below.) These are all elementary results for the ideal gas so that, given n and T, then P, E, and μ immediately follow.

3.4 The Saha Equation

In many situations the number densities of some species cannot be set *a priori* because "chemical" reactions are taking place. This is the problem referred to in §1.4 where mean molecular weights were computed. If the system is in thermodynamic equilibrium, however, then the chemical potentials of the reacting constituents depend on one another and this additional constraint is sufficient to determine the number densities.

As an example, consider the ionization–recombination reaction brought up earlier:

$$\mathrm{H}^+ + \mathrm{e}^- \longleftrightarrow \mathrm{H}^0 + \chi_\mathrm{H} \tag{3.29}$$

where $\chi_\mathrm{H} = 13.6$ eV is the ionization potential from the ground state of hydrogen (still assumed to have only one bound level). We assume that no other reactions are taking place that involve the above constituents and, in particular, that the gas is pure hydrogen. Reference to the photon in (3.29) has again been deleted because its chemical potential is zero and does not appear in the equilibrium condition (3.6), which will be invoked shortly.

To obtain the LTE number densities of the electrons and neutral and ionized versions of hydrogen, assume that all gases are in a Boltzmann distribution so that (3.23) applies. The reference energy levels for all species are established by taking the zero of energy as the just-ionized $\mathrm{H}^+ + \mathrm{e}^-$ state. (Other choices are possible of course.) Thus $\mathcal{E}_0$ for electrons and H^+ is zero, whereas for H^0 it is $-\chi_\mathrm{H} = -13.6$ eV lower on the energy scale. That is, we need 13.6 eV to convert H^0 to a free electron and a proton. The ground state of hydrogen has two near-degenerate states corresponding to spin-up or spin-down of the electron relative to the proton spin. For our purposes regard those states as having the same energy (but of course they do not, otherwise 21-cm HI radiation would not exist). Thus the degeneracy factor for H^0 is $g^0 = 2$. The situation for the free electron and H^+ is a bit more complicated because of the possible problem of double counting. If the spin axis of the proton is taken to be a fixed reference direction, then the free electron may have two spin directions relative to the free proton. Thus, $g^- = 2$ and $g^+ = 1$. The argument could be reversed without having any effect on the following.

With μ^-, μ^+, and μ^0 denoting the chemical potentials of the components in (3.29), equation (3.23) then yields

$$n_{\rm e} = \frac{2\left[2\pi m_{\rm e} kT\right]^{3/2}}{h^3} e^{\mu^-/kT} \tag{3.30}$$

$$n^+ = \frac{\left[2\pi m kT\right]^{3/2}}{h^3} e^{\mu^+/kT} \tag{3.31}$$

$$n^0 = \frac{2[2\pi(m_{\rm e}+m)kT]^{3/2}}{h^3} e^{\mu^0/kT}\, e^{\chi_{\rm H}/kT} \tag{3.32}$$

where $m_{\rm e}$ and $n_{\rm e}$ denote, respectively, the electron mass and number density, m is the proton mass and the neutral atom mass is set to $m_{\rm e}+m$.

Now form the ratio $n^+ n_{\rm e}/n^0$ and find

$$\frac{n^+ n_{\rm e}}{n^0} = \frac{(2\pi kT)^{3/2}}{h^3}\left(\frac{m_{\rm e} m}{m_{\rm e}+m}\right)^{3/2} e^{\left(\mu^- + \mu^+ - \mu^0\right)/kT}\, e^{-\chi_H/kT} .$$

But $\mu^- + \mu^+ - \mu^0 = 0$ for equilibrium by application of (3.6) so that we obtain the *Saha equation* for the single-level pure hydrogen gas

$$\frac{n^+ n_{\rm e}}{n^0} = \left(\frac{2\pi m_{\rm e} kT}{h^2}\right)^{3/2} e^{-\chi_H/kT} \tag{3.33}$$

where the reduced mass approximation $[m_e m/(m_{\rm e}+m)] \approx m_e$ has been used. A numerical version of part of this equation is

$$\left(\frac{2\pi m_e kT}{h^2}\right)^{3/2} = 2.415 \times 10^{15}\, T^{3/2} \ \ \mathrm{cm}^{-3} \tag{3.34}$$

and note that

$$kT = 8.617 \times 10^{-5}\, T \ \ \mathrm{eV} \tag{3.35}$$

where the eV units are handy for energies on the atomic scale.

To find the number densities, and not just ratios, further constraints must be placed on the system. A reasonable constraint is that of electrical neutrality, which requires that $n_e = n^+$ for a gas of pure hydrogen. Furthermore, nucleon number must be conserved so that $n^+ + n^0 = n$ where n is a constant if the density (ρ) is kept fixed.

We now define the degree of ionization (as in §1.4 and eq. 1.44)

$$y = \frac{n^+}{n} = \frac{n_{\rm e}}{n} \tag{3.36}$$

so that y is the fraction of all hydrogen that is ionized. The Saha equation (3.33) then becomes

$$\frac{y^2}{1-y} = \frac{1}{n}\left(\frac{2\pi m_e kT}{h^2}\right)^{3/2} e^{-\chi_H/kT} . \tag{3.37}$$

For sufficiently high temperatures, with fixed density, we expect the radiation field or collisions to effectively ionize all the hydrogen. This is indeed the case because we see that as $T \to \infty$, then $y \to 1$. Similarly, low temperatures mean less intense radiation fields and recombination wins with $y \to 0$.

For the pure hydrogen mixture $n = \rho N_A$ and equation (3.37) becomes

$$\frac{y^2}{1-y} = \frac{4.01 \times 10^{-9}}{\rho} T^{3/2} e^{-1.578\times10^5/T} . \tag{3.38}$$

The half-ionized ($y = 1/2$) path in the ρ-T plane for this mixture is then

$$\rho = 8.02 \times 10^{-9} T^{3/2} e^{-1.578\times10^5/T} \quad \text{g cm}^{-3}; \tag{3.39}$$

this is shown in Figure 3.1 as a very shallow curve for a range of what are interesting densities.

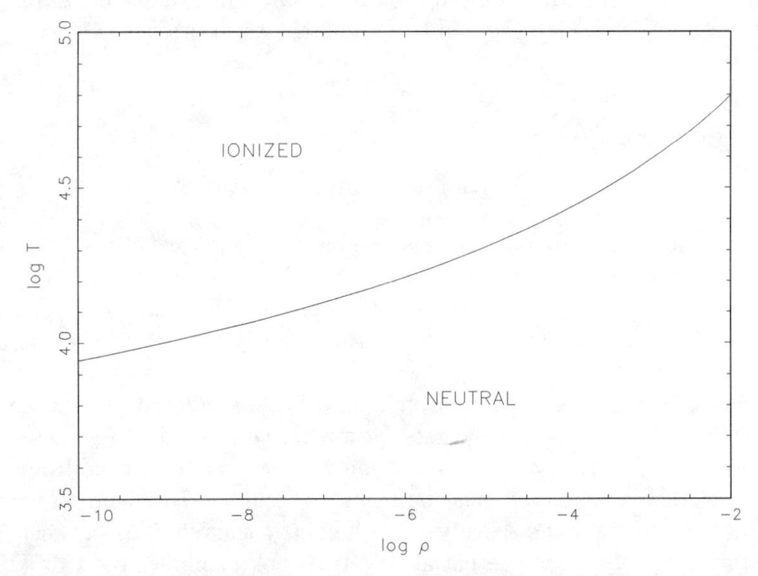

FIGURE 3.1. The half-ionization curve for a mixture of pure hydrogen undergoing the recombination–ionization reaction $H^+ + e^- \longleftrightarrow H^0 + \chi_H$ (ground state only).

The dominant factor in (3.38) and (3.39) is the exponential and this is what causes the half-ionization point to depend only weakly on density.

For hydrogen ionization from the ground state, the characteristic temperature for ionization-recombination is around 10^4 K and you may readily check that the transition from $y = 0$ to $y = 1$ takes place very rapidly as the temperature scans across that value (or, more precisely, at the temperature corresponding to $y = 1/2$ at a particular density). This is typical of ionization phenomena and a rough rule of thumb is that the transition temperature (where $y \approx 1/2$) is such that $\chi/kT \sim 10$ to within a factor of three or so depending on density. Thus, for example, the ionization potentials for removing the first and second electrons of helium are 24.6 eV and 54.4 eV, which correspond to transition temperatures of about 3×10^4 and 6×10^4 K. As we shall see, the presence of these zones of ionization have profound consequences for the structure of a star. You may wish to consider at this point a mixture of single-level hydrogen and helium (with two stages of ionization) and go through an analysis corresponding to the above to see how the various ions compete for electrons and to find out what the transition temperatures are for the three ionization stages involved. Even for this very practical but simple problem you will find that a computer is essential for your sanity.[3]

If the temperature and density of the hydrogen mixture are fixed, then equation (3.38) yields the ionization fraction y. The total hydrogen number density is clearly $n = \rho N_A$ and thus $n^+ = n_e = yn$ from (3.36). Chemical potentials, if required, follow from (3.30–3.32). The partial pressures and internal energies, which are additive, yield the total pressure

$$P = n(1+y)kT \tag{3.40}$$

and total internal energy

$$E = \tfrac{3}{2}n(1+y)kT + y\,n\chi_H \ \ \text{erg cm}^{-3}. \tag{3.41}$$

The last term in E appears because we have to take care of the ionization energy. If we wish to completely ionize the gas ($y \to 1$), then $(3nkT/2 + n\chi_H)$ erg cm^{-3} must be added to the system. Of this amount, $n\chi_H$ strips off the electrons, and the remainder brings the system up to the common temperature T.

The real calculation of ionization equilibria is as difficult as that for real equations of state (and the two are intimately connected). In principle, all species, energy levels, and reactions must be considered. In addition, the effects of real interactions must be included (and these depend on composition, temperature, and density), which change the relations between concentration and chemical potential. For textbook examples see Cox (1968, §15.3) and Kippenhahn and Weigert (1990, Chapter 14) with the warning that, in practice, accurate analytic or semianalytic solutions are seldom

---

[3] A pure helium version of this problem, along with other material, is the first exercise at the end of this chapter.

possible: you are usually faced with computer-generated tables of pressure and the like and the task is to use them intelligently.

3.5 Fermi-Dirac Equations of State

The most commonly encountered Fermi-Dirac elementary particles of stellar astrophysics are electrons, protons, and neutrons; all have spin one-half. (Neutrinos also appear but in contexts not usually connected with equations of state.) The emphasis here will be on electrons, but (almost) all that follows may apply to the other particles as well.

The number density of Fermi-Dirac particles is given by (3.8) and (3.9) with the choice of $+1$ in (3.8) and an energy reference level of $\mathcal{E}_0 = mc^2$. (Other choices are indeed possible for $\mathcal{E}_0$. They lead to an additive constant in the definition on the chemical potential and you have to watch out for this in the literature.) For these spin one-half particles, the statistical weight $g = 2$. Transcribing these statements then means that the number density is

$$n = \frac{8\pi}{h^3} \int_0^\infty \frac{p^2\, dp}{\exp\left\{\left[-\mu + mc^2 + \mathcal{E}(p)\right]/kT\right\} + 1} \tag{3.42}$$

where, in general,

$$\mathcal{E}(p) = mc^2 \left[\sqrt{1 + \left(\frac{p}{mc}\right)^2} - 1\right] \tag{3.43}$$

and

$$v(p) = \frac{\partial \mathcal{E}}{\partial p} = \frac{p}{m}\left[1 + \left(\frac{p}{mc}\right)^2\right]^{-1/2}. \tag{3.44}$$

We now explore some consequences of the above.

3.5.1 *The Completely Degenerate Gas*

The "completely degenerate" part of the title of this subsection refers to the unrealistic assumption that the temperature of the gas is absolute zero. In practice this does not happen but, under some circumstances, the gas effectively behaves as if it were at zero temperature and, for fermions in stars, these unusual circumstances are very important. So, in (3.42), note the peculiar behavior of the integrand as $T \to 0$. The exponential tends either to zero or infinity depending on, respectively, whether $-\mu + mc^2 + \mathcal{E}$ is < 0 or > 0. Therefore consider the interesting part of (3.8)

$$F(\mathcal{E}) = \frac{1}{\exp\left\{\left[\mathcal{E} - (\mu - mc^2)\right]/kT\right\} + 1} \tag{3.45}$$

where, as $T \to 0$, $F(\mathcal{E})$ approaches either zero or unity depending on whether $\mathcal{E}$ is greater or less than $\mu - mc^2$.

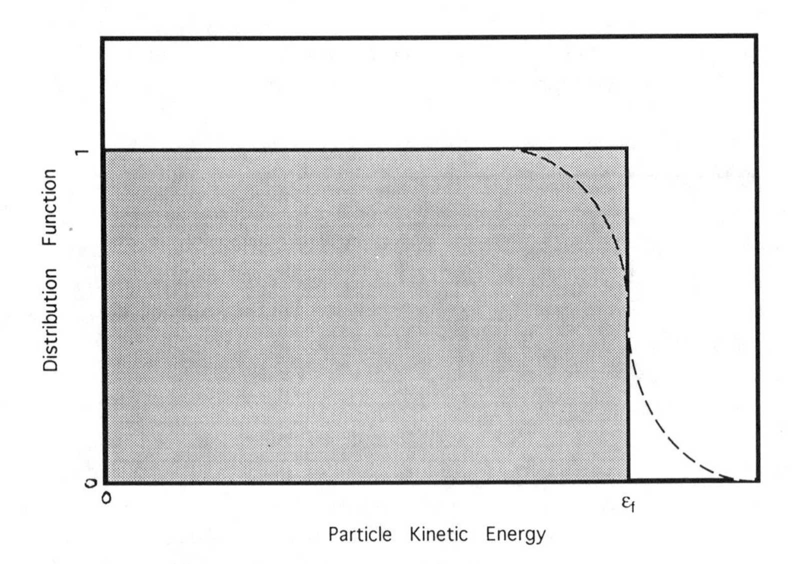

FIGURE 3.2. The function $F(\mathcal{E})$ of (3.45) versus particle kinetic energy for zero temperature. Fermions are restricted to the boxed-in area of unit height and do not have energies greater than the Fermi energy $\mathcal{E}_F$. The dashed line shows how $F(\mathcal{E})$ is changed by raising the temperature slightly. (In this case $\mathcal{E}_F/kT = 20$.)

The critical kinetic energy at which $F(\mathcal{E})$ is discontinuous is called the "Fermi energy" and it is denoted by $\mathcal{E}_F$. The situation is depicted in Figure 3.2 where in the shaded area, corresponding to particle energies $0 \leq \mathcal{E} \leq \mathcal{E}_F$, $F(\mathcal{E})$ is unity. Fermions are contained only in that energy range and not at energies greater than $\mathcal{E}_F$ where the distribution function is zero. (Ignore the dashed line for the moment. It shows what happens if the temperature is raised slightly above zero. See §3.5.3.)

The momentum corresponding to the Fermi energy is the Fermi momentum p_F. It is usually reduced to dimensionless form by setting $x = p/mc$ and defining $x_F = p_F/mc$. Then, from (3.43), we have

$$\mathcal{E}_F = mc^2 \left[\left(1 + x_F^2\right)^{1/2} - 1 \right] . \tag{3.46}$$

In this language, the chemical potential of the system is $\mu_F = \mathcal{E}_F + mc^2$ and it is the total energy, including rest mass energy, of the most energetic particle (or particles) in the system. If the spin is $1/2$ ($g = 2$), then all the rest of the particles are locked in pair-wise with spin-up and spin-down paired at each lower energy level by the Pauli exclusion principle. This "Fermi sea" is then capped by the "Fermi surface" at $\mathcal{E}_F$.

The relation between particle number density and the Fermi energy is found as follows. Because $F(\mathcal{E})$ is in the form of a unit step, (3.42) need only be integrated up to p_F. Hence

$$n = \frac{8\pi}{h^3}\int_0^{p_F} p^2\,dp = 8\pi\left(\frac{h}{mc}\right)^{-3}\int_0^{x_F} x^2\,dx = \frac{8\pi}{3}\left(\frac{h}{mc}\right)^{-3} x_F^3 . \tag{3.47}$$

To deal with astrophysically interesting numbers we shall, from this point on, deal exclusively with electrons unless otherwise noted.

It is traditional, but admittedly confusing, to delete the F subscript on x_F so that (3.47) is written

$$n_\mathrm{e} = \frac{8\pi}{3}\left(\frac{h}{m_\mathrm{e}c}\right)^{-3} x^3 = 5.865\times 10^{29}\,x^3 \quad \mathrm{cm}^{-3} \tag{3.48}$$

for electrons where $(h/m_\mathrm{e}c)$ is the electron Compton wavelength equal to 2.426×10^{-10} cm. The transcription to other spin 1/2 fermions is accomplished merely by changing the mass in (3.48).

To convert this to density units we reintroduce the electron mean molecular weight, μ_e, of (1.45–1.46) with $n_\mathrm{e} = \rho N_\mathrm{A}/\mu_\mathrm{e}$. Thus

$$\frac{\rho}{\mu_\mathrm{e}} = \frac{8\pi}{3N_\mathrm{A}}\left(\frac{h}{m_\mathrm{e}c}\right)^{-3} x^3 = B\,x^3 \tag{3.49}$$

with $B = 9.739\times 10^5$ g cm^{-3} for electrons. This may be looked upon as a relation that yields x (i.e., x_F) and, hence, $\mathcal{E}_F$ and p_F, once ρ/μ_e is given.

Note that the demarcation between nonrelativistic and relativistic mechanics occurs when $p_F \approx m_\mathrm{e}c$ or $x = x_F \approx 1$. The corresponding density is $\rho/\mu_\mathrm{e} \approx 10^6$ g cm^{-3} which, incidentally, is a typical central density for white dwarfs and is near the density at which the helium flash takes place. It remains to be shown, however, that temperatures in these contexts are sufficiently low to be effectively zero as far as electrons are concerned.

Looking ahead to neutron star matter, the numerical constant B in (3.49) is $B(\text{neutrons}) = 6.05\times 10^{15}$ g cm^{-3} and μ_e in that expression is set to unity; that is, we must replace μ_e by the amu weight of the neutron (essentially unity). For typical densities in a neutron star (comparable to nuclear densities of $\rho \approx 2.7\times 10^{14}$ g cm^{-3}), $x \approx 0.35$ and $\mathcal{E}_F \approx 57$ MeV. This implies that the neutrons are nonrelativistic because the neutron rest mass energy is 939.57 MeV.

The pressure of a completely degenerate electron gas is treated in the same way as that for the number density. It is the integral in (3.12) truncated at the Fermi momentum with $F(\mathcal{E})$ of (3.45) set to unity. A little work yields

$$P_\mathrm{e} = \frac{8\pi}{3}\frac{m_\mathrm{e}^4c^5}{h^3}\int_0^{x_F}\frac{x^4\,dx}{(1+x^2)^{1/2}} = A\,f(x) \tag{3.50}$$

where

$$A = \frac{\pi}{3}\left(\frac{h}{m_\mathrm{e}c}\right)^{-3} m_\mathrm{e}c^2 = 6.002\times 10^{22}\ \mathrm{dyne\ cm}^{-2}$$

for electrons (note the grouping of terms on the left-hand side) and

$$f(x) = x(2x^2 - 3)(1 + x^2)^{1/2} + 3\sinh^{-1} x\,. \tag{3.51}$$

In similar fashion the internal energy is given by the integral

$$E_\mathrm{e} = 8\pi \left(\frac{h}{m_\mathrm{e} c}\right)^{-3} m_\mathrm{e} c^2 \int_0^{x_F} x^2 \left[(1 + x^2)^{1/2} - 1\right] dx = A g(x) \tag{3.52}$$

with

$$g(x) = 8x^3 \left[(1 + x^2)^{1/2} - 1\right] - f(x)\,. \tag{3.53}$$

The units for E_e are erg cm^{-3}.

It will often prove useful to have limiting forms for $f(x)$ and $g(x)$ that correspond to the limits of relativistic or nonrelativistic electrons. These are:

$$f(x) \rightarrow \begin{cases} \frac{8}{5}x^5 - \frac{4}{7}x^7 + \cdots, & x \ll 1 \\ 2x^4 - 2x^2 + \cdots, & x \gg 1 \end{cases} \tag{3.54}$$

and,

$$g(x) \rightarrow \begin{cases} \frac{12}{5}x^5 - \frac{3}{7}x^7 + \cdots, & x \ll 1 \\ 6x^4 - 8x^3 + \cdots, & x \gg 1\,. \end{cases} \tag{3.55}$$

Note that $x \ll 1$ implies nonrelativistic particles, and $x \gg 1$ is the extreme relativistic limit. Also observe that

$$P_\mathrm{e} \propto E_\mathrm{e} \propto \begin{cases} (\rho/\mu_\mathrm{e})^{5/3}, & x \ll 1 \\ (\rho/\mu_\mathrm{e})^{4/3}, & x \gg 1 \end{cases} \tag{3.56}$$

and the limiting ratios of E_e to P_e are

$$\frac{E_\mathrm{e}}{P_\mathrm{e}} = \frac{g(x)}{f(x)} = \begin{cases} 3/2 \;(\gamma = 5/3), & x \ll 1 \\ 3 \;\;\; (\gamma = 4/3), & x \gg 1\,. \end{cases} \tag{3.57}$$

The values for γ are included as a reminder that for a γ-law equation of state the completely degenerate nonrelativistic electron gas acts like a monatomic ideal gas whereas, in the extreme relativistic limit, it behaves like a photon gas.

3.5.2 Application to White Dwarfs

As a simple, but important, application of completely degenerate fermion statistics, consider zero temperature stars in hydrostatic equilibrium whose internal pressures are due solely to electron degenerate material and whose densities and composition are constant throughout.

The easiest way to look at this is to apply the virial theorem in the hydrostatic form $3(\gamma - 1)U = -\Omega$ from (1.23). Because the star is assumed to have constant density, $\Omega = -(3/5)(G\mathcal{M}^2/\mathcal{R})$. If E_e is the volumetric energy density (with no contribution from the zero temperature ions), then $U = VE_e$ where V is the total stellar volume $V = (4\pi/3)\mathcal{R}^3$. In the nonrelativistic limit $E_e = 12Ax^5/5$ from (3.52) and (3.55), and x may be expressed in terms of ρ/μ_e via (3.49) and ρ, in turn, may be eliminated in favor of $\mathcal{M}$ and $\mathcal{R}$ by $\rho = \mathcal{M}/(4\pi\mathcal{R}^3/3)$. If the entire virial theorem is also cast in a form containing only $\mathcal{M}$ and $\mathcal{R}$, and if the constants A and B of (3.49–3.50) are given in terms of fundamental constants, then a little algebra yields the nonrelativistic mass-radius relation

$$\mathcal{M} = \frac{1}{4}\left(\frac{3}{4\pi}\right)^4 \left(\frac{h^2 N_A}{m_e G}\right)^3 \frac{N_A^2}{\mu_e{}^5}\frac{1}{\mathcal{R}^3}. \tag{3.58}$$

This relation has the remarkable property that as mass increases, radius decreases and is quite unlike the homology result for main sequence stars discussed in the first chapter.

For electrons, this yields the numeric expression

$$\frac{\mathcal{M}}{\mathcal{M}_\odot} \approx 10^{-6}\left(\frac{\mathcal{R}}{\mathcal{R}_\odot}\right)^{-3}\left(\frac{2}{\mu_e}\right)^5. \tag{3.59}$$

We state, without proof for now, that the interiors of white dwarf stars are almost entirely supported by electron degeneracy pressure, and that they typically have masses around $0.6\,\mathcal{M}_\odot$. If the electrons are nonrelativistic, then (3.59) yields a typical radius of $\mathcal{R} \approx 0.01\,\mathcal{R}_\odot$ for $\mu_e = 2$ (completely ionized ^{4}He, ^{12}C, ^{16}O, etc.). This radius is very close to that of the earth's with $\mathcal{R}_\oplus = 6.38 \times 10^8$ cm. An exact analysis involving integration of the hydrostatic equation using the nonrelativistic equation of state shows that (3.59) gives the correct result provided that the numerical coefficient is increased by (only!) a factor of two.

If μ_e in (3.58) is replaced by unity and the particle mass is taken to be that of the neutron, then the neutron star equivalent of (3.59) becomes

$$\frac{\mathcal{M}}{\mathcal{M}_\odot} \approx 5 \times 10^{-15}\left(\frac{\mathcal{R}}{\mathcal{R}_\odot}\right)^{-3} \qquad \text{(neutron stars)} \tag{3.60}$$

in the nonrelativistic limit. For $\mathcal{M} = \mathcal{M}_\odot$, $\mathcal{R} \approx 11$ km, which is in the right ballpark. (Note that general relativistic effects have been completely ignored, but this is the least of our sins because the nuclear force makes our noninteracting equation of state inaccurate.)

You will have realized by now that the simple arguments outlined above for making degenerate stellar models contain a serious flaw. The nonrelativistic degenerate electron pressure depends solely on density and composition (through μ_e); that is, in numeric form and using (3.49), (3.50), and (3.54)

$$P_e = 1.004 \times 10^{13}\left(\frac{\rho}{\mu_e}\right)^{5/3} \quad \text{dyne cm}^{-2} \tag{3.61}$$

and, as may easily be verified, the corresponding extreme relativistic expression is

$$P_e = 1.243 \times 10^{15} \left(\frac{\rho}{\mu_e}\right)^{4/3} \quad \text{dyne cm}^{-2}. \tag{3.62}$$

Thus if ρ and μ_e are constant, then so is P_e by virtue of the equation of state. But a constant pressure is inconsistent with hydrostatic equilibrium and, in fact, (1.38) is the correct solution for the pressure through a constant-density star. Thus P_e is not a constant and neither is E_e as assumed above. The trouble is that we have overconstrained the problem by insisting on the constancy of ρ combined with the degenerate equation of state.

The correct way to construct equilibrium degenerate models is to use the general expression for the pressure given by (3.50) along with the relation between ρ/μ_e and dimensionless Fermi momentum of (3.49). This yields a pressure-density relation, which is then put into the equation of hydrostatic equilibrium. The resulting equation is then combined with the equation of mass conservation yielding a second-order differential equation that must be integrated numerically. We shall not go into the tedious details here because more than adequate discussions are given in Chandrasekhar (1939, Chapter 11) and Cox (1968, §25.1). (However, some of these details involve the use of polytropes, which we discuss in Chapter 7.) Important results are summarized below.

In the limit of extreme relativistic degeneracy, where (3.62) is relevant, you may easily convince yourself by using dimensional analysis that the total stellar mass depends only on μ_e and not on radius. An exact analysis yields

$$\frac{\mathcal{M}}{\mathcal{M}_\odot} = \frac{\mathcal{M}_\infty}{\mathcal{M}_\odot} = 1.456 \left(\frac{2}{\mu_e}\right)^2 \tag{3.63}$$

where $\mathcal{M}_\infty$ is the *Chandrasekhar limiting mass.*[4] A virial analysis similar to that used to find (3.58), but done in the relativistic limit, yields a result differing from the above by only a change in the constant (a 1.75 instead of 1.456). We assume you will try to verify this and, if you do, you should also find that the full virial expression (1.23) implies d^2I/dt^2 becomes negative if the total mass exceeds $\mathcal{M}_\infty$. The interpretation is that electron degenerate objects (of fixed μ_e) cannot have masses exceeding the Chandrasekhar limit without collapsing the object. Increased densities and pressure cannot halt the collapse because the relativistic limit has already been reached. In

[4]The exact value of this limiting mass depends on physics we have not included in our analysis. Hamada and Salpeter (1961), for example, consider the effects of electrostatic interactions and electron captures on various nuclei. For single white dwarfs with normal masses and compositions, these effects are not that significant. However, we can imagine massive objects formed by various means in binary systems where such effects could well give a stable maximum mass less than the Chandrasekhar limiting mass.

the nonrelativistic limit, on the other hand, a new configuration may be reached by decreasing the radius as indicated by (3.59). Extreme relativistic equations of state, including that for photons, are too "soft" compared to the effects of self-gravity. (You can't make the particles exceed the speed of light to try to increase pressures!) This conclusion might have been anticipated because extreme relativistic effects imply $\gamma \to 4/3$.

The astrophysical significance of the Chandrasekhar limiting mass is as follows. If electron degenerate configurations are good representations of white dwarfs, and if those objects are the final end product of evolution for most stars, then the late stages of evolution are severely constrained. That is, if a star does not finally rid itself of enough mass to eventually leave a white dwarf with $\mathcal{M} \lesssim 1.46\mathcal{M}_\odot$ (assuming a reasonable value of $\mu_e = 2$), then something catastrophic will happen at some time in its life. Since there are so many white dwarfs in the sky, a large fraction of stars either start off with sufficiently low masses, or they manage to rid themselves of the excess mass.

The regime intermediate between nonrelativistic and full relativistic degeneracy is intractable using simple means, and full-scale models must be calculated. The following useful and quite accurate mass-radius relation bridging the two regimes (fit to actual calculations) is based on one given by Eggleton (1982) for electrons:

$$\frac{\mathcal{R}}{\mathcal{R}_1} = 2.02 \left[1 - \left(\frac{\mathcal{M}}{\mathcal{M}_\infty}\right)^{4/3}\right]^{1/2} \left(\frac{\mathcal{M}}{\mathcal{M}_\infty}\right)^{-1/3} . \tag{3.64}$$

Here, $\mathcal{M}/\mathcal{M}_\infty$ is given by (3.63), and $\mathcal{R}_1$ is defined by

$$\mathcal{R}_1 = \frac{1.117 \times 10^{-2}}{\mu_e} \mathcal{R}_\odot . \tag{3.65}$$

This radius is a typical scale length for electron degenerate objects. The relativistic and nonrelativistic limits of (3.64) go to the correct values as $\mathcal{R} \to 0$ (relativistic) or $\mathcal{M}$ becomes small (nonrelativistic). It is shown plotted in Figure 3.3.

We shall have more to say about white dwarfs in Chapter 9. One crucial item that has not been addressed here, and that pertains to these objects, is the effect of temperature on degeneracy. After all, if white dwarfs were really at zero temperature we would not see them.

3.5.3 Effects of Temperature on Degeneracy

The crucial step in deriving some of the thermodynamics of the completely degenerate zero temperature fermion gas was the realization that the distribution function becomes a unit step function at a kinetic energy equal to $\mu - mc^2$. If the zero temperature condition is relaxed, the distribution

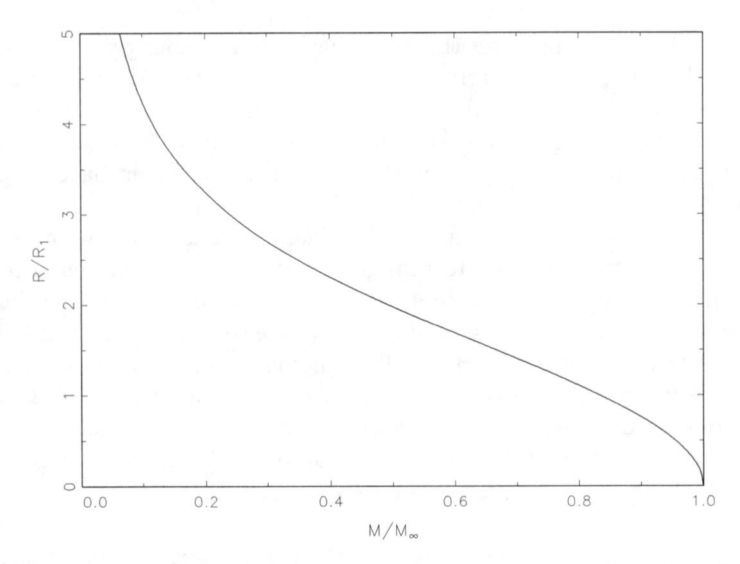

FIGURE 3.3. The mass-radius relation for zero temperature white dwarfs with constant μ_e. (See eqs. 3.64–3.65.)

function follows suit. Suppose the temperature is low (on some scale yet to determined) but not zero. Fermions deep in the Fermi sea, at energies much less than $\mathcal{E}_F$, need roughly an additional $\mathcal{E}_F$ energy units to move around in energy. That is, if the energy input to the system, as measured by kT, is much smaller than $\mathcal{E}_F$, then low-energy particles are excluded from promotion to already occupied upper energy levels by the Pauli exclusion principle. Fermions near the top of the Fermi sea do not have that difficulty and they may find themselves elevated into states with energies greater than $\mathcal{E}_F$. Thus as temperature is raised from zero, the stepped end of the distribution function smooths out to higher energies. This is the effect shown in Figure 3.2 by the dashed line. If temperatures rise high enough, we expect the effects of Fermi-Dirac statistics to be washed out completely and the gas should merge into a Maxwell-Boltzmann distribution. With this discussion as a guide, it should be apparent that a rough criterion for the transition from degeneracy to near- or nondegeneracy is $\mathcal{E}_F \approx kT$. As an example, we apply this criterion to nonrelativistic electrons.

The Fermi energy of a nonrelativistic electron gas is $\mathcal{E}_F = mc^2 x_F^2/2$, which is easily obtained by expanding the radical in (3.46) for small x_F. The dimensionless Fermi momentum x_F is then converted to ρ/μ_e using (3.49). After this is applied to $\mathcal{E}_F \approx kT$, and numbers put in, the criterion becomes

$$\frac{\rho}{\mu_e} \approx 6.0 \times 10^{-9} T^{3/2} \quad \text{g cm}^{-3}. \tag{3.66}$$

If ρ/μ_e exceeds the value implied by the right-hand side of (3.66) for a given temperature, then the gas is considered degenerate. Realize though that this is a rough statement: there is no clean demarcation line on the T–ρ/μ_e plane that distinguishes degenerate from nondegenerate electrons.

The extreme relativistic analogue to (3.66) is

$$\frac{\rho}{\mu_e} \approx 4.6 \times 10^{-24}\, T^3 \quad \text{g cm}^{-3}\,. \tag{3.67}$$

The density near which special relativistic effects become important has been estimated earlier as $\rho/\mu_e \approx 10^6$ g cm^{-3}. Equations (3.66) and (3.67) are illustrated in Figure 3.4 where the transition near 10^6 g cm^{-3} has been smoothed. Note that the center of the present-day sun, as indicated in the figure, is nondegenerate but close enough to the transition line that good solar models include the effects of Fermi-Dirac statistics.

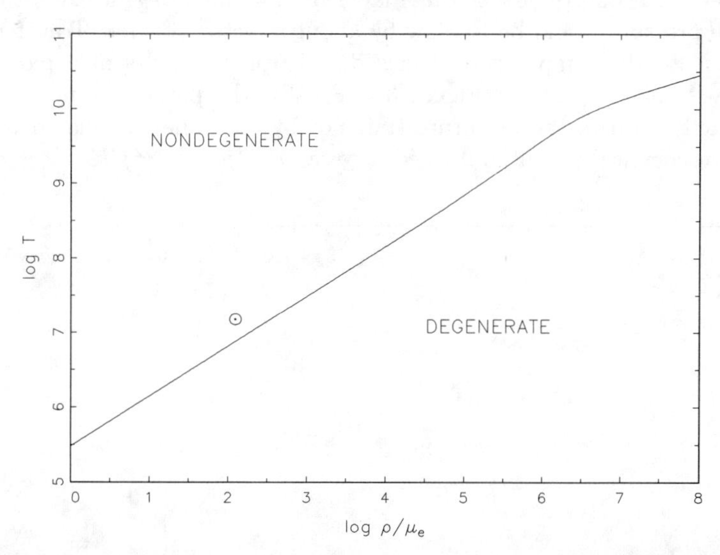

FIGURE 3.4. The domains of nondegenerate and degenerate electrons in the T–ρ/μ_e plane. The location of the center of the present-day sun in these coordinates is indicated by the $\odot$ sign.

A better idea of how the transition from degeneracy to nondegeneracy takes place with respect to temperature and ρ/μ_e requires explicit evaluation of the Fermi-Dirac integrals. In general, this involves numeric integration, although there are some useful series expansions and we shall discuss one of these in a bit. The reader is referred to Cox (1968) and other references at the end of this chapter for a full discussion but the results are summarized in Figure 3.5, which is derived from the numeric tabulations in Appendix A2 of Cox and his §24.4. Cloutman (1989) discusses some techniques for computing the Fermi–Dirac integrals and includes a `FORTRAN` program listing (see also Eggleton et al. 1973 and Antia 1993). Plotted versus ρ/μ_e in Figure 3.5 is the ratio of electron pressure at nonzero temperature, $P_e(T, \rho/\mu_e)$, to the electron pressure for complete degeneracy at zero temperature, $P_e(T = 0, \rho/\mu_e)$. Values near unity for this ratio imply strong degeneracy for $P_e(T, \rho/\mu_e)$ whereas large values mean that the gas is nondegenerate and, if large enough, the Maxwell-Boltzmann expression may be used. Note that the effects of electron-positron pairs created by the radiation field are not included here. These become important if temperatures approach or exceed $kT \approx m_e c^2$ (i.e., $T \gtrsim 6 \times 10^9$ K). We shall discuss pair-created electrons briefly in Chapter 6, where they play a role in creating neutrinos.

A parameter called η is plotted as dashed lines on the figure and an η of five, for example, corresponds to the situation where the true pressure is only about 15% greater than we would find if the gas were completely degenerate. Along the dashed line labeled "$\eta = 0$," a degenerate estimate for the pressure would be too low by about a factor of three. Transferring this line to the temperature versus density plane results in a plot that is very similar to that of Figure 3.4. Finally, the parameter η, which is commonly used in the literature (but not by everyone), is related to the electron chemical potential defined here by $\eta = (\mu - m_e c^2)/kT$.

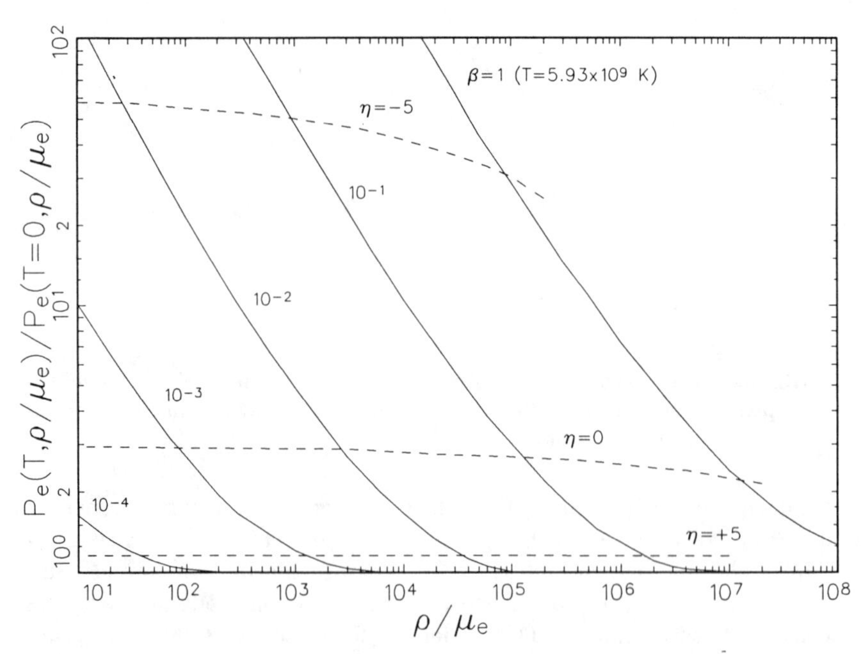

FIGURE 3.5. The domains of nondegenerate and degenerate electrons in temperature and density as expressed by the ratio $P_e(T, \rho/\mu_e)/P_e(T = 0, \rho/\mu_e)$. Temperatures are given in units of $\beta = kT/m_e c^2$ where $\beta = 1$ corresponds to 5.93×10^9 K. The dashed lines are lines of constant η, which is sometimes called the "degeneracy parameter" and is related to the chemical potential (see text).

For strongly, but not completely, degenerate gases, there are useful expansions for number density, pressure, and internal energy that are often quoted in the literature. We shall not derive those expansions here (see the references) but they all depend on the mild relaxation of the shape of the distribution function near $\mathcal{E}_F$. Following Landau and Lifshitz (1958, §57) we write any of the Fermi-Dirac integrals (for number density, etc.) in the kinetic energy-dependent form

$$I(\mu, T) = \int_0^\infty \frac{G(\mathcal{E})\, d\mathcal{E}}{\exp\left[\left(-\mu + mc^2 + \mathcal{E}\right)/kT\right] + 1} . \tag{3.68}$$

The integral I may be expressed as an asymptotic (but not necessarily convergent) series whose leading terms are

$$I(\mu, T) = \int_0^{\mu'} G(\mathcal{E})\, d\mathcal{E} + \frac{\pi^2}{6}\frac{\partial G}{\partial \mathcal{E}}(kT)^2 + \frac{7\pi^4}{360}\frac{\partial^3 G}{\partial \mathcal{E}^3}(kT)^4 + \cdots , \tag{3.69}$$

where $\mu' = \mu - mc^2$ and all the partials are evaluated at μ'. It is assumed that μ'/kT is much larger than unity.

It is a simple, but tedious, exercise to transform the integrals for n, P, and E of, respectively, (3.9), (3.12), and (3.13), into their energy space counterparts and then to find $G(\mathcal{E})$. Another way, however, is to transform all of the elements in the expansion (3.69) into $x = p/mc$–space using (3.43); that is, $\mathcal{E} = mc^2\left[(1+x^2)^{1/2} - 1)\right]$. A big part of this was done when the expressions for the completely degenerate electron gas were written down in the equations for n_e (3.47), P_e (3.50), and E_e (3.52). Thus, for example, the leading term in the expansion of (3.69) for n_e is simply (neglecting constants)

$$n_e \text{ (first term)} \propto \int_0^{x_f} x^2\, dx .$$

Here x_f takes the place of $\mu' = \mu - mc^2$ and, since we have converted from energy to x-space, it should be obvious that the relation between x_f and μ' is

$$\mu' = \mu - m_e c^2 = m_e c^2\left[\left(1 + x_f^2\right)^{1/2} - 1\right] . \tag{3.70}$$

This relation is given in the same spirit as was done for the completely degenerate case where the Fermi energy was related to the chemical potential by $\mathcal{E}_F = \mu - m_e c^2$ and $\mathcal{E}_F$ was given in terms of x_F through (3.46). In that instance, x_F and, hence, μ were found by fixing the number density n_e and using (3.47). The same sort of thing can be done here except there is an additional complication because temperature also appears in the thermodynamics; that is, n_e must be a function of both x_f (or μ) and T. This all can be accomplished by performing the indicated operations in the expansion (3.69). Carrying out this enterprise is left to you as an exercise in elementary calculus but the result, to second order in temperature, is,

$$n_e = \frac{8\pi}{3}\left(\frac{h}{m_e c}\right)^{-3} x_f^3 \left[1 + \pi^2 \frac{1+2x_f^2}{2x_f^4}\left(\frac{kT}{m_e c^2}\right)^2 + \cdots\right] \text{ cm}^{-3}. \quad (3.71)$$

This expansion is useful only if the second term in the brackets is small compared to unity. A useful rule of thumb is to be wary if it exceeds 0.1 to 0.2. In any case, given any two of n_e (or ρ/μ_e), T, or x_f (or μ), the third follows. Looked at another way (and we shall use this shortly), (3.71) may be used to find out how the chemical potential changes with respect to temperature for fixed n_e or ρ/μ_e. Note that as $T \to 0$, the number density approaches the completely degenerate expression (3.47) with $x_f \to x_F$, and $\mu' \to \mathcal{E}_F$.

The corresponding expansions for pressure and internal energy are as follows:

$$P_e = Af(x_f)\left[1 + 4\pi^2 \frac{x_f(1+x_f^2)^{1/2}}{f(x_f)}\left(\frac{kT}{m_e c^2}\right)^2 + \cdots\right] \quad (3.72)$$

$$E_e = Ag(x_f)\left[1 + 4\pi^2 \frac{(1+3x_f^2)(1+x_f^2)^{1/2} - (1+2x_f^2)}{x_f\, g(x_f)}\left(\frac{kT}{m_e c^2}\right)^2 + \cdots\right] \quad (3.73)$$

where $f(x_f)$ and $g(x_f)$ are given, respectively, by (3.51) and (3.53). Note that P_e is in dyne cm^{-2} and E_e is the volumetric energy density in erg cm^{-3} (and *not* specific energy density in erg g^{-1}).

These equations will be used to find such things as specific heats and temperature exponents for the almost completely degenerate electron gas. **Note**: as a matter of practicality, x_f is often computed as if the gas were completely degenerate. Thus if the correction term for temperature is very small, then x (or x_F) of (3.48) is used instead of x_f as a good approximation for direct calculation of n_e, P_e, and E_e in (3.71–3.73).

3.6 "Almost Perfect" Equations of State

In real gases, interactions must be taken into account that modify the "perfect" results given above. In addition, a stellar equation of state might consist of many components with radiation, Maxwell-Boltzmann, and degenerate gases competing in importance. This short section will not attempt to show how imperfections are treated in detail but will indicate where some are important in practical situations. The results of this discussion are summarized in Figure 3.6 for a hypothetical gas composed of pure hydrogen.

In an almost-ideal gas, a measure of the interaction energy between ions is the Coulomb potential between two ions. If the ionic charge is Z, then

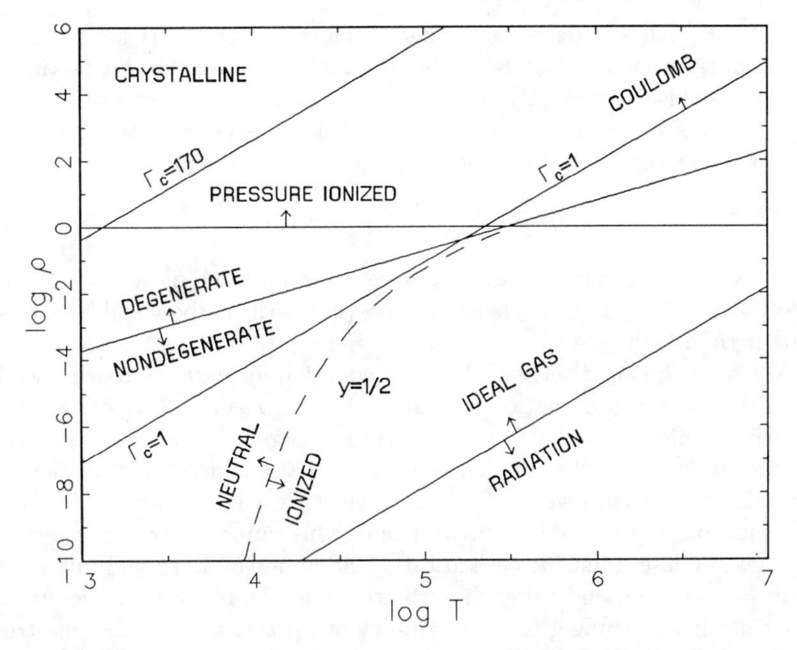

FIGURE 3.6. A composite showing how the ρ–T plane is broken up into regions dominated by pressure ionization, degeneracy, radiation, ideal gas, crystallization, and ionization-recombination. The gas is assumed to be pure hydrogen.

the potential is Z^2e^2/a, where a is some typical separation between the ions. Coulomb effects are expected to become important when this energy is comparable to kT. Thus form the ratio

$$\Gamma_C \equiv \frac{Z^2e^2}{akT} \tag{3.74}$$

where $\Gamma_C = 1$ is the rough demarcation between where Coulomb effects might be important or not. A $\Gamma_C > 1$ implies they probably are important. The distance a is usually taken as the radius of a Wigner-Seitz sphere whereby $(4\pi a^3/3) = (1/n_{\rm I})$ and $n_{\rm I}$ is the ion number density. If the gas consists of pure ionized hydrogen and $\Gamma_C = 1$, then (3.74) becomes

$$\rho = 8.49 \times 10^{-17}\, T^3 \quad \text{g cm}^{-3}. \tag{3.75}$$

If the density is greater than that implied by (3.75) for a given temperature, then you can be reasonably certain that a perfect gas is not as perfect as could be desired. This line is shown in Figure 3.6. You may check, from the material given previously, that the centers of very low mass ZAMS stars are encroaching upon both this line and the one for degeneracy effects. Carefully done stellar models of these stars contain corrections for these effects.

If Γ_C becomes large enough, then Coulomb effects overwhelm those of thermal agitation and the gas settles down into a crystal. The best estimates as to how this takes place yield a Γ_C of around 170 for the transition. With this value of Γ_C in a hydrogen gas (which is unrealistic for a crystallizing composition but fine for talking purposes), (3.74) becomes

$$\rho = 4.2 \times 10^{-10}\, T^3 \quad \text{g cm}^{-3}. \tag{3.76}$$

This is not an academic issue because some portions of very cool white dwarfs are thought to turn crystalline (but with carbon and/or oxygen rather than hydrogen).

We have already discussed the Saha equation for pure hydrogen and the density-temperature relation for half-ionization was given by (3.39). That relation is also shown in Figure 3.6. In deriving the Saha equation it was implicitly assumed that the energy levels of the hydrogen atom (had we included all of them) were known and that their energies were independent of conditions in the ambient environment. This cannot be true in general. If the gas is dense, then the electrostatic field of one atom should influence a neighboring atom and hence disturb atomic levels. In the extreme, we can imagine this continuing until electron clouds practically rub and electrons are ionized off the parent atoms. This is a crude description of *pressure ionization.* To estimate under what conditions this occurs, take the rubbing picture seriously and find at what density the Wigner-Seitz radius equals the radius of the first Bohr orbit of hydrogen (0.5×10^{-8} cm). A very easy calculation says that this takes place when

$$\rho \approx 1 \quad \text{g cm}^{-3}. \tag{3.77}$$

This density is shown in Figure 3.6 as the line that terminates ordinary Saha ionization. Such densities are commonplace in stellar interiors and lead to the statement that the larger bulk of those interiors are ionized as far as the lighter elements are concerned independent of the effects of the radiation field.[5]

We finally ask under what conditions radiation pressure dominates over ideal gas pressure or the other way round. That is, where does $aT^4/3 = \rho N_A kT/\mu$? With the assumption of complete ionization in hydrogen this becomes

$$\rho = 1.5 \times 10^{-23}\, T^3 \quad \text{g cm}^{-3} \tag{3.78}$$

as shown in the figure. This ends the discussion of the major elements determining pressures and internal energies in simple environments.[6]

---

[5] As a side comment, note that several lines in the figure cross at $T \approx 3 \times 10^5$ K and $\rho \approx 1$ g cm^{-3}. You can be assured that computing accurate equations of state in that region of the T–ρ plane is a nightmare.

[6] We have purposely ignored equations of state at ultrahigh densities such

3.7 Adiabatic Exponents and Other Derivatives

For the most part, all we need in the way of thermodynamic variables to construct a simplified stellar model is the internal energy and pressure as a function of density, temperature, and composition (as was done in Chapter 1). To construct realistic models, and to evolve them in time, however, we need several thermodynamic derivatives. We shall assume, at first, that the detailed composition, including concentrations of ions, etc., has been determined and that chemical reactions are not taking place. We also assume that you have some facility in transforming thermodynamic functions under reversible conditions and that you are familiar with their properties.

3.7.1 Keeping the Composition Fixed

If changes in temperature and density (or volume) do not cause corresponding changes in the relative concentrations of various species of atoms or ions in the stellar mixture, then the calculation of thermodynamic derivatives is not particularly difficult. We now examine this situation and ignore until later those complications arising from chemical reactions.

Specific Heats

The first derivatives encountered in elementary thermodynamics are specific heats. In general form these are defined by

$$c_\alpha = \left(\frac{dQ}{dT}\right)_\alpha \tag{3.79}$$

where α is kept fixed as T changes. In the following, Q will have the units of erg g^{-1} and thus the specific heats will have units of erg g^{-1} K^{-1}. The most useful variables for α for much of our later discussion are P, ρ, or the specific volume $V_\rho = 1/\rho$. (We shall also have occasion to use the ordinary volume, V.) From the first law for a reversible process (and see eq. 1.10)

$$dQ = dE + P\,dV_\rho = dE + P\,d\left(\frac{1}{\rho}\right) = dE - \frac{P}{\rho^2}\,d\rho$$

so that

$$c_{V_\rho} = \left(\frac{dQ}{dT}\right)_\rho = \left(\frac{\partial E}{\partial T}\right)_\rho \quad \text{erg g}^{-1}\ \text{K}^{-1}. \tag{3.80}$$

For an ideal monatomic gas $E = 3N_{\mathrm{A}}kT/2\mu$ erg g^{-1} (from 3.28) so that $c_{V_\rho} = 3N_{\mathrm{A}}k/2\mu$ and $E = c_{V_\rho}T$. Note that the composition has not been

---

as are found in neutron stars and the collapsing cores of supernovae. This is a difficult subject itself worthy of a monograph. For further reading we suggest Chapters 2 and 8 of Shapiro and Teukolsky (1983), and Bethe (1990, §§3-4).

mentioned here except in the mean molecular weight μ: it is kept fixed by assumption.

To find c_P, recall (from any of many thermodynamic texts) that c_P and c_{V_ρ} (or c_V) are related by

$$c_P - c_{V_\rho} = -T\left(\frac{\partial P}{\partial T}\right)^2_{(\rho \text{ or } V_\rho)}\left(\frac{\partial P}{\partial V_\rho}\right)^{-1}_T .$$

To cast this in a form that will prove more suitable for later purposes we reintroduce the power law expression for the equation of state given in Chapter 1 by (1.64):

$$P = P_0 \rho^{\chi_\rho} T^{\chi_T} \tag{3.81}$$

where P_0, χ_ρ, and χ_T are constants. This means that the last two are also defined by

$$\chi_T = \left(\frac{\partial \ln P}{\partial \ln T}\right)_{(\rho \text{ or } V_\rho)} = \frac{T}{P}\left(\frac{\partial P}{\partial T}\right)_{(\rho \text{ or } V_\rho)} \tag{3.82}$$

and

$$\chi_\rho = \left(\frac{\partial \ln P}{\partial \ln \rho}\right)_T = -\left(\frac{\partial \ln P}{\partial \ln V_\rho}\right)_T = \frac{\rho}{P}\left(\frac{\partial P}{\partial \rho}\right)_T = -\frac{1}{\rho P}\left(\frac{\partial P}{\partial V_\rho}\right)_T . \tag{3.83}$$

Thus

$$c_P - c_{V_\rho} = \frac{P}{\rho T}\frac{{\chi_T}^2}{\chi_\rho} \quad \text{erg g}^{-1}\text{ K}^{-1}. \tag{3.84}$$

For an ideal monatomic gas $\chi_\rho = \chi_T = 1$ and

$$c_P - c_{V_\rho} = \frac{N_\text{A} k}{\mu} \quad \text{erg g}^{-1}\text{ K}^{-1} \tag{3.85}$$

which gives $c_P = 5N_\text{A}k/2\mu$.

We also define γ, the ratio of specific heats, to be

$$\gamma = \frac{c_P}{c_{V_\rho}} = 1 + \frac{P}{\rho T c_{V_\rho}}\frac{{\chi_T}^2}{\chi_\rho} \tag{3.86}$$

which will be discussed shortly. (This need not be the γ of the γ-law equation of state—see later.)

Adiabatic Exponents

The dimensionless adiabatic exponents, the "Γs," measure the thermodynamic response of the system to adiabatic changes and will be used extensively. (Two of them, Γ_1 and Γ_2, were already introduced in Chapter 1.) They are defined as follows:

$$\Gamma_1 = \left(\frac{\partial \ln P}{\partial \ln \rho}\right)_\text{ad} = -\left(\frac{\partial \ln P}{\partial \ln V_\rho}\right)_\text{ad} \tag{3.87}$$

$$\frac{\Gamma_2}{\Gamma_2 - 1} = \left(\frac{\partial \ln P}{\partial \ln T}\right)_{\rm ad} = \frac{1}{\nabla_{\rm ad}} \tag{3.88}$$

which also defines $\nabla_{\rm ad}$, and

$$\Gamma_3 - 1 = \left(\frac{\partial \ln T}{\partial \ln \rho}\right)_{\rm ad} = -\left(\frac{\partial \ln T}{\partial \ln V_\rho}\right)_{\rm ad} . \tag{3.89}$$

As in Chapter 1, the subscript "ad" means that the indicated partials are to be evaluated at constant entropy. (We shall not need it directly, but extensive use will be made of $\nabla_{\rm ad}$ in later chapters.) It will shortly become clear why the Γ_i appear in such curious combinations in the definitions, but first note that not all the Γ_i are independent. You may easily show that

$$\frac{\Gamma_3 - 1}{\Gamma_1} = \frac{\Gamma_2 - 1}{\Gamma_2} = \nabla_{\rm ad} . \tag{3.90}$$

Computation of the Γ_i is tedious but not particularly enlightening. Complete and clear derivations may be found in Cox (1968) but we suggest you try to derive the expressions that follow using the more compact methods given in Landau and Lifshitz (1958). They start from fundamentals and then use powerful yet simple Jacobian transformations to derive what is needed. All you need watch out for is the distinction between V and V_ρ. When you get done, realize that there are many variations in the ways that the Γ_i may be expressed and the following may not always be the most efficient to use (i.e, you may wish to rearrange things). The adiabatic exponents are:

$$\Gamma_3 - 1 = \frac{P}{\rho T}\frac{\chi_T}{c_{V_\rho}} = \frac{1}{\rho}\left(\frac{\partial P}{\partial E}\right)_\rho \tag{3.91}$$

$$\Gamma_1 = \chi_T\left(\Gamma_3 - 1\right) + \chi_\rho = \frac{\chi_\rho}{1 - \chi_T \nabla_{\rm ad}} \tag{3.92}$$

$$\frac{\Gamma_2}{\Gamma_2 - 1} = \nabla_{\rm ad}{}^{-1} = c_P \frac{\rho T}{P}\frac{\chi_\rho}{\chi_T} = \frac{\chi_\rho}{\Gamma_3 - 1} + \chi_T . \tag{3.93}$$

The last exponent, γ, is given by,

$$\gamma = \frac{c_P}{c_{V_\rho}} = \frac{\Gamma_1}{\chi_\rho} = 1 + \frac{\chi_T}{\chi_\rho}\left(\Gamma_3 - 1\right) = \frac{\Gamma_3 - 1}{\chi_\rho}\nabla_{\rm ad}{}^{-1} . \tag{3.94}$$

Note that the right-hand side result for Γ_3 implies that $P = (\Gamma_3 - 1)\rho E$ so that the γ in the γ-law equation of state of (1.22) is Γ_3 and, generally, not one of the other gammas. Lay the blame for any possible confusion here to the quirks of historical nomenclature.

Explicit values for all the exponents and specific heats, etc., for interesting gases follow below. Remember, however, that chemical reactions are still absent so that the relative concentrations of ions and electrons are fixed despite changes in thermodynamic quantities.

Mixtures of Ideal Gases and Radiation

For a monatomic ideal gas χ_ρ and χ_T are equal to unity and $\Gamma_1 = \Gamma_2 = \Gamma_3 = \gamma = 5/3$. A pure radiation "gas" has $\chi_\rho = 0$, $\chi_T = 4$, and $\Gamma_1 = \Gamma_2 = \Gamma_3 = 4/3$. Note that $\gamma = \Gamma_1/\chi_\rho \to \infty$ in this case.

If $\gamma = \Gamma_1 = \Gamma_2 = \Gamma_3$ of the same constant value, as can be satisfied by an ideal gas, then

$$P \propto \rho^\gamma \tag{3.95}$$
$$P \propto T^{\gamma/(\gamma-1)} \tag{3.96}$$
$$T \propto \rho^{(\gamma-1)} \tag{3.97}$$

along adiabats. This is the result usually quoted in elementary physics texts for adiabatic behavior: it is collectively true only if the exponents satisfy the above equality.

In modeling simple stars, it often turns out that an equation of state consisting of a mixture of ideal gas and radiation suffices:

$$P = \frac{\rho N_A kT}{\mu} + \frac{aT^4}{3} = P_g + P_{rad} \quad \text{dyne cm}^{-2} \tag{3.98}$$

and

$$E = \frac{3N_A kT}{2\mu} + \frac{aT^4}{\rho} \quad \text{erg g}^{-1}. \tag{3.99}$$

The density and temperature exponents may be obtained almost by inspection so that

$$\chi_\rho = \frac{P_g}{P} \equiv \beta \tag{3.100}$$

which also defines β, the ratio of gas to total pressure, and

$$\chi_T = 4 - 3\beta\,. \tag{3.101}$$

(This β is not to be confused with $\beta = kT/m_e c^2$ introduced earlier.) Further analysis, using the general expressions given previously, yields

$$c_{V_\rho} = \frac{3N_A k}{2\mu}\left(\frac{8-7\beta}{\beta}\right) \quad \text{erg g}^{-1}\ \text{K}^{-1} \tag{3.102}$$

$$\Gamma_3 - 1 = \frac{2}{3}\left(\frac{4-3\beta}{8-7\beta}\right) \tag{3.103}$$

$$\Gamma_1 = \beta + (4-3\beta)(\Gamma_3 - 1) \tag{3.104}$$

$$\frac{\Gamma_2}{\Gamma_2 - 1} = \frac{32 - 24\beta - 3\beta^2}{2(4-3\beta)} \tag{3.105}$$

and, finally,

$$\gamma = \frac{\Gamma_1}{\beta}\,. \tag{3.106}$$

It is easy to confirm that all quantities go to their proper limits as $\beta \to 1$ (ideal gas) or $\beta \to 0$ (pure radiation) and that all quantities are intermediate between their pure gas and radiation values for intermediate β.

Mixtures of Degenerate and Ideal Gases

The first thing we shall find is the specific heat at constant volume for an almost completely degenerate electron gas. Recall our earlier discussion of the temperature corrections to such a gas where the number density, n_e, was given as a function of T and x_f in (3.71). If the volume or density of the gas is fixed while temperature is varied, then n_e does not change but x_f must. Thus $(\partial n_e/\partial T)_\rho = 0$. If this operation is performed on (3.71), then the right-hand side of the resulting equation contains $(\partial x_f/\partial T)_\rho$, which may be solved to first order in T as

$$\left(\frac{\partial x_f}{\partial T}\right)_\rho = -\frac{\pi^2 k^2}{m_e^2 c^4}\frac{1+2x_f^2}{3x_f^3}\, T. \tag{3.107}$$

When you derive this you will find that it is missing a denominator of the form $\left[1+\mathcal{O}\left(T^2\right)\right]$ where $\mathcal{O}\left(T^2\right)$ contains terms that are of order T^2. Those terms must be ignored because they are of the same order as other correction terms that would have appeared if the equation for n_e had been carried out to higher order in temperature. Thus (3.107) is correct to first order in T and T is small on some appropriate scale.

To find the specific heat we have to differentiate E_e of (3.73) with respect to T while keeping density fixed. This operation yields, through the chain rule, nasty terms such as $[dg(x_f)/dx_f]\,(\partial x_f/\partial T)_\rho$. When these are all straightened out (see Chandrasekhar 1939, Chapter 10, §6), we find

$$c_{V_\rho\,(\mathrm{e})} = \frac{8\pi^3 m_e^4 c^5}{3h^3 T\rho}\left(\frac{kT}{m_e c^2}\right)^2 x_f\left(1+x_f^2\right)^{1/2} \tag{3.108}$$

or

$$c_{V_\rho\,(\mathrm{e})} = \frac{1.35\times 10^5}{\rho}\, T\, x_f\left(1+x_f^2\right)^{1/2} \quad \mathrm{erg\ g^{-1}\ K^{-1}}. \tag{3.109}$$

Note the presence of ρ in the above. It is required because this specific heat is a specific specific heat (from the units). As before, it is reasonable to replace x_f with x_F using (3.47–3.48) provided that temperature correction terms are small in all of n_e, P_e, and E_e.

From here on, we have to make some reasonable physical assumptions about the nature of the stellar gas. Because of pressure ionization, we expect all or most of the nuclear species to be completely ionized so that all electrons are free to swim in the Fermi sea. Thus pressure and energy, as additive quantities, are determined by bare ions and the free electrons. Radiation should play no significant role because, if it did, the temperatures would be so high that electrons would no longer be nearly degenerate—which we assumed at the onset. Thus the total pressure consists of $P = P_e + P_I$ where "I" means "ions." Internal energies and specific heats are also additive. The reason we bring this up is because the rest of the thermodynamic derivatives are, for the most part, logarithmic (like the Γs) and we cannot simply add them together. It is best to give an example.

The temperature exponent of pressure, χ_T, is $(T/P)\left(\partial P/\partial T\right)_\rho$ from (3.82) where P is the *total* pressure. We cannot separate χ_T into components describing just the electrons or just the ions. We had the same problem when treating the gas and radiation mixture of the previous section but the calculations there were fairly straightforward. Here, however, the complexity of the electron gas equation of state makes things computationally more difficult. Nevertheless, we can compute all the derivatives fairly easily if we assume that temperatures are very low. If this is the case, then electron degeneracy pressure greatly exceeds that of the ions and $P_\mathrm{e} \gg P_\mathrm{I}$. The same is not true for the partials of pressure with respect to temperature. By following the same course of analysis as was outlined above for the specific heat, you should verify that $\left(\partial P_\mathrm{e}/\partial T\right)_\rho \propto T$. On the other hand, $\left(\partial P_\mathrm{I}/\partial T\right)_\rho = N_\mathrm{A} k\rho/\mu_\mathrm{I}$ where μ_I is the ion mean molecular weight. (The ions are still assumed to be ideal.) Thus for low enough temperatures the temperature derivative of electron pressure may be neglected compared to that of the ions. The net result is that for low temperatures

$$\chi_T \rightarrow \frac{N_\mathrm{A} k}{\mu_\mathrm{I}} \frac{\rho T}{P_\mathrm{e}} \tag{3.110}$$

and, as $T \rightarrow 0$, so does χ_T. The electrons have nothing to say in the matter.

The density exponent $\chi_\rho = (\rho/P)\left(\partial P/\partial \rho\right)_T$ of (3.83) is easier. The electron pressure dominates both terms for low temperatures so that

$$\chi_\rho \rightarrow \frac{\rho}{P_\mathrm{e}}\left(\frac{\partial P_\mathrm{e}}{\partial \rho}\right)_T \rightarrow \begin{cases} 5/3 & \text{nonrelativistic} \\ 4/3 & \text{relativistic.} \end{cases} \tag{3.111}$$

The limiting forms come directly from the pressure-density relations (3.56) for the degenerate gas.

The rest of the derivatives require that the specific heats be found. We already have $c_{V_\rho\,(\mathrm{e})}$ (from 3.108) and we know that the ion specific heat is $3N_\mathrm{A}k/2\mu_\mathrm{I}$ (from, e.g., 3.80) and it is a constant. Therefore, for sufficiently low temperatures

$$c_{V_\rho} \rightarrow c_{V_\rho\,(\mathrm{I})} = \frac{3N_\mathrm{A}k}{2\mu_\mathrm{I}} = \frac{1.247 \times 10^8}{\mu_\mathrm{I}} \quad \text{erg g}^{-1}\ \text{K}^{-1} \tag{3.112}$$

and the electrons do not matter. The combination of pressure dominance by electrons, low sensitivity of pressure to temperature (small χ_T), and low specific heats (only the ions matter), all add up to a potentially explosive situation when very reactant nuclear fuels are present, as in the helium flash.

Having found the above, it should be a simple matter for you to verify the following: $c_P = c_{V_\rho\,(\mathrm{I})}$, $\Gamma_3 - 1 = 2/3$, $\Gamma_1 = \chi_\rho$, and $\nabla_\mathrm{ad} = 2/3\chi_\rho$.

3.7.2 Allowing for Chemical Reactions

We now give an example of how the thermodynamic derivatives are found when chemical reactions are allowed to take place. For simplicity, the ideal gas, one-state hydrogen atom will again be used, and radiation in the equation of state will be ignored. As usual, real calculations are very difficult and you are referred to Cox (1968, §9.18) for a fuller discussion. As you will see, even in the simple example given here, the analysis is made difficult because relative concentrations of particles vary as temperatures and densities change.

Because we assume that all changes in the system take place along paths in thermodynamic equilibrium, the Saha equation of (3.33) holds and

$$\frac{n^+ n_e}{n^0} = \mathcal{B} T^{3/2} e^{-\chi_H/kT} \tag{3.113}$$

where $\mathcal{B}$ is

$$\mathcal{B} = \left(\frac{2\pi m_e k}{h^2}\right)^{3/2} = 2.415 \times 10^{15} \ \text{cm}^{-3}\ \text{K}^{-3/2}$$

and the other symbols are the same as defined in §3.4. Define N (as in §3.1) so that $N\rho \equiv n = n^+ + n^0$. Thus N is the total ion plus neutral atom number density per unit mass and it is independent of density and will not change as the system is compressed or expanded. With the usual definition of $y = n^+/n = n_e/n$, the pressure may be written

$$P = (n_e + n^+ + n^0)kT = (1+y)N\rho kT \quad \text{dyne cm}^{-2} \tag{3.114}$$

and the specific internal energy is (see 3.41)

$$E = (1+y)\frac{n}{\rho}\frac{3kT}{2} + y\frac{n}{\rho}\chi_H \quad \text{erg g}^{-1} \tag{3.115}$$

or

$$E = (1+y)N\frac{3kT}{2} + yN\chi_H \quad \text{erg g}^{-1}. \tag{3.116}$$

Having the pressure and internal energy now allows us to compute the thermodynamic derivatives. First note that the analysis leading to the determination of those derivatives in the previous discussion involved only taking partials with respect to either temperature or density with the other kept fixed: concentrations were never mentioned in that analysis. But this implies that partials with respect to concentrations (i.e., the N_i) were never needed. Thus the general expressions derived for the specific heats, the Γs, etc., are formally correct and all we need do is put in the correct pressures and internal energies that contain the information about chemical equilibrium. To carry this out in detail, however, still requires some effort. We start with easier quantities, χ_T and χ_ρ, and leave most of the rest of the work to you.

The ionization fraction y is given by a slightly rewritten version of (3.37):

$$\frac{y^2}{1-y} = \frac{\mathcal{B}}{N\rho} T^{3/2} e^{-\chi_{\rm H}/kT} . \tag{3.117}$$

We now have the three relations $P = P(\rho, T, y)$, $E = E(\rho, T, y)$, and the Saha equation. Take total differentials of the first two to find

$$dP = P\left[\frac{dT}{T} + \frac{d\rho}{\rho} + \frac{dy}{1+y}\right]$$

and

$$dE = \tfrac{3}{2} NkT(1+y)\left[\frac{dT}{T} + \frac{2}{3}\left(\frac{3}{2} + \frac{\chi_{\rm H}}{kT}\right)\frac{dy}{1+y}\right] .$$

Note that N remains fixed because it is the number of hydrogen nuclei per gram and cannot change with temperature, density, or volume.

Also take the differential of the Saha equation and divide the result by the Saha equation to find

$$\frac{dy}{1+y} = \mathcal{D}(y)\left[\left(\frac{3}{2} + \frac{\chi_{\rm H}}{kT}\right)\frac{dT}{T} - \frac{d\rho}{\rho}\right]$$

where

$$\mathcal{D}(y) = \frac{y(1-y)}{(2-y)(1+y)} . \tag{3.118}$$

Note that $\mathcal{D}(1) = \mathcal{D}(0) = 0$ and, for general $0 \le y \le 1$, $\mathcal{D}(y) \ge 0$. It reaches a maximum at the half-ionization point $y = 1/2$ where $\mathcal{D}(1/2) = 1/9$.

The left-hand side of the differentiated Saha equation appears explicitly in the expressions for dP and dE. Therefore, use that equation to eliminate any reference to dy in dP and dE and find, for dE,

$$dE = \frac{3}{2} NkT(1+y)\left\{\left[1 + \mathcal{D}\frac{2}{3}\left(\frac{3}{2} + \frac{\chi_{\rm H}}{kT}\right)^2\right]\frac{dT}{T} - \mathcal{D}\frac{2}{3}\left(\frac{3}{2} + \frac{\chi_{\rm H}}{kT}\right)\frac{d\rho}{\rho}\right\} .$$

From this find directly

$$c_{V_\rho} = \frac{3}{2} Nk(1+y)\left[1 + \mathcal{D}\frac{2}{3}\left(\frac{3}{2} + \frac{\chi_{\rm H}}{kT}\right)^2\right] \quad \text{erg g}^{-1}\ \text{K}^{-1} . \tag{3.119}$$

Note that $Nk = N_{\rm A}k/\mu_{\rm I}$ and $Nky = N_{\rm A}k/\mu_{\rm e}$ from which may be found $\mu_{\rm I}$ of (1.42) and $\mu_{\rm e}$ of (1.45).

By treating the pressure differential in like fashion we find

$$\frac{dP}{P} = \left[1 + \mathcal{D}\left(\frac{3}{2} + \frac{\chi_{\rm H}}{kT}\right)\right]\frac{dT}{T} + (1 - \mathcal{D})\frac{d\rho}{\rho}$$

so that

$$\chi_\rho = 1 - \mathcal{D} \tag{3.120}$$

and

$$\chi_T = 1 + \mathcal{D}\left(\frac{3}{2} + \frac{\chi_{\rm H}}{kT}\right) . \tag{3.121}$$

Because $\mathcal{D} \geq 0$, we have $\chi_\rho \leq 1$ and $\chi_T \geq 1$.

The Γ_i may now be calculated using equations (3.91) through (3.93) in the forms that contain χ_ρ, χ_T, and c_{V_ρ}. After a bit of algebra the results are:

$$\Gamma_3 - 1 = \frac{2 + 2\mathcal{D}\left(3/2 + \chi_{\rm H}/kT\right)}{3 + 2\mathcal{D}\left(3/2 + \chi_{\rm H}/kT\right)^2} \tag{3.122}$$

$$\frac{\Gamma_2}{\Gamma_2 - 1} = \frac{5 + 2\mathcal{D}\left[\chi_{\rm H}/kT + \left(3/2 + \chi_{\rm H}/kT\right)\left(5/2 + \chi_{\rm H}/kT\right)\right]}{2 + 2\mathcal{D}\left(3/2 + \chi_{\rm H}/kT\right)} \tag{3.123}$$

and Γ_1 follows from (3.90).

Note that as y approaches zero or unity (so that $\mathcal{D} \to 0$) all the Γ_i approach their ideal gas values of 5/3. This is as it should be. If the gas is completely neutral or totally ionized, then the equation of state is of its usual ideal gas form since y is not changing. It is the intermediate case that is of interest.

To compute the Γ_i the scheme is: choose ρ (or n) and T, find y from the Saha equation (3.117) (and $\mathcal{D}$ by means of 3.118), and then apply the above expressions. A typical result is shown in Figure 3.7 where $\Gamma_3 - 1$ is plotted as a function of temperature for three densities. The half-ionization point, $y = 1/2$, is indicated. Note that if T is near the typical hydrogen ionization temperature of 10^4 K, Γ_3 drops rapidly from its value of 5/3 to much lower values. Even the dangerous 4/3 may be passed by in the process.

The reason why Γ_3 (and the other Γs) behaves the way it does when ionization is taking place is quite simple. First suppose no ionization or recombination processes can operate in an almost completely neutral gas so that concentrations remain constant as the system is compressed adiabatically. In that case $\Gamma_3 = 5/3$, $T \sim \rho^{2/3}$, (as in 3.97) and the gas heats up. If, however, we allow ionization to take place, then compression may still heat up the gas, but ionization is much more sensitive to temperature changes than to changes in density. Hence, ionization is accelerated. But this takes energy and that energy is paid for at the expense of the thermal motion in the gas. Thus the temperature tends not to rise as rapidly as $\rho^{2/3}$ and Γ_3 is smaller than its value with no ionization.

As we shall see shortly, all the Γ_i are important in some respect or another: Γ_3 says something about how the heat content of the gas responds to compression; Γ_1 is intimately tied up with dynamics (partially through the sound speed); the behavior of Γ_2 may be a deciding factor in whether convection may take place.

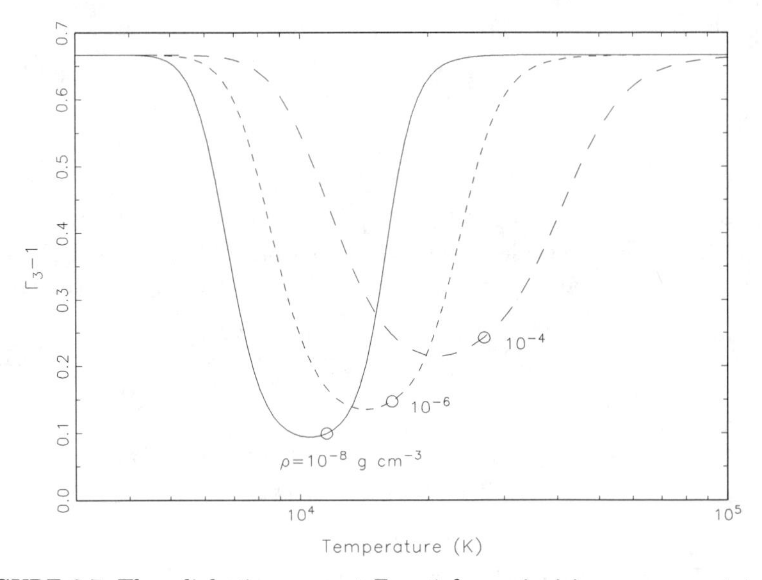

FIGURE 3.7. The adiabatic exponent $\Gamma_3 - 1$ for an ionizing pure one-state hydrogen gas is plotted as a function of temperature. Results are shown for three densities. The $\circ$ indicates the half-ionization point.

The effects of radiation or other ionizing species and energy levels are included in more complete analyses than what we have done here (see Cox 1968). In addition, the effects of pressure ionization (among other things) have to be included in many situations. Even though the results we have obtained are useful for many calculations in stellar structure, you should be aware that real models are usually constructed using tabular equations of state with P, E, and, sometimes, various derivatives given as functions of temperature and density for a fixed nuclear composition. Very often these are included in tabulations of opacities—which we discuss in the next chapter.

3.8 Exercises

1. We have already explored the Saha equation using a pure hydrogen gas as an example. Now consider the more complicated ^{4}He atom with its two electrons. Assume, as in the hydrogenic example, that the neutral atom and first ionized ion are in their respective ground states. The ionization potentials are: to remove the first (second) electron requires $\chi_1 = 24.587$ eV ($\chi_2 = 54.416$ eV). To agree on a common nomenclature, let n_e, n_0, n_1, and n_2 be the number densities of, respectively, electrons, neutral atoms, and first and second ionized ions. The total number density of atoms plus ions of the pure helium

gas is denoted by n. Furthermore, define z_e as the ratio n_e/n and, in like manner, let z_i be n_i/n where $i = 0, 1, 2$. The gas is assumed to be electrically neutral. For the following you will also need the degeneracy factors for the atoms and ions and these are to be found on page 35 of Allen (1973).

(a) Following the hydrogenic case, construct the ratios $n_e n_1/n_0$ and $n_e n_2/n_1$. In doing so you must take care in establishing the zero points of energy for the various constituents. The final form you obtain should not contain chemical potentials (and you must show why this is true).

(b) Apply $n = n_0 + n_1 + n_2$, overall charge neutrality, and recast the above Saha equations so that only z_1 and z_2 appear as unknowns. The resulting two equations have temperature and n or, equivalently, $\rho = 4n/N_A$ as independent parameters.

(c) Simultaneously solve the two Saha equations for z_1 and z_2 for temperatures in the range $4 \times 10^4 \leq T \leq 2 \times 10^5$ K with a fixed value of density from among the choices $\rho = 10^{-4}$, 10^{-6}, or 10^{-8} g cm^{-3}. Choose a dense grid in temperature because you will soon plot the results. (These cases will prove useful when discussing pure helium opacities in Chapter 4.) Once you have found z_1 and z_2, also find z_e and z_0 for the same range of temperature. Note that this is a numerical exercise and use of a computer is strongly advised.

(d) Now plot all your zs as a function of temperature for your chosen value of ρ. (Plot z_0, z_1, and z_2 on the same graph.) This is an essential step because it will make clear how the ionization responds to temperature changes.

(e) Find the half-ionization points on your plot. The two temperatures you obtain (for fixed density) will correspond to the single half-ionization point for pure hydrogen.

2. We earlier established that photon mean free paths were very short in a star except in the very outermost layers. This means that photons must follow a tortuous path to eventually escape from a star and must take a long time in doing so. To estimate this time, assume that a photon is created at the center of a star and thereafter undergoes a long series of scatterings off electrons until it finally reaches the surface. The mean free path associated with each scattering is $\lambda_{\text{phot}} = (n_e \sigma_e)^{-1}$ where σ_e is the Thomson scattering cross section $\sigma_e \approx 0.7 \times 10^{-24}$ cm^2 (see §4.4.1). For simplicity, assume that the star has a constant density so that λ_{phot} is also constant. This is an order-of-magnitude problem so don't worry too much about constants of order unity. (Real diffusion in stars is much more complicated and your estimate for the time will be an underestimate.)

(a) Using one-dimensional random walk arguments, show that $L \approx \mathcal{R}^2/\lambda_{\text{phot}}$ is the *total* distance a photon must travel if it starts its scattering career at stellar center and eventually ends up at the surface at $\mathcal{R}$.

(b) Since the photons travel at the speed of light, c, find the time, τ_{phot}, required for the photon to travel from the stellar center to the surface. (Assume that any scattering process takes place instantaneously.)

(c) Give an estimate for τ_{phot}, in the units of years, for a star of mass $\mathcal{M}/\mathcal{M}_\odot$ and radius $\mathcal{R}/\mathcal{R}_\odot$.

3. Neutron stars are assumed to be objects with $\mathcal{M} \sim \mathcal{M}_\odot$, $\mathcal{R} \sim 10$ km ($\langle\rho\rangle \sim 10^{14}$ g cm^{-3}) where internal temperatures (kT) are small compared to the Fermi energies of electrons, protons, and neutrons (which are assumed to be the only particles present). To demonstrate that the name neutron star is apt, consider the following. Assume that the stellar temperature is zero and that chemical equilibrium exists between electrons, protons, and neutrons. The reaction connecting them is

$$n \Longleftrightarrow p + e^- + Q$$

where $Q = 0.782$ MeV and we are neglecting the electron antineutrino, which should appear on the right-side of the reaction. Further assume that the electrons are completely relativistic but that protons and neutrons are nonrelativistic.

(a) Convince yourself that the "Saha" equation is

$$\mathcal{E}_n + Q = \mathcal{E}_p + \mathcal{E}_e$$

where the $\mathcal{E}$s are the Fermi energies of the respective particles. Do your "self-convincing" two ways: (*i*) argue from the chemical potential equation of the reaction; (*ii*) make a physical argument based on the energetics of the reaction and the Pauli exclusion principle.

(b) Now to find the number densities of the particles as a function of density. Assume charge neutrality, so that $n_e = n_p$, and use the Saha equation to find n_e, n_p, and n_n for densities in the range $10^{13} \lesssim \rho \lesssim 2 \times 10^{14}$ g cm^{-3}. You may take the density as being $\rho = (n_p + n_n)m$ where m is the mass of either proton or neutron.

(c) Plot your number density results as a function of density and, if possible, compare to what you might find in the literature.

4. Show for the ideal gas ($\mu/kT \ll 0$) that $P = nkT$ is a general result independent of whether the particles are relativistic, nonrelativistic, or anything in between.

3.9 References and Suggested Readings

Introductory Remarks

The place to go for general information on stellar equations of state is

▹ Cox, J.P. 1968, *Principles of Stellar Structure*, in two volumes (New York: Gordon and Breach).

In particular, see his Chapters 9–11, 15, and 24. We also recommend Part III of

▹ Kippenhahn, R., and Weigert, A. 1990, *Stellar Structure and Evolution* (Berlin: Springer–Verlag).

A favorite text of ours is

▹ Landau, L.D., and Lifshitz, E.M. 1958, *Statistical Physics* (London: Pergamon)

and its later editions. We recommend it for its clarity (but it is not easy) and wealth of practical applications. You will even find material about neutron stars in it. A classic, though brief, text in classical thermodynamics is

▹ Pippard, A.B. 1957, *The Elements of Classical Thermodynamics* (Cambridge: Cambridge University Press).

A complete discussion of what conditions must be met to sensibly use the approximation of LTE may be found in

▹ Mihalas, D. 1978, *Stellar Atmospheres*, 2d ed. (San Francisco: Freeman).

Anyone thinking seriously about studying stars should try to find a copy. The last we heard, it is out of print but permission may be granted by the publisher to reproduce it (but check for royalty fees).

§**3.3**: Ideal Monatomic Gas

A complete textbook discussion of Fermi-Dirac equations of state for use in stars was first published by

▹ Chandrasekhar S. 1939, *An Introduction to the Study of Stellar Structure* (Chicago: University of Chicago Press).

It is available in paperback Dover editions and is well worth buying at modest cost. We shall refer to this work quite often. Other versions may be found in §3.5 of

▹ Chiu, H.-Y. 1968, *Stellar Physics*, Vol. 1. (Waltham, MA: Blaisdell)

and Chapter 24 of

▹ Cox, J.P. 1968, *Principles of Stellar Structure*, in two volumes (New York: Gordon and Breach).

§**3.4**: The Saha Equation

Systematic application of the Saha equation to multicomponent mixtures is not easy. The bookkeeping required to keep track of all the energy levels is a daunting task to say nothing of getting information on level parameters. See Chapter 15 of

▷ Cox, J.P. 1968, *Principles of Stellar Structure*, in two volumes (New York: Gordon and Breach)

and Chapter 14 of

▷ Kippenhahn, R., and Weigert, A. 1990, *Stellar Structure and Evolution* (Berlin: Springer–Verlag).

§**3.4**: Fermi-Dirac Equations of State

Chandrasekhar (1939) and Cox (1968) (see above) are standard references.

The reference to Peter Eggleton (1982) is given as a private communication in

▷ Truran, J.T., and Livio, M. 1986, *Ap.J.*, **308**, 721

who use it in some work concerning nova systems. We have extended Eggleton's mass-radius fit for white dwarfs to accommodate general μ_e.

▷ Cloutman, L.D. 1989, *Ap.J. Suppl.*, **71**, 677

is a good source of numerical techniques for computing the Fermi–Dirac integrals. See also

▷ Eggleton, P., Faulkner, J., and Flannery, B. (1973), *Astron. Astrophys.*, **23**, 325

for a thermodynamically self-consistent and efficient computation of the equation of state for arbitrarily degenerate and arbitrarily relativistic ionized gases.

▷ Antia, H.M. 1993, *Ap.J. Suppl.*, **84**, 101

gives rational expansions for the Fermi-Dirac integrals. Early work on realistic corrections to the perfect Fermi-Dirac gas includes

▷ Hamada, T., and Salpeter, E.E. 1961, *Ap.J.*, **134**, 683.

§**3.4**: "Almost Perfect" Equations of State

Our Figure 3.6 is our version of Figure 1 of

▷ Fontaine, G., Graboske, H.C., and Van Horn, H.M. 1977, *Ap.J. Suppl.*, **35**, 293.

This paper has an excellent discussion of the problems that arise when ionization (including pressure effects) and electron degeneracy must be accounted for. Their results are in the form of tables.

We have not discussed nuclear equations of state. To get an idea of what may be involved see Chapters 2, 8, and 9 of

▷ Shapiro, S.L., and Teukolsky, S.A. 1983, *Black Holes, White Dwarfs, and Neutron Stars* (New York: John Wiley & Sons)

and the review article by

▷ Bethe, H.A. 1990, *Rev. Mod. Phys.*, **62**, 801.

§**3.8**: Exercises

▷ Allen, C.W. 1973, *Astrophysical Quantities* 3d ed. (London: Athlone)

4 Radiative and Conductive Heat Transfer

In an intuitive picture of diffusion,
one usually conceives of a slow leakage
from a reservoir of large capacity by means
of a seeping action. These ideas apply in the
radiative diffusion limit as well.
— Dimitri Mihalas in *Stellar Atmospheres* (1978)
OK, this plus a little math and I suppose we're done.

In this chapter we discuss two ways by which heat can be transported through stars: diffusive radiative transfer by photons and heat conduction. The third mode of transport, which is by means of convective mixing of hot and cool material, will wait until Chapter 5. For references on the theory and application of energy transfer in stars, we recommend the following excellent texts by Mihalas (1978) and Mihalas and Mihalas (1984). Cox (1968) and Rybicki and Lightman (1979) also contain some very useful material. The discussion here will barely scratch the surface of this complex subject and will be directed toward the specific end of finding approximations suitable for the stellar interior.

4.1 Radiative Transfer

In discussing blackbody radiation and equations of state we assumed LTE as a very good approximation. We do know, however, that LTE implies complete isotropy of the radiation field and this, in turn, means that radiant energy cannot be transported through the stellar medium. Anisotropy in the field is *required* for that to happen. On the other hand, it is easy to demonstrate that only a small degree of anisotropy is needed to drive photons through most of the stellar interior. Another way to phrase this is that even small gradients in temperature can do the job. For example, a crude estimate of the overall temperature gradient in the sun is given by the ratio $(T_c/\mathcal{R}_\odot) \approx 10^{-4}$ K cm^{-1}. Although convection might augment heat transport in parts of the star, that small gradient is usually sufficient. At the solar photosphere, however, we shall see that gradients are large and, in any case, the radiation field must eventually become very anisotropic since radiation only leaves the star at the surface while none enters.

What we shall examine is what near-isotropy in the radiation field implies for the stellar interior. In the end, we shall find that the diffusion equation discussed in the first chapter (§1.5) is more than adequate for our purpose. Consideration of the very surface will be deferred until later. For a start, assume that all photons have the same frequency (ν). Amends will be made shortly.

Central to a discussion of radiative transfer is the *specific intensity*, $I(\vartheta)$. It is so defined that the product $I(\vartheta)\, d\Omega$ is the radiative energy flux (in erg cm^{-2} s^{-1}) passing through a solid angle $d\Omega$ (in sr=steradians) around a colatitude angle ϑ (in spherical coordinates ϑ and φ) at some position r or, in plane parallel geometry, z. We delete, for now, reference to r, z, and φ in the intensity and we make the important assumption that the energy transfer does not depend on time. The picture is that of a thin cone of radiation starting from r and passing through $d\Omega$ as shown in Figure 4.1.

The quantity $u(\vartheta)\, d\Omega$ is the corresponding energy density (in the units of erg cm^{-3}). This may be related to $I(\vartheta)$ by considering how much radiant energy is contained in a tube of unit cross section and length 1 sec $\times\, c$ along a thin cone in the direction ϑ. Thus

$$u(\vartheta)\, d\Omega = \frac{I(\vartheta)}{c}\, d\Omega. \tag{4.1}$$

The total energy density U is obtained by integrating (4.1) over 4π steradians with

$$U = \int_{4\pi} u(\vartheta)\, d\Omega = \frac{2\pi}{c} \int_{-1}^{1} I(\mu)\, d\mu \tag{4.2}$$

where azimuthal symmetry (for a round star or flat plane) in φ has been assumed and $\mu = \cos\vartheta$ with $-1 \le \mu \le 1$.

Because we are eventually interested in the net outflow (or inflow) of energy along z or r, define the *total flux*, $\mathcal{F}$, as follows. Place a unit area (one cm^2) perpendicular to the z-direction (in planar geometry to simplify matters). The projection of $I(\vartheta)$ onto this area is then $I(\vartheta)\cos\vartheta \times 1$ cm^2. If this last is integrated over $d\Omega$ and the result is divided by $1 \times$ cm^2 we

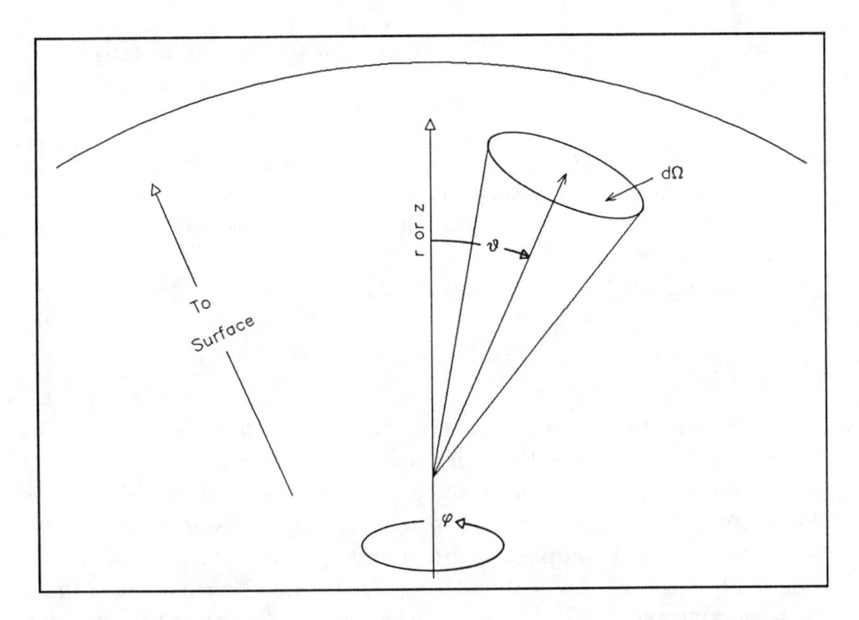

FIGURE 4.1. The geometry associated with the specific intensity $I(\vartheta)$. The position coordinate may either be radius r, in a spherically symmetric star or vertical distance z, in a plane parallel "star." In the latter case, symmetry in the transverse x and y coordinates is assumed. The properties of the stellar medium are then independent of azimuthal angle φ for either choice of geometry.

then obtain the total flux in the z-direction

$$\mathcal{F} = \int_{4\pi} I(\vartheta) \cos\vartheta \, d\Omega = 2\pi \int_{-1}^{1} I(\mu)\mu \, d\mu \quad \text{erg s}^{-1} \text{ cm}^{-2}. \tag{4.3}$$

Note that if I is a constant, then the total flux is zero because the same amount of radiation comes in as goes out. Hence I must vary with μ (or ϑ) for radiant energy to be transported.

What are the sources and sinks for $I(\vartheta)$? At any location z or r, $I(\vartheta)$ may be fed by radiation being scattered from other directions into ϑ or by direct emission from local atoms. We are not going to go into the subtleties of what these different kinds of processes mean for radiative transfer, but will rather lump them together into the *mass emission coefficient*, $j(\vartheta)$, constructed as follows. If ds is a distance directed along $I(\vartheta)$ over which $I(\vartheta)$ is augmented by the amount $dI(\vartheta)$ due to scattering or emission, then $j(\vartheta)$ is defined by

$$dI(\text{put into } \vartheta) = j(\vartheta)\rho \, ds.$$

Photons can also be removed from the beam by absorption and scattering. These processes are accounted for by the *opacity* (or mass absorption coefficient), κ (in cm^2 g^{-1}), so that the amount removed from $I(\vartheta)$ is

$$dI(\text{taken out of } \vartheta) = -\kappa\rho I(\vartheta)\, ds.$$

Note that if j is zero and κ and ρ are both constant, then $I \propto e^{-\kappa\rho s}$ as in simple attenuation. Recall that the product $(\kappa\rho)^{-1}$ was previously used to compute a typical mean free path in (3.2). We see now that it is an e–folding length for attenuation.

The net change in $I(\vartheta)$ per unit path length is then

$$\frac{1}{\rho}\frac{dI(\vartheta)}{ds} = j - \kappa I(\vartheta) \tag{4.4}$$

which is the *equation of transfer*. Note that in the form we give it, the transfer equation really only holds in planar geometry: a rigorous derivation in spherical geometry would have to contain curvature terms. Because we shall finally get the diffusion equation, such niceties are unnecessary here.

If we had LTE and complete isotropy and spatial uniformity of the radiation field, then $(dI/ds) = 0$ and $I = (j/\kappa)$ would be constant with no radiant energy transported. In that case the energy density of the radiation field is, from (4.2),

$$U = \frac{2\pi}{c}\int_{-1}^{1} I\, d\mu = \frac{4\pi}{c} I. \tag{4.5}$$

But in LTE, U follows from the blackbody result $U = aT^4$ (as in 3.17 with $E_{\rm rad}$ replaced by U) so that

$$I = \frac{j}{\kappa} = \frac{c}{4\pi} aT^4 = \frac{\sigma}{\pi} T^4 = B(T) \quad \text{(in LTE)} \tag{4.6}$$

where $B(T)$, the integrated Planck function, was introduced in Chapter 3 as equation (3.21). The frequency–dependent Planck function is

$$B_\nu(T) = \frac{2h\nu^3}{c^2}\left(e^{h\nu/kT} - 1\right)^{-1} \tag{4.7}$$

as may be deduced from (3.18) and (3.20).

If, as we suppose, the radiation field is nearly isotropic through most of the star, then the intensity should closely resemble $B(T)$. The question is, by how much? To answer this requires a little more work.

At this point we introduce frequency into all expressions and realize that quantities such as I, j, and κ must all depend on ν so that we can talk about photons of a given frequency being added to or subtracted from a beam in direction ϑ, etc. In addition, we introduce the *source function* (which, for us, will be merely a computational device), $S_\nu(\vartheta) = j_\nu/\kappa_\nu$, and the *optical depth*, τ_ν, with

$$d\tau_\nu = -\kappa_\nu \rho\, dz \ \text{(or } dr). \tag{4.8}$$

The integrated version of (4.8) is

$$\tau_\nu(z) = \tau_{\nu,0} - \int_{z_0}^{z} \kappa_\nu \rho \, dz \tag{4.9}$$

where z_0 is some spatial reference level and $\tau_{\nu,0}$ is the optical depth evaluated at that level. If z_0 corresponds to the "true surface" of the star where density and pressure presumably go to zero, then $\tau_{\nu,0}$ is taken to be zero. We shall make this choice and thus $\tau_\nu(z)$ measures depth from the surface in curious, but dimensionless, units.

If we now recall that ds is measured along the direction of $I(\vartheta)$, then $dz = \cos\vartheta \, ds = \mu \, ds$ and, putting all together, the equation of transfer becomes

$$\mu \frac{dI_\nu(\tau,\mu)}{d\tau_\nu} = I_\nu - S_\nu. \tag{4.10}$$

The solution of the transfer equation is easy to pose but difficult to carry out in practice. As a first step, note that (4.10) admits of the integrating factor $\exp(-\tau/\mu)$ where here (and often elsewhere) the subscript ν and the arguments ϑ or μ will be deleted for visual clarity. Thus multiply through by that factor, recognize a perfect differential, and find

$$\frac{d}{d\tau}\left[e^{-\tau/\mu} I\right] = -e^{-\tau/\mu}\frac{S}{\mu}.$$

If we formally integrate from some reference level τ_0 to a general level τ, then the solution is

$$I(\tau,\mu) = e^{-(\tau_0-\tau)/\mu} I(\tau_0,\mu) + \int_{\tau}^{\tau_0} e^{-(t-\tau)/\mu} \frac{S(t)}{\mu} \, dt \tag{4.11}$$

where t is a dummy integration variable.

Depending on the range of μ (or ϑ), different values for τ_0 are chosen in seeking solutions for $I(\tau,\mu)$. For forward-directed radiation (heading out toward the surface) with $\mu \geq 0$ $(0 \leq \vartheta \leq \pi/2)$, choose τ_0 to be very large and positive so that the reference level lies deep (at least with respect to optical depth) within the star. Thus with $\tau_0 \rightarrow \infty$

$$I(\tau,\mu \geq 0) = \int_{\tau}^{\infty} e^{-(t-\tau)/\mu} \frac{S(t)}{\mu} \, dt \,. \tag{4.12}$$

If $\mu < 0$, signifying inwardly directed radiation, use $\tau_0 = 0$ so that

$$I(\tau,\mu < 0) = \int_{\tau}^{0} e^{-(t-\tau)/\mu} \frac{S(t)}{\mu} \, dt \,. \tag{4.13}$$

In the last expression, advantage has been taken of the fact that the level $\tau_0 = 0$ has been chosen to be the true surface of the star, where it is required that $I(0, \mu < 0) = 0$; that is, there is no incoming radiation at the surface.[1]

If the deep interior is to be nearly in LTE, we then expect the source function $S_\nu = j_\nu / \kappa_\nu$ to be almost independent of angle and, from (4.6), to be near its Planckian value $B_\nu(T)$ at depth τ_ν (assumed to be appropriately large). If this is so, then it seems reasonable to expand $S(t)$ in a Taylor series which, to first order, is

$$S(t) = B(\tau) + (t - \tau)\left(\frac{\partial B}{\partial \tau}\right)_\tau \tag{4.14}$$

where $B(\tau)$ stands for $B[T(\tau)]$. (A more exacting discussion of this, and the approach to the diffusion equation, may be found in Mihalas 1978, §2–5.)

Inserting (4.14) into (4.12–4.13) yields

$$I(\tau, \mu \geq 0) = B(\tau) + \mu\left(\frac{\partial B}{\partial \tau}\right)_\tau \tag{4.15}$$

and,

$$I(\tau, \mu < 0) = B(\tau)\left[1 - e^{\tau/\mu}\right] + \mu\left(\frac{\partial B}{\partial \tau}\right)_\tau \left[e^{\tau/\mu}\left(\frac{\tau}{\mu} - 1\right) + 1\right] . \tag{4.16}$$

Since $\mu < 0$ in (4.16), we may neglect the exponential $e^{\tau/\mu}$ for τ large and find that (4.15) *is valid for all* μ. You may easily verify that higher-order terms in the Taylor series expansion for $S(t)$ in (4.14) lead to additional terms in $I(\tau, \mu)$ that go as $|\partial^n B / \partial \tau^n| \sim B/\tau^n$. Thus, roughly speaking, convergence is rapid if τ is greater than unity.

This looks promising. Since we expect temperature to increase inward, as does τ, then $\partial B / \partial \tau > 0$. Thus, because of the presence of the factor μ in (4.15) or (4.16), the outwardly directed intensity (with $\mu \geq 0$) is enhanced over its Planckian value, $B(\tau)$, whereas the intensity directed inward (with $\mu \leq 0$) is reduced. The net result is a flow of radiation outward when the two intensities are integrated over their respective angles and the two results are added. To compute the flux of this radiation use (4.3) and find

$$\mathcal{F}(\tau) = 2\pi \int_{-1}^{1} \left[B(\tau) + \mu \frac{\partial B(\tau)}{\partial \tau}\right] \mu \, d\mu = \frac{4\pi}{3} \frac{\partial B(\tau)}{\partial \tau} \tag{4.17}$$

in the units of erg cm^{-2} s^{-1}. Since the integrated Planck function is proportional to T^4, the amount of energy flux carried by radiation depends only on how rapidly temperature varies with optical depth.

---

[1]This statement would be inapplicable to a star in a close binary where its companion might bathe the stellar surface with radiation. A similar situation holds for stellar winds where the wind itself may radiate profusely.

If (4.17) represents the total flux (implying that it contains all the frequency-dependent fluxes integrated over frequency), then the *luminosity* for a spherical star at radius $r(\tau)$ is

$$\mathcal{L}(r) = \mathcal{L}_r = 4\pi r^2 \mathcal{F}(r) \quad \text{erg s}^{-1}. \tag{4.18}$$

A measure of the anisotropy in the intensity is the comparison of $I(\tau)$ to $\partial B(\tau)/\partial\tau$. We may calculate $\partial B(\tau)/\partial\tau$ from the flux by means of (4.17) and use $(\sigma/\pi)T^4(\tau)$ as an estimate for $I(\tau)$ from (4.6). As an example, consider the sun. Using global values for everything in sight, a typical flux may be found from $\mathcal{L}_\odot \sim 4\pi\mathcal{R}_\odot^2\mathcal{F}_\odot \sim 4\times 10^{33}$ erg s^{-1} or, $\mathcal{F}_\odot \sim 7\times 10^{10}$ erg cm^{-2} s^{-1}. Thus $\partial B(\tau)/\partial\tau_\odot \sim 2\times 10^{10}$ erg cm^{-2} s^{-1}. A typical solar temperature is $T_\odot \sim 10^7$ K so that $I_\odot = B_\odot \sim 2\times 10^{23}$ erg cm^{-2} s^{-1}. The measure of anisotropy is then $\left[\partial B(\tau)/\partial\tau\right]_\odot/I_\odot \sim 10^{-13}$! We have, of course, used estimates for various numbers here but the final result is quite representative of the true situation in the deep interior.

The truncation to first order of the expansion for $I(\tau,\mu)$ is reasonable provided that $\tau \gtrsim 1$. We shall find that an optical depth of unity lies very close below the physical surface in most stars. Hence the approximations used here will be valid for just about all of a star. The thin region above $\tau \approx 1$ we call the *atmosphere* and it is the region where radiation is processed so that we ultimately see it. Except for some simple calculations to be considered in Chapter 7, the atmosphere will be left to the specialists in that important subject.

4.2 The Diffusion Equation

To properly derive the diffusion approximation, we return to the expression for the flux but include the frequency dependence:

$$\mathcal{F}_\nu = \frac{4\pi}{3}\frac{\partial B_\nu}{\partial \tau_\nu} \tag{4.19}$$

which may be rewritten using the definition of $d\tau_\nu$ as

$$\mathcal{F}_\nu = -\frac{4\pi}{3}\frac{1}{\kappa_\nu\rho}\frac{\partial B_\nu}{\partial r}\,. \tag{4.20}$$

The derivative of B_ν is cast into more a convenient form by using the chain rule so that

$$\frac{\partial B_\nu}{\partial r} = \frac{\partial B_\nu}{\partial T}\frac{dT}{dr}$$

where, if desired, $\partial B_\nu/\partial T$ may be found using (4.7). The flux is then

$$\mathcal{F}_\nu = -\frac{4\pi}{3}\frac{1}{\rho}\frac{dT}{dr}\frac{1}{\kappa_\nu}\frac{\partial B_\nu}{\partial T}\,. \tag{4.21}$$

To obtain the total flux integrate over frequency and define the *Rosseland mean opacity*, κ, by

$$\frac{1}{\kappa} = \left[\int_0^\infty \frac{1}{\kappa_\nu}\frac{\partial B_\nu}{\partial T}\,d\nu\right]\left[\int_0^\infty \frac{\partial B_\nu}{\partial T}\,d\nu\right]^{-1} \tag{4.22}$$

so that the total flux is

$$\mathcal{F} = -\frac{4\pi}{3}\frac{1}{\kappa\rho}\frac{dT}{dr}\int_0^\infty \frac{\partial B_\nu}{\partial T}\,d\nu.$$

The last integral is eliminated by observing that

$$\int_0^\infty \frac{\partial B_\nu}{\partial T}\,d\nu = \frac{\partial}{\partial T}\int_0^\infty B_\nu(T)\,d\nu = \frac{\partial B}{\partial T} = \frac{ac}{\pi}T^3$$

where (4.6) and $\sigma = ac/4$ have been used.[2] Thus, finally,

$$\mathcal{F}(r) = -\frac{4ac}{3}\frac{1}{\kappa\rho}T^3\frac{dT}{dr} = -\frac{c}{3\kappa\rho}\frac{d(aT^4)}{dr}. \tag{4.23}$$

This version of $\mathcal{F}$ is in the Fick's law form introduced in the first chapter (§1.5) where the diffusion coefficient $\mathcal{D}$ is now identified as $\mathcal{D} = c/(3\kappa\rho)$. The factor of 1/3 that appears is usual in diffusion theory and the remainder represents a velocity (c) times a mean free path $\lambda = 1/(\kappa\rho)$. The derivative term in (4.23) implies that the "driving" is caused by spatial gradients in the energy density (aT^4) of the radiation field.

The total luminosity in the diffusion approximation to radiative transfer is simply $\mathcal{L} = 4\pi r^2\mathcal{F}$ or

$$\mathcal{L}(r) = \mathcal{L}_r = -\frac{16\pi acr^2}{3\kappa\rho}T^3\frac{dT}{dr} = -\frac{4\pi acr^2}{3\kappa\rho}\frac{dT^4}{dr} \tag{4.24}$$

which is what was stated in (1.57). There are several other ways of expressing $\mathcal{L}_r$ which will prove useful for future work. Among these are the following.

The Lagrangian form of (4.24) is obtained by using the mass equation (1.1) to convert the radial derivative to one of mass:

$$\mathcal{L}_r = -\frac{\left(4\pi r^2\right)^2 ac}{3\kappa}\frac{dT^4}{d\mathcal{M}_r} \tag{4.25}$$

and this was used in the dimensional arguments of §1.6. Absorbing the factor of a and recognizing $P_{\rm rad} = (1/3)aT^4$ also yields

$$\mathcal{L}_r = -\frac{\left(4\pi r^2\right)^2 c}{\kappa}\frac{dP_{\rm rad}}{d\mathcal{M}_r}. \tag{4.26}$$

---

[2]Note that we do not cover here the important effects of stimulated emission on opacities. These are treated in the references listed previously.

For still another version, introduce the equation of hydrostatic equilibrium (1.5) (in a slightly disguised form and note the presence of the pressure scale height of eq. 3.1 in the middle term) so that

$$-\frac{d\ln P}{d\ln r} = \frac{r}{\lambda_P} = \frac{G\mathcal{M}_r\rho}{rP}. \tag{4.27}$$

Then divide both sides by $(d\ln T/d\ln r)$ to find

$$\frac{(d\ln P/d\ln r)}{(d\ln T/d\ln r)} = \frac{d\ln P}{d\ln T} = -\frac{G\mathcal{M}_r\rho}{rP}\frac{1}{(d\ln T/d\ln r)}.$$

The reciprocal of the derivative in the middle of this expression is used to define a new quantity, ∇, called "del" with

$$\nabla \equiv \frac{d\ln T}{d\ln P} = -\frac{r^2P}{G\mathcal{M}_r\rho}\frac{1}{T}\frac{dT}{dr}. \tag{4.28}$$

Sometimes a subscript "act" is appended to ∇ to denote "actual." The implication is that ∇ represents the actual run, or logarithmic slope, of local temperature versus pressure in the star. If ∇ is known by some means or another, then a simple rearrangement of (4.24) yields

$$\mathcal{L}_r = \frac{16\pi acG}{3}\frac{T^4}{P\kappa}\mathcal{M}_r\nabla. \tag{4.29}$$

4.3 A Brief Diversion into "∇'s"

Besides the ∇ defined earlier, it is useful to define another logarithmic quantity, $\nabla_{\rm rad}$, called "delrad," as follows. Suppose $\mathcal{L}_r$ is the luminosity corresponding to an energy flux transported by *any* means and not necessarily just by radiation. Then define $\nabla_{\rm rad}$ by turning (4.29) around so that

$$\nabla_{\rm rad} \equiv \left(\frac{d\ln T}{d\ln P}\right)_{\rm rad} \equiv \frac{3}{16\pi acG}\frac{P\kappa}{T^4}\frac{\mathcal{L}_r}{\mathcal{M}_r}. \tag{4.30}$$

Thus $\nabla_{\rm rad}$ is the local logarithmic slope of temperature versus pressure which *would* be required *if* all the given luminosity were to be carried by radiation. This quantity will prove useful for future work although, at the moment, it may seem to be superfluous baggage. But, for example, suppose you were given the run of density, temperature, opacity, and energy generation rate in a star and the luminosity and ∇ and $\nabla_{\rm rad}$ as functions of radius. You have no a priori knowledge of how the energy is transported. It could well be that the luminosity, $\mathcal{L}$, at a given radius consists of a part that reflects diffusive transfer, $\mathcal{L}_{\rm rad}$, plus a contribution from other sources such as convection, $\mathcal{L}_{\rm conv}$, with $\mathcal{L} = \mathcal{L}_{\rm rad} + \mathcal{L}_{\rm conv}$. The luminosity of (4.29) is obviously $\mathcal{L}_{\rm rad}$ because it is that which is generated by Fick's law with, in the present nomenclature, a gradient term ∇; that is, ∇ is the actual driving gradient in the star and thus $\mathcal{L}_{\text{rad}}$ follows from it. However, ∇_{rad} derives from $\mathcal{L}$. Thus if $\nabla = \nabla_{\text{rad}}$ then all the luminosity must be radiative, $\mathcal{L} = \mathcal{L}_{\text{rad}}$, and $\mathcal{L}_{\text{conv}} = 0$. If, on the other hand, $\nabla_{\text{rad}} > \nabla$, then $\mathcal{L} > \mathcal{L}_{\text{rad}}$ and $\mathcal{L}_{\text{conv}}$ is not zero and radiation does not transport all of the energy. (Note that we have not yet mentioned conduction. We shortly shall find a way to incorporate it in with radiative transfer. This cannot be done for convection.)

The preceding analysis will turn out to be not all that abstract. Recall that yet another "del" was defined in (3.88) and (3.90) of the previous chapter as $\nabla_{\text{ad}} = (\partial \ln T / \partial \ln P)_{\text{ad}}$. It has the same T–ρ structure as ∇ but it is a thermodynamic derivative. All three "dels" consist of logarithmic derivatives of temperature with respect to pressure except they are computed under different circumstances. When convection is discussed in the next chapter, these three derivatives will serve to establish a description of how energy is transported in the stellar interior.

4.4 Radiative Opacity Sources

The calculation of realistic stellar opacities is easily among the most difficult of problems facing the stellar astrophysicist. At the present time, the most commonly used opacities for stellar mixtures are those generated at the Los Alamos National Laboratory (LANL) and at the Lawrence Livermore National Laboratory for both astronomers and non-astronomers although other groups are now engaged in producing these necessary numbers. (The original need for opacities at LANL was, needless to say, prompted not by astrophysical considerations but rather by those of fission and fusion bomb work.) Opacities are available in tabular form and include many stellar mixtures with opacities computed over wide ranges of density and temperature. The references at the end of this chapter include published sources and it would be a worthwhile exercise for you to plot up some opacities and get a feel for how they behave.

The following discussion is by no means complete and will give only sketches (if even that) of what goes into the calculation of opacities. A physically clear, and not terribly difficult, description of the ingredients of the calculations may be found in Chapter 3 of Clayton (1968). Cox (1968, chap. 16) also contains some very useful material. The aim is to construct a total Rosseland mean opacity, κ_{rad}, which is the sum of contributions from the following sources. We shall start with the simplest, which is electron scattering.

4.4.1 Electron Scattering

Equation (4.4) gave a prescription for calculating how much intensity is removed from a beam when an opacity source is present. In the instance where the opacity is independent of frequency, a simple relation may be found between the opacity and the cross section of the process responsible for beam attenuation. A cross section for a process may be defined quite generally as in this example of low-energy electron scattering. If a beam of photons of a given flux—now defined as the number of photons per cm^2 per second—is incident upon a collection of stationary electron targets, then the rate at which a given event (a photon electron scattered out of the beam) takes place per target is related to the cross section, σ, by

$$\sigma = \frac{\text{number of events per unit time per target}}{\text{incident flux of photons}} \quad \text{cm}^2. \tag{4.31}$$

As we shall soon indicate, the cross section for low-energy electron scattering is independent of energy, and the transfer equation that describes how a beam is attenuated is (4.4) with j set to zero. Thus if n_e is the number density of free target electrons, then the product $I\sigma n_\text{e}\, ds$ is the number of scatterings in cm^{-2} s^{-1} erg over the path length ds (from the definition of σ) and this is to be equated to $I\kappa\rho\, ds$ of (4.4). The desired relation between κ and σ is then

$$\kappa = \frac{\sigma n_\text{e}}{\rho} \quad \text{cm}^2\ \text{g}^{-1}. \tag{4.32}$$

For electron or photon thermal energies well below the rest mass energy of the electron ($kT \ll m_\text{e}c^2$ or $T \ll 5.93 \times 10^9$ K), ordinary frequency-independent Thomson scattering describes the process very well and the cross section for that is

$$\sigma_\text{e} = \frac{8\pi}{3}\left(\frac{e^2}{mc^2}\right)^2 = 0.6652 \times 10^{-24} \quad \text{cm}^2 \tag{4.33}$$

where $\left(e^2/mc^2\right)$ is the classical electron radius. Because, as it will turn out, electron scattering is most important when stellar material is almost completely ionized, it is customary to compute n_e according to the prescription of (1.45) and (1.50). Thus $n_\text{e} = \rho N_\text{A}(1+X)/2$ where X is the hydrogen mass fraction. Folding this in with (4.32–4.33) we obtain the electron scattering opacity

$$\kappa_\text{e} = 0.2(1+X) \quad \text{cm}^2\ \text{g}^{-1}. \tag{4.34}$$

If heavy elements are very abundant or ionization is not complete, then n_e must be calculated in a more general way using the ionization fractions, etc., of (1.45). Note also that in a mixture consisting mostly of hydrogen, this opacity decreases rapidly from the value implied by (4.34) at temperatures below the ionization temperature of hydrogen of near 10^4 K: there are just too few free electrons left. The corresponding temperature for a gas consisting mostly of helium is around 5×10^4 K. (And see Figs. 4.2 and 4.3.)

As remarked upon in §1.5, this opacity depends neither on temperature nor density if ionization is complete and hence its temperature and density exponents s and n in $\kappa = \kappa_0 \rho^n T^{-s}$ (of 1.59) are $s = n = 0$.

Besides having to worry about exotic mixtures of elements and partial ionization, the electron scattering opacity presented above must be modified for high temperatures (relativistic effects with $kT \gtrsim m_e c^2$) and for the effects of electron degeneracy where electrons may be inhibited from scattering into already occupied energy states. It is fortunate that this is done for us in most opacity tables. Later on, we shall show some graphical illustrations of some of these tables and give useful analytic fits to some of the results.

4.4.2 Free–Free Absorption

As is well-known from elementary physics, a free electron cannot absorb a photon because conservation of energy and momentum cannot both be satisfied during the process. If, however, a charged ion is in the vicinity of the electron, then electromagnetic coupling between the ion and the electron can serve as a bridge to transfer momentum and energy making the absorption possible. It should be apparent that this absorption process is the inverse of normal bremsstrahlung wherein an electron passing by and interacting with an ion emits a photon.

A complete derivation will not be presented here (which would deal with the quantum mechanics of the absorption) but a rough estimate of the opacity may be found classically. We first compute the emission rate for bremsstrahlung, and then turn the problem around.

Imagine an electron of charge e moving nonrelativistically at velocity v past a stationary ion of charge $Z_c e$. As the electron goes past, it is accelerated and radiates power according to the Larmor result

$$P(t) = \frac{2}{3}\frac{e^2}{c^3}\,a^2(t)$$

where $a(t)$ is the time-dependent acceleration. If we naively assume that the electron trajectory is roughly a straight line, then it is easy to show (as an E&M problem in Landau and Lifshitz 1971, §73) that the time-integrated power, or energy, radiated is

$$E_s = \frac{Z_c^2 e^6 \pi}{3c^3 m_e^2}\frac{1}{vs^3}$$

where s is the impact parameter for the trajectory (i.e., the distance of closest approach were the trajectory to remain straight).

The maximum energy radiated during the scattering will be peaked in angular frequency around $\omega \approx v/s$. Thus if E_ω is the energy emitted per unit frequency, then E_ω must be simply related to E_s, which is the energy emitted per unit impact parameter. If $2\pi s\, ds$ is the area of an annular target that intercepts a uniform velocity beam of electrons, then

$$E_\omega\, d\omega = -E_s\, 2\pi s\, ds = \frac{2Z_c^2 e^6}{3c^3 m_e^2} \frac{\pi^2}{v^2}\, d\omega$$

where ω has been set to v/s and the minus sign comes about because $ds > 0$ implies $d\omega < 0$.

To get a rate of emission per unit frequency, assume that the electron distribution is Maxwell-Boltzmann so that (3.24) applies and

$$n_e(v)\, dv = 4\pi n_e \left(\frac{m_e}{2\pi kT}\right)^{3/2} e^{-m_e v^2/2kT}\, v^2\, dv$$

(after the transformation $p = m_e v$ is used in 3.24). The product $n_e(v)v$ is the flux of electrons per unit velocity so that $E_\omega n_e(v)v\, dv$ integrated over all permissible v is the desired rate per target ion per unit frequency. All that remains is to multiply by the ion number density, n_I, and to identify the result as being part of the mass emission coefficient j of (4.4). The total power emitted per unit frequency and volume is then

$$4\pi j_\omega \rho = n_I \int_v E_\omega n_e(v) v\, dv$$

where j_ω, assumed isotropic, has been integrated over 4π steradians.

The lower limit on the integral should correspond to the minimum velocity required to produce a photon of energy $\hbar\omega$, namely, $(1/2)v_{\rm min}^2 = \hbar\omega$. Even though we have assumed that the electrons are nonrelativistic, the upper limit is taken as infinity. (Unless temperatures are very high, the exponential in the Maxwell-Boltzmann distribution will serve as an effective cutoff.) The integral is elementary and yields

$$4\pi j_\omega \rho\, d\omega = \frac{2\pi}{3} \frac{Z_c^2 e^6}{m_e c^3} \left(\frac{2\pi}{m_e kT}\right)^{1/2} n_e n_I\, e^{-\hbar\omega/kT}\, d\omega .$$

Finally, integrate over ω and find

$$4\pi j \rho = \frac{2\pi}{3} \frac{Z_c^2 e^6}{m_e c^3 \hbar} \left(\frac{2\pi kT}{m_e}\right)^{1/2} n_e n_I \approx 10^{-27}\, Z_c^2 n_I n_e T^{1/2} \quad \text{erg cm}^{-3}\ \text{s}^{-1}.$$

This result is very nearly correct; the numerical coefficient should be 1.4×10^{-27} and an additional quantum mechanical "gaunt factor" ($g_{\rm ff}$), which is of order unity, should appear (as in, for example, Spitzer 1962, §5.6)

To get the absorption coefficient, we assume that the radiation field is in LTE with $j/\kappa = S = B(T)$ and that κ is due only to free-free absorption. Thus, $\kappa = j/B(T) = \pi j/\sigma T^4$, or, putting in the numbers,

$$\kappa_{\rm f-f} \approx 4 \times 10^{-24} \frac{Z_c^2 n_e n_I T^{-3.5}}{\rho} \propto \rho T^{-3.5} \quad \text{cm}^2\ \text{g}^{-1}$$

where the last proportionality arises from eliminating the number densities, both of which are proportional to density.

The functional relation of $\kappa_{\rm f-f}$ to ρ and T of the above is basically correct. The numerical coefficient is too high by a factor of ten. To use this opacity as a Rosseland mean, we must really perform the integration indicated in (4.22) and put in the relevant atomic physics. All this is done when constructing opacity tables, and we defer to them. There is, however, a fair approximation to the free-free opacity, which does prove useful in working with simplified stellar models, and it is

$$\kappa_{\rm f-f} \approx 10^{23} \frac{\rho}{\mu_{\rm e}} \frac{Z_{\rm c}^2}{\mu_{\rm I}} T^{-3.5} \quad {\rm cm}^2 \ {\rm g}^{-1} \tag{4.35}$$

where Z_c is an average nuclear charge and $\mu_{\rm e}$ and $\mu_{\rm I}$ are the mean atomic weights used previously on several occasions. Note that this opacity requires the presence of free electrons: if none are present then $\kappa_{\rm f-f}$ should be zero. This is effectively taken care of by $\mu_{\rm e}$ where, if all ions are neutral, then $\mu_{\rm e} \to \infty$ from the definition of $\mu_{\rm e}$ in (1.45–1.46, with $y_i = 0$). For a mixture composed of hydrogen and some helium (and traces of metals), we expect the free-free opacity to be negligible below temperatures of around 10^4 K (or perhaps a little higher if densities are relatively high: see the half-ionization curve for hydrogen of Fig. 3.1).

The main features of (4.35) are correct and the general form is that of a Kramers' opacity, which was used in Chapter 1 (§1.5). Recall from there that the opacity was written in power law form $\kappa = \kappa_0 \rho^n T^{-s}$ in equation (1.59) and please note the *sign* in the temperature dependence. Thus the free-free opacity may be characterized by $n = 1$ and $s = 3.5$.

The strongest dependence in $\kappa_{\rm f-f}$ is that of temperature. In our quick and dirty derivation, this comes about because j is a weak function of temperature ($j \sim T^{1/2}$) whereas $B(T) \sim T^4$. Another closely related approach is to construct directly the cross section for the free-free process. The contribution to this quantity from electrons in the velocity band dv is $\sigma \propto n_{\rm e}(v)\, dv/v\nu^3$ where ν is the frequency of the absorbed photon. (Several factors varying relatively slowly with frequency or velocity have been neglected here.) An average for this cross section over velocity introduces a temperature dependence going as $T^{-1/2}$ (from integrating $n_{\rm e}[v]\, dv/v$). The Rosseland mean integral of (4.22) weights most heavily those photons with frequencies near $\nu \approx 4kT/h$ (as you may easily verify). Thus ν^{-3} in the cross section gives a dependence of T^{-3} and this is folded in with the velocity average contribution to yield a factor of $T^{-3.5}$. The opacity is proportional to the cross section and, hence, $s = 3.5$.

4.4.3 Bound-Free Absorption

This is absorption of a photon by a bound electron where the photon energy is sufficient to remove the electron from the atom or ion altogether. To do a proper job of opacity calculation, the atomic physics of all the atoms and ions in the mixture must be handled with great care. However, it may be shown that the frequency dependence of the opacity κ_ν is again $1/\nu^3$ and that the total bound-free opacity is again of Kramers' form. A rough and ready estimate, permissible for simple stellar calculations, has been presented by Schwarzschild (1958) who gives (with some factors of order unity deleted)

$$\kappa_{\mathrm{b-f}} \approx 4 \times 10^{25} Z(1+X)\rho T^{-3.5} \quad \mathrm{cm}^2\ \mathrm{g}^{-1} \tag{4.36}$$

where X and Z are, respectively, the hydrogen and metal mass fractions discussed in §1.4. This expression should not be applied if temperatures are much below $T \approx 10^4$ K because, as only part of the story, most photons are not energetic enough to ionize the electrons.

Schwarzschild also gives an expression for the free-free opacity, which is quite useful although it makes some assumptions about the composition:

$$\kappa_{\mathrm{f-f}} \approx 4 \times 10^{22} (X+Y)(1+X)\rho T^{-3.5} \quad \mathrm{cm}^2\ \mathrm{g}^{-1}. \tag{4.37}$$

4.4.4 Bound-Bound Opacity

This opacity is associated with photon-induced transitions between bound levels in atoms or ions. The calculation is quite complex because it involves detailed description of absorption line profiles under a wide variety of conditions of line broadening, etc. The form of the opacity is of a Kramers' type and could be included in with the preceding expressions. Since, however, it is usually of magnitude less than f-f or b-f, we shall not give any estimates here. For the level of accuracy required in actual stellar modeling, on the other hand, all components of the opacity must be treated carefully. The job of computing opacities is not for the amateur!

4.4.5 H^- *Opacity and Others*

Among the more important sources of opacity in cooler stars is that resulting from free-free and bound-free transitions in the negative hydrogen ion, H^-. It is, for example, the most important opacity source for the solar atmosphere. Because of the large polarizability of the neutral hydrogen atom, it is possible to attach an extra electron to it with an ionization potential of 0.75 eV. But this implies that the resulting negative ion is very fragile and is readily ionized if temperatures exceed a few thousand degrees ($kT \approx 0.75$ eV). Making the ion is not an easy task either because it requires both neutral hydrogen and free electrons. This means that some electrons must be made available from any existing ionized hydrogen (helium will be neutral for $T \lesssim 10^4$ K) or from outer shell electrons contributed from abundant metals such as Na, K, Ca, or Al. In this respect, the H^- opacity is sensitive not only to temperature but also to metal abundance.

If temperatures are less than about 2500 K, or if metal donors have very low abundances, then insufficient numbers of free electrons are available to make H^- and the opacity becomes very small.

An estimate of the opacity contributed by H^- may be obtained by using existing tabulations (to be discussed shortly). The following power law fit gives reasonable results (give or take a factor of ten) for temperatures in the range $3000 \lesssim T \lesssim 6000$ K, densities $10^{-10} \lesssim \rho \lesssim 10^{-5}$ gm cm^{-3}, a hydrogen mass fraction of around $X \approx 0.7$ (corresponding to main sequence atmospheric hydrogen abundances), and a metal mass fraction $0.001 \lesssim Z \lesssim 0.03$ assuming a solar mix of individual metals:

$$\kappa_{\mathrm{H}^-} \approx 2.5 \times 10^{-31}\,(Z/0.02)\,\rho^{1/2}T^{9} \quad \mathrm{cm}^2\ \mathrm{g}^{-1}. \tag{4.38}$$

This expression should only be used for estimates when tabulated opacities are not available. On the other hand, it does give the flavor of how this opacity operates and it will prove useful when we examine some properties of cool stars. Note that its power law exponents are $n = 1/2$ and $s = -9$. Unlike Kramers', it increases strongly with temperature until about 10^4 K, above which Kramers' and electron scattering take over. This will be apparent when curves of realistic opacities are presented later.

For very cool stars with effective temperatures of less than about 3000 K, opacity sources due to the presence of molecules or small grains become important. Because of the proliferation of complex molecules in cool stars and the difficulty in modeling their abundances and opacities, there is still a great deal of uncertainty about how the atmospheres of cool stars really work. This situation is liable to be with us for several more years.

This ends our discussion of opacities derived from atomic processes. In the interiors of dense objects, however, there are other processes that control the flow of energy.

4.5 Heat Transfer by Conduction

We have already stated that the structural support of the deep interior of a white dwarf or of some red supergiants is due to the presence of degenerate electrons. Not only do these electrons prevent the interior from collapsing, they also are the major means by which energy is transported outward (or, in some instances, inward). The mechanism is by means of electron heat conduction down a temperature gradient—as in a metal—and it is at this point that we must do a little solid-state physics (and that's why dealing with stellar interiors is so much fun: you get to do almost everything).

A good approximation to heat transfer in metals is, again, Fick's law of diffusion:

$$\mathcal{F}_{\mathrm{cond}} = -\mathcal{D}_{\mathrm{e}} \frac{dT}{dr}\,. \tag{4.39}$$

Here, $\mathcal{D}_\mathrm{e}$ is a diffusion coefficient with "e" standing for electron. It is convenient to recast (4.39) into a form identical to that used in diffusive radiative transfer (i.e., eq. 4.23 or 4.24) by defining a "conductive opacity," κ_cond, with

$$\kappa_\mathrm{cond} = \frac{4acT^3}{3\mathcal{D}_\mathrm{e}\rho} . \tag{4.40}$$

The conductive flux is then

$$\mathcal{F}_\mathrm{cond} = -\frac{4ac}{3\kappa_\mathrm{cond}\rho} T^3 \frac{dT}{dr} . \tag{4.41}$$

Assuming, for the moment, that we already know how to compute κ_cond, how do we combine this opacity with atomic opacities (since, if we have a temperature gradient, photons should also flow)? The total energy flux, from radiation and conduction electrons combined, is additive. Thus, calling the radiative component $\mathcal{F}_\mathrm{rad}$, the total is $\mathcal{F}_\mathrm{tot} = \mathcal{F}_\mathrm{rad} + \mathcal{F}_\mathrm{cond}$ if convection is ignored. By inspection (see 4.24), the opacities are additive as in a parallel resistive circuit or

$$\frac{1}{\kappa_\mathrm{tot}} = \frac{1}{\kappa_\mathrm{rad}} + \frac{1}{\kappa_\mathrm{cond}} \tag{4.42}$$

with

$$\mathcal{F}_\mathrm{tot} = -\frac{4ac}{3\kappa_\mathrm{tot}\rho} T^3 \frac{dT}{dr} . \tag{4.43}$$

Note that whichever opacity is the *smaller* of κ_rad or κ_cond it is also the more important in determining the total opacity and hence the heat flow (as in a current flowing through a parallel circuit). In normal stellar material κ_cond is large (conduction is negligible) compared to radiative opacities and, in those situations, only the latter need be considered. The opposite is usually true in dense degenerate material.

The diffusion coefficient, $\mathcal{D}_\mathrm{e}$, has the general form (see, e.g., Kittel 1968) $\mathcal{D}_\mathrm{e} \approx c_V v_\mathrm{e} \lambda/3$ where c_V is the specific heat at constant volume of the degenerate electrons, v_e is some typical (or relevant) electron velocity, and λ is an electron collisional mean free path. In the following, we shall derive the diffusion coefficient for nonrelativistic electrons.

The specific heat of a nearly completely degenerate electron gas was given by equation (3.108) in the last chapter. The momentum parameter x_f in that equation is very much less than unity for nonrelativistic electrons so that

$$c_V \approx \frac{8\pi^3 m_e^2 c}{3h^3} k^2 T x_f \quad \mathrm{erg\ cm^{-3}\ K^{-1}} \tag{4.44}$$

where a factor of $1/\rho$ has been deleted from (3.108) to convert to the indicated units. Since $x \propto (\rho/\mu_\mathrm{e})^{1/3}$ from (3.49) is a good approximation to x_f at low temperatures, it is easy to see that $c_V \propto (\rho/\mu_\mathrm{e})^{1/3} T$.

For v_e and λ, we must recall an important fact of degenerate life: any collisional process involving a degenerate electron cannot result in that electron being scattered into an already filled energy state. What this means is that only electrons near the top of the Fermi sea can participate effectively in the conduction process. Thus the velocity v_e should satisfy $m_\mathrm{e} v_\mathrm{e} \approx p_F \propto x \propto (\rho/\mu_\mathrm{e})^{1/3}$. The most efficient means of scattering these electrons is via Coulomb interactions with the surrounding ion gas. Thus write $\lambda = 1/(\sigma_\mathrm{C} n_\mathrm{I})$ where n_I is the ion number density and σ_C is the Coulomb scattering cross section. A typical way to estimate σ_C is to consider what electron-ion impact parameters result in a "significant" degree of scattering. Following arguments similar to those in Spitzer (1962), an encounter in which the electron kinetic energy is about the same as the electron-ion electrostatic potential will result in a significant scatter. Thus consider electrons for which $m_\mathrm{e} v_\mathrm{e}^2 \approx Z_c e^2/s$. The significant impact parameter is then $s \propto 1/v_\mathrm{e}^2 \propto (\rho/\mu_\mathrm{e})^{-2/3}$. The cross section, in simplest terms, is $\sigma_\mathrm{C} \approx \pi s^2 \propto (\rho/\mu_\mathrm{e})^{-4/3}$. Thus $\lambda \propto (\rho/\mu_\mathrm{e})^{4/3}/n_\mathrm{I}$ or, after introducing the ion mean molecular weight μ_I, $\lambda \propto (\rho/\mu_\mathrm{e})^{4/3}(\mu_\mathrm{I}/\rho)$, and $\mathcal{D}_\mathrm{e} \propto (\mu_\mathrm{I}/{\mu_\mathrm{e}}^2)\,\rho T$. Inserting these results into (4.40) and accounting for all the numerical factors previously ignored, we find that the conductive opacity is

$$\kappa_\mathrm{cond} \approx 4 \times 10^{-8} \frac{{\mu_\mathrm{e}}^2}{\mu_\mathrm{I}} Z_c^2 \left(\frac{T}{\rho}\right)^2 \quad \mathrm{cm}^2\ \mathrm{g}^{-1} \tag{4.45}$$

(and remember that Z_c is the ion charge). As crude as this derivation has been, the final result is not so bad when compared to accurate calculations. The temperature and density exponents are about right ($s \approx n \approx -2$) and the coefficient is correct to within an order of magnitude (or so).

As an example of where conductive opacities are important, consider the deep interior of a typical cool white dwarf with $\rho \approx 10^6$ g cm^{-3}, $T \approx 10^7$ K, and a composition of carbon (which is close enough). The results of the last chapter imply that the gas is certainly degenerate and the material pressure ionized. This implies that the radiative opacity is electron scattering with $\kappa_\mathrm{e} \approx 0.2$ cm^2 g^{-1}. Equation (4.45) yields $\kappa_\mathrm{cond} \approx 5 \times 10^{-5}$ cm^2 g^{-1} with $\mu_\mathrm{e} = 2$, $\mu_\mathrm{I} = 12$, and $Z_c = 6$. Because $\kappa_\mathrm{cond} \ll \kappa_\mathrm{rad}$, the total opacity is $\kappa_\mathrm{tot} \approx \kappa_\mathrm{cond}$ after applying (4.42).

4.6 Tabulated Opacities

As has been emphasized repeatedly here, modern stellar structure and evolution studies almost never use the simple kinds of expressions quoted here for opacities expect, perhaps, for pedagogic purposes. In practice, extensive tables or, sometimes, analytic fits to these tables[3] are used that give radiative and conductive opacities over wide ranges of temperature and density for various compositions of interest. A specific opacity is obtained by a multidimensional interpolation in tables. Sources for such tables are listed

[3]One such fit is used in the ZAMS model maker `ZAMS.FOR` of Appendix C.

in the end of chapter references. Note that despite all the effort that has gone into the calculation of stellar opacities, it is commonly conceded that they are far from perfect. For some purposes in which the minor details are of importance, such as for variable star calculations, it may be said that they are adequate—albeit barely in some instances.

Figures 4.2 and 4.3 show two sets of older radiative opacities from LANL (both from Cox and Tabor 1976) plotted as functions of temperature and density for two different compositions.[4] The first is the "King IVa" set in which the composition is $X = 0.70$, $Y = 0.28$, and $Z = 0.02$ with a solar mix of metals (see Chapter 8). The name "King IVa" given to the tabulated set of opacities from which this figure was generated means that it was the IVth (sub a) set requested by David King (University of New Mexico). This kind of nomenclature is often found in the published LANL tables. (These opacities were used by King in his, and his collaborators', studies of Cepheid variables.) The most obvious feature is the pronounced hump around $T \approx 10^4$ K. As density increases, the location of the peak of the hump moves out to slightly higher temperatures. This behavior reflects the temperature versus density relation of the Saha equation for the half-ionization point of hydrogen as given by (3.39) and that relation is indicated on the figure. (The relation should actually be modified because the mixture used for the figure is not pure hydrogen but it is close enough.) The sharp drop in opacity to the left of $T \approx 10^4$ K signals the demise of free-free and bound-free transitions as hydrogen becomes neutral and the radiation field cools to lower energies, but H^- prevents the opacity from disappearing altogether. Features at higher temperatures include the effects of first and second helium ionization, which can be detected as mild increases in opacity at temperatures a little over 10^4 K and near 10^5 K at the lower densities. These features, although seemingly minor, are important for many variable stars, which are "driven" by helium ionization (see Chapter 10). Apart from such irregularities, the opacities follow a Kramers' law and fall off in temperature until high temperatures are reached whereupon electron scattering takes over ($\kappa_e \approx 0.34$). At the highest temperatures, the opacity dips below the Thompson scattering level and this is due to relativistic effects.

Figure 4.2 also illustrates some problems with using tabulated opacities. First of all, they do not completely cover the temperature–density plane (that would be impossible) but rather include just enough information to be of use for modeling certain classes of stars. If you wish to study stars whose properties are very different from those for which the given table was computed, then you have to extend the table or make a new one. Never extrapolate off a table (if possible). Secondly, you will note that the lines

[4]We only show the LANL opacities. A new set from Lawrence Livermore National Laboratory is also available and you may wish to compare the two. See Rogers and Iglesias (1992) for further details.

in the figure do not always look smooth—they are not—and we have made no attempt at smoothing but have just connected the tabulated points by straight lines. This is where intelligent interpolation is needed.

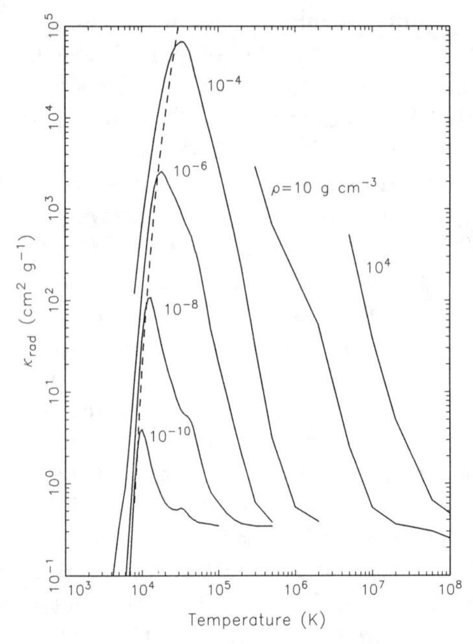

FIGURE 4.2. Plotted are the LANL radiative opacities for the King IVa mixture $X = 0.7$, $Y = 0.28$, and $Z = 0.02$. The mix of metals comprising Z corresponds to those seen in the solar atmosphere. Material for this figure comes from the tabulations of Cox and Tabor (1976). The dashed line shows the half-ionization curve for pure hydrogen.

The second figure shows the results for an almost pure helium mix (with $X = 0$, $Y = 0.97$, $Z = 0.03$) opacity set requested by Morris Aizenman (National Science Foundation). (This is the Aizenman IV table and it has been used in modeling the deep interiors of evolved stars.) The scales are the same as in the first figure. Here the first and second helium ionization stages are well-marked by the double-humped peaks. Also note that the opacities are about an order of magnitude lower (for a given T and ρ) than the hydrogen-rich mixture before the electron scattering threshold is reached. Apart from the difference in Z and the possible effects of conduction, both of which we ignore for the moment, you could use the radiative opacities in both figures to compute opacities in an evolutionary sequence starting from the ZAMS and continuing on to hydrogen exhaustion. Interpolation (usually logarithmic) would be used to find the opacity as a function of, say, Y. Note, however, that you might expect problems in interpolation because of the large differences in opacity between the two tabulated mixes. The correct procedure would be to calculate tables for intermediate values of Y

but, even in these days of the supercomputer, this can be expensive.

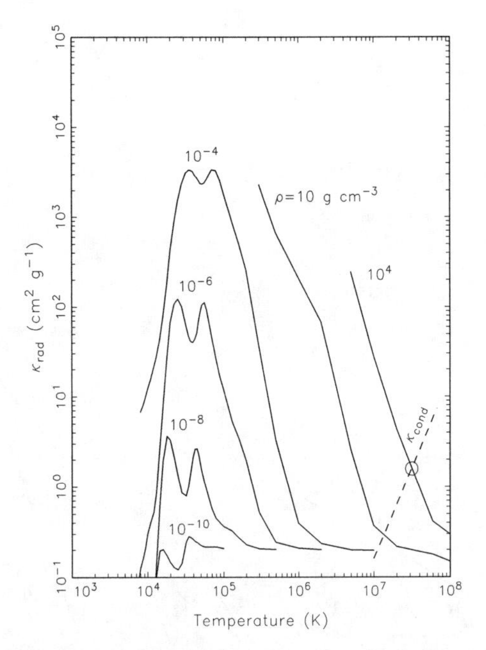

FIGURE 4.3. These are the radiative opacities for the helium-rich Aizenman IV mix $X = 0$, $Y = 0.97$, and $Z = 0.03$ from Cox and Tabor (1976). Also shown (as a dashed line) is the conductive opacity for $\log \rho$ of +4 from (4.45).

In Figure 4.3 we also show the conductive opacity for pure helium at a density of 10^4 g cm^{-3} (from eq. 4.45). The intersection of this opacity with the radiative opacity at the same density and temperature is indicated by a circle. If density is kept fixed, then a reduction in temperature causes κ_{cond} to decrease as T^2 (see 4.45). At the same time the radiative opacity increases (see figure) so that $\kappa_{\text{cond}} < \kappa_{\text{rad}}$. This means, in accordance with our previous arguments, that the total opacity becomes more like the conductive opacity and the radiative opacity begins not to count. Conversely, a rise in temperature makes κ_{rad} more important. Thus if the *total* opacity were plotted on the figure, the opacity contours would be very different in some regions of ρ and T and especially where densities are high and temperatures are low.

Figure 4.4 illustrates part of the point just made. This figure, from Hayashi, Hōshi, and Sugimoto (1962), shows what regions of the $\log \rho$-$\log T$ plane are dominated by various kinds of opacity. (The composition is typical Population I.) The line labeled $\psi = 0$ denotes the onset of degeneracy and, as you may verify, corresponds roughly to the transition line where conduction takes over.

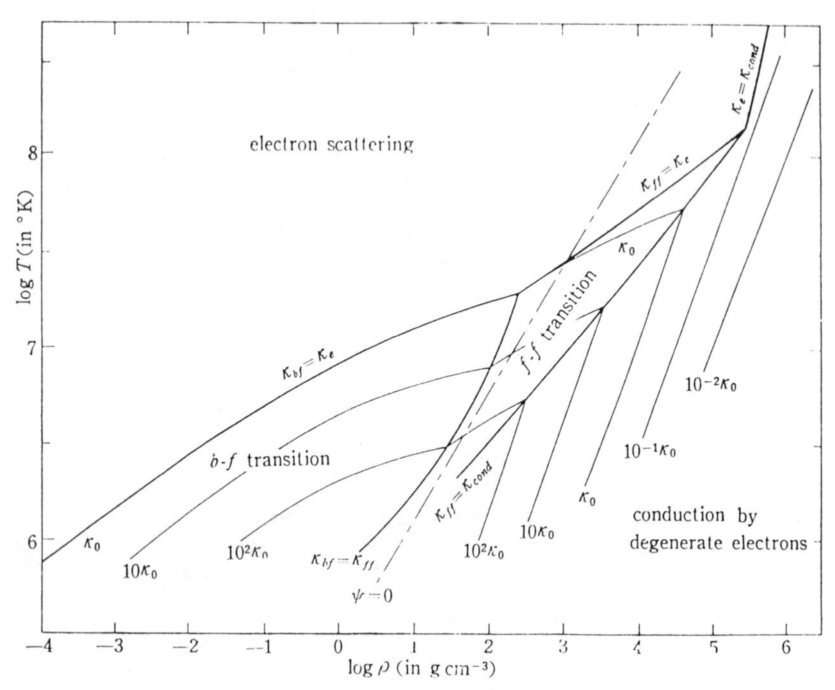

FIGURE 4.4. This figure, from Hayashi *et al.* (1962), illustrates where various opacities are most important as functions of temperature and density. The mixture is Population I. The opacity nomenclature is almost the same as in the text except that the lines are labeled in units of the electron scattering opacity here denoted by $\kappa_0 (= \kappa_e) = 0.2(1 + X)$. Reproduced with permission.

4.7 References and Suggested Readings

Introductory Remarks and §**4.1**: Radiative Transfer

We recommend the texts

▹ Mihalas, D. 1978, *Stellar Atmospheres*, 2d ed. (San Francisco: Freeman)

▹ Mihalas, D., and Mihalas, B.W. 1984, *Foundations of Radiative Hydrodynamics* (Oxford: Oxford University Press).

The emphasis of these two is different, but complementary, and both contain modern and practical material. Chapters 4–8 of

▹ Cox, J.P. 1968, *Principles of Stellar Structure*, in two volumes (New York: Gordon and Breach)

discusses stellar atmospheres more from the viewpoint of applications to stellar interiors than do the Mihalas references.

▹ Rybicki, G.B., and Lightman, A.P. 1979, *Radiative Processes in Astrophysics* (New York: Wiley & Sons)

also contains useful material.

§**4.4**: Radiative Opacity Sources

The material in

▹ Clayton, D.D. 1968 *Principles of Stellar Evolution and Nucleosynthesis*, (New York: McGraw-Hill)

is presented from a physicist's point of view and we recommend it highly. Cox, J.P. 1968, *Principles of Stellar Structure*, bases his exposition primarily on the LANL method of calculating opacities. The LANL method (in an older but still good discussion) is given by

▹ Cox, A.N. (1965) in Chapter 3 of *Stellar Structure*, Eds. L.H. Aller and D.B. McLaughlin (Chicago: University of Chicago Press).

The reference to

▹ Landau, L.D., and Lifshitz, E.M. 1971, *Classical Theory of Fields* (Oxford: Pergamon Press)

can be found, as with other volumes in this classic series, in more recent editions. The monograph by

▹ Spitzer, L. 1962, *Physics of Fully Ionized Gases* 2d ed. (New York: Interscience)

contains much of interest for the astrophysicist. Another newer work of his is

▹ Spitzer, L. Jr. 1978, *Physical Processes in the Interstellar Medium* (New York: Wiley & Sons).

The text by

▹ Schwarzschild, M. 1958, *Structure and Evolution of the Stars* (Princeton: Princeton University Press)

is counted as the first modern work describing how stars evolve. It is now out of date but still worth perusing.

§4.5: Heat Transfer by Conduction

Several undergraduate solid-state (now "condensed matter") texts give the basic material on thermal conduction by electrons. The text by

▷ Kittel, C. 1968, *Introduction to Solid State Physics* (New York: Wiley & Sons)

(or later editions) is particularly clear.

Conductive opacities are discussed in

▷ Hubbard, W.B., and Lampe, M. 1969, *Ap.J. Suppl.*, **18**, 297

▷ Lamb, D.Q., and Van Horn, H.M. 1975, *Ap.J.*, **200**, 306

▷ Itoh, N., Mitake, S., Iyetomi, H., and Ichimaru, S. 1983, *Ap.J.*, **273**, 774

▷ Itoh, N., Kahyama, Y. Matsumoto, N., and Seki, M. 1984, *Ap.J.*, **285**, 758.

The reference to Spitzer (1962) is given above.

§4.6: Tabulated Opacities

Extensive tabulations of radiative opacities in the literature may be found in

▷ Cox, A.N., and Stewart, J.N. 1970, *Ap.J. Suppl.*, **19**, pp. 243, 261

▷ Cox, A.N., and Tabor, J.E. 1976, *Ap.J. Suppl.*, **31**, 271

▷ Weiss, A., Keady, J.J., and Magee, N.H. Jr. 1990, *Atomic Data and Nuclear Data Tables*, **45**, 209.

You should also consult

▷ Mihalas, D., Hummer, D.G., Mihalas, B.W., and Däppen, W. 1990, *Ap.J.*, **350**, 300

for a discussion of equations of state. The papers in this series (and there will be more of them) describe a project to compute opacities independently of those of LANL. Also in competition with LANL is the group at Lawrence Livermore National Laboratory. You may wish to refer to

▷ Rogers, F.J., and Iglesias, C.A. 1992, *Ap.J. Suppl.*, **79**, 507

who also supply tables. A common result of the new opacities is an increase of opacity over the older LANL calculations due to metallic lines. The consequences of these increases are still being explored.

We shall have other occasions to refer to the classic article by

▷ Hayashi, C., Hōshi, R., and Sugimoto, D. 1962, *Suppl. Progress of Theoretical Physics, Japan*, **22**.

It is now outdated by modern standards but contains a particularly clear development of the ingredients of stellar structure.

You may wish to find out how difficult it is to produce accurate analytic fits to opacities. Some attempts are due to

▹ Christy, R.F. 1966, *Ap.J.*, **144**, 108

▹ Iben, I. Jr. 1975, *Ap.J.*, **196**, 525

▹ Stellingwerf, R.F. 1975, *Ap.J.*, **195**, 441

and

▹ Stellingwerf, R.F. 1975, *Ap.J.*, **199**, 705

contains a footnote correction. The last two fits are used in the ZAMS computer code `ZAMS` of Appendix C.

5

Heat Transfer by Convection

Double, double toil and trouble;
Fire burn and cauldron bubble.
— W. Shakespeare (*Macbeth*)
That about sums it up.

The major portion of this chapter will be devoted to a discussion of the "mixing length theory," or "MLT," of convective heat transport in stars. Although this theory has many faults, it has served as a useful phenomenological model for a description of stellar convection for more than thirty years. Almost all numerical simulations of stellar evolution use it in one guise or another. Near the end of the chapter we shall briefly discuss alternatives to the MLT and why a realistic description is so difficult.

Our discussion of the MLT will differ considerably from Cox (1968)—where details of the usual derivation of the MLT are the most completely laid out in the textbook literature—and, because of this, the chapter will be rather long. (The connection between the usual treatments and ours shall be made clear, however.) We have chosen this path because we believe that it gives a better insight into the underlying assumptions of the MLT so that you will better appreciate its strengths and weaknesses. We shall also have to state that stellar convection is still one of the major hurdles in modeling stars and will probably remain so for several years to come.[1]

---

[1]We wish to thank John I. Castor for much of the material in this chapter. He distilled a good deal of the work referenced here into a palatable form and gave

5.1 The Mixing Length Theory

The mixing length theory was originally formulated in its "stellar" form by Biermann (1951), Vitense (1953), and Böhm-Vitense (1958) based on earlier 1925 work of Prandtl (see Prandtl 1952). Since then, it has been elaborated on and modified in many ways and one should no longer call it just the MLT without citing exactly which version is being referred to. A "classic" derivation of one version of the theory may be found in Cox (1968, chap. 14).

Before giving this text's version of the MLT, it is best to attempt a brief description of the general idea behind the theory. Imagine that the stellar fluid is composed of readily identifiable "eddies," "parcels," or "elements" (or, in more colloquial terms, "bubbles" and "blobs"), which can move from regions of high heat content to regions of lower heat content, or conversely; that is, they are capable of transporting or convecting heat through the fluid. These parcels arise from unspecified instabilities in the fluid but have properties not drastically different from their surroundings. If conditions are ripe, then buoyancy effects cause the parcels to, say, rise in the star through some characteristic distance ℓ, the *mixing length*, before they lose their identity as separate parcels and break up and merge with the surrounding fluid. As they rise, they maintain pressure equilibrium with their surroundings. Since these particular parcels start their rise in an environment having a higher heat content (higher temperature) than where they break up, heat is thereby transported from the starting position up to the level at the additional height ℓ. To complicate matters, the parcel may release heat to its surroundings as it rises. At the same time, cool parcels at a higher level sink a distance ℓ, and they too break up. The net effect is heat transport directed outward in the star. The rate of transfer is established by the parcel formation rate, velocity (w) of rise, ℓ, the heat content of the star as a function of depth, and by how "leaky" the parcels are as they rise. This sounds relatively simple and, in fact, it is—in the context of MLT. We shall also see that most formulations of the theory have a major virtue for computation: all that matters is that temperature, density, and other stellar quantities be known at a single radius of interest. If so, then a convective heat flux may be computed at that point. The MLT is thus a *local* theory.

The sequence we shall follow in discussing the MLT is first to derive the criterion for buoyancy, and then to estimate the heat leakage from a parcel. This will give us the equations of motion. Finally, we shall find expressions for the convective flux and discuss how they are used.[2]

---

it as part of a course in stellar atmospheres at the University of Colorado in the early 1970s.

[2]Some of you may wish to short-circuit much of the following material to get to an end result and then return to fill in the details. To do so, browse through

5.1.1 Criteria for Convection

First of all, certain general assumptions are made that should be explicitly set forth. Besides neglecting magnetic fields, rotation, and the like, we assume the following (with comments).

1. A readily identifiable parcel has a characteristic dimension of the same order of size as the mixing length ℓ.

2. The mixing length is much shorter than any scale length associated with the structure of the star. Examples of such lengths are the pressure scale height, λ_P of (3.1), and similar scale heights for temperature and density. Another is the stellar radius or some nonnegligible fraction thereof.

3. The parcel always has the same internal pressure as that of its surroundings. This means that however the convective processes work, the time scales associated with them are always long enough that pressure equilibrium is maintained. Thus, for example, if v_s is the local sound speed in the parcel, then the sound traversal time across the parcel, ℓ/v_s, is short compared to, say, the ascent or descent time of the parcel through the distance ℓ.

4. Acoustic phenomena may be ignored altogether, as may shocks, etc.

5. Temperatures and densities within and outside a parcel differ by only a small amount.

The combination of these assumptions constitutes the "Boussinesq" approximation. What it implies is that the fluid is *almost* incompressible and that density variations (which may give rise to buoyancy effects) and temperature variations in the fluid are very small. The Boussinesq approximation usually works very well in the laboratory, where scale heights are large compared to container sizes (which roughly set the maximum size of a convective cell). In its application to stars, however, we shall see that the mixing length must be near the size of λ_P or one of the other scale heights for reasonable results to be obtained. Thus, in practice, the MLT will turn out to violate one of its internal assumptions. Furthermore, it is unfortunate that laboratory-derived constraints on the MLT are essentially nonexistent because of the following. The dimensionless Rayleigh number **Ra** (see later) associated with laboratory fluids is usually less than 10^{11} but stellar convection is characterized by high values, $\mathbf{Ra} \sim 10^{20}$—give or take a few orders of magnitude. The same situation applies to the Prandtl

---

§5.1.1 and then skip to §5.1.5 where, after some preliminary discussion, equation (5.50) gives an expression for the convective flux in the case where fluid motions are adiabatic. The quantity $\mathcal{Q}$ in (5.50) is of order unity and the various "dels" have been defined in §4.3.

number, **Pr**, where, in stars, **Pr** $\sim 10^{-9}$, but in the lab, **Pr** ~ 1. Note also, in passing, the troublesome consequences of (1) in the above: how can the parcel get anywhere if its dimensions are of the same order as the distance it travels (ℓ)?

With the above alerts in mind, consider a plane parallel fluid under the influence of gravity where z measures the height up through the fluid. Inside a typical parcel created by some unspecified process, denote the interior temperature, pressure, and density by T', P, and ρ', respectively. Outside the parcel, the corresponding quantities are denoted by T, P, and ρ. Note that the pressures inside and out are the same by virtue of assumption (3). Suppose that $T' > T$ (but not by much) so that the parcel is hotter than its surroundings. Normally this implies that $\rho' < \rho$ because of the interior versus exterior pressure equilibration. If the volume of the parcel is $V \sim \ell^3$, then Archimedes' principle states that the parcel will experience a net upward buoyancy force of

$$\rho V g - \rho' V g \tag{5.1}$$

where g is the local gravity. Note that we have not specified exactly what the volume of the parcel is in terms of ℓ. It could be spherical ($4\pi[\ell/2]^3/3$), a cube (ℓ^3), or what have you. These fine distinctions involving constants of order unity give rise to some of the variants in mixing length theory and we shall ignore them. In any case, the parcel now commences to rise.

We must now determine what is the mean velocity of the parcel as it rises through the mixing length distance, ℓ, and what is its temperature compared to the ambient temperature when it merges into the surrounding fluid. The latter comparison will tell us how much energy the parcel will release when it loses its identity. For all this we shall need information about temperature gradients.

We denote β to be the *negative* of the ambient temperature gradient

$$\beta = -\frac{dT}{dz} \tag{5.2}$$

where (almost always) $\beta > 0$. This gradient is assumed known despite the fact that heat transported by rising and descending parcels may very well establish just what that gradient is. We can relate β to other known quantities by observing that

$$\frac{dT}{dz} = \frac{dT}{dP}\frac{dP}{dz} = T\frac{d\ln T}{d\ln P}\frac{d\ln P}{dz} = -\frac{T}{\lambda_P}\nabla = -\beta \tag{5.3}$$

where ∇ is the "actual del" introduced in the preceding chapter as equation (4.28). The pressure scale height in the above may be recast in terms of the local sound speed with the aid of (1.35):

$$\lambda_\text{P} = -\left(\frac{d\ln P}{dz}\right)^{-1} = \frac{P}{g\rho} = \frac{v_\text{s}^2}{g\Gamma_1} \tag{5.4}$$

where Γ_1 is the adiabatic exponent defined by (3.87). Thus

$$\beta = \frac{T}{\lambda_P}\nabla = \frac{g\Gamma_1 T}{v_s^2}\nabla. \tag{5.5}$$

To describe how the temperature inside the parcel varies as the parcel rises, first write

$$\frac{dT'}{dz} = T'\frac{d\ln T'}{d\ln P'}\frac{d\ln P'}{dz}.$$

If we assume, as a start, that the rising parcel exchanges no heat with its surroundings, then the term $d\ln T'/d\ln P'$ must describe adiabatic variations of temperature with pressure. This is the thermodynamic derivative $\nabla_{\rm ad} = (d\ln T/d\ln P)_{\rm ad}$ introduced earlier (as in eqs. 3.88 and 3.90). Because all fluctuations are assumed to be small, it is appropriate to replace the lone factor of T' by T in the right-hand side of the above. (We cannot do the same with the temperature derivatives because they will drive the motions.) In addition, we replace P' with P, because of pressure equilibration, so that the last factor may be turned into a pressure scale height. Finally we may append an "ad" subscript to dT'/dz because of the adiabaticity assumption and write

$$\left(\frac{dT'}{dz}\right)_{\rm ad} = -\frac{T}{\lambda_P}\nabla_{\rm ad} = -\beta_{\rm ad}. \tag{5.6}$$

The question is now whether the parcel, once having commenced to rise adiabatically, will continue to rise. It may well be that as the parcel rises to greater heights and its internal pressure drops, its interior temperature may also have decreased adiabatically to a level where the parcel is cooler than its surroundings. In that case it has negative buoyancy and it tends to sink back down before transversing a mixing length. It is the other possibility, in which the parcel's temperature continues to exceed that of the surroundings, that is of major interest. Here, the fluid is said to be *convectively unstable* and the perturbation that causes the parcel to rise takes place in an environment that encourages further rising until a mixing length is traversed. This latter condition may be expressed as follows.

First observe that both dT'/dz and dT/dz are assumed to be *negative*. Thus the condition that T' decreases *more slowly* than T with height is expressed as

$$\left(\frac{dT'}{dz}\right)_{\rm ad} > \left(\frac{dT}{dz}\right) \quad \text{(convectively unstable).} \tag{5.7}$$

This convectively unstable situation is illustrated in Figure 5.1. Another way to express this is to use (5.2), (5.5), and (5.6) to write

$$\beta > \beta_{\rm ad} \quad \text{(convectively unstable), or} \tag{5.8}$$

$$\nabla > \nabla_{\rm ad} \quad \text{(convectively unstable).} \tag{5.9}$$

The three conditions (5.7–5.9) are equivalent and, in the stellar context, are often called the *Schwarzschild criteria* (K. Schwarzschild 1906). These are *local* criteria and thus require information from only the one height (or radius) of interest in the star.

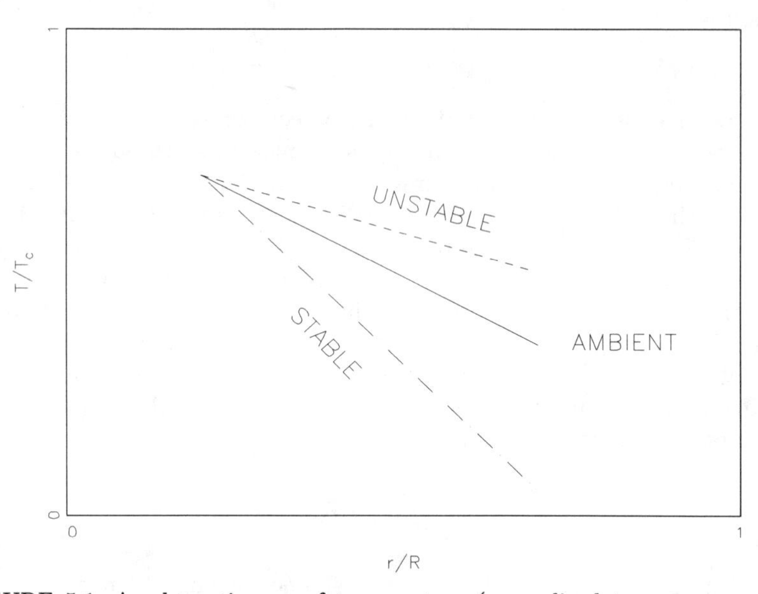

FIGURE 5.1. A schematic run of temperature (normalized to some common temperature T_c) versus radius for: (a) the ambient medium (solid line); (b) a convectively unstable parcel (dashed line) with $(dT'/dr)_{\rm ad} > (dT/dr)$; (c) a stable parcel (dashed-dotted line) with $(dT'/dr)_{\rm ad} < (dT/dr)$.

The criteria (5.7–5.9) are also equivalent to the statement that if entropy decreases outward at some point ($dS/dr < 0$), then the fluid is convectively unstable (Cox 1968, §13.4 or our §7.3.3). Put another way, convection does not take place in hydrostatic stars where the entropy increases outward. It will turn out that in regions where convection is very efficient, ∇ is only very slightly greater than $\nabla_{\rm ad}$. In such regions the entropy is very nearly constant with height.

We finally begin to see what role the "dels" play in convection. If, by some mischance, adiabatic perturbations arise where the local run of temperature versus pressure is such that the local value of ∇ is greater than the local thermodynamic equivalent, $\nabla_{\rm ad}$, then convection should be present. However, if this is really so, then convection must change the thermal structure and, hence, ∇. We shall soon explore what this implies. First, however, an examination of the assumption of adiabaticity.

5.1.2 *Radiative Leakage*

Real life is never really adiabatic. Because our parcel is either cooler or hotter than its surroundings, heat must be exchanged between the two. Thus consider the energy equation

$$\frac{dQ'}{dt} = -\nabla \bullet \mathcal{F}_{\text{rad}} \tag{5.10}$$

where Q' is the heat content per unit volume in the parcel and $\mathcal{F}_{\text{rad}}$ is the radiant flux from the parcel out to the ambient medium. (Don't confuse the gradient or divergence operators used here and the various "dels.") Now recall that the Boussinesq conditions of no acoustic phenomena and small fluctuations imply that the only time density perturbations are to be taken into account is when they are coupled to gravity to cause buoyancy. Thus in considering heat balance, $P\,dV$ work terms are neglected so that (5.10) reduces to

$$\left(\frac{\partial T'}{\partial t}\right) = -\frac{1}{\rho c_P} \nabla \bullet \mathcal{F}_{\text{rad}} \tag{5.11}$$

where c_P is in erg g^{-1} K^{-1}.

For the radiative flux we again choose a diffusion approximation

$$\mathcal{F}_{\text{rad}} = -\mathcal{K}\nabla T \tag{5.12}$$

where the diffusion constant $\mathcal{K}$ is assumed constant over the mixing length ℓ. The gradient term on the right-hand side of this equation and subsequent right-hand sides of the energy equation will be taken to be some linear combination of T and T' (as will be done shortly). We now introduce the opacity by using a relation identical to that used in the calculation of conductive opacities (see 4.40); namely,

$$\mathcal{K} = \frac{4acT^3}{3\kappa\rho} . \tag{5.13}$$

Equation (5.11) then becomes

$$\left(\frac{\partial T'}{\partial t}\right) = \frac{\mathcal{K}}{\rho c_P} \nabla^2 T = \frac{4acT^3}{3\kappa\rho^2 c_P} \nabla^2 T. \tag{5.14}$$

The ratio $\mathcal{K}/\left(\rho c_P\right)$ is the thermal diffusivity (or conductivity), ν_T, with

$$\nu_T = \frac{4acT^3}{3\kappa\rho^2 c_P} \tag{5.15}$$

and it has the units of cm^2 s^{-1}. For the characteristic length in these units choose ℓ itself. Therefore a characteristic radiative cooling time associated with radiation from, or into, the parcel is ℓ^2/ν_T. The energy equation then becomes

$$\left(\frac{\partial T'}{\partial t}\right) = \nu_T \nabla^2 T \tag{5.16}$$

and, in this form, it is sometimes known as Fourier's equation (see Landau and Lifshitz 1959, §50).

Since we have no way to determine the precise structure of our parcel, we should model $\nabla^2 T$ as simply as we can. The obvious way is to replace it by $(T-T')/\ell^2$. The order of the temperatures is correct because, for example, if $T' > T$ then $\nabla^2 T$ and $(\partial T'/\partial t)$ are both negative as they should be: the parcel loses heat because it is hotter than its surroundings.

At this point we have to amend the energy equation somewhat because, as it stands, it is in Eulerian form and it describes the change in heat content at a fixed position. To follow what happens in the parcel as it moves, convert to Lagrangian coordinates by the following well-known transformation, which is discussed in any text in hydrodynamics (and see eq. 10.70):

$$\frac{DT'}{Dt} = \left(\frac{\partial T'}{\partial t}\right) + \mathbf{w} \bullet \nabla T' \tag{5.17}$$

where D denotes the Lagrangian (or Stokes') operator and $\mathbf{w}$ is the (vector) parcel velocity. Since only vertical movement is contemplated here, $\mathbf{w} \bullet \nabla T' = w(\partial T'/\partial z)$. The term $\partial T'/\partial t$ in (5.17) takes care of the instantaneous heat loss so that the remaining advective term, $w\,(\partial T'/\partial z)$, should describe how T' behaves without such loses. Thus we identify the derivative in the last term as

$$\left(\frac{\partial T'}{\partial z}\right) = \left(\frac{dT'}{dz}\right)_{\text{ad}} = -\beta_{\text{ad}}$$

from (5.6), and (5.17) becomes

$$\frac{DT'}{Dt} = \frac{\nu_T}{\ell^2}(T - T') - \beta_{\text{ad}} w \tag{5.18}$$

after replacing the Laplacian in (5.16) by its numerical difference analogue.

Now compare T to T' by introducing

$$\Delta T = T' - T \tag{5.19}$$

which depends on z and t. By combining (5.18) and (5.19) and realizing that the vertical rate of change of the ambient temperature as seen by the moving parcel is

$$\frac{DT}{Dt} = \frac{dT}{dz}\frac{dz}{dt} = -\beta w$$

we arrive at

$$\frac{D\Delta T}{Dt} = (\beta - \beta_{\text{ad}})\, w - \frac{\nu_T}{\ell^2}\Delta T\,. \tag{5.20}$$

This equation describes the time-dependent temperature contrast between the parcel and its immediate surroundings as the parcel moves.

5.1.3 The Equation of Motion

Can we now say something about w? Because buoyancy forces are responsible for the motion of the parcel,

$$\frac{dw}{dt} = \frac{(\rho - \rho')}{\rho} g \tag{5.21}$$

in a first approximation to the acceleration implicit in (5.1). To interject a little more realism, however, introduce a viscous drag on the parcel as follows. In the Navier-Stokes equation of motion for an incompressible fluid (which is consistent with the Boussinesq approximation) we find a drag term $\nu \nabla^2 \mathbf{w}$, where ν is the kinematic viscosity having the same units as ν_T. (See, for example, Landau and Lifshitz 1959, §15.) We replace the Laplacian by w/ℓ^2 and amend (5.21) to find

$$\frac{dw}{dt} = \frac{(\rho - \rho')}{\rho} g - \frac{\nu}{\ell^2} w. \tag{5.22}$$

The additional term always acts to decelerate the parcel. We wish now to convert this equation of motion into one containing temperatures and not densities.

The small relative density contrast $(\rho - \rho')/\rho$ is related to that in temperature through the coefficient of thermal expansion, $-\mathcal{Q}$, taken at constant pressure; that is,

$$-\mathcal{Q} = \left(\frac{d \ln \rho}{d \ln T} \right)_P .$$

(The constraint is required by pressure equilibrium.) If the density is written as a function of temperature, pressure, and composition (denoted by μ) with $\rho = \rho(T, P, \mu)$, then $-\mathcal{Q}$ is

$$-\mathcal{Q} = \left(\frac{\partial \ln \rho}{\partial \ln T} \right)_{\mu,P} + \left(\frac{\partial \ln \rho}{\partial \ln \mu} \right)_{P,T} \left(\frac{\partial \ln \mu}{\partial \ln T} \right)_P . \tag{5.23}$$

It follows that

$$\frac{(\rho - \rho')}{\rho} = -\mathcal{Q} \frac{(T - T')}{T} = \frac{\mathcal{Q}}{T} \Delta T \tag{5.24}$$

is the desired relation.

The coefficient $\mathcal{Q}$ is generally clumsy to compute because of the composition dependence, which should be kept in to allow for composition changes either within the parcel (such as ionization) or in the surroundings. If composition changes may be neglected, then it is easy to show that

$$\mathcal{Q} = \frac{\chi_T}{\chi_\rho} \tag{5.25}$$

where χ_T and χ_ρ were discussed in Chapter 3 (§3.7.1). For a fluid composed of a mixture of ideal gas and radiation, $\mathcal{Q} = (4 - 3\beta)/\beta$ where β is the ratio of gas to total pressure (as in Eqs. 3.100 and 3.101).

The equation of motion is then

$$\frac{dw}{dt} = \frac{Qg}{T}\Delta T - \frac{\nu w}{\ell^2}, \tag{5.26}$$

which is to be considered along with the energy equation (5.20).

5.1.4 Convective Efficiencies and Time Scales

Neutral stability against convection applies when both $D\Delta T/Dt$ and dw/dt are zero. Setting these conditions for (5.20) and (5.26) yields the criterion for neutral stability

$$\mathbf{Ra} = \frac{Q}{T} g\left(\beta - \beta_{\text{ad}}\right) \frac{\ell^4}{\nu_T \nu} = 1\,. \tag{5.27}$$

The dimensionless quantity **Ra** is the Rayleigh number and it is a measure of how well the driving of convection, indicated by terms such as $g\left(\beta - \beta_{\text{ad}}\right)$, compares to damping processes (ν_T and ν). We shall demonstrate shortly that if $\mathbf{Ra} > 1$, then both $D\Delta T/Dt$ and dw/dt have exponentially growing solutions. That is, a parcel is accelerated and becomes (even) hotter than its surroundings and thus any fluctuation is amplified in importance. In the laboratory, nature seems to require a critical Rayleigh number of about 10^3 before convection sets in. We shall see shortly that convective regions (or "zones") in stars have much larger Rayleigh numbers than this.

The other dimensionless number mentioned previously was the Prandtl number, which is defined as

$$\mathbf{Pr} = \frac{\nu}{\nu_T}\,. \tag{5.28}$$

It measures the relative importance of viscous damping effects to those of thermal diffusivity and is a property of the material but not the flow. Earlier we claimed that this number was very small in stars. To confirm this, consider a fully ionized ideal gas. The kinematic viscosity is estimated by Spitzer (1962) as

$$\nu \approx \frac{2 \times 10^{-15}}{\rho} \frac{T^{5/2} A^{1/2}}{Z^4 \ln \Lambda_c} \quad \text{cm}^2\ \text{s}^{-1} \tag{5.29}$$

where

$$\Lambda_c \approx 10^4 \frac{T^{3/2}}{n_e^{1/2}} \tag{5.30}$$

and A and Z are the atomic weight and charge of the ions. To compute a representative value use typical values from the sun of $T_\odot \approx 10^7$ K, $\rho_\odot \approx A \approx Z \approx 1$, and find that $\nu \sim 100$ cm^2 s^{-1}. A similar calculation for the thermal diffusivity yields (using an ideal gas and electron scattering opacity) $\nu_T \sim 10^{10}$ cm^2 s^{-1} from which we infer that $\mathbf{Pr} \sim 10^{-8}$.

For the Rayleigh number we need ℓ and $\beta - \beta_{\rm ad}$. Let the former be $\ell \sim \mathcal{R}_\odot/100$ (give or take an order of magnitude) and, for the latter, $|\beta - \beta_{\rm ad}| \sim \beta \sim T_\odot/\mathcal{R}_\odot$. Also let $\mathcal{Q} \approx 1$ and $g \approx G\mathcal{M}_\odot/\mathcal{R}_\odot^2$. This all yields $\mathbf{Ra} \sim 10^{17}$ with considerable uncertainty, but it is very large.

Yet another dimensionless number of interest is the *Péclet* number, defined as $w\ell/\nu_T$. It compares the turbulent diffusivity to the thermal diffusivity. (Incidently, the Reynolds number is defined as $w\ell/\nu$ so that the Péclet number is the product of the Reynolds number and $\mathbf{Pr}$.) Using estimates of w to be found later, the Péclet number is approximately $(\mathbf{Ra} \times \mathbf{Pr})^{1/2}$ if convection is "efficient" but if "inefficient" then it is $\mathbf{Ra} \times \mathbf{Pr}$. The dimensionless product $\mathbf{Ra} \times \mathbf{Pr}$ will appear frequently so we might as well give it a name now:

$$\Lambda \equiv \mathbf{Ra} \times \mathbf{Pr} = \frac{\mathcal{Q} g \ell^4}{T \nu_T^2} \left(\beta - \beta_{\rm ad}\right) . \tag{5.31}$$

We shall find that Λ is a direct measure of the time scale for convection as compared to the time scale for radiative cooling. In that sense it is a measure of convective efficiency (or inefficiency).

Consider the differential equations for ΔT and w given as (5.20) and (5.26). If we suppose that all the coefficients in those equations are constant not only over the distance ℓ but for all time, then it is an easy matter to solve for ΔT and w as functions of time. The solutions will be of the form ΔT or $w \propto \exp(\sigma t)$ where σ is a complex angular frequency. We shall find that $1/\Re(\sigma)$ defines a characteristic time scale for the growth (decay) of convection if $\mathbf{Ra} > 1$ ("$\Re$" means "real part of").

To carry out this analysis, combine the two differential equations into a single second-order equation for either w or ΔT, substitute $\exp(\sigma t)$, and find that σ must satisfy the characteristic equation

$$\sigma^2 + \sigma \frac{\nu_T}{\ell^2} \left[1 + \mathbf{Pr}\right] + \frac{\nu_T \nu}{\ell^4} \left[1 - \mathbf{Ra}\right] = 0 . \tag{5.32}$$

Note first that $\mathbf{Ra} < 0$ implies no convection and we thus consider only positive values of $\mathbf{Ra}$. An inspection of the properties of the roots of this equation reveals the following. If $\mathbf{Ra} < 1$, then the real part of σ is negative. This means that both w and ΔT decrease exponentially with time and the convective dies out. $\mathbf{Ra} > 1$, on the other hand, always yields at least one positive real root and the temperature contrast and velocity both increase exponentially.

We can save ourselves some labor at this point by observing that the first bracket in (5.32) contains the Prandtl number, which is small compared to unity in stars, and we shall neglect it. The second bracket contains $\mathbf{Ra}$, which is large when stellar convection is present, and we thus discard the

"1." We are left with

$$\sigma^2 + \frac{\nu_T}{\ell^2}\sigma - \frac{Q}{T}g\,(\beta - \beta_{\rm ad}) = 0 \tag{5.33}$$

as the equation describing the time development of the convective parcel.

The last term in (5.33) is important not only here but in the theory of variable stars (see Chapter 10), planetary atmospheres, and other fields, and it deserves a name. Define the *Brunt-Väisälä frequency*, N, by

$$N^2 = -\frac{Q}{T}g\,(\beta - \beta_{\rm ad}) \; . \tag{5.34}$$

This may be cast into another form by introducing the expressions for Q of (5.23) or (5.25) and converting βs to ∇s. In the simplest case of no composition gradients N^2 becomes

$$N^2 = -\frac{\chi_T}{\chi_\rho}\left(\nabla - \nabla_{\rm ad}\right)\frac{g}{\lambda_{\rm P}} \; . \tag{5.35}$$

If there is a gradient in the mean molecular weight because the composition of nuclear species changes with height, then the parenthesis should contain an additional term

$$+\frac{\chi_\mu}{\chi_T}\frac{d\ln\mu}{d\ln P} \quad \text{with } \chi_\mu = \left(\frac{\partial \ln P}{\partial \ln \mu}\right)_{\rho,T} .$$

Note that the effects of ionization on μ are already automatically accounted for through the equation of state and (5.35) should not be modified if μ changes solely due to those effects. (See Cox 1968, §13.3 for an analysis.)

Related to N^2 is the *Schwarzschild discriminant*

$$A(r) = \frac{d\ln\rho}{dr} - \frac{1}{\Gamma_1}\frac{d\ln P}{dr} \tag{5.36}$$

where the logarithmic derivatives are over the actual run of ambient pressure and density in the star. With a little effort, this can be converted into

$$A = \frac{\chi_T}{\chi_\rho}\left(\nabla - \nabla_{\rm ad}\right)\frac{1}{\lambda_{\rm P}} \tag{5.37}$$

if composition gradients are again not present. It should be clear from (5.9) that either $N^2 < 0$ or $A > 0$ implies convective instability. The relation between N^2 and A is

$$N^2 = -Ag = -\Sigma^2 \tag{5.38}$$

where Σ has been introduced as a convenience. Because the Brunt–Väisälä frequency contains direct dynamic information concerning whether convection is present or not, then $1/|N|$ should be directly related to a convective time scale (as we shall see).

The convective efficiency, Λ, of (5.31) is now

$$\Lambda = -\frac{N^2\ell^4}{\nu_T^2} = \frac{\Sigma^2\ell^4}{\nu_T^2} = \frac{Qg}{\lambda_P}\left(\nabla - \nabla_{\rm ad}\right)\frac{\ell^4}{\nu_T^2} \tag{5.39}$$

and the quadratic for σ of (5.33) becomes

$$\sigma^2 + \frac{\nu_T}{\ell^2}\sigma - \Sigma^2 = 0, \quad \text{or} \tag{5.40}$$

$$\left(\frac{\sigma}{\Sigma}\right)^2 + \frac{1}{\Lambda^{1/2}}\left(\frac{\sigma}{\Sigma}\right) - 1 = 0\,. \tag{5.41}$$

The roots of the characteristic equation are now

$$\frac{\sigma}{\Sigma} = -\frac{1}{2\Lambda^{1/2}} + \frac{1}{2}\left(\frac{1}{\Lambda} + 4\right)^{1/2} \tag{5.42}$$

where the plus sign has been chosen so that with $\Lambda > 0$ (convective instability) σ is real and positive and both w and ΔT have positive exponential growth rates.

Since ℓ^4/ν_T^2 is the square of a radiative cooling time (see discussion after 5.15), then

$$\Lambda \sim \left(|N| \times \text{ cooling time}\right)^2 .$$

Thus $\Lambda \gg 1$ implies that the radiative cooling time is long compared to some kind of a dynamic convection time $1/|N|$ and the convection should be regarded as efficient: whatever else the parcel does, it does not lose much heat in its travels. The opposite is true for $\Lambda \ll 1$ and the convection is said to be inefficient. It is instructive to examine σ/Σ in these two limits.

1. If $\Lambda \gg 1$ (efficient convection), then

$$\frac{\sigma}{\Sigma} \approx 1\,. \tag{5.43}$$

There are two subcases here, depending on whether N^2 is positive or negative. If $N^2 < 0$, implying convective instability, then $\Sigma = |N|$ (see 5.38) and σ are both real and w and ΔT grow exponentially with a time scale set solely by $1/|N|$. In the case of convective stability with $N^2 > 0$, Σ and σ are pure imaginary. This corresponds to oscillatory behavior in both w and ΔT with frequency N and with no growth or decay of the motion. Adiabatic behavior such as this is possible because the cooling time is essentially infinite and no radiation is lost from the parcel. It thus moves up and down, waxing and waning periodically in temperature, but does little else. Because gravity is the restoring force in such motions (as opposed to pressure), this kind of fluid behavior is associated with "gravity waves" (see Chapter 10).

2. If $\Lambda \ll 1$ (inefficient convection), then

$$\frac{\sigma}{\Sigma} \approx \Lambda^{1/2} \ll 1 \,. \tag{5.44}$$

In the convective case where Λ and Σ are both positive, w and ΔT again grow exponentially but at a reduced rate (compared to the above) because the parcel easily leaks heat by radiation. With no convection, w and ΔT are exponentially damped.

3. Intermediate values of Λ involve semiperiodic behavior with growing or damped exponentials depending on the sign of N^2.

Finally, we estimate the magnitudes of w and ΔT. It is most convenient to measure these two quantities in terms of σ since both $w = dz/dt$ and ΔT vary as $\exp(\sigma t)$. From this we have $w = \sigma z$. Taking $z = \ell$, a characteristic (or, perhaps, terminal) velocity of a convecting parcel is

$$w = \sigma \ell \,. \tag{5.45}$$

Similarly,

$$\frac{d\Delta T}{dt} = \sigma \Delta T = (\beta - \beta_{\rm ad})\, w - \frac{\nu_T}{\ell^2} \Delta T$$

or, after substituting for w and solving for ΔT, the temperature contrast between the parcel and its surroundings is (in various guises)

$$\Delta T = \frac{\sigma \ell \,(\beta - \beta_{\rm ad})}{\sigma + \nu_T/\ell^2} = \frac{\sigma^2 \ell \,(\beta - \beta_{\rm ad})}{\Sigma^2} = \frac{\sigma^2 \ell T}{\mathcal{Q} g} \,. \tag{5.46}$$

From the material given previously, note that the convective velocity becomes

$$w = \Sigma \ell \quad \text{for efficient convection,} \tag{5.47}$$

and

$$w = \Sigma \Lambda^{1/2} \ell \quad \text{for inefficient convection.} \tag{5.48}$$

You may also check that the Péclet number, $w\ell/\nu_T$, is indeed equal to $\Lambda^{1/2} = (\mathbf{Ra} \times \mathbf{Pr})^{1/2}$ in the efficient regime and to Λ if the convection is inefficient.

5.1.5 Convective Fluxes

We are now prepared to find convective fluxes in this version of MLT. We have assumed that a convectively unstable parcel rises a total distance ℓ with a characteristic velocity w, while the typical temperature contrast between the parcel and its surroundings is ΔT. The parcel merges into the surroundings at ℓ. Realize that we have assumed that various ambient quantities in the star, such as those contained in the coefficients of equations

(5.20) and (5.26), are constant over the vertical distance ℓ. This assumption is only reasonable if assumption (1) we started out with really is satisfied; namely, if ℓ is smaller than all other scale heights. Granting this, then the heat released by the parcel upon dissolution is $\rho c_P \Delta T$ erg cm^{-3} where c_P is used here because there is still pressure equilibration. The *rate* at which heat is released is then $\rho w c_P \Delta T$ erg cm^{-2} s^{-1} and this is the convective flux. Thus

$$\mathcal{F}_{\text{conv}} = \rho w c_P \Delta T = \frac{\rho c_P T \sigma^3 \ell^2}{\mathcal{Q} g} = \frac{\rho c_P T \left(\nabla - \nabla_{\text{ad}}\right) \sigma^3 \ell^2}{\lambda_P \Sigma^2} \tag{5.49}$$

in a few of many forms. The convective luminosity is then $\mathcal{L}_{\text{conv}}(r) = 4\pi r^2 \mathcal{F}_{\text{conv}}(r)$.

Limiting expressions for $\mathcal{F}_{\text{conv}}$ arising from extremes of convective efficiency, Λ, are (assuming $\Lambda > 0$ in the first place) as follows.

1. With efficient convection $\Lambda \gg 1$, $\sigma \approx \Sigma$, and

$$\mathcal{F}_{\text{conv}} \approx \frac{\rho c_P T \Sigma^3 \ell^2}{\mathcal{Q} g} = \frac{\rho c_P \ell^2 T g^{1/2} \mathcal{Q}^{1/2} \left(\nabla - \nabla_{\text{ad}}\right)^{3/2}}{\lambda_P^{3/2}} . \tag{5.50}$$

   Note: For those of you who have not followed all of the previous material leading to this equation, this is the expression to use when computing convective fluxes if you do not want to work too hard. Most stellar convection is efficient and adiabatic. What it also implies (see later), is that in most convection zones ∇ is *very* close to ∇_{ad} in absolute value.

2. For inefficient convection, $\Lambda \ll 1$ and $\sigma \approx \Sigma \Lambda^{1/2}$. Equation (5.50) again applies but with an additional small factor of $\Lambda^{3/2}$. Thus, with all other things being equal, $\mathcal{F}_{\text{conv}}$ is reduced by the factor $\Lambda^{3/2}$ below the efficient regime. Note also that the opacity enters because of the presence of ν_T in Λ (from 5.39 and 5.15).

The above results are the same as those given by Cox (1968, eqs. 14.119 and 14.122) to within factors of order unity. To show this, however, requires some algebra and a transcription of his variables to those given here (see §5.1.7).

The flux $\mathcal{F}_{\text{conv}}$ may also be phrased to contain the Mach number of the convective parcels. Using (5.49) and $\sigma = w/\ell$

$$\mathcal{F}_{\text{conv}} = \frac{\rho c_P T}{\ell \mathcal{Q} g} v_s^3 \left(\frac{w}{v_s}\right)^3 . \tag{5.51}$$

The Mach number, w/v_s, comes in as a high power in $\mathcal{F}_{\text{conv}}$ and it spells possible trouble. If w/v_s approaches unity, then assumption (4) of the MLT has clearly been violated because acoustic effects are no longer ignorable. What is usually done in practice when the Mach number approaches unity is to limit the convective flux by setting w/v_s in the above to, say, 1/2. What this has to do with reality remains a mystery.

5.1.6 Calculations in the MLT

Thus far we have assumed that ∇ (or β) and ℓ are known and that, from this, we can then compute Λ, σ, etc., and, finally, $\mathcal{F}_{\rm conv}$. There are two major practical difficulties with this. The first is that there is no guarantee the system is consistent; that is, $\mathcal{F}_{\rm conv}$, if not zero, must surely help determine the structure of the star and, hence, ∇, $\nabla_{\rm ad}$, β, and so on. Some sort of iterative process must then be required. The second objection is that in stellar structure calculations it is very often true that it is the total flux that is relatively well determined at some stage in the calculation and not ∇ (or, worse yet, $[\nabla - \nabla_{\rm ad}]$). This situation is common when examining the outer stellar regions where the luminosity is nearly constant and is primarily determined by nuclear burning deeper within the star. The issue in also complicated by the fact that the total flux is compounded from the convective flux (if present) and the radiative flux (which is always there). How are these situations handled?

For various reasons, which should become apparent later, we introduce the *Nusselt number*, **Nu**. The usual literature definition derives from laboratory practice where, for example, a fluid is confined between surfaces held a distance ℓ apart and a temperature differential ΔT is maintained across the fluid. The Nusselt number is then defined as $(\mathcal{F}_{\rm tot}\ell/\nu_T\Delta T)$ where $\mathcal{F}_{\rm tot}$ is the total flux, $\mathcal{F}_{\rm tot} = \mathcal{F}_{\rm conv} + \mathcal{F}_{\rm rad}$. From previous work (as, for example, the discussion leading up to 5.15) it should be clear that the combination $\nu_T\Delta T/\ell$ is a measure of $\mathcal{F}_{\rm rad}$ across the fluid. Thus if the Nusselt number goes to unity, then convection is not present. The larger the Nusselt number, on the other hand, the more convection contributes to the total. In laboratory situations, **Nu** goes as $\Lambda^{1/3} = (\mathbf{Ra} \times \mathbf{Pr})^{1/3}$ for $\Lambda \gg 1$ (efficient convection). We shall find that our Nusselt number, having the same spirit in definition, goes as $\Lambda^{1/2}$. The MLT and laboratory convective versus radiative fluxes do not have quite the same behavior when convection is efficient. We construct our Nusselt number as follows.

The radiative flux is given by (see 5.2, 5.12, 5.13, and 5.15)

$$\mathcal{F}_{\rm rad} = \rho c_P \nu_T \beta = \frac{\rho c_P \nu_T T}{\lambda_P} \nabla. \tag{5.52}$$

Now define a new flux, $\mathcal{F}_{\rm rad}^{\rm ad}$, by breaking up the radiative flux in the following way. Let

$$\mathcal{F}_{\rm rad} = \mathcal{F}_{\rm rad}^{\rm ad} + \frac{\rho c_P \nu_T T}{\lambda_P} (\nabla - \nabla_{\rm ad}) \tag{5.53}$$

with

$$\mathcal{F}_{\rm rad}^{\rm ad} = \frac{\rho c_P \nu_T T}{\lambda_P} \nabla_{\rm ad} = \rho c_P \nu_T \beta_{\rm ad}. \tag{5.54}$$

$\mathcal{F}_{\rm rad}^{\rm ad}$ is then the radiative flux that would be present were $\nabla = \nabla_{\rm ad}$. Contained in the fluxes we have thus far are ∇ and $\nabla_{\rm ad}$. One other "del" remains and that is $\nabla_{\rm rad}$. From the discussion leading up to (4.30), this last "del" is related to the total flux by

$$\mathcal{F}_{\rm tot} = \mathcal{F}_{\rm rad} + \mathcal{F}_{\rm conv} = \frac{\rho c_P \nu_T T}{\lambda_P} \nabla_{\rm rad} \, . \tag{5.55}$$

We now assume that $\mathcal{F}_{\rm tot}$ is known and thus so is $\nabla_{\rm rad}$. The Nusselt number is then defined in terms of the various fluxes just mentioned:

$$\mathbf{Nu} = 1 + \frac{\mathcal{F}_{\rm conv}}{\mathcal{F}_{\rm rad} - \mathcal{F}_{\rm rad}^{\rm ad}} = \frac{\mathcal{F}_{\rm tot} - \mathcal{F}_{\rm rad}^{\rm ad}}{\mathcal{F}_{\rm rad} - \mathcal{F}_{\rm rad}^{\rm ad}} \, . \tag{5.56}$$

This Nusselt number has some of the same properties mentioned earlier in that it goes to unity when convection is not present, and it becomes large when convection carries most of the flux.

Easy substitutions yield other ways of expressing **Nu**, such as

$$\mathbf{Nu} = 1 + \frac{\Sigma \ell^2}{\nu_T} \left(\frac{\sigma}{\Sigma} \right)^3 = 1 + \Lambda^{1/2} \left(\frac{\sigma}{\Sigma} \right)^3 \, . \tag{5.57}$$

Limits of **Nu** for extremes in Λ are:

1. For $\Lambda \gg 1$ ($\sigma \approx \Sigma$) and efficient convection, $\mathbf{Nu} \approx \Lambda^{1/2}$ as indicated earlier.

2. If convection is inefficient ($\Lambda^{1/2} \approx \sigma/\Sigma \ll 1$), then $\mathbf{Nu} \approx 1$ and convection carries little or none of the flux.

From the definitions of the "dels," it is also true that

$$\mathbf{Nu} = \frac{\nabla_{\rm rad} - \nabla_{\rm ad}}{\nabla - \nabla_{\rm ad}} \, . \tag{5.58}$$

If this is multiplied through by the last version of Λ in (5.39), then an easy result is

$$\Lambda \, \mathbf{Nu} = \frac{Qg}{\lambda_P} \left(\nabla_{\rm rad} - \nabla_{\rm ad} \right) \frac{\ell^4}{\nu_T^2} = C \tag{5.59}$$

where the quantity $C = \Lambda \, \mathbf{Nu}$ is known because $\mathcal{F}_{\rm tot}$ (and, hence, $\nabla_{\rm rad}$) and $\nabla_{\rm ad}$ are assumed known. Note that C is of the same form as the efficiency parameter Λ except that ∇ has been replaced by $\nabla_{\rm rad}$.

The left-hand side of equation (5.59) may be transformed into a function of Λ only by substituting (5.42) into (5.57) to yield

$$\Lambda \, \mathbf{Nu} = \Lambda + \Lambda^{3/2} \left(\frac{\sigma}{\Sigma} \right)^3 = \Lambda + \left[(\Lambda + 1/4)^{1/2} - 1/2 \right]^3 \, . \tag{5.60}$$

We now define a new variable x by

$$\Lambda = x^2 + x \, , \tag{5.61}$$

so that (5.59) and (5.60) become

$$x^3 + x^2 + x - C = 0 . \tag{5.62}$$

The scheme to find $\mathcal{F}_{\text{conv}}$ and $\mathcal{F}_{\text{rad}}$ at a particular point in the star is as follows. We assume that the total flux, $\mathcal{F}_{\text{tot}}$, temperature, opacity, ℓ, ∇_{ad}, ∇_{rad} (from $\mathcal{F}_{\text{tot}}$ of 5.55), etc., are known at that point. We do not know the value of the gradient ∇ (because if we did, then we would have all the fluxes). With what we do know, C may be found from (5.59), x from the solution of the cubic in (5.62), and Λ from (5.61). Equation (5.39) then yields ∇, and $\mathcal{F}_{\text{conv}}$ and $\mathcal{F}_{\text{rad}}$ finally follow from the application of (5.49) and (5.52).

An alternative, and quicker, path is to compute x and Λ and then apply

$$\frac{\mathcal{F}_{\text{conv}}}{\mathcal{F}_{\text{tot}}} = \frac{\nabla_{\text{rad}} - \nabla}{\nabla_{\text{rad}}} = \frac{\nabla_{\text{rad}} - \nabla_{\text{ad}}}{\nabla_{\text{rad}}} \left(1 - \frac{1}{\mathbf{Nu}} \right) \tag{5.63}$$

which may be derived from (5.52–5.56) and (5.58), and where $\mathbf{Nu} = C/\Lambda$ from (5.59). $\mathcal{F}_{\text{tot}}$ is, of course, assumed known.

An important thing to note here is that *only local quantities are required*; the properties of the rest of the star are, in effect, of no consequence for this particular calculation. Hence, the MLT is a local theory in this regard and this is one of its major virtues. The tough part for the stellar model builder is, of course, to make the model consistent in the framework of the MLT and the stellar structure equations.

It is easy to verify the following for limits of C:

1. As $C \to 0$ for $\Lambda > 0$, then $x \to C$, $\Lambda \approx x \ll 1$, and $\mathbf{Nu} \to 1$. This implies that $\mathcal{F}_{\text{tot}} \to \mathcal{F}_{\text{rad}}$, and $\nabla \to \nabla_{\text{rad}}$. Thus small values of C mean feeble convection in which the actual ∇ is very close to ∇_{rad}. Also remember (from §4.3) that $\nabla_{\text{rad}} \geq \nabla$ because $\mathcal{F}_{\text{tot}} \geq \mathcal{F}_{\text{rad}}$. Therefore, if there is any possibility of convection at all and $\nabla > \nabla_{\text{ad}}$, then we must have $\nabla_{\text{rad}} > \nabla > \nabla_{\text{ad}}$. The case of very inefficient convection then corresponds to $\nabla_{\text{rad}} \to \nabla > \nabla_{\text{ad}}$. For *no* convection, $\mathcal{F}_{\text{tot}} = \mathcal{F}_{\text{rad}}$ implies $\nabla = \nabla_{\text{rad}} \leq \nabla_{\text{ad}}$.

2. As $C \to \infty$, then $x^3 \to C$, $\Lambda \approx x^2 \approx C^{2/3} \gg 1$, $\Lambda \mathbf{Nu} = C$. The Nusselt number is then $\mathbf{Nu} \approx C^{1/3} \gg 1$ so that $\mathcal{F}_{\text{tot}} \to \mathcal{F}_{\text{conv}}$ and convection essentially carries all the flux. Since $\mathbf{Nu} \gg 1$, then $\nabla \to \nabla_{\text{ad}}$. The last can also be expressed as $\nabla_{\text{rad}} > \nabla \to \nabla_{\text{ad}}$ for vigorous and efficient convection. Convection in this instance is very often called *adiabatic* because the actual run of temperature versus pressure in the convective region is very close to adiabatic.

5.1.7 Another Version of the MLT

For those of you who have been following Cox (1968, Chapter 14), this may be useful. Our main variables are Λ, C, x, and $\mathbf{Nu}$. Cox (1968) uses variables named ζ, A, and B, among others. Note that his mixing length is denoted by Λ whereas ours is ℓ. We include in the following relations the numerical factors of order unity that Cox has used in his derivations but that we have elected to ignore (because most of them are arbitrary to some degree in any case). Aside from these factors, the procedure is very similar and a little work should reveal correspondences between the following variables and the ones we have chosen to use.

First define

$$\zeta = 1 - \frac{1}{\mathbf{Nu}} \tag{5.64}$$

where **Nu** is our Nusselt number. This means (see 5.63) that

$$\frac{\mathcal{F}_{\text{conv}}}{\mathcal{F}_{\text{tot}}} = \frac{\nabla_{\text{rad}} - \nabla}{\nabla_{\text{rad}}} = \frac{\nabla_{\text{rad}} - \nabla_{\text{ad}}}{\nabla_{\text{rad}}}\zeta . \tag{5.65}$$

The quantity ζ is a solution of the equation

$$\zeta^{1/3} + B\zeta^{2/3} + a_0 B^2 \zeta - a_0 B^2 = 0 \tag{5.66}$$

where

$$B^3 = \frac{A^2}{a_0}\left(\nabla_{\text{rad}} - \nabla_{\text{ad}}\right) \tag{5.67}$$

$$A = \frac{1}{9\sqrt{2}} \frac{Q^{1/2} g^{1/2} \ell^2}{\lambda_P^{1/2} \nu_T} , \tag{5.68}$$

and a_0 is a constant which, for most of his discussion, Cox sets to 9/4 but that may take on other values depending on exactly what version of the MLT you prefer. If *all* the numeric constants in the above were set to unity then we would have the correspondences

$$\mathit{\Lambda} = A^2\left(\nabla - \nabla_{\text{ad}}\right), \quad B^3 = C = A^2\left(\nabla_{\text{rad}} - \nabla_{\text{ad}}\right)$$

and $x^3 = B^3\zeta = C\zeta = C - \mathit{\Lambda}$.

As emphasized before, factors of order unity usually arise from geometrical considerations (is the parcel round or a cube?), or different ways of averaging quantities (is the average velocity w really $\sigma\ell$ or should it be $\sigma\ell/2$?). Well, it doesn't really matter at this level. Just be consistent and be sure to let others know exactly what are your ground rules.[3]

5.1.8 *Numeric Examples*

Plots of the logarithms of $\mathit{\Lambda}$ and **Nu** versus $\log C$ are shown in Figure 5.2. It is apparent that the limiting behaviors for $\mathit{\Lambda}$ and **Nu** given above

---

[3]Our ground rules for calculating convective fluxes in `ZAMS.FOR` of Appendix C are those described in the previous sections.

take place in the immediate vicinity of $C \approx 1$ and this value of C may be considered the critical value for efficient versus inefficient convection.

Now the question is what are interesting values of C in stellar interiors. Before this can be answered, however, we note that C contains the mixing length ℓ to the (unfortunately) high fourth power. As mentioned in the introduction, laboratory experiments cannot guide us in determining how ℓ should be treated. In the stellar context, prejudice lies in the direction of supposing that in a convective layer that is, say, a few or many scale heights (of one sort or another) deep, the shear produced by expansion and contraction in the compressible fluid results in eddies or parcels that only retain their identity over lengths comparable to a scale height. That is, we imagine parcels being continually formed and destroyed with typical lengths ℓ equal roughly to a scale height, and that many such scale heights may fit into a convective layer. As remarked upon earlier, this picture contradicts the assumption that scale lengths in the ambient fluid be long compared to parcel sizes—but so be it.

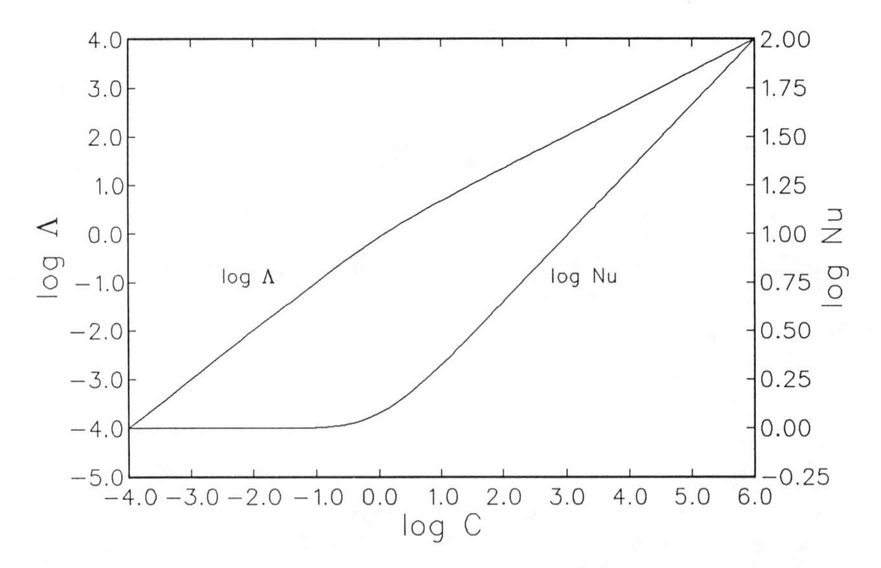

FIGURE 5.2. The efficiency parameter Λ and the Nusselt number **Nu** are plotted against the quantity C of equation (5.59). Note the qualitative change in the behavior of both Λ and **Nu** near C of unity.

The most popular scale chosen for stellar calculations is the pressure scale height λ_P. To allow some flexibility we set

$$\ell = \alpha\lambda_P \tag{5.69}$$

where α, assumed to be constant, is of order unity. In this context α is sometimes called the *mixing length parameter*. With this definition of ℓ, the crucial quantity C becomes

$$C = \frac{\mathcal{Q}g}{\nu_T^2} \alpha^4 \lambda_P^3 \left(\nabla_{\text{rad}} - \nabla_{\text{ad}} \right) . \tag{5.70}$$

A test of the utility (but not necessarily the validity) of the MLT is whether the foregoing analysis actually works in practice. A good testbed is the sun. We shall go into more detail about solar evolutionary models in Chapter 8, but it appears that in order to match the age, size, and luminosity of the present-day sun, an α of about unity is required. The use of the MLT does seem to work reasonably well in modeling a nearby star and our prejudices about what to use for the mixing length appear to be justified. This seems to be true, as far as can be ascertained at this time, for modeling of other kinds of stars. However, enough progress has been made in the observation and theory of stars, that *details* of convection are now important. For example, the consequence of subsurface convective motions can easily be detected at the sun's photosphere, but how are these to be modeled and what does such modeling mean for convection deeper down in the sun?

Let us then assume that the MLT with $\alpha \approx 1$ is a provisionally acceptable model for convection. In what situations may this model be applied? Some examples are as follows.

Deep Convection in Upper Main Sequence Stars

From the discussion of the preceding section, the criterion for local convective instability is that ∇_{rad} be greater than ∇_{ad}. The gradient ∇_{rad} is defined by (4.30), which we reproduce here for convenience:

$$\nabla_{\text{rad}} = \frac{3}{16\pi acG} \frac{P\kappa}{T^4} \frac{\mathcal{L}_r}{\mathcal{M}_r} \tag{5.71}$$

where $\mathcal{L}_r = 4\pi r^2 \mathcal{F}_{\text{tot}}$. The other gradient of interest is $\nabla_{\text{ad}} = (\Gamma_2 - 1)/\Gamma_2$. Only in special circumstances can the adiabatic exponent Γ_2 approach unity and values near 5/3 are more normal. Thus, for practical purposes, ∇_{ad} is a number near 0.4 or a little less. Convection then requires ∇_{rad} to be greater than this value (see 5.63). This can be accomplished in several ways. One of these arises from the behavior of $\mathcal{L}_r/\mathcal{M}_r$. In particular, let us examine what happens near the center of a massive ($\mathcal{M} \gtrsim 1.5\mathcal{M}_\odot$) upper ZAMS star. From previous remarks, the CNO cycles dominate over the pp chains in these stars because of the temperature sensitivity of the former. By the same token, the energy generation rate should be sharply peaked at the stellar center and then drop rapidly outward as temperature decreases. For stars in thermal balance this also implies that $\mathcal{L}_r$ should increase rapidly from zero as r increases away from the center. The consequences for the ratio $\mathcal{L}_r/\mathcal{M}_r$ may be seen by examining the behavior of the energy equation $d\mathcal{L}_r/d\mathcal{M}_r = \varepsilon$ of (1.55) as r or $\mathcal{M}_r$ becomes very small. It should be obvious that, for small $\mathcal{M}_r$, $\mathcal{L}_r/\mathcal{M}_r \to \varepsilon_c$, where ε_c is the energy generation rate computed at $r = 0$. (This and other expansions of the stellar structure equations near the stellar center will be examined more closely in Chapter 7.) Thus one way to get a large $\nabla_{\rm rad}$ might be to have a large energy generation rate. The physical picture is that with a large ε in some region, the flux required for thermal balance is so large that energy transport by radiation is insufficient and convection must eventually carry some of the load.

As a specific example taken almost randomly from the literature (or you can use the ZAMS code discussed in Appendix C), consider a 30 $\mathcal{M}_\odot$ ZAMS model. The composition is typical Pop. I with $X = 0.7$, $Z = 0.03$, and the mass fraction of CNO nuclei is $X_{\rm CNO} = Z/2$. The central temperature and density are $T_c = 3.6 \times 10^7$ K and $\rho_c = 3$ g cm^{-3}. The total luminosity and radius are $\mathcal{L} = 5.5 \times 10^{38}$ erg s^{-1} and $\mathcal{R} = 4.6 \times 10^{11}$ cm. From this information we can readily find (as discussed in Chapter 6) the energy generation rate $\varepsilon_c \approx 2 \times 10^5$ erg g^{-1} s^{-1}. Even though this is a large number, we have to compute the rest of the factors in (5.71) to get a final result for $\nabla_{\rm rad}$. These will also be found for the center.

From the discussion of opacity sources in the previous chapter it is clear that the opacity at stellar center derives primarily from electron scattering so that (see 4.34) $\kappa_c \approx 0.34$ cm^2 g^{-1}. The total pressure, including the effects of radiation, is $P_c \approx 1.88 \times 10^{16}$ dyne cm^{-2}, and the ratio of gas to total pressure is $\beta_c \approx 0.775$. With these results we may apply the techniques of Chapter 3 (specifically Eq. 3.105) to find $\Gamma_2 = 1.41$, or $\nabla_{\rm ad} = 0.29$, and, finally, $\nabla_{\rm rad} = 3.0$. Thus $\nabla_{\rm rad} > \nabla_{\rm ad}$ by a factor of ten and the stellar center is easily convective.

To compute the flux carried by convection, we need a few more quantities which are parts of C in (5.59). The thermal diffusivity, given by (5.15) is readily found to be $\nu_T \approx 5 \times 10^9$ cm^2 s^{-1} at stellar center. The mixing length is more difficult. Using a mixing length parameter of $\alpha = 1$ means $\ell = \lambda_P$. But what is the pressure scale height? As in the discussion in the introductory remarks of Chapter 3, the pressure scale height at stellar center is infinite, which makes for a nonsensical mixing length. An upper limit is the radius of the star itself, so take $\lambda_P \approx \mathcal{R} \approx 4 \times 10^{11}$ cm. From ν_T and the mixing length we may readily compute the cooling time, λ_P^2/ν_T, for a parcel to be about 10^6 years. This long cooling time implies that the convection is very adiabatic.

The central gravity that appears in C presents a problem because it is formally zero; for small r, $g \approx 4\pi G r^3 \rho_c/3$. Since, as will be evident shortly, C will be large with almost any reasonable choice of g, choose $g \approx 1.4 \times 10^4$ cm s^{-2}, which is the gravity at the point where $\mathcal{M}_r \approx 0.1\,\mathcal{M}$.

Putting this all together yields $C \approx 10^{20}$—obviously a number far in excess of unity—and this does not even fit on the scale of Figure 5.2. We do know, however, the behavior of various quantities with large C; namely, $\Lambda \approx C^{2/3} \approx 2 \times 10^{13}$, and $\mathbf{Nu} \approx C^{1/3} \approx 5 \times 10^6$. Since, from (5.58), $\mathbf{Nu} = (\nabla_{\rm rad} - \nabla_{\rm ad}) / (\nabla - \nabla_{\rm ad})$, it follows that $(\nabla - \nabla_{\rm ad}) \approx 5 \times 10^{-7}$.

For all practical purposes, the structure at and near the stellar center is $\nabla = \nabla_{\rm ad}$ and is, in a sense, determined solely by the thermodynamics. This is typical of deep-seated convection in stars: convection is very nearly adiabatic. It also demonstrates some other points. Because C is so large, some leeway is permitted in how the mixing length is chosen. If we had changed ℓ in this example by even a factor of ten either way, ∇ would still equal $\nabla_{\rm ad}$ to quite a few significant places. Thus neglected "factors of order unity" really are of little consequence in the analysis.

You might be tempted at this juncture to try and compute the convective flux by using the value of $(\nabla - \nabla_{\rm ad})$ just found and inserting it into (5.49). Don't! The difference in the "dels" is not well determined and may change by an order of magnitude or more depending on just how mixing lengths, etc., are treated. All you really know is that ∇ is very nearly equal to $\nabla_{\rm ad}$. The sensible way is to either use (5.63) or the following to find $\mathcal{F}_{\rm conv}/\mathcal{F}_{\rm tot}$ or $\mathcal{L}_{\rm rad}/\mathcal{L}_{\rm tot}$. Note that since $\mathcal{L}_{\rm rad}(r) \propto \nabla(r) = \nabla_{\rm ad}(r)$, and $\mathcal{L}_{\rm tot}(r) \propto \nabla_{\rm rad}(r)$, then $\mathcal{L}_{\rm rad}/\mathcal{L}_{\rm tot} \approx 0.1$. That is, convection carries about 90% of the total flux near the stellar center.

Finally, to find the convective velocity, $w = \sigma\ell \approx \sigma\lambda_P$, realize that $\Lambda \gg 1$ implies $\sigma \approx \Sigma$. But, from (5.39) $\Sigma = \left(\nu_T/\lambda_P^2\right)\Lambda^{1/2}$ and thus $\sigma \approx 10^{-7}$ s. The convective time scale is then $\tau_{\rm conv} = 1/\sigma \approx 10^7$ s $\approx$ one year, and $w \approx 5\times10^4$ cm s^{-1}. The sound velocity is $v_s \approx 10^8$ cm s^{-1} so the convective motions are comfortably subsonic. All this is again typical of deep interior convection: velocities are slow compared to sound and time scales are fairly long. Even though convection transports a lot of luminosity, it can do so at its leisure because of the high heat content of the dense interior.

Envelope Convective Zones

The situation can be quite different for convection in the outer layers of stars. As an example consider a star on the RGB or AGB. From a model (actually of a Mira variable) with $\mathcal{M} = 0.7\mathcal{M}_\odot$, $\mathcal{L} \approx 2\times10^{37}$ erg s^{-1}, $\mathcal{R} \approx 3\times10^{13}$ cm, and $T_{\rm eff} \approx 2500$ K, we find the following parameters at $r/\mathcal{R} = 0.7$: $\nabla_{\rm rad} \approx 1000$, $\nabla_{\rm ad} \approx 0.1$, $\nabla \approx 0.3$, $\mathcal{M}_r \approx 0.53\mathcal{M}_\odot$, $\mathcal{L}_r \approx 2\times10^{37}$ erg s^{-1}, $T \approx 10^4$ K, $\rho \approx 10^{-8}$ g cm^{-3}, $\lambda_P \approx 3\times10^{12}$ cm, $\kappa \approx 100$ cm^2 g^{-1}, $r \approx 2\times10^{13}$ cm, $g \approx 0.1$ cm s^{-2}, $P \approx 500$ dyne cm^{-2}, and no thermonuclear energy generation. This point is in the middle of the hydrogen ionization zone and hence the low value of $\nabla_{\rm ad}$. In red supergiants, the ionization zones (hydrogen, helium, and perhaps H_2) are very broad. This is because, among other things, the temperature and density profiles are relatively flat. At the same time, luminosities are very high, as are opacities, hence the large value of $\nabla_{\rm rad}$. Obviously this ionization zone is convective.

From the above figures you may estimate that $C \approx 10^{11}$, $\mathbf{Nu} \approx 4\times10^3$, $\Lambda \approx 2\times10^7$, $w \approx 2.5\times10^5$ cm s^{-1}, $v_s \approx 5\times10^5$ cm s^{-1} (Mach 1/2), and $\tau_{\rm conv} \sim$ one year.

For this kind of star, convection is efficient and transports most of the power but the convection has to work a little harder because of the low densities. In fact, the convective flow is near supersonic. The temperature versus pressure structure is no longer adiabatic as in the previous example because $\nabla \approx 3\nabla_{\rm ad}$.

As a side remark about this particular model, it turns out that it is a reasonable representation of a Mira variable. (A "pulsational stability analysis" shows this to be the case.) However, the period of this model variable is about one year, which is about the same as the convective time scale from the MLT analysis. If convection is difficult to model for static stars, be assured that coupling it with dynamic motions in a variable star model compounds the difficulty manyfold.

The convective envelopes of stars hotter than the one just considered tend to be less efficient and smaller in size. Finally, for stars whose effective temperatures are quite high, convection zones associated with envelope ionization processes either are not present or are of negligible importance in transporting heat. In the latter case, C tends to be small or, at most, near unity. For the borderline efficiency case of $C \approx 1$, those annoying factors of near unity that differ among various versions of MLT become important in one respect: they can significantly change the value of C and the convective flux.

Figure 5.3 shows the H–R diagram divided into two regions where, in the hotter region, convection is inefficient or hardly present in the outer stellar layers, whereas the cooler stars have active and efficient convection zones. The primary reason for this is that the ionization zones in the cooler stars lie deeper in the star where opacities are high (not electron scattering as in the hot deep interior of upper main sequence) and $\nabla_{\rm ad}$ is relatively small.

5.2 Variations on the MLT

Several times in this chapter mention has been made of attempts to better model parcels and their motions by introducing geometrical factors, changing the method of averaging various quantities, etc. We shall not describe these attempts because all they tend to do is change the efficiency of convection somewhat. As a practical matter, one has to be very careful when reading the literature to figure out precisely what version (or version of a version) is being used. Besides simple geometrical factors, the choice of scale height used to construct the mixing length differs among authors. Variations here include computing the mixing length using the pressure scale height unless that scale is longer than the distance from the stellar position in question to the upper boundary of the convection zone. In that case the latter distance is used (as in Böhm and Cassinelli 1971). Each of the variations sounds reasonable but they are still carried out in the framework of the MLT. One recent study that explores the effect on stellar

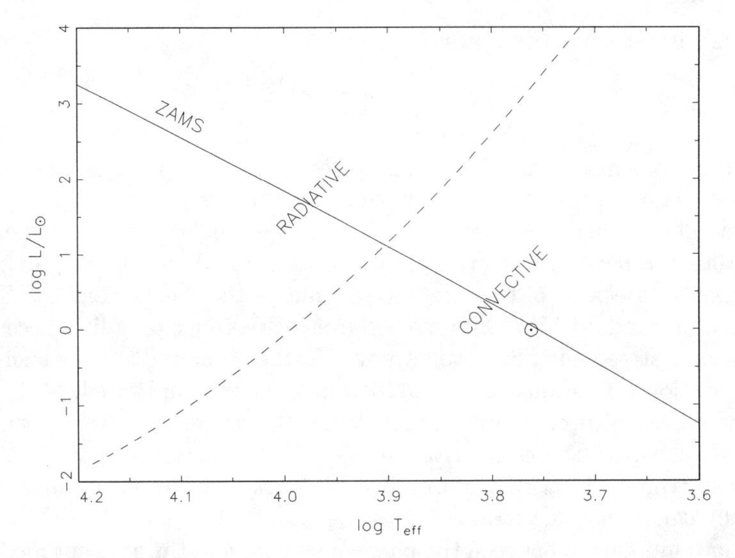

FIGURE 5.3. The dashed line on this schematic H–R diagram divides those stars with active and efficient outer convection zones to the right of the line from those to the left that have feeble and inefficient convection. There is, of course, a gradual transition across the line. The sun is indicated by the $\odot$ sign. Adapted from Cox (1968).

evolution of varying mixing length parameters is Pedersen *et al.* (1990).

Some attempts at modeling stellar convection that are certainly not part of generic formulations of the MLT include the following. We have imagined our parcel rising (for example) at some characteristic velocity w and then merging with the ambient medium and releasing heat once ℓ has been traversed. At the top of the convection zone, where ∇ finally becomes smaller than $\nabla_{\rm ad}$, all these parcels are supposed to be able to stop and not penetrate into the stable layer above. This does not even sound reasonable. We expect some fluid elements to *overshoot* into the overlying stably stratified medium. Below the convection zone we might also expect downward flowing elements to penetrate into the underlying stable medium to some extent. The question is how to model this behavior without really working too hard. Descriptions of how this is done in practice by some researchers may be found in the references and especially see the review by Trimble (1992).

One way to attack this problem, however, is to consider a rising parcel having velocity w_0 at the top of a convection zone at radius r_0. Assuming that the parcel does not come to a complete halt at r_0 as it begins to enter the convectively stable region above r_0, it must still experience a deceleration because of negative buoyancy. The work done against the parcel by the negative buoyancy may be estimated by combining (5.1), (5.19), and (5.24). Consideration of the kinetic energy lost by the parcel then yields the velocity at some position $r \geq r_0$

$$w^2(r) = w_0^2 + 2\int_{r_0}^{r} \frac{gQ\Delta T}{T}\, dr' \tag{5.72}$$

where viscous dissipation has been neglected and the fluid is assumed to be essentially incompressible over some distance less than a density scale height. The temperature contrast $\Delta T = T' - T$ is negative in the stable fluid and thus the parcel must eventually slow down and stop (over a distance less than ℓ). Because of the integral, this makes the theory nonlocal. Thus, unlike the standard MLT, information from some range of radii is necessary to make a statement about some point further removed. This makes the computation (of various things) a good deal more complicated.

The effects of overshooting are twofold: it may mix matter of varying composition past the convective interface and it may transport heat. To calculate these effects requires even more work than we have implied. To be fully consistent, the term ΔT must be handled correctly. It is the difference in temperature between the parcel's interior and the ambient medium. Presumably T is known but T' is another matter. If we are going to treat ΔT as something that can vary with height for a *particular* parcel, then consider the following. Two parcels start rising from different positions, r_1 and r_2, within the convection zone. When they reach $r_3 > r_1$, r_2, their interior temperatures will not be the same in general. What then is the meaning of $T' - T$ at r_3? A little thought reveals that to do this sort of thing also requires a nonlocal treatment of handling temperatures, etc.

There are different implementations of the above and not all give the same kind of answer for the same kind of stellar model. For an upper main sequence model of the sort discussed in the last section, it appears that typical overshooting lengths are a little bit less than a pressure scale height. How much less (zero to 70% of a scale height?) is a matter of debate and the consequences are important. Overshooting in the convective cores of massive stars affects the lifetime (by mixing fuel) of that stage, and may easily cause the products of nuclear burning to emerge at the stellar surface where they might become visible (as in Wolf-Rayet stars). The caution here is that all these modifications, as well intentioned as they may be, are still based primarily on the MLT with all its faults.

Another consideration is whether the Boussinesq assumption of no acoustic effects is reasonable. The Reynolds number $\mathbf{Re} = w\ell/\nu$ is easily computed from the material already presented. For the upper ZAMS core $\mathbf{Re} \sim 10^{13}$ and it is even larger for the Mira envelope. Because laboratory values of $\mathbf{Re}$ exceeding ~ 100 imply a transition from smooth to turbulent flow, we expect stellar convection to be characterized by turbulent eddies of many size scales. (What does a parcel size of the order of a "mixing length" mean in this context?) It has been suggested that the energy flux carried by acoustic noise due to turbulence in some convection zones may actually be comparable to that carried by mean flows. One model, due to

Lighthill (1952, 1954) and Proudman (1952), yields an estimate for this flux of $\mathcal{F}_{\text{turb}} \approx a \int \left(\rho w^3 M^5/\ell\right) dr$ where the integration is to be carried out over the extent of the convection zone, M is the Mach number, and a is a number of order 10. If this result is really applicable to stars, then the computation of convective fluxes for high-velocity flows must be far more difficult than what has been outlined in this chapter.

5.2.1 *Beyond the MLT*

Any attempt to model convection in all its glory must be, a priori, highly nonlocal and nonlinear because the full equations of hydrodynamics, including turbulence, must be considered. This may turn out to be an impossible program for the near future. There has, however, been some progress, which we now discuss very briefly. An excellent introduction to what follows may be found in the first few chapters of Chandrasekhar (1981). References to the works of those involved with this program in the stellar context are listed at the end of this chapter.

The analytical and computational thrust of this research consists in considering the equations of hydrodynamics but where relevant variables are separated into two parts. One corresponds to a horizontal mean of a variable while the other part deals with fluctuations. For example, the spatial and temporal behavior of the density (in plane parallel geometry) is $\rho(z,t) = \langle\rho(z,t)\rangle + \rho'(z,t)$. It is the second, and fluctuating, term on the right that will describe how mass, energy, etc., are transported. The next step is to introduce the anelastic approximation in which $\partial\rho'/\partial t$ is neglected in the equation of continuity

$$\left(\frac{\partial \rho}{\partial t}\right) + \boldsymbol{\nabla} \bullet (\rho \mathbf{v}) = 0.$$

The effect of this is to filter out acoustic waves (as in the MLT). The reason this is done is that acoustic waves have higher frequencies than gravity waves and deleting the former allows the time evolution of the convection to be followed more efficiently on the computer. It also assumes that acoustic fluxes are relatively unimportant in transporting energy. This may well restrict the validity of the model to those envelopes in which convection is not too vigorous.

The next step is to expand the horizontal structure of fluctuating quantities into a finite number of horizontal "planforms" or "modes." These planforms may be of various shapes. For example, a hexagonal planform might be used where fluid may rise in the center of the hexagon and sink at the edges (or the other way around). Such structures are seen in the laboratory under the right conditions in Bénard cells, or on the surface of the sun (as a direct consequence of the underlying convective layer). Another form might be that of a "roll". In any case, such planforms of various shapes and sizes may be added together to model the convective motions. In the calculations reported below, only one planform was used.

These techniques were applied to a model spectral type A star of 1.8 $\mathcal{M}_\odot$ on the main sequence with $T_{\text{eff}} = 8000$ K. Such stars are thought to have two thin and relatively inefficient (from an MLT point of view) convection zones near the stellar surface, which are associated with H^+ and He^{++} ionization. Since these zones are inefficient, any variations in the convection should have a minor effect on the stellar model as a whole. Furthermore, the zones are sufficiently thin that non-Boussinesq effects of stratification should also be minor; that is, they should be amenable to an MLT approach so that comparisons with non-MLT treatments will make more sense. Another reason for choosing an A star is that a subclass of these, the "metallic line A stars," show wild abundance anomalies (about a factor of ten compared to the sun) in such elements as vanadium, nickel, copper, zinc (all overabundant), and calcium and scandium (underabundant). The reason for the anomalies probably lies in the competition between settling of heavy elements and their mixing by convection.

The solutions obtained differ dramatically (in one sense) from those using a version of the MLT similar to that described earlier. The maximum convective flux in the He^{++} zone contributes some 6% to the total in the planform calculation and this is about two orders of magnitude greater a contribution than the MLT gives. Furthermore, severe overshooting takes place into the stably stratified region between the two convection zones. Thus the two zones become coupled in a non-MLT way.

Thus far this analysis has not been applied to convection zones that are important for the general structure of a star.[4] One reason for this is that, even in the calculation described above, the convection part of the computer program was larger and considerably more complex than the rest of the program, which went on to handle the rest of the star. We shall have to wait a while before more definitive comparisons with the MLT and its relatives are completed.

5.2.2 Semiconvection

This important topic is related in some respects to overshooting. Briefly, the process of semiconvection comes about as indicated in the following example. In a pioneering evolutionary study, Schwarzschild and Härm (1958) found that the convective cores of massive ($\mathcal{M} > 10\mathcal{M}_\odot$) stars behaved in a curious fashion. It turns out that the convective core of such a star tends to be larger than the region of active CNO cycle burning and, because of the contribution of radiation pressure to the equation of state, the convective

---

[4]Exceptions are the work by Latour *et al.* (1983) or Nordlund and Stein (1990) on the sun. Another way of dealing with real convection in stellar models is reported in Lydon *et al.* (1992, 1993).

core tends to move outward as evolution proceeds. This means that if we assume convective motions can efficiently mix material in such cores, then snapshots of the hydrogen content of the star as it evolves might resemble that shown in Figure 5.4. The evolutionary stages are labeled as discontinuous "steps" in hydrogen mass fraction starting from step zero on the main sequence when the star is homogeneous and has an initial hydrogen mass fraction $X = 0.7$. (Incidently, all special effects such as mass loss or simple overshooting are ignored here.) What is the effect of these discontinuities on the model and are they consistent with the equations of stellar structure developed thus far? Consider the following.

If the standard MLT has been used to describe convection, then, as discussed previously, $\nabla = \nabla_{\text{ad}}$ to high precision in the core. Outside the core, $\nabla = \nabla_{\text{rad}}$. Now ∇_{ad} should be roughly continuous across the outer edge of the composition discontinuity because Γ_2 is roughly independent of composition provided that ionization is complete. We expect the latter to be true since we are deep within the star. What about ∇_{rad}. Is it continuous? That "del" is given by equation (5.71) and, in that expression, P cannot be discontinuous because that would introduce infinite radial derivatives of pressure and, hence, infinite forces. Similarly, temperature cannot be discontinuous because the radiative luminosity is proportional to its gradient. Lastly, $\mathcal{M}_r$ and $\mathcal{L}(r)$ are continuous provided that density does not do something bizarre and if we stay outside the energy generation region. This leaves the opacity.

In the deep interior $\kappa \approx 0.2(1 + X)$ from electron scattering. Since X is discontinuous, so is κ and, thus, ∇_{rad}. However, we do see that the ratio $\nabla_{\text{rad}}/\kappa$ is continuous.

If quantities interior to the composition discontinuity are indicated by an "i" subscript, and those exterior by an "e," then we must have

$$\frac{\nabla_{\text{rad,i}}}{\kappa_i} = \frac{\nabla_{\text{rad,e}}}{\kappa_e}.$$

For the hydrogen profile evolution shown in Figure 5.4, $X_{\text{i}} \leq X_{\text{e}}$ implies $\kappa_{\text{i}} \leq \kappa_{\text{e}}$, or

$$\nabla_{\text{rad,e}} \geq \nabla_{\text{rad,i}}$$

across the discontinuity in X. An illustrative run of the ratio $\nabla_{\text{rad}}/\nabla_{\text{ad}}$ with radius is shown in Figure 5.5 for the ZAMS stage and one later stage. Note that ∇_{rad} must be greater than ∇_{ad} for points within the convection zone (otherwise there is no convection). Because of the discontinuity in ∇_{rad}, however, there is a small region outside the convection zone that is not radiative; that is, this region must also be convective. If this is really so, then we have a contradictory situation. Just what happens in this region is still a matter of some debate. It is supposed that some mixing takes place in this region so that some composition gradients are smoothed out. Exactly how this is accomplished has not been established. But, however it is done, it is referred to as "semiconvective mixing" and it may have a strong

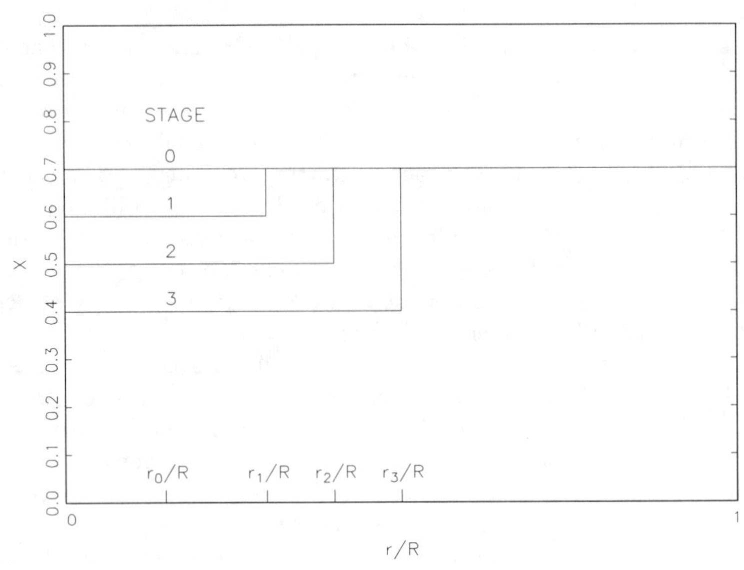

FIGURE 5.4. Schematic profiles of the hydrogen mass fraction, X, for a massive star as evolution proceeds from the ZAMS (step 0). As the star evolves (steps 1 through 3), the convection zone moves outward and mixing causes the hydrogen, which is gradually being depleted by nuclear burning, to have constant abundance out to larger radii. The convection zone is assumed to extend out to the discontinuity in X. The radii r_0, r_1, etc., are the outermost radii of the convection zones and hydrogen discontinuities in the various evolutionary stages.

effect on the later stages of evolution when shell sources are effective. Brief summaries of this general problem may be found in Chiosi and Maeder (1986) in the context of massive stars, in the general review by Trimble (1992), and in Hansen (1978) for other evolved objects where chemical discontinuities play a role in structure.

To summarize the contents of this chapter we need only say that the treatment of stellar convection is still one of the weak points in stellar structure and evolution theory. The MLT and its variants are not necessarily bad, and do give what appear to be reasonable results, but a future (far future?) implementation of a proper treatment will probably yield some enlightening surprises for the field.

5.3 References and Suggested Readings

§**5.1**: The Mixing Length Theory

Chapter 14 of

▹ Cox, J.P. 1968, *Principles of Stellar Structure*, in two volumes (New York: Gordon and Breach)

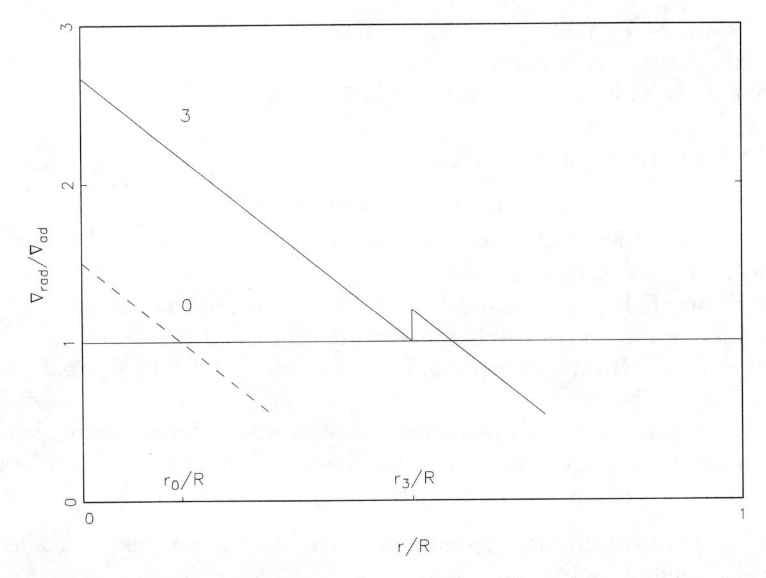

FIGURE 5.5. An illustration of how the ratio $\nabla_{\rm rad}/\nabla_{\rm ad}$ might vary as a function of radius for two of the evolutionary stages pictured in Figure 5.4. The radii r_0 and r_3 are the same as in that figure.

contains a good description of the MLT as usually applied and should be read after our chapter. Section 5.1.7 gives a brief review of the difference between our nomenclature and that of Cox (1968).

The development of the MLT may be traced through the following papers:

▹ Schwarzschild, K. 1906, *Gott. Nach.*, **1**, 41

▹ Biermann, L. 1951, *Z. Astrophys.*, **28**, 304

▹ Prandtl, L. 1952, *Essentials of Fluid Dynamics* (London: Blakie)

▹ Vitense, E. 1953, *Zs. f. Ap.*, **32**, 135

▹ Böhm-Vitense, E. 1958, *Zs. f. Ap.*, **46**, 108.

An accessible and modern commentary on Boussinesq convection and the MLT may be found in

▹ Spiegel, E.A. 1971, *Ann. Rev. Astron. Ap.*, **9**, 323.

As usual, we recommend

▹ Landau, L.D., and Lifshitz, E.M. 1959, *Fluid Mechanics* (Reading: Addison-Wesley)

(or later edition) for their physical style.

Chapter 5 of

▹ Spitzer, L. 1962, *Physics of Fully Ionized Gases* 2d ed. (New York: Interscience)

is our source for the calculation of kinematic viscosity.

The parameters for the 30 $\mathcal{M}_\odot$ core convection are from

▷ Stothers, R. 1963, *Ap.J.*, **138**, 1074
and the Mira material is from
▷ Langer, G.E. 1971, *M.N.R.A.S.*, **155**, 199.

§5.2: Variations on the MLT

▷ Chan, K.L., and Sofia, S. 1987, *Science*, **235**, 465
give an interesting discussion on some validity tests for the MLT in deep convection zones. The paper by
▷ Böhm, K.H., and Cassinelli, J. 1971 *Astron. Ap.*, **12**, 21
concerns the convective envelopes of white dwarfs.
▷ Pedersen, B.B., VandenBerg, D.A., and Irwin, A.W. 1990, *Ap.J.*, **352**, 279
have constructed series of evolutionary models with different mixing length parameters. You may wish to consult this paper to see how these series differ from one to the other.

Early estimates of the rate of production of acoustic noise due to turbulence may be found in
▷ Lighthill, M.J. 1952, *Proc. Roy. Soc. London*, **A221**, 564
▷ *Ibid* 1954, *Proc. Roy. Soc. London*, **A222**, 1
▷ Proudman, J. 1952, *Proc. Roy. Soc. London*, **A214**, 119.

The following papers by A. Maeder and collaborators describe one model for overshooting in a version of the MLT and give results for the evolution of upper main sequence stars:
▷ Maeder, A., 1975, *Astron. Ap.*, **40**, 303
▷ Maeder, A. 1975, *Astron. Ap.*, **43**, 61; Maeder, A. 1976, *Astron. Ap.*, **47**, 389
▷ Maeder, A. 1982, *Astron. Ap.*, **105**, 149
▷ Maeder, A., and Bouvier, P. 1976, *Astron. Ap.*, **50**, 309
▷ Maeder, A., and Mermilliod, J.C. 1981, *Astron. Ap.*, **93**, 136.
Additional material, including a brief appraisal of methods used in handling semiconvection, is given by
▷ Chiosi, C., and Maeder, A. 1986, *Ann. Rev. Astron. Ap.*, **24**, 329.
A recent compilation of evolutionary results that summarizes 40,000 models is
▷ Maeder, A. 1990, *Astron. Ap. Suppl.*, **84**, 139.
Marked differences in structure may result from varying degrees of discontinuities in models. Some results are discussed in
▷ Hansen, C.J. 1978, *Ann. Rev. Astron. Ap.*, **16**, 15.

An excellent short review containing up-to-date material on semiconvection and overshoot is due to
▷ Trimble, V. 1992, *P.A.S.P.*, **104**, 1.

The text by

▹ Chandrasekhar, S. 1981, *Hydrodynamic and Hydromagnetic Stability* (New York: Dover)

should be on every theorists' bookshelf. It can be heavily mathematical at times, but it is an indispensable introduction to modern developments in convection theory.

Planform methods, with applications, are given in the following series of papers involving many of the same names as collaborators:

▹ Gough, D.O., Spiegel, E.A., and Toomre, J. 1975, *J. Fluid Mech.*, **68**, 695

▹ Latour, J., Spiegel, E.A., Toomre, J., and Zahn, J.-P. 1976, *Ap.J.*, **207**, 233

▹ Toomre, J., Zahn, J.-P., Latour, J., and Spiegel, E.A. 1976, *Ap.J.*, **207**, 545

▹ Latour, J., Toomre, J., and Zahn, J.-P. 1981, *Ap.J.*, **248**, 1081

▹ Latour, J., Toomre, J., and Zahn, J.-P. 1983, *Solar Phys.*, **82**, 387.

The role of penetrative overshooting in compressible flow has been examined for simple models by

▹ Hurlburt, N.E., Toomre, J., and Massaguer, J.M. 1986, *Ap.J.*, **311**, 563.

The use of massively parallel supercomputers is now enabling investigators to model compressible convection in two and three dimensions. For a taste of what is being done you might look into

▹ Nordlund, Å, and Stein, R.F. 1990, *Comp. Phys. Com.*, **59**, 119.

Other examples of dealing with real convection are due to

▹ Lydon, T.J., Fox, P.A., and Sofia, S. 1992, *Ap.J.*, **397**, 701

▹ *Ibid* 1993, *Ap.J. Letters*, **403**, L79.

The upper main sequence calculations, which led to our discussion of stellar semiconvection are those of

▹ Schwarzschild, M., and Härm, R. 1958, *Ap.J.*, **128**, 348.

6
Stellar Energy Sources

Energy is Eternal Delight
— William Blake (1757–1827)

Don't get smart alecksy
with the galaxy
leave the atom alone.
— E.Y. Harburg (on the "Bomb")

Now that we have established how energy is transported in the stellar interior, we shall backtrack and see how that energy is generated. This chapter will discuss energy production by the conversion of gravitational energy into internal energy and by thermonuclear processes. One section will deal with neutrinos.

6.1 Gravitational Energy Sources

In the first chapter (§1.3.2) we described how the virial theorem could be used to estimate the Kelvin-Helmholtz time scale for maintaining the luminosity of a star by means of gravitational contraction. What we obtained was a global property of the system because the virial deals only with integrated quantities. We now examine how this works on a local scale.

The local condition for thermal balance was discussed earlier (in §1.5). It described how the local rate of thermonuclear energy generation, ε, is balanced by the mass divergence of luminosity with

$$\frac{\partial \mathcal{L}_r}{\partial \mathcal{M}_r} = \varepsilon \tag{6.1}$$

for a given gram of material (with ε in erg g^{-1} s^{-1}). If this equality does not hold, then the energy content decreases (increases) for $\partial \mathcal{L}_r / \partial \mathcal{M}_r$ greater (less) than ε. The difference $\varepsilon - (\partial \mathcal{L}_r / \partial \mathcal{M}_r)$ is then the rate at which heat is added to, or removed from, each gram. This difference is just dQ/dt where Q is the heat content in erg g^{-1}. Combining this with the first law of thermodynamics yields

$$\frac{dQ}{dt} = \frac{\partial E}{\partial t} + P \frac{\partial}{\partial t}\left(\frac{1}{\rho}\right) = \varepsilon - \frac{\partial \mathcal{L}_r}{\partial \mathcal{M}_r} \tag{6.2}$$

where it is understood that the partial time derivatives are applied in a Lagrangian sense so that a particular gram of matter is followed in time.

This expression may be rewritten by defining the *gravitational* energy generation rate

$$\varepsilon_{\text{grav}} = -\left[\frac{\partial E}{\partial t} + P \frac{\partial}{\partial t}\left(\frac{1}{\rho}\right)\right] \tag{6.3}$$

so that (6.2) becomes

$$\frac{\partial \mathcal{L}_r}{\partial \mathcal{M}_r} = \varepsilon + \varepsilon_{\text{grav}} \, . \tag{6.4}$$

Note that $\varepsilon_{\text{grav}}$ may be positive, negative, or zero.

Another way of expressing $\varepsilon_{\text{grav}}$ is to cast it in a more useful form containing time derivatives of density and pressure rather the internal energy. This involves a little work using thermodynamic identities and some of the thermodynamic derivatives discussed in Chapter 3 (see, e.g., Cox 1968, §17.6). The result we shall use is

$$\varepsilon_{\text{grav}} = -\frac{P}{\rho\,(\Gamma_3 - 1)}\left[\frac{\partial \ln P}{\partial t} - \Gamma_1 \frac{\partial \ln \rho}{\partial t}\right] \tag{6.5}$$

where, for simplicity, we assume Γ_1 is constant in time. Note that $\varepsilon_{\text{grav}}$ is zero for adiabatic processes wherein $P \propto \rho^{\Gamma_1}$. Energy release in gravitational sources thus arises from departures from adiabaticity during contraction or expansion. This may also be seen by rewriting (6.5) in the form

$$\varepsilon_{\text{grav}} = -\frac{P}{\rho\,(\Gamma_3 - 1)} \frac{\partial}{\partial t}\left[\ln\left(P/\rho^{\Gamma_1}\right)\right] . \tag{6.6}$$

Thus, for example, if the pressure rises less rapidly than adiabatic upon compression so that

$$P \sim \rho^{\Gamma_1 - \delta}$$

with δ small but positive, then

$$\varepsilon_{\text{grav}} = \delta \frac{P}{\rho\left(\Gamma_3 - 1\right)} \frac{\partial \ln \rho}{\partial t} > 0$$

where the greater than sign is used for compression. This result is reasonable because a less than adiabatic rise of pressure upon compression implies that energy is being released.

We may now improve on our earlier estimate of the Kelvin-Helmholtz time scale. As in the discussion of §1.3.2, we ignore thermonuclear energy sources so that (6.4) becomes $(\partial \mathcal{L}_r / \partial \mathcal{M}_r) = \varepsilon_{\text{grav}}$. If we assume homologous contraction, which is often fairly close to the truth, then the results of §1.6, with mass held constant, yield

$$\frac{\rho}{\rho_0} = \left(\frac{\mathcal{R}}{\mathcal{R}_0}\right)^{-3} \tag{6.7}$$

and

$$\frac{P}{P_0} = \left(\frac{\mathcal{R}}{\mathcal{R}_0}\right)^{-4}. \tag{6.8}$$

Hydrostatic equilibrium has been used in the derivation of these relations and, hence, time scales must be long compared to dynamic times.

If we take the "0" star to be some initial configuration at time zero, then the time-dependent source $\varepsilon_{\text{grav}}$ may be found by applying (6.5) and then expressing the result in terms of the initial values P_0, ρ_0, the adiabatic exponents Γ_1 and Γ_3, and the rate of change of $\mathcal{R}$. The total luminosity is then found as a function of time by integrating over all mass. This sequence yields

$$\mathcal{L}(t) = -\int_{\mathcal{M}} \frac{P_0}{\rho_0} \frac{\mathcal{R}_0}{\mathcal{R}^2(t)} \frac{d\mathcal{R}}{dt} \left[\frac{3\Gamma_1 - 4}{\Gamma_3 - 1}\right] d\mathcal{M}_r . \tag{6.9}$$

The time-independent factors in the integral may be reexpressed using the virial result (1.21) with $\ddot{I} = 0$. Thus, for example, eliminate the initial pressure and density using

$$-\Omega_0 = 3 \int_{\mathcal{M}} \frac{P_0}{\rho_0} \, d\mathcal{M}_r .$$

If the adiabatic exponents are constant in space as well as time, then the integration is easy and

$$\mathcal{L}(t) = -\frac{1}{3} \frac{\Omega_0 \mathcal{R}_0}{\mathcal{R}^2} \frac{d\mathcal{R}}{dt} \left[\frac{4 - 3\Gamma_1}{\Gamma_3 - 1}\right].$$

This may be further simplified using (1.7) where, as you may recall, q is a dimensionless constant of order unity with $\Omega_0 = -q\left(G\mathcal{M}^2/\mathcal{R}_0\right)$. Thus

$$\mathcal{L}(t) = -q \frac{G\mathcal{M}^2}{\mathcal{R}^2} \frac{d\mathcal{R}}{dt} \left[\frac{\Gamma_1 - 4/3}{\Gamma_3 - 1}\right]. \tag{6.10}$$

If $\Gamma_1 = \Gamma_3 = 5/3$, then the luminosity relation of (1.28) is regained. Note that here, however, we have specified a bit more carefully just which Γs are involved in the luminosity. Before it was a generic γ from a γ-law equation of state.

Real stars sometimes contract in a near-homologous fashion. It is more usual, however, for evolving stars to contract in some regions while other regions expand. The most familiar example of the latter is during post–main sequence evolution where the core contracts and the outer regions expand. In those real situations, $\varepsilon_{\rm grav}$ must be treated as the local quantity it is.

6.2 Thermonuclear Energy Sources

The major text reference for this section is the excellent book by Donald Clayton (1968), which emphasizes nuclear astrophysics. Although some of the numerical results in that work are out of date, the chapters dealing with how to use nuclear physics to find thermonuclear energy generation rates certainly are not. We advise referring to that book for many of the details we shall be forced to leave out in the following.

After we first consider how reaction rates are found using experimental data supplemented by theory, we shall examine hydrogen, helium, and advanced stages of thermonuclear burning.

6.2.1 Preliminaries

We shall concern ourselves, for the moment at least, only with reactions initiated by charged particles. Neutron-induced reactions, which are of great importance for nucleosynthesis, will be discussed later.

Most thermonuclear reactions in stars proceed through an intermediate nuclear state called the compound nucleus. That is, if α represents some *projectile* (say a proton), and X is a *target* nucleus, and these react to finally give rise to nuclear products β and Y, then the compound nucleus Z^* is the intermediary state. In reaction equation language this statement reads

$$\alpha + X \rightarrow Z^* \rightarrow Y + \beta \, . \tag{6.11}$$

The "$*$" appended to Z implies that the compound nucleus is (almost always) in an excited state. Reaction (6.11) is often written as $X(\alpha, \beta)Y$.

A basic assumption here is that Z^* forgets how it was formed and may decay or break up by any means consistent with conservation laws and selection rules. One permitted breakup "channel" is that consisting of the particles that originally formed Z^*—in our example, $\alpha + X$. As an example, suppose we had produced an excited state of ^{12}C by way of proton capture on ^{11}B with

$$p + {}^{11}\mathrm{B} \rightarrow {}^{12}\mathrm{C}^*.$$

The compound state may then break up in a number of ways ("exit channels") such as

$$\begin{aligned} {}^{12}\mathrm{C}^* &\rightarrow {}^{12}\mathrm{C}^{**} + \gamma \\ &\rightarrow {}^{11}\mathrm{B} + \mathrm{p} \\ &\rightarrow {}^{11}\mathrm{C} + \mathrm{n} \\ &\rightarrow {}^{12}\mathrm{N} + \mathrm{e}^- + \overline{\nu}_e \\ &\rightarrow {}^{8}\mathrm{Be} + \alpha, \quad \textit{etc.} \end{aligned}$$

Here, a "$**$" may mean another excited state of ^{12}C and the other symbols, γ, p, n, e^-, $\overline{\nu}_e$, and α refer, respectively, to gamma–ray photon, proton, neutron, electron, electron antineutrino, and alpha particle (^{4}He nucleus).

The state $^{12}\mathrm{C}^*$ may be thought of as some combination of the states of all the possible exit channels or decay modes. With each of these modes we can associate a mean-life for decay, τ_i, and, through the uncertainty principle, a width in energy, Γ_i, where

$$\Gamma_i \tau_i = \hbar\,. \tag{6.12}$$

Here, $\hbar$ is Planck's constant divided by 2π. Thus the more long-lived the state is, the less likely it is that the decay will take place and the smaller is the width in energy. The probability that $^{12}\mathrm{C}^*$ will decay through the channel i is measured by the comparison of τ_i to the sum of all the other lifetimes. Put more precisely, the probability $\mathcal{P}_i$ is given by

$$\mathcal{P}_i = \frac{1/\tau_i}{\sum_j (1/\tau_j)} = \frac{\tau}{\tau_i} \tag{6.13}$$

where τ, the total mean life of $^{12}\mathrm{C}^*$, is

$$\tau = \left(\sum_j \frac{1}{\tau_j} \right)^{-1}. \tag{6.14}$$

If Γ is defined as the *total energy width*, $\sum_j \Gamma_j$, then

$$\mathcal{P}_i = \frac{\Gamma_i}{\Gamma}\,. \tag{6.15}$$

Among the possible exit channels in our example is the channel by which $^{12}\mathrm{C}^*$ was formed in the first place. We call the latter the entrance channel and denote the width associated with that state as Γ_{entr}. If we could turn the clock backward then, in a crude sense, the ratio $\Gamma_{\mathrm{entr}}/\Gamma$ should be a measure of forming $^{12}\mathrm{C}^*$ through that channel.

For charged particle channels the Γ_i are broken down into two factors. There is the intrinsic probability that the state Z^* is composed of the charged constituents of the ith channel in a common nuclear potential. Here is where the nuclear physics really comes in. We shall just give it a name and a symbol: that factor is the "reduced width" γ_i^2. The second factor describes the probability that the separate constituents of the compound state ($\alpha+X$ or $Y+\beta$ of 6.11) can dissociate from one another and become distinct particles far-removed from one another. For charged particles this means that not only must their combined energy within the nuclear potential well be positive relative to the total energy (including rest mass energy) of the separated particles but, further, that they be able to overcome the Coulomb barrier between them. This last probability will be denoted by P_ℓ, the *Coulomb penetration factor*. To clarify what this means, we first digress into some nuclear energetics.

6.2.2 Nuclear Energetics

The total binding energy, B, of a nucleus is defined as the energy required to break up and disperse to infinity all the constituent nucleons (protons and neutrons) in that nucleus. Using "mc^2" arguments this is

$$B = (\text{mass of constituent nucleons} \;-\; \text{mass of bound nucleus})\, c^2 \quad (6.16)$$

where B is usually expressed in units of MeV. (Note that we do not worry here about some refinements concerning just how "mass" is defined such as whether electronic binding energies are included or not.) The average binding energy per nucleon, B/A, is defined as B divided by the total nucleon mass number, A. It is roughly a measure of the energy required to remove the most energetic nucleon from a given nucleus in its ground state. Figure 6.1 shows a schematic of the experimentally derived B/A for the most stable isobar of nuclei with atomic mass A.

The important feature of the binding energy per nucleon curve is its rapid rise from low mass number nuclei to a plateau around A of 60, and then a gradual decline thereafter. The plateau around $A \approx 60$ includes the relatively common element iron, and thus that region is commonly referred to as the "iron-peak" region. The significance of this curve is well known: it, unfortunately, makes fission and fusion weaponry possible. The fusion branch of the B/A curve is that region below $A \approx 60$. Thus if we fuse two light nuclei on that branch, energy is released. In particular, the requirement for energy release in fusion is that $B_1 + B_2 < B_3$ where nuclei 1 and 2 combine to give nucleus 3. That is, if we tear the first two nuclei apart and then reconstitute the remains to form the third, then energy is left over.

As an example, consider the reaction that fuses three ^{4}He nuclei (or α-particles) to form one nucleus (or atom) of ^{12}C. Reading roughly from Figure 6.1, B/A for ^{4}He is ≈ 7.1 MeV per nucleon. For carbon, $B/A \approx 7.7$.

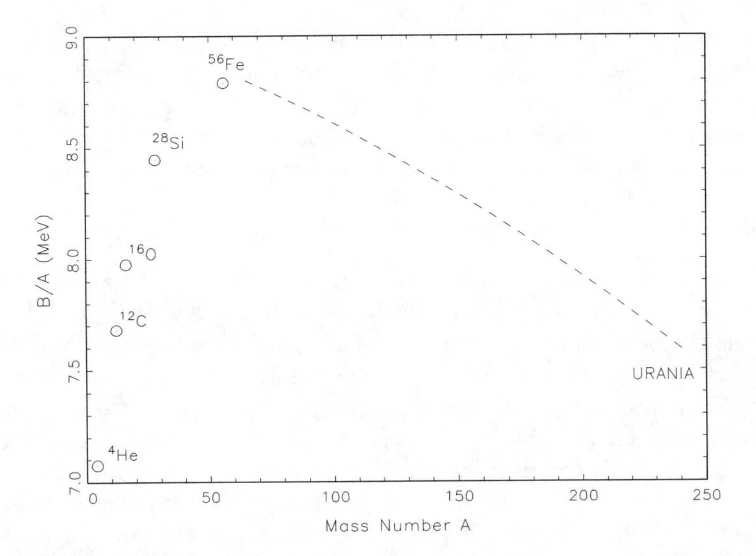

FIGURE 6.1. A schematic of the binding energy per nucleon, B/A, as a function of atomic mass number A for the most stable isobar of A. "Urania," who here represents the elements around uranium, is the Muse of Astronomy.

(More exact numbers are 7.074 and 7.6802 MeV.) The total binding energies are then $4 \times 7.1 \approx 28.4$ and $12 \times 7.7 \approx 92.4$ MeV respectively. Taking apart three α-particles then requires $3 \times 28.4 \approx 85.2$ MeV. Reconstituting the resulting protons and neutrons into ^{12}C gives back 92.4 Mev leaving an excess, or "Q-value," for the reaction of about 7.2 MeV (really 7.275 MeV). Since a gram of pure ^{4}He consists of $N_{\rm A}/4 \approx 1.5 \times 10^{23}$ atoms and it requires three ^{4}He nuclei per reaction, then a complete conversion to ^{12}C yields approximately 6×10^{17} erg g^{-1} after the MeV have been converted to ergs (by multiplying by 1.602×10^{-6} erg MeV^{-1}). In practice, nuclear mass or, better yet, mass excess ($\Delta = [M - A]c^2$) tables are used to find the Q-values of reactions (as in Wapstra and Bos 1977).

Fission on the branch with A greater than about 60 achieves the same end as the above. What this means is that nuclei around the iron peak, which are the most tightly bound of all nuclei (per constituent nucleon), are not of much use as an energy source. Thus any star that ends up with nuclei in the iron peak has lost potential fuel; this is a matter of grave consequence for the star.

Once enough nucleons have been packed into a nucleus, B/A approaches its saturation value of around 8 MeV. To remove one nucleon in a high-lying state from such a nucleus then requires about 8 MeV. How much deeper is the nuclear well below 8 MeV? An estimate may be gained from the observation that nucleons in the nuclear well behave, to zeroth order, as independent fermions of spin 1/2 in a zero temperature sea. That is, the nuclear protons (or neutrons) are stacked in energy in pairs from the bottom of the potential well on up. The Fermi energy of the most energetic proton may be found from the number density of protons in the nucleus as we did with electrons in a Fermi sea in Chapter 3. For this, assume that the nucleus is spherical with radius R. A perfectly good estimate for the nuclear radius is

$$R \approx 1.4 \times 10^{-13} A^{1/3} \text{ cm}. \tag{6.17}$$

With A nucleons (each with a mass of about 1.67×10^{-24} g) packed into the sphere, this yields a nuclear density of around 1.5×10^{14} g cm^{-3}.

Since the proton and neutron rest mass energies are near 940 MeV, and all other nuclear energies are in the MeV range, we may treat the nucleons as nonrelativistic particles. If the protons and neutrons are considered independently, and if $Z \approx N \approx A/2$, then a simple calculation yields (see §3.5.1), for either neutrons or protons, a Fermi momentum of $p_F \approx 1.15 \times 10^{-14}$ g cm s^{-1} and a Fermi energy $\mathcal{E}_F = p_F^2/2m$ of close to 25 MeV. Thus the potential depth of the typical massive nucleus is about $25 + 8 \approx 30$ MeV.

For a charged particle seeking either to leave or enter a nucleus, there is another important energy, the maximum height of the Coulomb barrier between the interacting nuclei. If, for example, the charges of the target and projectile are Z_X and Z_α, then that height is $B_\text{C} = Z_\alpha Z_X e^2/R$, where R is now taken to be the minimum interparticle separation

$$R = 1.4 \left(A_\alpha^{1/3} + A_X^{1/3} \right) \text{ fm} \tag{6.18}$$

and "fm" denotes "femtometer" or "fermi" and is 10^{-13} cm. The Coulomb barrier height is then

$$B_\text{C} \approx 1.44 \frac{Z_\alpha Z_X}{R} \text{ MeV} \tag{6.19}$$

with R in fm. For typical fusioning nuclei with, say, $Z_\alpha \approx Z_X \sim 2$, and $A_\alpha \approx A_X \sim 4$, $B_\text{C} \sim$ MeV. The potential well energetics just described are summarized in Figure 6.2.

The barrier penetration factor mentioned previously is the probability that a particle may quantum mechanically tunnel through the Coulomb barrier shown in Figure 6.2. Two situations are of interest here. A projectile, such as α of reaction (6.11), with initial kinetic energy $\mathcal{E}$, must tunnel through the Coulomb barrier to reach the target X or, in the exit channel, the compound state Z^* breaks up internally into $Y + \beta$ and β must have enough energy to tunnel its way out to freedom from Y. It is usually the case in stellar charged particle reactions that particles β of the exit channel have energies comparable to, or greater than, B_C whereas entrance channel particles α have much lower energies than B_C.

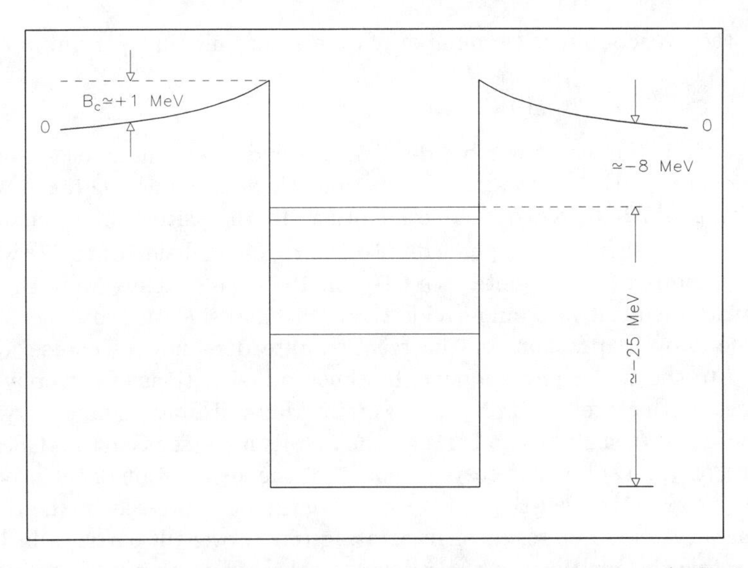

FIGURE 6.2. A sketch of a nuclear potential well including the $1/r$ Coulomb barrier between the target nucleus and a hypothetical charged projectile. (Not to scale.)

The relation between the barrier penetration factor P_ℓ, reduced width (γ_i^2), and Γ_i is

$$\Gamma_i = 2P_\ell \gamma_i^2 \tag{6.20}$$

where, as we shall see shortly, P_ℓ can depend sensitively on the ratio of kinetic energy to Coulomb barrier height. The subscript "ℓ" refers to the angular momentum quantum number ℓ and it implies that the effectiveness of barrier penetration also depends on the relative angular momenta of the particles involved in the reaction.

6.2.3 Astrophysical Thermonuclear Cross Sections and Reaction Rates

To derive a rate for a particular reaction, we need the cross section for that reaction. For the moment, imagine that we are in the laboratory, and the target nucleus X is stationary while being bombarded by a beam of monoenergetic particles, α, of velocity v. Calling $\sigma_{\alpha\beta}$ the cross section for the reaction (6.11), where the subscripts tell us what are the in and out particles, we have (as in the definition 4.31)

$$\sigma_{\alpha\beta}(v) = \frac{\text{number of reactions per unit time per target } X}{\text{incident flux of projectiles } \alpha} \ \text{cm}^2. \tag{6.21}$$

The incident projectile flux is $n_\alpha v$ where n_α is the number density of projectiles in the beam so that the reaction rate per target nucleus is $n_\alpha v\, \sigma_{\alpha\beta}$.

The *total* reaction rate (in number of events per unit time per unit volume of target) is

$$r_{\alpha\beta}(v) = n_\alpha n_X \, \sigma_{\alpha\beta}(v) \, v \quad \text{cm}^{-3}\ \text{s}^{-1} \tag{6.22}$$

where n_X is the target number density. If α and X are the same species of particle (as in the proton-proton reaction) then care must be taken when applying (6.22) to avoid double counting. In that situation, a factor of $(1+\delta_{\alpha X})^{-1}$ should be appended onto the right-hand side of (6.22) where $\delta_{\alpha X}$ is the Kroenecker delta (see Clayton 1968, §4-2). Thus, were the two particles identical we would divide the right-hand side by two.

The above expression for the reaction rate does not, of course, quite apply to the stellar environment. In almost all situations of astrophysical interest both targets and projectiles are in Maxwell-Boltzmann energy distributions. We shall not go through the details here (see Clayton, Chapter 4, or Cox §17.12) but it is easy to show that the distribution of the product $n_\alpha n_X$ is also Maxwell-Boltzmann but it must be expressed in the center of mass system. The result of integrating (6.22) over all particles in their respective distributions is the rate, $r_{\alpha\beta}$, phrased in terms of a suitably averaged product of cross section and velocity:

$$r_{\alpha\beta} = n_\alpha n_X \, \langle \sigma v \rangle_{\alpha\beta} \quad \text{cm}^{-3}\ \text{s}^{-1} . \tag{6.23}$$

The total number densities appear in this expression as n_α and n_X and $\langle \sigma v \rangle_{\alpha\beta}$ is constructed by the weighting of cross section and velocity with the differential Maxwell-Boltzmann distribution, $\Psi(\mathcal{E})$, in energy space. The latter was given in Chapter 3 as (3.25); that is,

$$\Psi(\mathcal{E}) = \frac{2}{\pi^{1/2}} \frac{1}{(kT)^{3/2}} \, e^{-\mathcal{E}/kT} \, \mathcal{E}^{1/2} . \tag{6.24}$$

Note that here, and hereafter in this section, $\mathcal{E}$ will be the center of mass energy of an α–X pair. With this weighting $\langle \sigma v \rangle_{\alpha\beta}$ is given by

$$\langle \sigma v \rangle_{\alpha\beta} = \frac{\int_0^\infty \Psi(\mathcal{E}) \, \sigma_{\alpha\beta}(\mathcal{E}) \, v \, d\mathcal{E}}{\int_0^\infty \Psi(\mathcal{E}) \, d\mathcal{E}} \tag{6.25}$$

where the velocity is in the center of mass. Note that the integral in the denominator is unity because of normalization.

Putting in Ψ explicitly and using $v = (2\mathcal{E}/m)^{1/2}$, $\langle \sigma v \rangle_{\alpha\beta}$ becomes

$$\langle \sigma v \rangle_{\alpha\beta} = \left(\frac{8}{\pi m} \right)^{1/2} (kT)^{-3/2} \int_0^\infty \sigma_{\alpha\beta}(\mathcal{E}) \, e^{-\mathcal{E}/kT} \, \mathcal{E} \, d\mathcal{E} \quad \text{cm}^3\ \text{s}^{-1} . \tag{6.26}$$

The mass, m, in this expression is the reduced mass,

$$m = m_\alpha m_X / (m_\alpha + m_X) . \tag{6.27}$$

The problem is now reduced to finding $\sigma_{\alpha\beta}(\mathcal{E})$ in the center of mass. It turns out that the cross section can almost always be written in the form

$$\sigma_{\alpha\beta}(\mathcal{E}) = \pi\lambda\!\!\!^{-2} g \frac{\Gamma_\alpha \Gamma_\beta}{\Gamma^2} f(\mathcal{E}) . \tag{6.28}$$

Here the widths are as defined previously with Γ_α being the width of the entrance channel $(\alpha + X)$, Γ_β is that of the exit channel $(\beta + Y)$, and Γ the total width. All are functions of energy. The wavelength $\lambda\!\!\!^-$ is the reduced DeBroglie wavelength

$$\pi\lambda\!\!\!^{-2} = \frac{\pi\hbar^2}{2\mathcal{E}m} = \frac{0.657}{\mathcal{E}(\text{MeV})}\frac{1}{\mu} \quad \text{barns } (10^{-24} \text{ cm}^2) \tag{6.29}$$

where $\mathcal{E}$ on the right-hand side is in MeV. The quantity μ (not to be confused with other μs we have defined) is the reduced mass in amu units (with one amu equal to 931.494 MeV in μc^2 units) and it is given by $\mu = A_\alpha A_X/(A_\alpha + A_X)$. The factor g is statistical and contains information on the spins of the target, projectile, and compound nucleus. It is just a number of order unity for our purposes. Finally, the factor $f(\mathcal{E})$; we shall refer to it as a "shape factor."

The various important pieces of the cross section come about as follows. For the factor containing the DeBroglie wavelength: suppose we have a nucleus that is a perfect absorber. If a particle α with linear momentum p comes within an impact parameter distance s, then its quantized angular momentum is $sp = \ell\hbar$. Thus each ℓ has associated with it an impact parameter s_ℓ. For angular momenta between ℓ and $\ell + 1$, the target area, or fractional cross section, of a ring bounded by s_ℓ and $s_{\ell+1}$ is $\sigma_{\ell,\ell+1} = \pi\left(s_{\ell+1}^2 - s_\ell^2\right) = \pi\lambda\!\!\!^{-2}(2\ell + 1)$, where $\lambda\!\!\!^- = \hbar/p$.

The factor $\Gamma_\alpha\Gamma_\beta/\Gamma\Gamma$ is the joint probability of forming $\alpha + X$ and then $\beta + Y$ through the compound state Z^*, as was indicated by our earlier discussion.

The shape factor, $f(\mathcal{E})$, hides many physical effects that we can only allude to. We shall give it one of two forms: either the "resonant" or "nonresonant" $f(\mathcal{E})$. The first form varies rapidly with energy over some interesting energy range whereas the second form is always slowly varying.

1. The *resonant* form of f is that of an isolated Breit–Wigner resonance (in its most simple guise) whereby

$$f(\mathcal{E}) = \frac{\Gamma^2}{(\mathcal{E} - \mathcal{E}_r)^2 + (\Gamma/2)^2} . \tag{6.30}$$

This form of f is strongly peaked at, or near, the *resonance energy* $\mathcal{E}_r$ and reflects the fact that the compound nucleus has a discrete state at an energy $\mathcal{E}^*$ corresponding to

$$m_{Z^*} c^2 + \mathcal{E}^* = m_\alpha c^2 + m_X c^2 + \mathcal{E}_r . \tag{6.31}$$

If Γ does not vary appreciably over the interval $\mathcal{E}_r - \Gamma/2 \leq \mathcal{E} \leq \mathcal{E}_r + \Gamma/2$, then the width of the state at half maximum is Γ. The Breit-Wigner resonant cross section is then

$$\sigma_{\alpha,\beta} = \pi \lambda\!\!\!^{-2} g \frac{\Gamma_\alpha \Gamma_\beta}{(\mathcal{E} - \mathcal{E}_r)^2 + (\Gamma/2)^2} \, . \tag{6.32}$$

For energies near $\mathcal{E}_r$, the last factor dominates and the cross section has a sharp peak around $\mathcal{E}_r$.

2. The *nonresonant* case arises when $f(\mathcal{E})$ is a constant or is slowly varying compared to other factors in the cross section. This situation commonly arises when $\mathcal{E}$ is far removed from $\mathcal{E}_r$; that is, when the reaction takes place far in the tail of a resonance. It may also arise when the reaction is intrinsically nonresonant (as we shall see in the example of the proton–proton reaction). This case will be treated first because it contains features common to many types of charged particle reactions.

6.2.4 Nonresonant Reaction Rates

These rates require further discussion of the Coulomb barrier penetration factor because it is part of the widths contained in $\sigma_{\alpha\beta}$ (and see 6.20). We first note that nuclear reactions of major astrophysical interest are exothermic: they produce energy and the Q-value is positive in the nuclear energy equation

$$m_\alpha c^2 + m_X c^2 = m_\beta c^2 + m_Y c^2 + Q \, . \tag{6.33}$$

The Q-value is usually of order MeV and that is shared as kinetic energy among the exit channel particles if they are in their ground states. In a real reaction, the entrance channel kinetic energy is added on to the left-hand side of (6.33) but reappears in the exit channel and hence does not change what energy is added to the system—namely, Q per reaction. The entrance channel energy is, however, very important for the cross section. In terms of kT, entrance channel energies are typically $kT = 8.6174 \times 10^{-8}\, T(K)$ keV (see Eq. 3.35). The total temperature range spanning the H–, He–, and C(carbon)-burning stages is about $10^7 \lesssim T(K) \lesssim 10^9$ under normal circumstances. In terms of kT this is $1 \lesssim kT \lesssim 100$ keV. Thus input channel energies are usually considerably less than those of the exit channel.

The Coulomb barrier height, B_C, of (6.19) is also very large in comparison to input channel energies. From a classical perspective, this means that the target and projectile can never combine under these circumstances because the Coulomb barrier of Figure 6.2 cannot be penetrated. On the quantum level, on the other hand, this is possible. As in other such problems involving tunneling through a relatively high barrier, however, we expect the barrier penetrability factor, P_ℓ, to be very sensitive to energy. We again refer the reader to Clayton (or almost any text on nuclear physics) where the following form for P_ℓ in the entrance channel is derived using WKBJ methods and which is valid for $B_C \gg \mathcal{E}$:

$$P_\ell(\mathcal{E}) \propto e^{-2\pi\eta} \tag{6.34}$$

where η is the dimensionless Sommerfeld factor

$$\eta = \frac{Z_\alpha Z_X e^2}{\hbar v} = 0.1574 Z_\alpha Z_X \left(\frac{\mu}{\mathcal{E}}\right)^{1/2} . \tag{6.35}$$

Here, $\mathcal{E}$ is the entrance channel kinetic energy (CM) in MeV and μ is the reduced mass of (6.27). For $\mathcal{E} \sim kT \lesssim 100$ keV, $\mu \sim 1$, and $Z_\alpha Z_X \gtrsim 2$, find that $2\pi\eta \gtrsim 12$. With this factor in the exponential of P_ℓ it is clear that the latter is a very sensitive function of $\mathcal{E}$. Other energy- and angular momentum–dependent terms enter into a complete formulation of P_ℓ but, unless energies become comparable to the barrier height, these terms are not nearly as important as the exponential.[1] For energies small compared to B_C we shall use (6.34) because it is more than adequate. Any proportionality factors we have left out of that expression will be absorbed in various elements of the cross section.

The situation for the exit channel is different. Here, for charged particle channels, stellar exothermic reactions with Q-values in excess of a couple of MeV mean that the Coulomb barrier is relatively easy to tunnel through near or above its top and P_ℓ becomes insensitive to $\mathcal{E}$. This is also true when the exit channel is electromagnetic and a γ-ray is produced along with a charged ion or when the exiting particle is a neutron. Thus the exit channel width Γ_β in expression (6.28) for the cross section is taken to be a constant independent of the entrance channel energy $\mathcal{E}$ or, at worst, it may vary slowly with energy. The total width, Γ, is also assumed to vary slowly with $\mathcal{E}$ by the same arguments.

Putting the above elements together, and remembering that $\lambda\!\!\!^{-2}$ goes as $1/\mathcal{E}$, the nonresonant form for the cross section of (6.28) becomes

$$\sigma_{\alpha\beta}(\mathcal{E}) = \frac{S(\mathcal{E})}{\mathcal{E}}\, e^{-2\pi\eta} . \tag{6.36}$$

Here $S(\mathcal{E})$ is a slowly varying function of $\mathcal{E}$ and contains all the energy dependencies not contained in $\lambda\!\!\!^-$ or in the exponential of P_ℓ.

Since the form of η is known, a common procedure is to extract $\sigma_{\alpha\beta}$ at experimentally accessible laboratory energies (usually, and unfortunately, at energies not much below 100 keV) and plot $S(\mathcal{E}) = \sigma_{\alpha\beta}\, \mathcal{E}\, e^{2\pi\eta}$ as a

---

[1] To find the exact dependence of the barrier penetrability on ℓ and $\mathcal{E}$ requires calculating the Coulomb wave functions. These are discussed in most nuclear physics texts and in some mathematical physics texts.

function of $\mathcal{E}$. If necessary $S(\mathcal{E})$ is then extrapolated, with perhaps some help from theory, down to astrophysically interesting energies. We shall give an example shortly. Some of the major uncertainties in reaction rates derive from this procedure because low experimental energies usually involve low cross sections and, hence, relatively large experimental errors.

Assuming that $\sigma_{\alpha\beta}(\mathcal{E})$ is known either from experiment and/or by extrapolation of $S(\mathcal{E})$, we can now introduce (6.36) into the expression (6.26) for $\langle\sigma v\rangle_{\alpha\beta}$ and find

$$\langle\sigma v\rangle_{\alpha\beta} = \left(\frac{8}{\pi m}\right)^{1/2} (kT)^{-3/2} \int_0^\infty S(\mathcal{E}) \exp\left[-\left(\frac{\mathcal{E}}{kT} + \frac{b}{\mathcal{E}^{1/2}}\right)\right] d\mathcal{E} \,. \quad (6.37)$$

Here, m is the reduced mass in grams, S is in erg–cm, and $b/\mathcal{E}^{1/2}$ replaces $2\pi\eta$ with

$$b = 0.99 Z_\alpha Z_X \mu^{1/2} \quad (\text{MeV})^{1/2} \quad (6.38)$$

and μ is in amu. As a first step in evaluating the integral in (6.37) $S(\mathcal{E})$ is either evaluated at some typical energy where most reactions take place, or it is extrapolated to zero energy, yielding $S(0)$, and a first-order constant derivative, $dS/d\mathcal{E}|_0$, is added on to that. These are refinements (albeit often necessary) and, for our purposes, it will be sufficient to assume that S is some experimentally determined constant that may be taken out from inside the integral.

With S assumed constant, the numerical form of $\langle\sigma v\rangle_{\alpha\beta}$ is then

$$\langle\sigma v\rangle_{\alpha\beta} = \frac{1.6\times 10^{-15}}{\mu^{1/2}(kT)^{3/2}} S \int_0^\infty \exp\left[-\left(\frac{\mathcal{E}}{kT} + \frac{b}{\mathcal{E}^{1/2}}\right)\right] d\mathcal{E} \ \text{cm}^3\ \text{s}^{-1} \quad (6.39)$$

where S is now in MeV-barns and $\mathcal{E}$ and kT are in MeV.[2]

The structure of the integrand in (6.39) reflects the combination of two strongly competing factors. The barrier penetration factor contributes the factor $\exp\left(-b/\mathcal{E}^{1/2}\right)$, which increases rapidly with increasing energy, whereas the Maxwell-Boltzmann exponential decreases rapidly as energy increases. The integrand thus increases as energy increases because the Coulomb barrier becomes more penetrable but, to offset that, the number of pairs of particles available for the reaction decreases in the exponential tail of the distribution. What results is a compromise between the two competing factors. This is illustrated in Figure 6.3, where the integrand of (6.39) is plotted for two temperatures. The reactants are protons and ^{12}C nuclei.

The integrand of (6.39) is aptly called the "Gamow peak" because of its shape and in honor of George Gamow who early on investigated the problems of quantum mechanical transmission though barriers (and who made

[2]Our continual switching of units is done not to be capricious: you will see all these combinations in the literature.

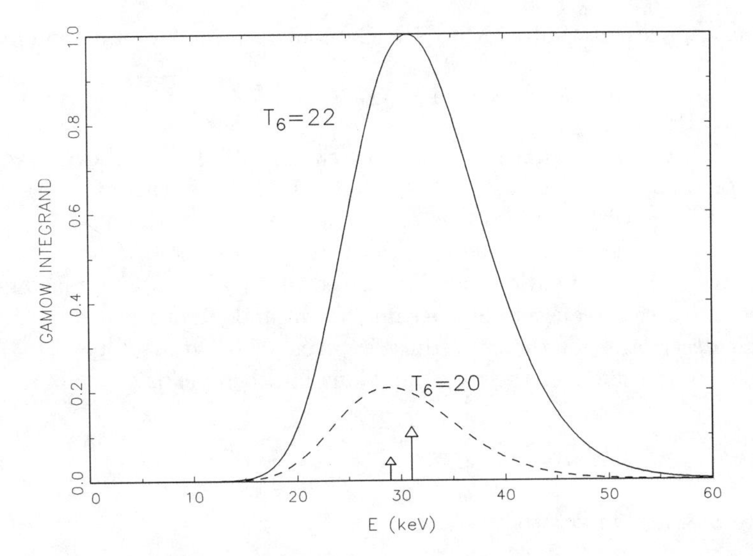

FIGURE 6.3. The integrand of (6.39) is plotted against center of mass energy (in keV) for the temperatures $T_6 = 20$ and $T_6 = 22$. The input channel is protons on ^{12}C. The figure has been scaled by dividing both integrands by 6.6×10^{-22}, which is the maximum (prescaled) of the integrand for $T_6 = 22$. The small (large) vertical arrow indicates the summit of the Gamow peak for $T_6 = 20$ ($T_6 = 22$) (see text).

many contributions to astrophysics). It is easy to show that the summit of the peak lies at an energy of

$$\mathcal{E}_0 = 1.22 \left(Z_\alpha^2 Z_X^2 \mu T_6^2 \right)^{1/3} \text{ keV}. \tag{6.40}$$

Here, T_6 is the temperature in units of 10^6 K. Thus, for example, at a temperature of 2.2×10^7 K ($T_6 = 22$), $kT = 1.896$ keV and, for the $p + {}^{12}$C reaction, $\mathcal{E}_0 \approx 30.8$ keV. (The $p + {}^{12}$C reaction will be used as a prototype reaction for both resonant and nonresonant rates here.) The location of the summit in energy for these conditions is indicated by the large arrow in Figure 6.3. Note that this result depends only on the reaction being nonresonant; the details of the nuclear physics are almost irrelevant. Also shown in Figure 6.3 is the integrand for a slightly lower temperature of $T_6 =$ 20 where the height of the peak is lower by a factor of almost five compared to the higher temperature. This is characteristic of low energy charged particle reactions where the rate is a sensitive function of temperature. If you compute logarithmic derivatives of these numbers, you will find that the height of the peak varies as roughly the 17th power of the temperature. This should come as no surprise because we are dealing with one of the reactions in the CNO cycles (see §1.5).

The approximate full-width of the Gamow peak (at $1/e$ of maximum) is

$$\Delta \approx 2.3\,(\mathcal{E}_0\, kT)^{1/2} \tag{6.41}$$

in whatever common units are used for $\mathcal{E}_0$ and kT. In the above example with $T_6 = 22$, $\Delta \approx 17.6$ keV so that roughly half of the reactions arise from reactant pairs with energies $22 \lesssim \mathcal{E} \lesssim 40$ keV $>> kT$.

A closed expression for the integral in (6.39) does not exist, but a perfectly useful approximation may be derived that follows from replacing the integrand with a Gaussian of the same height and curvature at maximum. A full description of the procedure is given in Clayton (1968, §4–3). A simple integration over the resulting Gaussian then yields

$$\langle \sigma v \rangle_{\alpha\beta} = \frac{0.72 \times 10^{-18} S}{\mu Z_\alpha Z_X}\, e^{-\tau}\, \tau^2 \;\; \text{cm}^3\ \text{s}^{-1} \tag{6.42}$$

where S is now in keV–barns, and

$$\tau = \frac{3\mathcal{E}_0}{kT}\,. \tag{6.43}$$

Correction terms that improve on the Gaussian consist of multiplying (6.42) by $1+(5/12\tau)+\cdots$. Because $\mathcal{E}_0$ is usually much greater than kT, we shall neglect this correction (and while we have not included anything fancy for possible slow energy variations in S, it can be expressed as a Taylor series with knowledge of $dS/d\mathcal{E}$, as in Clayton, §4-3).

We can express (6.42) in terms of temperature by unwinding τ to find

$$\langle \sigma v \rangle_{\alpha\beta} = \frac{0.72 \times 10^{-18} S a^2}{\mu Z_\alpha Z_X}\, \frac{e^{-a T_6^{-1/3}}}{T_6^{2/3}} \;\; \text{cm}^3\ \text{s}^{-1} \tag{6.44}$$

where $a = 42.49\left(Z_\alpha^2 Z_X^2 \mu\right)^{1/3}$ and S is in keV-barns. The temperature exponents in the exponential and the denominator are characteristic of nonresonant reactions.

To give an example of how the above results are applied, consider the well-studied reaction $^{12}\text{C}\,(\text{p},\gamma)\,^{13}\text{N}$. (Clayton also uses this reaction as a prototype.) At typical hydrogen-burning temperatures on the main sequence, this reaction proceeds primarily through the low energy tail of a resonance in ^{13}N (at 2.37 MeV) which, in the laboratory frame, is directly accessed by a proton with an energy of 0.46 Mev (and remember, the ^{12}C nuclei are stationary in the laboratory frame). The cross section for this reaction is shown in Figure 6.4 (taken from Fowler et al. 1967) where the abscissa is the laboratory energy of the proton.

The laboratory to center of mass conversion of the total kinetic energy of a projectile-target pair is $\mathcal{E}(\text{CM}) = \mathcal{E}(\text{Lab})[m_X/(m_\alpha + m_X)]$ where, in our example, α refers to the proton and X to ^{12}C. For $T_6 = 22$, the Gamow

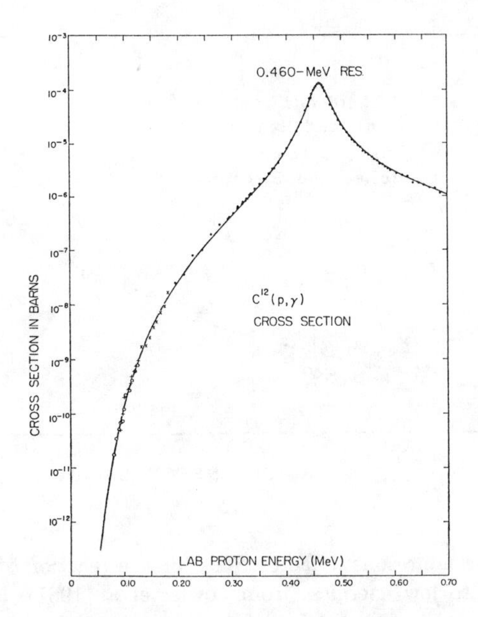

FIGURE 6.4. The experimental cross section for the reaction ^{12}C (p, γ) ^{13}N from Fowler et al. (1967). The solid curve is based on theory. (See also Figure 6.5.) Reproduced, with permission, from the *Annual Review of Astronomy and Astrophysics*, Volume 5, ©1967 by Annual Reviews, Inc.

peak (center of mass) energy of $\mathcal{E}_0 = 30.8$ keV corresponds to a proton laboratory energy of 33.4 keV = 0.0334 MeV. It is clear that the information contained in Figure 6.4 must be extrapolated down to energies well below the experimental data in order that $\langle \sigma v \rangle$ be computed. This is done, as discussed earlier, by removing the penetration factor and $1/\mathcal{E}$ dependence (of 6.36) from the data and then plotting $S(\mathcal{E})$. The results of this procedure are shown in Figure 6.5 (also from Fowler et al. 1967). Extrapolation to low energies yields $S(\mathcal{E}=0) = 1.4$ Mev-barns for the reaction. The rate may now be computed using (6.42) and ancillary equations.

The result we now quote for the nonresonant reaction ^{12}C(p, γ)^{13}N is from Fowler et al. (1975) where all their corrections for the energy dependence in $S(\mathcal{E})$ and adjustments to the Gaussian approximation to the shape of the Gamow peak are included for completeness. (Note that in their notation $\langle \sigma v \rangle_{\mathrm{p},\gamma}$ becomes $\langle ^{12}\mathrm{C\,p}\rangle_\gamma$.) Their result is

$$\begin{aligned} \langle \sigma v \rangle_{\mathrm{p},\gamma} &= 3.39 \times 10^{-17}(1 + 0.0304\, T_9^{1/3} + 1.19\, T_9^{2/3} + 0.254\, T_9 \\ &\quad + 2.06\, T_9^{4/3} + 1.12\, T_9^{5/3}) \times T_9^{-2/3} \\ &\quad \times \exp\left[-13.69/T_9^{1/3} - (T_9/1.5)^2\right] \ \mathrm{cm}^3\ \mathrm{s}^{-1}. \end{aligned} \tag{6.45}$$

The correction terms alluded to in the comments following equation (6.43)

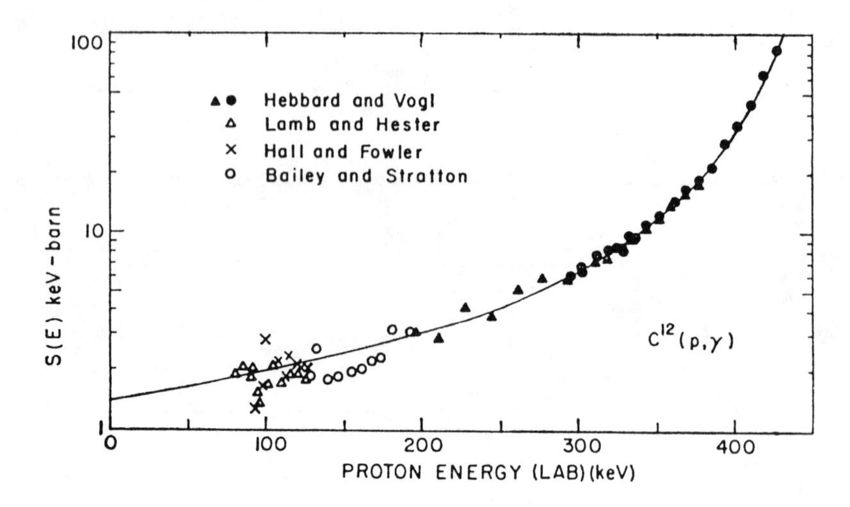

FIGURE 6.5. The nonresonant factor $S(\mathcal{E})$ for the reaction $^{12}\mathrm{C}\,(\mathrm{p},\gamma)\,^{13}\mathrm{N}$ with an extrapolation to low energies (from Fowler et al. 1967). Reproduced, with permission, from the *Annual Review of Astronomy and Astrophysics*, Volume 5, ©1967 by Annual Reviews, Inc.

are those in the parenthesis following the unit term and in the T_9^2 term of the exponential. You may easily verify that the rest of the expression follows from the information already given. Note that the units of temperature are in billions of degrees, which is generally the case for the reaction rate material published by Fowler and his coauthors (for a listing of those papers, see the references at the end of this chapter). The quoted limits of applicability of (6.45) are $0 \leq T_9 \leq 0.55$ for reasons to be made apparent shortly.

Note: From now on in this chapter we shall usually not include all correction terms or information on all resonances for a reaction (nor electron screening terms—see later). Instead, only those terms necessary for quick and dirty calculations will be given. If you have to do better than that, we strongly urge that you consult the details given in the references and check the recent literature; rates have a way of changing through the years.

Once $\langle\sigma v\rangle$ has been found then the total reaction rate per unit volume is given by (see 6.23 and 1.42)

$$r_{\alpha\beta} = \rho^2 N_A^2 \frac{X_\alpha X_X}{A_\alpha A_X} \langle\sigma v\rangle_{\alpha\beta} \ \mathrm{cm}^{-3}\ \mathrm{s}^{-1} \tag{6.46}$$

where the X_i are the mass fractions of particles α and X.[3]

The energy generation rate per gram, $\varepsilon_{\alpha\beta}$, is simply the reaction rate multiplied by the Q-value (in ergs) of the reaction divided by the density. The units of $\varepsilon_{\alpha\beta}$ are then erg g^{-1} s^{-1} in accordance with the discussion in §1.5. Thus

$$\varepsilon_{\alpha\beta} = \frac{r_{\alpha\beta} Q}{\rho} \ \text{erg g}^{-1}\ \text{s}^{-1}. \tag{6.47}$$

A little care must be exercised if a neutrino is produced as a result of the reaction. Under all but the most unusual circumstances, matter is essentially transparent to these particles and thus the energy associated with the neutrino is lost from the star. That energy must then be subtracted from Q. Neutrinos play no role in the $^{12}\text{C}\,(\text{p},\gamma)\,^{13}\text{N}$ reaction and its Q is 1.944 MeV=3.115×10^{-6} ergs.

The functional dependence of ε on temperature is, of course, the same as that of the reaction rate: both go as $\exp\left(-aT^{-1/3}\right)/T^{2/3}$ using (6.44). The density dependence for ε is obviously linear. From these considerations it is easy to derive the temperature and density exponents used in previous chapters. With

$$\varepsilon = \varepsilon_0 \rho^{\lambda} T^{\nu} \tag{6.48}$$

the logarithmic derivatives ν and λ for nonresonant reactions are

$$\lambda = \left(\frac{\partial \ln \varepsilon}{\partial \ln \rho}\right)_T = 1 \tag{6.49}$$

$$\nu = \left(\frac{\partial \ln \varepsilon}{\partial \ln T}\right)_\rho = \frac{a}{3T_6^{1/3}} - \tfrac{2}{3}\,. \tag{6.50}$$

Note again that these expressions do not include any of the correction terms discussed above (and given in 6.45). In addition, it is clear that ν depends on temperature. For $^{12}\text{C}\,(\text{p},\gamma)\,^{13}\text{N}$ at $T_6 = 20$, $\nu \approx 16$. This is a number characteristic of the CNO cycles, as was stated in Chapter 1, and it derives primarily from the effect of the Coulomb barrier.

What happens if we now systematically raise the temperature of the gas containing protons and ^{12}C? At some point the Gamow peak will begin to encroach upon the resonance at 0.424 MeV (in the center of mass) and our notions about how the peak is formed will break down. This temperature is easy to estimate. If $\mathcal{E}_0 + \Delta/2 \approx \mathcal{E}_r$ then the peak begins to overlap the resonance. For this reaction $\mathcal{E}_0 + \Delta/2 = 0.393 T_9^{2/3} + 0.213 T_9^{5/6}$, which equals 0.42 MeV when $T_9 \approx 0.6$. This is the upper temperature limit quoted for (6.45). (The width of the resonance must also be taken into account and that can lower the temperature calculated for the limit.) Therefore, on to resonant reactions.

---

[3] If α and X are the same, then the right-hand side should be divided by two as indicated in the discussion following (6.22).

6.2.5 *Resonant Reaction Rates*

The form of the cross section for a resonant reaction is dominated by the factor $\left[(\mathcal{E}-\mathcal{E}_r)^2+(\Gamma/2)^2\right]^{-1}$ of (6.32). Because $\Psi(\mathcal{E})$, the Maxwell-Boltzmann distribution function given by (6.24), and the Γs vary slowly over a resonance (at least as long as $\mathcal{E}_r \gg \Gamma/2$), what is usually done is to evaluate $\Psi(\mathcal{E})$ and Γ_α at $\mathcal{E}_r$ thereby letting the resonant form act like a delta function.[4] Using this approximation $\langle\sigma v\rangle_{\alpha\beta}$ becomes

$$\begin{aligned}\langle\sigma v\rangle_{\alpha\beta} &= \frac{\pi\hbar^2 g}{2m}\left(\frac{8}{\pi m}\right)^{1/2}(kT)^{-3/2}\,e^{-\mathcal{E}_r/kT}\,\Gamma_\alpha(\mathcal{E}_r)\Gamma_\beta(\mathcal{E}_r)\\ &\times\int_0^\infty \frac{d\mathcal{E}}{(\mathcal{E}-\mathcal{E}_r)^2+(\Gamma/2)^2}\end{aligned} \tag{6.51}$$

where we have used (6.26), (6.29), and (6.32). All that is left to evaluate is the integral over the resonance denominator. Because the integrand peaks sharply at $\mathcal{E}_r$ and nowhere else, including negative energies, it is customary to extend the lower limit of the integral to $-\infty$. The integral is elementary if Γ is taken constant and yields

$$\langle\sigma v\rangle_{\alpha\beta}.= \hbar^2\left(\frac{2\pi}{mkT}\right)^{3/2} g\frac{\Gamma_\alpha\Gamma_\beta}{\Gamma}\,e^{-\mathcal{E}_r/kT}\,. \tag{6.52}$$

This form is particularly useful because sometimes a poor resolution experiment only yields an integrated cross section, $\int_{\text{res}}\sigma(\mathcal{E})\,d\mathcal{E}$, where the integral is only over the resonance. The same sort of delta-function trick used above may be used here to yield

$$\int_{\text{res}}\sigma(\mathcal{E})\,d\mathcal{E} = \frac{\hbar^2\pi^2}{m\mathcal{E}_r}\left\{g\frac{\Gamma_\alpha\Gamma_\beta}{\Gamma}\right\}. \tag{6.53}$$

The term in braces is called $(\omega\gamma)_r$ by Fowler et al. (1967) and these are tabulated for many reactions by them. In these terms,

$$\begin{aligned}\langle\sigma v\rangle_{\alpha\beta} &= \left(\frac{2\pi\hbar^2}{mkT}\right)^{3/2}\frac{(\omega\gamma)_r}{\hbar}\,e^{-\mathcal{E}_r/kT}\\ &= 2.56\times10^{-13}\,\frac{(\omega\gamma)_r}{(\mu\,T_9)^{3/2}}\,e^{-11.605\,\mathcal{E}_r/T_9}\end{aligned} \tag{6.54}$$

where $(\omega\gamma)_r$ and $\mathcal{E}_r$ are in MeV.

---

[4]Note that this is one way to do it. Often, in practice, $\langle\sigma v\rangle_{\alpha\beta}$ is numerically integrated over the resonance using the experimental data. This is especially true if the resonance is broad or $\mathcal{E}_r$ is at low energy or at a negative energy with respect to where the input channel produces the compound state.

Fowler et al. (1967, 1975) give $(\omega\gamma)_r = 6.29 \times 10^{-7}$ MeV, $\Gamma(\mathcal{E}_r) = .0325$ MeV, and $\mathcal{E}_r = 0.424$ MeV for the 0.46 MeV (lab) state in ^{13}N. When inserted into (6.54) these yield

$$\langle \sigma v \rangle_{\mathrm{p},\gamma} = \frac{1.8 \times 10^{-19}}{T_9^{3/2}} \, e^{-4.925/T_9} \ \mathrm{cm}^3 \ \mathrm{s}^{-1} \tag{6.55}$$

for the resonant contribution to the ^{12}C(p,γ)^{13}N reaction. This result is applicable for $0.25 \leq T_9 \leq 7$ and is to be added onto the nonresonant expression (6.45) with due regard for the appropriate temperature limits. You will also want to consult Caughlan et al. (1985, 1988) for numerical tabulations of this rate (and others). The temperature exponent for a resonant rate of the form (6.52 or 6.54) is

$$\nu = \frac{11.61\mathcal{E}_r}{T_9} - \frac{3}{2} \tag{6.56}$$

with $\mathcal{E}_r$ in MeV and, of course, $\lambda = 1$.

6.2.6 Other Forms of Reaction Rates

Thus far we have only considered reactions initiated by protons (in particular) or charged nuclei such as alpha particles (to be dealt with in detail later). We now consider other types of reactions that either do not fall into this category or require special handling (such as the proton-proton reaction).

Neutron Capture Cross Sections

It is now believed that most of the elements with $A > 60$ were formed as the result of successions of neutron capture reactions and electron decays. These reactions can take place during some of the more normal stages of evolution or in supernovae. As an example of the former in helium burning, neutrons are formed by the reaction $^{13}\mathrm{C}(\alpha, \mathrm{n})^{16}\mathrm{O}$ and these neutrons are then captured on "seed" nuclei in the iron range of elements. This is an example of "s–process" nucleosynthesis where "s" stands for "slow" (as a time scale). The rapid "r–process" is usually associated with the fast time scale of supernovae, where a myriad of reactions take place involving many nuclei. In neither process does the production of very heavy nuclei represent an energy source—and thus has little direct effect on evolution—but both are important for our understanding of heavy element abundances found in nature.

The experimental determination of neutron capture cross sections is difficult to come by because neutral particles are hard to control and the neutron has a relatively short lifetime. (The determination by Mampe et al. 1989 gives $\tau = 887.6 \pm 3$ s for the e–folding life). It is fortunate that the form of the cross section is relatively simple and, for low energies, varies as v^{-1} (and see Clayton 1968, §7–3). Because they are unaffected by a Coulomb barrier there are no strong energy dependencies. Thus, at the lowest level of approximation, $\langle\sigma v\rangle$ is constant for any given reaction. For a summary of experiment versus theory see Boa and Käppeler (1987).

Weak Interactions

As an aftermath of the reaction $^{12}C(p,\gamma)^{13}N$, the nucleus ^{13}N is left in the ground state after emission of the γ-ray. Several things may now happen to that nucleus. One possibility is the reaction $^{13}N(p,\gamma)^{14}O$ (Q=4.628 MeV). The $\langle\sigma v\rangle$ for this reaction is

$$\begin{aligned}\langle\sigma v\rangle_{p,\gamma} &= \frac{6.71\times10^{-17}}{T_9^{2/3}}\,e^{-15.202/T_9^{1/3}} \\ &+\frac{4.04\times10^{-19}}{T_9^{3/2}}\,e^{-6.348/T_9}\ \text{cm}^3\ \text{s}^{-1} \qquad (6.57)\end{aligned}$$

for temperatures appropriate to main sequence CNO cycling. You should be able to identify both nonresonant and resonant contributions in this expression. The resonant term is only important at higher temperatures.

The time scale for destruction of ^{13}N is clearly the number density of that nucleus divided by the rate of the reaction (i.e., the rate of destruction of ^{13}N). Thus define

$$\tau_{p,\gamma} = \frac{n(^{13}N)}{r_{p,\gamma}} = \frac{1}{n_p\langle\sigma v\rangle_{p,\gamma}}\ \text{s} \qquad (6.58)$$

where n_p is the proton number density. (We assume, in these sorts of arguments, that volume is constant with time so that number density does not change for that reason. You could better describe number density as number per gram.) The number density of protons is $n_p = N_A\rho X$ where X is the hydrogen mass fraction. For $T_6 = 20$, $X \approx 1$, and $\rho \approx 10$, find that $n_p \approx 6\times10^{24}$ cm^{-3}, and $\langle\sigma v\rangle_{p,\gamma} \approx 4\times10^{-40}$ cm^3 s^{-1}. This yields a time scale $\tau_{p,\gamma} \approx 10^7$ years.

There is, however, a complication here because ^{13}N is an unstable nucleus even in its ground state and positron decays into ^{13}C with a half-life of only 10 minutes.[5] The reaction is

$$^{13}N \rightarrow {}^{13}C + e^+ + \nu_e,\ \ \tau_{1/2} = 10\ \text{min}\,. \qquad (6.59)$$

---

[5]The Q-value for the reaction is 2.22 MeV but the neutrino carries away an average of 0.71 MeV leaving 1.51 MeV for the positron. This last energy is eventually returned to the stellar gas when the positron annihilates with an ambient electron.

We thus encounter a typical situation in nuclear astrophysics: a choice must always be made regarding what reactions are important in any given situation. Here it appears that the (p,γ) reaction may safely be ignored because the ^{13}N nuclei are whisked away by positron decay before the protons can get at them. On the other hand, were the temperature and density higher, say 2.5×10^8 K and 10^3 g cm^{-3}, then the resonant contribution reduces $\tau_{p,\gamma}$ to about one minute so that the capture reaction competes with the beta-decay. Such is the case in the explosive burning of hydrogen near the surface of a classical nova.

Another interesting feature arises because of the rapidity of the positron of ^{13}N. Before the start of hydrogen-burning in a Pop I star the abundance of ^{12}C is about 0.5% by mass of the total but the concentration of ^{13}N is, of course, zero because it is unstable. When burning does commence, the concentration of ^{13}N begins to be built up by the $^{12}\mathrm{C}(\mathrm{p},\gamma)^{13}\mathrm{N}$ reaction. If temperatures are not too high, proton captures on ^{13}N may be neglected (as in the above) and the rate of change of abundance for ^{13}N is given by

$$\frac{d^{13}\mathrm{N}}{dt} = \mathrm{n_p}\ {}^{12}\mathrm{C}\ \langle\sigma v\rangle_{p,\gamma} - \lambda\ {}^{13}\mathrm{N} \quad \mathrm{cm}^{-3}\ \mathrm{s}^{-1} \tag{6.60}$$

where ^{13}N and ^{12}C represent the number densities of the respective nuclei. The beta-decay constant, λ, is related to the half-life by $\lambda = 0.693/\tau_{1/2}$. The time development of ^{13}N under conditions of constant temperature and density and the assumption that elapsed times are sufficiently short that $\mathrm{n_p}$ and ^{12}C also remain constant is

$$^{13}\mathrm{N}(t) = \frac{\mathrm{n_p}{}^{12}\mathrm{C}\langle\sigma v\rangle_{p,\gamma}}{\lambda}\left(1 - e^{-\lambda t}\right) \quad \mathrm{cm}^{-3} . \tag{6.61}$$

This means that the concentration of ^{13}N rapidly approaches an equilibrium value and that it can just as well be computed by setting the time derivative of ^{13}N in (6.60) to zero and solving for ^{13}N. This situation is also common in nuclear astrophysics where the concentration of a nuclide involved in a comparatively rapid reaction may often easily be computed. Other examples arise in reaction chains where other considerations apply (as in the proton-proton chains to be discussed shortly).

Another kind of reaction, which results in the emission of an electron neutrino, is the capture by a nucleus of either a free electron or one in an atomic orbital. An example from the proton-proton chains is

$$\mathrm{e}^- + {}^7\mathrm{Be} \rightarrow {}^7\mathrm{Li} + \nu_\mathrm{e} . \tag{6.62}$$

In the laboratory, neutral atoms of ^{7}Be capture K-shell electrons with a half-life of about 53 days for the capture. In the hydrogen-burning stellar interior, however, temperatures are high enough to completely ionize essentially all of the ^{7}Be present and the reaction must proceed using the free electrons in the stellar plasma. The rate of the reaction is determined, in effect, by how well the wave functions of the electrons overlap the nucleus and by the intrinsic strength of the weak interaction process (see Chiu 1968, chap. 6, for example). An effective $\langle \sigma v \rangle$ is

$$\langle \sigma v \rangle_{e^- \nu_e} = \frac{2.23 \times 10^{-34}}{T_9^{1/2}} \quad \text{cm}^3 \text{ s}^{-1} \tag{6.63}$$

exclusive of some correction terms and restrictions for $T_6 \leq 1$ (see Caughlin et al. 1985). This expression has a form entirely different from those found thus far.

To find the rate for $^7\text{Be}(e^-, \nu_e)^7\text{Li}$, (6.63) must be multiplied by n_e, the free electron number density, and the number density of ^{7}Be. The energy generation rate follows, as usual, by dividing by the density and multiplying by the Q–value. In this reaction the total Q is 0.862 MeV but the neutrinos (of two energies depending on what state of ^{7}Li is produced) carry away all but 0.046 MeV of that figure. Even though this eventually means that the reaction is a minor direct contributor to the energy generation rate in the proton-proton chains, the neutrinos so produced are among those presumably seen by neutrino detectors "focused" on the sun (of which more in Chapter 8).

Electron captures are also important in high-density situations, where electron Fermi energies range into the MeVs—such as in at least one kind of supernova—and we shall touch upon this in §6.8 when we discuss neutrino emission mechanisms. Pertinent references are Fuller et al. (1982, 1985).

The Proton-Proton Reaction

The proton-proton chains are initiated by the reaction

$$^1\text{H} + {}^1\text{H} \rightarrow {}^2\text{H} + e^+ + \nu_e \tag{6.64}$$

where ^{2}H is a deuteron (often given the designation ^{2}D). This crucial but, as it turns out, unlikely reaction requires that two protons form a coupled system (the "diproton") while flashing past one another and, at practically that same instant, one of these protons must undergo a weak decay by emitting a positron and electron neutrino. The two remaining massive particles, proton and neutron, are then left together as the rather fragile deuteron (2.22 MeV binding energy). This sequence of events is so unlikely that it probably will never be measurable with any certainty in the laboratory. However, the theory—for once—appears to be quite reliable.

We shall not derive the rate for this reaction here (see Chiu 1968, Clayton 1968, and, further back in time, the pioneering work of Bethe, Critchfield, and Salpeter listed in the references). It turns out that one of the major uncertainties is the beta-decay lifetime for the neutron, which is needed to compute the reverse process of proton decay. (Other problems may arise because of unusual conditions in the stellar plasma but these are not of a fundamental nature.)

The reaction is nonresonant and the energy dependence of the cross section arises mostly from the Coulomb barrier between the initial proton pair. The Q-value is 1.192 MeV if the energy carried away by the electron neutrino is discarded. From Caughlan and Fowler (1988) we find that for the ${}^1\mathrm{H}({}^1\mathrm{H}, \mathrm{e}^+ + \nu_\mathrm{e})\,{}^2\mathrm{H}$ reaction

$$\begin{aligned} \langle \sigma v \rangle_{\mathrm{pp}} &= \frac{6.34 \times 10^{-39}}{T_9^{2/3}} \left(1 + 0.123 T_9^{1/3} + 1.09 T_9^{2/3} + 0.938 T_9\right) \\ &\quad \times \exp\left(-3.380/T_9^{1/3}\right) \ \mathrm{cm}^3\ \mathrm{s}^{-1} . \end{aligned} \tag{6.65}$$

(New measurements of the neutron lifetime and other corrections result in an increase of the multiplicative factor by a small number of percent. We have not included these here. See Gould and Guessoum 1990.)

The reaction rate is obtained by multiplying by $n_\mathrm{p}^2/2$ where the factor of 1/2 comes about because of the double-counting problem for identical initial particles discussed earlier. The result, excluding correction terms, is

$$r_{\mathrm{p-p\ reaction}} = \frac{1.15 \times 10^9}{T_9^{2/3}} X^2 \rho^2 \, e^{-3.380/T_9^{1/3}} \ \mathrm{cm}^{-3}\ \mathrm{s}^{-1} \tag{6.66}$$

where X is the hydrogen mass fraction. The temperature exponent for the reaction rate and energy generation rate is

$$\nu_{\mathrm{p-p\ reaction}} = \frac{11.3}{T_6^{1/3}} - \frac{2}{3} \tag{6.67}$$

which, for a solar center temperature of about $T_6 = 15$, is $\nu \approx 4$.

It is easy to compute the mean life of a proton against destruction by the p-p reaction; namely,

$$\tau_\mathrm{p} = -\frac{n_\mathrm{p}}{dn_\mathrm{p}/dt} = \frac{n_p}{2 r_{\mathrm{pp}}} . \tag{6.68}$$

(Note that a factor of 2 appears because each reaction destroys two protons.) For $T_6 \approx 15$, $\rho \approx 100$ g cm^{-3}, and $X \approx 0.7$, find that $\tau_\mathrm{p} \approx 6 \times 10^9$ years. That this time scale is close to the nuclear time scale given by (1.88) is no accident: the p-p reaction is so slow that it effectively controls the rate at which the p-p chains operate as a whole.

6.2.7 Special Effects

A major modification to normal reaction rates discussed above has to do with alterations to the Coulomb potential between reactants due to the presence of intervening electrons. This is the problem of "electron screening." Here we only treat the regime where the effects are "weak" and, even then, only approximately. The case of "strong" screening is beyond the scope of this text and, in any event, many questions regarding this regime have not been satisfactorily resolved. For a very readable first paper on the subject, see Salpeter (1954).

Consider two completely ionized identical nuclear reactants of nuclear charge Z. It is assumed that the medium consists solely of these species and free electrons. We introduced the Wigner-Seitz radius, a, in Chapter 3, where it was defined by $(4\pi a^3/3) = (1/n_{\rm I})$ where $n_{\rm I}$ was the ion number density. If $Z^2e^2/a \ll kT$, then a simple exercise in Debye-Hückel theory yields the following expression for the electrostatic potential of one ion surrounded by a cloud of electrons (see, for example, Landau and Lifshitz 1958, §74):

$$\phi(r) = \frac{Ze}{r}\, e^{-\kappa_d r} \tag{6.69}$$

where κ_d, the inverse of the Debye radius, is

$$\kappa_d = \left[\frac{4\pi e^2}{kT}\left(Z^2 n_{\rm I} + n_{\rm e}\right)\right]^{1/2} . \tag{6.70}$$

The net effect of the exponential is to reduce the potential barrier below its pure Coulomb value of Ze/r at a given radius. In other words, the electrons screen the ions from one another to some extent.

Since we are interested in how this modified potential affects the barrier penetrability, the radii of interest for nuclear reactions are those roughly equal to, or less than, the classical turning point of the motion which, for zero angular momentum, is given by $r_t = Z^2e^2/\mathcal{E}$ where $\mathcal{E}$ is the kinetic energy at infinite separation. For $\mathcal{E} \approx \mathcal{E}_0 \sim 10$ keV at the Gamow peak, and Z of unity, the turning point radius is about 10^{-11} cm. We can then approximate $\phi(r)$ for $r \leq r_t$ by

$$\phi(r) \approx \frac{Ze}{r}(1 - \kappa_d r) \tag{6.71}$$

if $\kappa_d r_t \ll 1$ or, in this numerical example, if $(n_{\rm I}/kT) \ll 10^{39}$ cm^{-3} erg^{-1}. For a solar type main sequence star with $\rho_c \approx 100$ and $T_c \approx 10^7$ K, $n_{\rm I}/kT \approx 10^{34}$, which seems safe enough.

The above implies that the electrostatic potential energy, $U = Ze\phi$, has been reduced by an amount $U_0 \approx Z^2e^2\kappa_d$ because of the screening presence of the electron cloud surrounding the ions. This, in turn, implies that the interacting charged particles effectively have their center of mass kinetic energy enhanced by an amount U_0; that is, they do not have to use up as much kinetic energy in approaching one another because the Coulomb barrier has, in effect, been reduced in height. Therefore, as a first go at seeing what this means, replace $\sigma_{\alpha\beta}(\mathcal{E})$ by $\sigma_{\alpha\beta}(\mathcal{E}+U_0)$ in expression (6.26). The other kinetic energies appearing in that expression are not altered in this approximation. We then transform the variable of integration in (6.26) from $\mathcal{E}$ to $\mathcal{E} - U_0$ with the result

$$\langle \sigma v \rangle_{\alpha\beta} \propto \int_{U_0}^{\infty} (\mathcal{E} - U_0) \sigma_{\alpha\beta}(\mathcal{E}) \, e^{-\mathcal{E}/kT} \, e^{U_0/kT} \, d\mathcal{E} \, .$$

Because the dominant contribution to the rate occurs for $\mathcal{E}$ equal to either $\mathcal{E}_0$ or $\mathcal{E}_r$ (in the nonresonant and resonant forms, respectively), and both of these are greater than U_0 in cases where all these approximations will apply, then notice that the major change made in the transform of the integral is the introduction of the factor $e^{U_0/kT}$. The lower limit of the integral is now U_0 but this may be replaced by zero since very low energies (compared to $\mathcal{E}_0$ or $\mathcal{E}_r$) contribute little to the total and, similarly, the linear term in energy may be replaced by $\mathcal{E}$. The result is an integral identical to the original except for the factor $e^{U_0/kT}$. In other words,

$$\langle \sigma v \rangle_{\alpha\beta} \text{ (with screening)} = \langle \sigma v \rangle_{\alpha\beta} \text{ (unscreened)} \times e^{U_0/kT} \tag{6.72}$$

with $e^{U_0/kT} \geq 1$. Thus the rate is increased by the screening.

If this is done consistently, with due account made for differences of charge of the reactants, etc. (see Clayton 1968, §4–8, or Cox 1968, §17.15), then, as an example, for protons on ^{12}C in a Pop I mix at $T_6 = 20$ and $\rho = 100$ g cm^{-3}, you will find that $\exp(U_0/kT) \approx 1.25$. [6] The effect of screening may thus be significant even at relatively low densities and high temperatures. The above formalism falls apart badly, however, when U_0/kT approaches anything like unity and other steps must be taken. Some of this will be brought up again later in the context of helium-burning reactions. For now, note the curious quantum mechanical fact that reactions may occur even at very low or "zero" temperatures because of zero-point energy vibrations in a lattice as *the* extreme in screening.

We shall not explicitly indicate that screening corrections should be applied to many reactions discussed in this chapter but keep them in mind because nuclear burning at the higher densities may be effected strongly by these corrections.

6.3 The Proton-Proton Chains

The major reaction sequences in the proton-proton chains are given in Table 6.1. By "major" we mean that some minor reactions have been left out of this tabulation and that the reactions given are those appropriate to hydrogen-burning at normal main sequence temperatures.

[6] In doing this simple calculation you will find that our U_0 is of the opposite sign from that used by Clayton but is consistent with Cox. This is merely a pedagogical preference.

TABLE 6.1. The proton-proton chains

PP–I	$^1H + ^1H$	$\longrightarrow$	$^2H + e^+ + \nu_e$
	$^2H + ^1H$	$\longrightarrow$	$^3He + \gamma$
	$^3He + ^3He$	$\longrightarrow$	$^4He + ^1H + ^1H$
PP–II	$^3He + ^4He$	$\longrightarrow$	$^7Be + \gamma$
	$^7Be + e^-$	$\longrightarrow$	$^7Li + \nu_e(+\gamma)$
	$^7Li + ^1H$	$\longrightarrow$	$^4He + ^4He$
PP–III	$^7Be + ^1H$	$\longrightarrow$	$^8B + \gamma$
	^8B	$\longrightarrow$	$^8Be + e^+ + \nu_e$
	^8Be	$\longrightarrow$	$^4He + ^4He$

There are three "chains," denoted by "PP-I," "PP-II," and "PP-III," and these are accessed by alternative reaction paths as indicated. The end products of each chain are ^{4}He nuclei. Generally speaking, these chains become more important in the order I, II, and III as temperature increases. Starting at the p-p reaction itself and going to the end of any of these chains eventually involves using four protons to make each α-particle. To do so, two of the protons must be converted to neutrons and this is done by means of some combination of positron decays or electron captures.

The various reactions in the p-p chains may proceed at wildly different rates and this is illustrated in Figure 6.6, where $N_A\langle\sigma v\rangle$ is plotted versus temperature for all capture reactions. The intrinsically slowest is the p-p reaction and its run of $N_A\langle\sigma v\rangle$ has been multiplied by 10^{18} just so it could appear on the figure. The next reaction in the PP–I chain, $^2H(p,\gamma)^3He$, is so fast that the abundance of the fragile nucleus ^{2}H is kept at a very small level, which may be computed using an equilibrium argument similar to that discussed for ^{13}N. The last reaction in the PP–I chain operates much more slowly then the preceding but, again, it is very fast compared to the p-p reaction and a long-term equilibrium abundance of ^{3}He may be calculated (see Clayton 1968, §§5-2, 5-3, for a full discussion of equilibration). As the temperature is raised, the equilibrium abundance of ^{3}He decreases until the first reaction in the PP–II chain begins to compete. Note that ^{4}He is always present in main sequence hydrogen-burning because of its production in Big Bang nucleosynthesis and the rate for the $^3He + ^3He$ reaction depends on the square of the number density $n(^3He)$.

The PP-II chain continues with an electron capture on ^{7}Be. Compared to ion capture reactions its rate is comparatively constant with temperature (see 6.63 and recall that computation of its rate requires multiplying $\langle\sigma v\rangle$

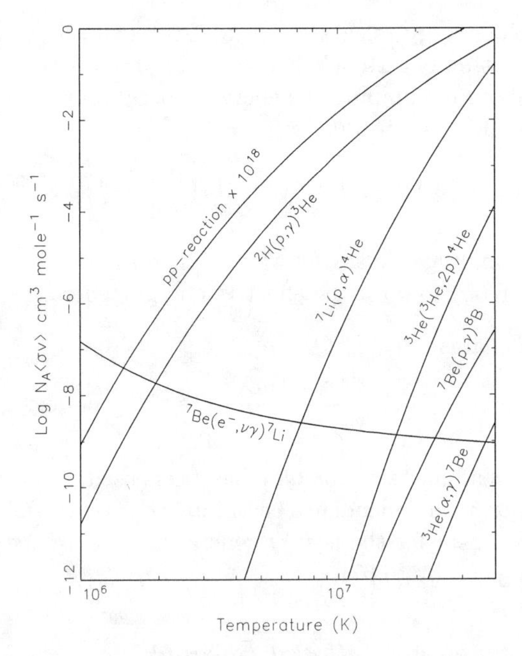

FIGURE 6.6. Shown as a function of temperature are $N_{\mathrm{A}}\langle\sigma v\rangle$ for all the capture reactions in the proton-proton chains of Table 6.1. Note that the curve for the proton-proton reaction has been multiplied by a factor of 10^{18}. Material for this figure comes from Caughlan and Fowler (1988).

by both electron and ^{7}Be number densities). The alternative reaction in (PP-III) is proton capture on ^{7}Be. Since the number density of electrons is roughly the same as that of protons, the crossing point of the curves for $^7\mathrm{Be}(\mathrm{e}^-, \nu)^7\mathrm{Li}$ versus $^7\mathrm{Be}(\mathrm{p},\gamma)^8\mathrm{B}$ yields the temperature at which the PP-III chain begins to compete with PP-II. This occurs at a temperature of around $T_6 \approx 24$. The ^{8}B nucleus produced by the proton capture is unstable to positron decay with a half-life of about 0.8 s. The accompanying neutrino, along with that from the ^{7}Be electron capture, is being detected in solar neutrino experiments (see Chapter 8).

The final nuclear event in the PP–III chain, which is the decay of ^{8}Be into two α-particles, is not only of great importance because it terminates the p-p chains, but it, and its inverse, is also one of the key reactions in helium-burning (see §6.5). The ^{8}Be nucleus is spectacularly unstable with a mean lifetime of only 9.7×10^{-17} s.

Because the slow p-p reaction starts off the p-p chains, the rate of processing to helium is controlled by that reaction. Thus the energy generation rate for the p-p chains must be proportional to $\langle\sigma v\rangle_{pp}$ of (6.65). The overall Q-value for the chains depends, however, on the weighted contributions of the three subchains to the rate of processing. Each of these contributes differently to the energy release because of the quantities and energies of the neutrinos lost among the chains. An overall effective Q-value may be estimated and used to compute the energy generation rate. From Fowler et al. (1975), this effective Q-value is

$$Q_{\text{eff}}(\text{p-p chains}) = 13.116 \left[1 + 1.412 \times 10^8 (1/X - 1)\, e^{-4.998/T_9^{1/3}}\right] \quad \text{MeV} \tag{6.73}$$

where X is the hydrogen mass fraction. This may be used in conjunction with (6.65) or (6.66) to form the effective energy generation rate

$$\begin{aligned} \varepsilon_{\text{eff}}(\text{p-p chains}) &= r_{\text{pp}}\, Q_{\text{eff}}/\rho \\ &\approx \frac{2.4 \times 10^4 \rho X^2}{T_9^{2/3}}\, e^{-3.380/T_9^{1/3}} \quad \text{erg g}^{-1}\ \text{s}^{-1} \end{aligned} \tag{6.74}$$

where only the leading term for Q_{eff} has been used. Note that since the dominant temperature dependence is still in the exponential of $\langle \sigma v \rangle_{\text{pp}}$, the temperature exponent for the energy generation rate for the combined p-p chains is again given by (6.67).

6.3.1 Deuterium and Lithium Burning

We include this short section on deuterium and lithium not only because their burning plays a special role in some stars but also because of the cosmological implications of their abundances. For reviews, see Boesgaard and Steigman (1985) and Steigman (1985). A complete compilation of cosmological results may be found in Yang et al. (1984), for example.

Deuterium is produced in "standard models" of the Big Bang by the reaction $\text{p} + \text{n} \rightarrow {}^2\text{H} + \gamma$. If temperatures are still very high, however, the reverse reaction destroys ^{2}H as rapidly as it is formed. Only when universal expansion has sufficiently cooled the radiation field does ^{2}H persist and p-p reactions can process nuclei to ^{4}He. The amount of ^{4}He left after this stage is done (expressed as a mass fraction) is $0.24 < Y_{\text{prim}} < 0.26$ and this may later be incorporated into stars. Observations of metal-poor (and, hence, old) galaxies indicate a mass fraction at the lower end of this range. The amount of ^{2}H and ^{3}He left over from the Big Bang is $1 < 10^5 \left[\left({}^2\text{H} + {}^3\text{He}\right)/{}^1\text{H}\right] < 20$ where the nuclear designations refer to number densities. Since ^{2}H is such a fragile nucleus, it is readily burned in stars and, in particular, in pre–main sequence evolution if temperatures exceed $T \gtrsim 6 \times 10^5$ K. It can then serve as an energy source (for a short time) to supplement gravitational contraction.

Lithium, as ^{7}Li, is produced and destroyed in the early universe by the same reactions given for the p-p chains. The final primordial amount left is $0.8 \lesssim 10^{10}({}^7\text{Li}/{}^1\text{H}) \lesssim 10$ from standard models. It too can be processed in stars by burning, mixing, etc. (and through cosmic rays), and we expect to see varying amounts in stellar atmospheres and the interstellar medium.

Pop I stars and their associated gas show a maximum abundance ratio by number density of $^7\mathrm{Li}/^1\mathrm{H} < 10^{-9}$ whereas Pop II stars generally have $^7\mathrm{Li}/^1\mathrm{H} \sim 10^{-10}$. Among the many puzzles in nucleosynthesis and stellar evolution, however, is the following—and it has to do with the sun. Among the oldest objects in the solar system are the meteorites. The abundance of $^7\mathrm{Li}$ has been measured in one class of these (the Type I Chondrites) to be $^7\mathrm{Li}/^1\mathrm{H} \sim 10^{-9}$ and this is consistent with the sun being a Pop I star. The "lithium problem" for the sun, however, is that the solar surface abundance of lithium is only $\left(^7\mathrm{Li}/^1\mathrm{H}\right) \sim 10^{-11}$, which is down by two orders of magnitude from what we expect. Standard evolutionary models for the sun cannot explain this and we raise this as a warning flag because the sun is our standard among stars.

6.4 The Carbon-Nitrogen-Oxygen Cycles

The major reactions comprising the CNO cycles at normally occurring hydrogen-burning temperatures are given in Table 6.2.

TABLE 6.2. The carbon-nitrogen-oxygen cycles

$$
\begin{array}{rcl}
^{12}\mathrm{C} + {}^1\mathrm{H} & \longrightarrow & ^{13}\mathrm{N} + \gamma \\
^{13}\mathrm{N} & \longrightarrow & ^{13}\mathrm{C} + \mathrm{e}^+ + \nu_\mathrm{e} \\
^{13}\mathrm{C} + {}^1\mathrm{H} & \longrightarrow & ^{14}\mathrm{N} + \gamma \\
^{14}\mathrm{N} + {}^1\mathrm{H} & \longrightarrow & ^{15}\mathrm{O} + \gamma \\
^{15}\mathrm{O} & \longrightarrow & ^{15}\mathrm{N} + \mathrm{e}^+ + \nu_\mathrm{e} \\
^{15}\mathrm{N} + {}^1\mathrm{H} & \longrightarrow & ^{12}\mathrm{C} + {}^4\mathrm{He} \\
 & \text{–or–} & \\
^{15}\mathrm{N} + {}^1\mathrm{H} & \longrightarrow & ^{16}\mathrm{O} + \gamma \\
^{16}\mathrm{O} + {}^1\mathrm{H} & \longrightarrow & ^{17}\mathrm{F} + \gamma \\
^{17}\mathrm{F} & \longrightarrow & ^{17}\mathrm{O} + \mathrm{e}^+ + \nu_\mathrm{e} \\
^{17}\mathrm{O} + {}^1\mathrm{H} & \longrightarrow & ^{14}\mathrm{N} + {}^4\mathrm{He}
\end{array}
$$

The general structure of the CNO cycles consists of a series of proton captures on isotopes of CNO interspersed with positron decays, and ending with a proton capture reaction yielding $^4\mathrm{He}$. The first set of reactions listed in Table 6.2 is called the CN cycle and the isotopes of carbon and nitrogen act as catalysts; that is, you can start almost anywhere in the cycle, destroy one of these isotopes, and, by looping around the cycle, eventually find a reaction that makes the same isotope. This does not mean that the concentrations will remain constant through time because that depends primarily on the relative rates of the reactions in the cycle.

The second set of reactions in Table 6.2, when combined with the first, constitute the complete CNO cycle (sometimes called a tricycle). It arises from a combination of two factors: either ^{16}O is (very likely) in the stellar mixture in the first place or, in any case, it will eventually be made by the reaction ^{15}N(p, γ)^{16}O. Note that the final reaction in the complete CNO cycle sends ^{14}N right back into the CN cycle.[7]

An exact description of just how the CNO cycles operate in hydrogen-burning is not a trivial matter because of the intricate cycling of isotopes. Both the rate of energy generation and the detailed abundances of all the isotopes depend on the initial concentrations of the catalytic nuclei, the mean lifetimes for the individual reactions as they depend on temperature, and how long the processing has been going on. It was recognized early on, however, that a key reaction in the cycles was ^{14}N(p, γ)^{15}O. It is relatively slow and involves the isotope ^{14}N, which appears in both sets of cycles (see Caughlan and Fowler 1962). As we saw in the last section, a slow reaction in a chain of reactions often sets the pace for the whole. We cannot go through the complete analysis here (see Clayton 1968, §5-4) but the important result is that if temperatures are high enough to initiate CN or CNO hydrogen-burning in main sequence stars, then the most abundant nucleus will end up being ^{14}N after long enough periods of time have elapsed. This means that the cycles eventually convert almost all of the original CN or CNO nuclei, depending on temperatures and time scales, to ^{14}N. It is thought that virtually all ^{14}N seen in nature has been produced in this way.

If not enough time is available to allow the CN or CNO cycles to reach equilibrium, then the above results have to be reevaluated. This means that the differential equations that govern the creation and destruction of individual isotopes must be followed explicitly in time. This is also usually necessary when detailed isotope ratios are desired. For example, many highly evolved stars show abundance ratios between ^{12}C, ^{13}C, and ^{14}N in their spectra that are anomalous compared to some sort of cosmic standard. It is highly likely that what is being seen here is the effect of CNO processed material having being brought to the stellar surface by mixing perhaps coupled with mass loss (in, e.g., red giant or asymptotic giant phases). Thus we see directly the products of nuclear burning. For a review of this important topic see, for example, Iben and Renzini (1984).

An estimate for the energy generation rate for the CN and CNO cycles may be obtained as follows. If enough time has elapsed so that the cycles are in equilibrium, then the reaction rate for the cycles is set by the rate of ^{14}N(p, γ)^{15}O. To find the energy generation rate we then need an overall Q-value. Fowler et al. (1975) recommend 24.97 MeV per proton capture on ^{14}N. We also need the number densities of protons (that is easy) and ^{14}N. The last is tricky because, if the cycles are in equilibrium, $n(^{14}\text{N})$ is the sum of the number densities of the original (before burning) CN or CNO

[7] Additional reactions often made part of the CNO cycles include ^{18}O(p, α)^{15}N, which uses up any ^{18}O that may be in the original mixture and feeds ^{15}N into the cycle.

nuclei. But, as discussed above, this depends on details of temperature and time scale history.

To get an idea of what errors might arise from making the wrong choice between CN and CNO, we should look at typical abundances of these nuclei in nature and a good place to look is the sun. Bahcall and Ulrich (1988), in a review on the status of solar models, quote the following relative abundances (from L. Aller) for C, N, and O in the solar atmosphere: $n(\mathrm{C}) = 0.28$, $n(\mathrm{N}) = 0.059$, and $n(\mathrm{O}) = 0.498$. These number densities are normalized so that the total number density of all metals (i.e, all elements except hydrogen and helium) is unity. Adding these up we find that CNO constitutes about 84% of all metals by number and that O makes up about 60% of CNO by number. Delving a little deeper into the tables in the review by Bahcall and Ulrich and multiplying number densities by atomic weights also reveals that the mass fraction of CNO is $X_{\mathrm{CNO}} \approx 0.74Z$ where Z is the metal mass fraction (see §1.4). Thus CNO elements comprise the majority of metals in the solar atmosphere and, by extension, of the atmospheres of other normal Pop I stars. In addition, we find that $X_{\mathrm{O}} \approx 0.67X_{\mathrm{CNO}}$. What this all means is that considerable caution should be exercised in choosing CN or CNO to represent $^{14}\mathrm{N}$ in the CNO energy generation rate. As a compromise, and after reviewing the above figures, a reasonable choice is to set the CN or CNO mass fraction to $Z/2$ with a possible error of about 25%.

Putting this together—and after consulting Fowler et al. (1975) for the reaction rate of $^{14}\mathrm{N}(\mathrm{p},\gamma)^{15}\mathrm{O}$—we find a very useful estimate for the energy generation rate for the CN or CNO cycles of

$$\varepsilon_{\mathrm{CNO}} \approx \frac{4.4 \times 10^{25} \rho X Z}{T_9^{2/3}} e^{-15.228/T_9^{1/3}} \quad \mathrm{erg\ g^{-1}\ s^{-1}}. \tag{6.75}$$

Detailed evolutionary calculations require more than this but it should suffice for making simple ZAMS models (as in `ZAMS.FOR` of Appendix C).

Figure 6.7 shows the pp chain and CNO cycle energy generation rates derived from (6.74) and (6.75) with density and either factors of X^2 (for the pp chains) or XZ (for CNO) removed. For a solar central temperature of $T_6 \approx 15$, $X = 0.7$, and $Z = 0.02$, find that $\varepsilon_{\mathrm{pp}} \approx 10 \times \varepsilon_{\mathrm{CNO}}$ so that the CNO contribution to the total energy generation rate is roughly 10%. However, it does not take much more massive a star on the main sequence with a higher than solar central temperature before the greater temperature sensitivity of the CNO cycles wins out and the pp chains lose their dominance. Note that the temperature exponent for the CNO cycles is

$$\nu(\mathrm{CNO}) = \frac{50.8}{T_6^{1/3}} - \tfrac{2}{3} \tag{6.76}$$

which is about 18 for $T_6 = 20$.

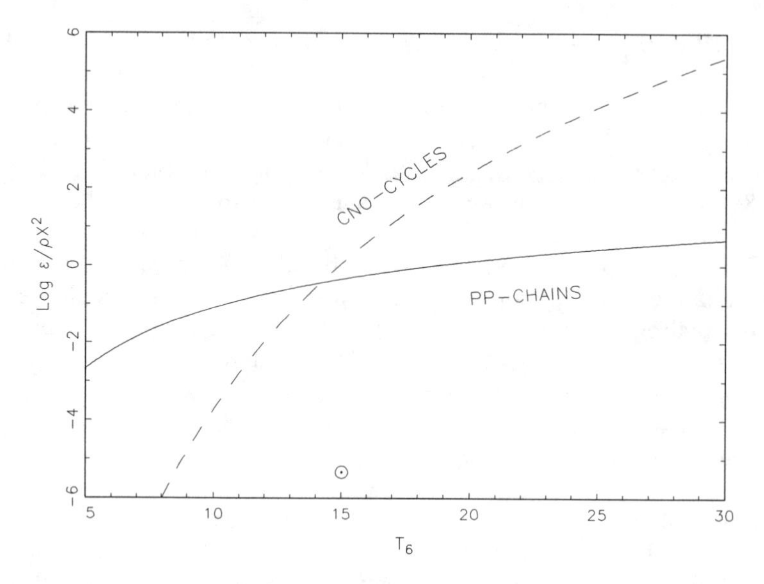

FIGURE 6.7. Plotted as a function of temperature are $\varepsilon_{\rm pp}/\rho X^2$ and $\varepsilon_{\rm CNO}/\rho XZ$. (The legend on the ordinate is generic: "$\varepsilon/\rho X^2$" refers to either depending on context.) To obtain the energy generation rates, you must multiply by the density and the appropriate mass fractions. The temperature of the present-day solar center is indicated by the Sun sign.

6.5 Helium-Burning Reactions

This section will deal with the triple-α and subsequent reactions in helium-burning. For the most part, the stellar environment is assumed to correspond to that of the cores of normal post–main sequence stars where temperatures do not greatly exceed 10^8 K. For an excellent historical review of the subject see Fowler (1986).

Helium-burning begins with the inverse of the $^8\text{Be} \rightarrow 2\,^4\text{He}$ decay that terminates the PP-III chain; that is, the first reaction is $^4\text{He} + {^4\text{He}} \rightarrow {^8\text{Be}}$, which is endothermic by 91.78 keV. We remarked earlier that ^{8}Be has a lifetime of only 10^{-16} s. Thus to produce ^{8}Be in any quantity whatsoever, the α-particles must have sufficient energy to gain access to the ground state of ^{8}Be and the formation rate of ^{8}Be must be sufficiently rapid to make up for its short lifetime. Since the ground state of ^{8}Be has a finite width ($\Gamma_{\alpha\alpha} = 6.8$ eV) we may ask at what temperature is it that the Gamow peak begins to encroach upon that resonance. In other words, if the reaction does not begin to look resonant, then the reaction rate for production may not catch up with the inverse decay. From (6.40), the location of the peak is at $\mathcal{E}_0 = 3.9T_6^{2/3}$ keV ($Z_\alpha = 2$, $\mu = 2$) and it equals 92 keV when $T = 1.2 \times 10^8$ K. If the effects of electron screening in high-density situations are ignored (for the present) then this roughly sets the minimum temperature for helium-burning.

Let us assume then that temperatures exceed 10^8 K and the ^{8}Be producing reaction proceeds rapidly. If rapid enough, the formation rate of ^{8}Be should begin to match the rate at which it is destroyed by decays; that is, the concentration of ^{8}Be should approach equilibrium. (This should, and can, be justified—as it is in the references.) One way to find the equilibrium concentration is to compute the rate of production by equating $n_\alpha^2 \langle \sigma v \rangle_{\alpha\alpha}/2$ (remember the factor of two for like particles) and equate this to $\lambda\, n(^8\text{Be})$ where λ is the decay constant for ^{8}Be. This unfortunately requires knowing $\langle \sigma v \rangle_{\alpha\alpha}$. But, there is an easier and more illuminating way to go about it. We may assume chemical equilibrium and use the Saha equation (3.33) except that we now have nuclei and not atoms, ions, and electrons as was the case for the hydrogen ionization reaction (3.29).

The reaction to be considered is

$$\alpha + \alpha \longleftrightarrow {}^8\text{Be}$$

and several easy modifications must be made to the Saha equation of (3.33). The first is to replace the number densities by $(n^+, n_e) \to n_\alpha$ and $n^0 \to n(^8\text{Be})$. The statistical factors, g, are unity for both ^{4}He and the ground state of ^{8}Be because both have zero spin. Instead of an ionization potential χ_H we now have the Q-value, which is –91.78 keV . Finally, the mass, m_e, is replaced by $m_\alpha^2/m(^8\text{Be}) \approx m_\alpha/2$ as the reduced mass. The "nuclear" Saha equation is then

$$\begin{aligned} \frac{n_\alpha^2}{n(^8\text{Be})} &= \left(\frac{\pi m_\alpha kT}{h^2} \right)^{3/2} e^{-Q/kT} \\ &= 1.69 \times 10^{34} T_9^{3/2}\, e^{1.065/T_9} . \end{aligned} \tag{6.77}$$

For typical conditions at, say, the start of the helium flash in lower mass stars where $\rho \approx 10^6$ g cm^{-3} ($n_\alpha \approx 1.5 \times 10^{29}$ cm^{-3} if the flash starts with pure helium) and $T_9 \approx 0.1$, find that the equilibrium concentration of ^{8}Be is about 10^{21} cm^{-3} or $n(^8\text{Be})/n_\alpha$ is only 7×10^{-9}.

With a seed of ^{8}Be nuclei now in place, however, the second stage of the triple-α reaction may now continue with the capture reaction $^8\text{Be}(\alpha,\gamma)^{12}\text{C}$. This is an exothermic resonant reaction, with $Q = 7.367$ MeV, which proceeds through an excited state $^{12}\text{C}^*$ with zero spin at 7.654 MeV. The emission of a γ-ray photon by $^{12}\text{C}^*$ does not come easily because once the compound excited state is formed it almost always decays right back to ^{8}Be and an α-particle. Yet, as in the first step of the triple-α described above, the forward reaction is sufficiently rapid (assuming a high enough temperature) that a small pool of ^{12}C nuclei in the excited state is built up and, again, the nuclear Saha equation may be used to find the concentration in the pool. It is not difficult to do this and it should be obvious that it finally results in an expression for $n(^{12}\mathrm{C}^*)/n_\alpha^3$ as a function of temperature after (6.77) is applied.

Once having found $n(^{12}\mathrm{C}^*)$, then the net rate of decay of $^{12}\mathrm{C}^*$ by γ-ray cascade (or electron-positron pair emission) rather than by an α-particle is $n(^{12}\mathrm{C}^*) \times \Gamma_{\rm rad}/\hbar$ where the combination $\Gamma_{\rm rad}/\hbar$ is the decay rate, $\lambda_{\rm rad}$, through the uncertainty relation. (The value of $\Gamma_{\rm rad}$ is only 3.67 meV.) The overall sequence of the triple-α is illustrated in Figure 6.7. It should be clear that the overall rate of the triple-α reaction is the same as the formation rate of the ground state of $^{12}\mathrm{C}$.

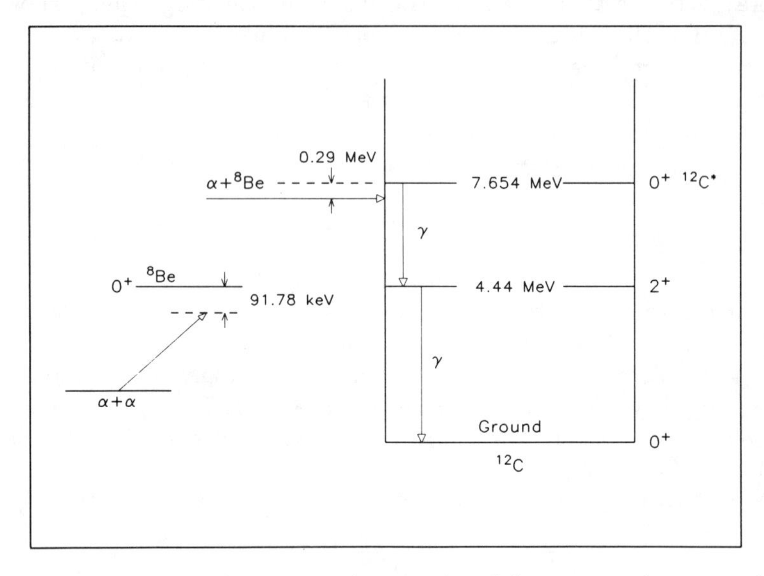

FIGURE 6.8. The level diagrams and energetics of the two reactions comprising the triple-α reaction (not to scale). The final result is the nucleus $^{12}\mathrm{C}$.

The above contains all the elements for computing the energy generation rate of the triple-α sequence. The final result we quote is taken from Harris et al. (1983, in their Table 1) where the quantity $N_{\rm A}^2\,\langle\alpha\alpha\alpha\rangle$ is to be found. This is multiplied by $\rho^2 Y^3 \mathrm{N_A} Q/6A_\alpha^3$ and Q=7.367–0.0918=7.275 MeV to yield

$$\varepsilon_{\alpha\alpha\alpha} = \varepsilon_{3\alpha} = \frac{5.1\times 10^8 \rho^2 Y^3}{T_9^3}\, e^{-4.4027/T_9} \quad \mathrm{erg\ g^{-1}\ s^{-1}}. \tag{6.78}$$

Here, $A_\alpha = 4$ (as part of n_α), and the division by 6 in the multiplying factor comes about because of triple counting of α-particles (as in dividing by two for double counting of protons in the pp chains). To verify (6.78) requires searching through some of the papers already referenced and we suggest you try to reproduce it to gain experience in how to use these references.[8] If intermediate rates are fast enough to satisfy the Saha equation, then the uncertainty in (6.78) is estimated to be only 15% (Fowler 1986).

It is easy to show from (6.78) that the temperature and density exponents for the triple-α reaction are

$$\lambda_{3\alpha} = 2, \quad \text{and} \quad \nu_{3\alpha} = \frac{4.4}{T_9} - 3. \tag{6.79}$$

For $T_8 = 1$, $\nu_{3\alpha} \approx 40$, which is considerably larger than the corresponding exponent for hydrogen-burning. This means that the helium fuel is potentially more explosive than hydrogen—a fact of considerable interest for the helium flash.

The effects of screening are difficult to assess for a reaction such as the triple-α and we shall not attempt to do so here. In addition, the above analysis is inappropriate for temperatures much below 10^8 K because $^4\text{He}(\alpha)^8\text{Be}$ no longer samples the resonance in ^{8}Be strongly. For an attempt to combine these various elements see Fushiki and Lamb (1987) who give general expressions for the energy generation rate including effects of weak and strong screening and, in the very high-density limit, "pycnonuclear" effects (a term coined by Cameron 1959; see also Ichimaru et al. 1992). These corrections can be very important.

The next step in helium-burning is α capture on ^{12}C to form ^{16}O. In the $^{12}\text{C}(\alpha,\gamma)^{16}\text{O}$ reaction, the $^{12}\text{C}+\alpha$ pair, at zero initial energy, enters ^{16}O at 7.162 Mev (which is the Q-value). The nearest resonance in ^{16}O, however, lies some 45 keV below that energy. Hence the reaction may proceed only in the upper tail of the resonance at temperatures near 10^8 K. Unfortunately the nuclear parameters for this resonance and the detailed behavior of the resonance tail are hard to come by experimentally (because the resonance cannot decay to ^{12}C plus an α) and, due to various factors, the theory is not easy either. The result given below is taken from Caughlan and Fowler (1988) and pertains only to that resonance. We omit contributions from higher energy resonances that are important for high-temperature burning. To quote from Fowler (1985): "If users find that their results in a given study are sensitive to the rate of the $^{12}\text{C}(\alpha,\gamma)^{16}\text{O}$ reaction, then they should repeat their calculations with 0.5 times and 2 times the values recommended here." That is, give a factor of two either way. This rate is continually being adjusted, so check the recent literature.[9]The energy

[8] In particular, you will need Fowler et al. (1967, 1975), and Harris et al. (1983). High-temperature correction factors and individual rates for the two parts of the triple-α may be found in Caughlan and Fowler (1988).

[9] New results indeed. Buchmann et al. (1993) report the results of new, and very subtle, experiments relevant to this rate. In addition, Weaver and Woosley (1993) discuss numerical nucleosynthesis experiments that suggest the rate should be (1.7 ± 0.5) times the rate given by (6.80). Keep posted.

generation rate is

$$\varepsilon(\alpha, {}^{12}\mathrm{C}) = \frac{1.5 \times 10^{25} Y X_{12}\, \rho}{T_9^2} \left(1 + 0.0489 T_9^{-2/3}\right)^{-2} \times \exp\left[-32.12 T_9^{-1/3} - (0.286 T_9)^2\right] \ \mathrm{erg\ g^{-1}\ s^{-1}} \quad (6.80)$$

excluding the effects of higher resonances and where X_{12} is the mass fraction of ^{12}C.

The importance of the uncertainties in this rate cannot be overemphasized for calculations of the nucleosynthesis of the elements carbon and oxygen. Since the next reaction in the helium-burning sequence, $^{16}\mathrm{O}(\alpha,\gamma)^{20}\mathrm{Ne}$, is rather slow at normal helium-burning temperatures, the competition between how fast ^{12}C is produced by the triple-α and how quickly it is converted to ^{16}O determines the final relative abundances of these two nuclei (and that is of concern for us H-, C- and O-based creatures). For the later evolutionary stages of lower mass stars, this may determine whether the final core, as in a white dwarf, is mostly carbon or oxygen. For your reference, the energy generation rate for the $^{16}\mathrm{O}(\alpha,\gamma)^{20}\mathrm{Ne}$ reaction ($Q = 4.734$ MeV) is, from Caughlan and Fowler (1988),

$$\varepsilon(\alpha, {}^{16}\mathrm{O}) = \frac{6.69 \times 10^{26} Y X_{16}\, \rho}{T_9^{2/3}} \times \exp\left[-39.757 T_9^{-1/3} - (0.631 T_9)^2\right] \ \mathrm{erg\ g^{-1}\ s^{-1}} \quad (6.81)$$

for not overly high temperatures.

Other capture reactions using α-particles that are of some importance to nucleosynthesis are those on various C, N, and O isotopes where one of the exit channel particles is a neutron—and we have discussed these briefly before.

6.6 Carbon, Neon, and Oxygen Burning

Once α–particles have been used up in helium-burning and if temperatures can rise to $T_9 \sim 0.5$–1, carbon burning commences and, at yet higher temperatures ($T_9 \gtrsim 1$), oxygen burning. Intermediate between these two burning stages is neon burning, which uses high-energy photons to break down ^{20}Ne by "photodisintegration" (see below) via $^{20}\mathrm{Ne}(\gamma, {}^{16}\mathrm{O})\alpha$.

The important branches of the reactions $^{12}\mathrm{C} + {}^{12}\mathrm{C}$ and $^{16}\mathrm{O} + {}^{16}\mathrm{O}$ are given in Table 6.3 where "yield" is the percentage of time the reaction results in the particular products on the right-hand side. The yield depends weakly on temperature and we ignore minor branches.

The $^{12}\mathrm{C}+{}^{12}\mathrm{C}$ reactions are followed by $^{23}\mathrm{Na}(\mathrm{p},\alpha)^{20}\mathrm{Ne}$ (Q=2.379 MeV) and $^{23}\mathrm{Na}(\mathrm{p},\gamma)^{24}\mathrm{Mg}$ (Q=11.691 MeV) using the protons released from the

TABLE 6.3. Carbon and oxygen burning reactions

Reaction	Yield	Q (MeV)
$^{12}C + ^{12}C \rightarrow ^{20}Ne + \alpha$	44%	4.621
$^{12}C + ^{12}C \rightarrow ^{23}Na + p$	56%	2.242
$^{16}O + ^{16}O \rightarrow ^{28}Si + \alpha$	21%	9.593
$^{16}O + ^{16}O \rightarrow ^{31}P + p$	61%	7.678
$^{16}O + ^{16}O \rightarrow ^{31}S + n$	18%	1.500

second reaction in Table 6.3. The α–particles can then be used on ^{16}O to form ^{20}Ne or on ^{20}Ne to yield ^{24}Mg. Depending somewhat on temperature and density, the net result of this chain of reactions is the formation of ^{20}Ne followed by lesser amounts of ^{23}Na and ^{24}Mg for quiescent carbon-burning. (For a good review of quiescent heavy ion burning see Thielemann and Arnett 1985.) The rate of energy generation for the two branches of the $^{12}C + ^{12}C$ reaction is, from Caughlan and Fowler (1988),

$$\varepsilon(^{12}C + ^{12}C) = \frac{1.43 \times 10^{42} Q \eta \rho X_{12}^2}{T_9^{3/2}} e^{-84.165 T_9^{-1/3}} \quad \text{erg g}^{-1} \text{ s}^{-1} \tag{6.82}$$

where the proper Q-value (in MeV) is to be used and η is the yield of Table 6.3 multiplied by 10^{-2}. If, as a convenience, you wish to make believe that the reaction ends up as ^{24}Mg, then use $\eta = 1$ and $Q = 13.933$ MeV.

You may easily check that the temperature and density exponents for (6.82) are $\nu = 28/T_9^{1/3} - 1.5$ and $\lambda = 1$. The large temperature exponent is, as usual, due to the large nuclear charge of the reactants. These heavy ion reactions are especially susceptible to electron screening effects (and often take place in dense environments) so take care if you require accurate rates.

Intermediate between carbon and oxygen burning are a set of reactions that use up the neon just produced and constitute the neon burning stage. The first of this set is new to us and is a result of the intensity of the radiation field as temperatures exceed $T_9 \gtrsim 1$. A temperature of T_9 of unity is about 0.1 MeV and there are substantial numbers of photons with energies exceeding that figure in the tail of the Planck distribution. These energies are in the range of those of low-lying nuclear states for some nuclei and it is now possible to excite those unstable states. The result is often the emission of particles from the nucleus in a "photodisintegration" reaction, which is the analogue of ionization in atoms. The relevant reaction for neon burning is $^{20}Ne(\gamma, \alpha)^{16}O$, which is the inverse of the last reaction for helium-burning. And, as in helium-burning, the α–particle produced can be captured right back by a ^{20}Ne nucleus but, more to the point, temperatures are sufficiently high to allow the sequence $^{20}Ne(\alpha, \gamma)^{24}Mg(\alpha, \gamma)^{28}Si$. (Note how, in the later burning stages, the reaction sequences get more convoluted with nucleons and α-particles being tossed around to make a great variety of heavy nuclei.) The net result is a pool of ^{16}O, ^{24}Mg, and ^{28}Si (see Figure 2.26).

The next stage is the burning of oxygen by $^{16}\text{O} + {}^{16}\text{O}$. (Note that $^{12}\text{C} + {}^{16}\text{O}$ is, in principle, possible at some point but ^{12}C is rapidly used up by carbon-burning and the rate of $^{12}\text{C} + {}^{16}\text{O}$ is intrinsically slow.) The three main reactions and their yields and Q-values are given in Table 6.3 with an energy generation of

$$\varepsilon(^{16}\text{O} + {}^{16}\text{O}) = \frac{1.3 \times 10^{52} Q \eta \rho X_{16}^2}{T_9^{2/3}} e^{-135.93 T_9^{-1/3}} \times e^{\left[-0.629 T_9^{2/3} - 0.445 T_9^{4/3} + 0.0103 T_9^2\right]}. \tag{6.83}$$

If all the reactions were somehow to proceed to ^{32}S the total Q would be 16.542 MeV.

Many reactions are possible after the last three in Table 6.3. Examples are: $^{31}\text{S} \to {}^{31}\text{P} + \text{e}^+ + \nu_\text{e}$, $^{31}\text{P}(\text{p},\alpha)^{28}\text{Si}(\alpha,\gamma)^{32}\text{S}$, etc. Completion of this stage of burning results in ^{28}Si, ^{30}Si, ^{34}S and, depending on conditions of temperature and density, ^{42}Ca and ^{46}Ti.

6.7 Silicon "Burning"

When temperatures begin to exceed some 3×10^9 K, a bewildering number of reactions are possible. We pointed out previously the effects of photodisintegration during neon burning where the radiation field was capable of "ionizing" nucleons from nuclei, which could then be used to build even more massive nuclei. As a relevant example consider $^{28}\text{Si}(\gamma,\alpha)^{24}\text{Mg}$ followed by $^{24}\text{Mg}(\alpha,\text{p})^{27}\text{Al}$ and, finally, $^{27}\text{Al}(\alpha,\text{p})^{30}\text{Si}$. Here nucleons have been recycled with the aid of photons to effectively add two neutrons to ^{28}Si to produce ^{30}Si. Amplify this to include many reactions that eventually lead up to nuclei in the iron peak—as those with the highest binding energy per nucleon—and you have the essentials of silicon burning (or perhaps "melting").

To follow all these reactions in detail is a daunting task and one that was first carried out in the pioneering calculations of Truran et al. (1966). What is required is consideration of many nuclei and the reactions that connect them (plus the cross sections for the reactions). One such reaction "network," along with the possible types of reactions, is shown in Figure 6.8. As the burning accelerates, the reactions proceed sufficiently rapidly that a state of "quasi-static equilibrium" begins to take hold. By this we mean that photodisintegration and particle capture reactions are nearly in equilibrium but with a bias toward the production of nuclei in the iron peak. Because of the rapidity of the reactions, the abundances of most nuclei may be approximated by a nuclear version of the Saha equation where, instead

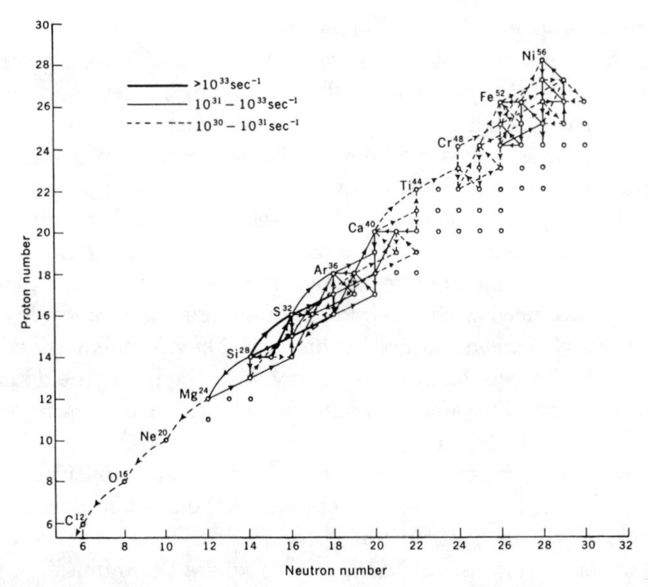

FIGURE 6.9. A sample reaction network for silicon burning that also shows the reactions possible between nuclei in the network. Adapted by Clayton (1968) from Truran et al. (1966). Copyright ©1968, 1983 by D.D. Clayton and reproduced by permission.

of ionization potentials, nuclear masses and energies are used. (See, for example, Bodansky, Clayton, and Fowler 1968.)

The result of silicon burning is production of nuclei in the iron peak. If enough time is allowed to elapse (as in quiescent burning) then the most abundant of these is ^{56}Fe. If, on the other hand, the time scales are short—as in a supernova—and electron or positron decays and electron captures do not have enough time to go to completion, then ^{56}Ni is the most abundant. This is crucial to our understanding of supernovae and was discussed in Chapter 2.

6.8 Neutrino Emission Mechanisms

Matter at ordinary temperatures and densities is extraordinarily transparent to neutrinos. Thus if neutrinos are produced in normal stellar interiors they are an energy loss mechanism because they carry off energy to space. But, there are situations where neutrinos may be intercepted by stellar material and either be absorbed or scattered. To get an idea of how extreme those conditions must be, consider a typical neutrino capture cross section of $\sigma_\nu \sim 10^{-44}\mathcal{E}_\nu^2$ cm^2 where $\mathcal{E}_\nu$ is the neutrino energy in MeV. The target for capture is unspecified here.[10] The mean free path is then $\lambda = 1/n\sigma_\nu$ where n is the number density of targets which, if they have a mean molecular weight near unity, yields $\lambda \sim 10^{20}\mathcal{E}_\nu^{-2}/\rho$ cm. To get short mean free paths we need very high densities or neutrino energies. These conditions for stars are met, as far as we know, only in the cores of supernovae. In the core collapse phase of Type II supernovae densities approach and then exceed those of nuclear densities (as discussed in §2.6.3). If a typical density is $\rho \sim 10^{14}$ g cm^{-3} then typical nucleon kinetic energies are in the 20 MeV range if we choose Fermi energies as being representative (as in §6.2.2). If neutrinos are produced with these kinds of energies then a crude estimate of λ_ν is 25 meters. Thus neutrinos in this sort of environment are effectively "trapped" and whatever energy they carry with them may be deposited in the collapsing core. This energy transfer mechanism is one ingredient in some supernovae calculations.

Aside from such extreme environments, we should consider neutrinos as an energy drain for stars. If ε_ν represents the specific energy rate (in erg g^{-1} s^{-1}) at which neutrinos are produced, then the energy equation, excluding other factors, is $d\mathcal{L}_r/d\mathcal{M}_r = -\varepsilon_\nu$ where the minus sign reminds us that ε_ν is a power drain. What we shall briefly explore here are some important neutrino producing mechanisms in the later stages of evolution.

The most familiar of these are electron (or positron) decay and electron capture involving nuclei. We have seen examples of these reactions in hydrogen-burning and more advanced stages. Usually the associated neutrino losses are rather modest. This is not the case, however, for the stages immediately prior to Type II supernova core collapse. The reason is that the Fermi energies of electrons in very dense environments are sufficiently high that electrons near the top of the Fermi sea are capable of being captured on protons in most nuclei. The result is not only copious neutrino production but also a shift to neutron-rich nuclei. This process can be a first step in the transformation of ordinary matter to neutron star matter. A brief overview is given by Bethe (1990). More esoteric mechanisms include the following.

Pair Annihilation Neutrinos

These neutrinos come about by the annihilation of an electron by a positron in the reaction

$$e^- + e^+ \longrightarrow \nu_e + \overline{\nu}_e. \tag{6.84}$$

But where can we get sufficient numbers of positrons to make this reaction at all interesting? This is not that difficult if temperatures are high enough ($kT \sim 2m_ec^2$) so that some fraction of ambient photons are capable of pair

[10]Much of this is discussed in Shapiro and Teukolsky (1983, Chapter 18), Bethe (1990), and references therein. We make no attempt to give derivations of this material or that which follows.

creation via

$$\gamma + \gamma \longleftrightarrow e^- + e^+. \tag{6.85}$$

If this reaction goes rapidly enough then the equilibrium number densities of both electrons and positrons can be calculated from the condition on their chemical potentials, $\mu(e^-) + \mu(e^+) = 0$ (see §3.1). Thus, in chemical equilibrium, $\mu(e^-) = -\mu(e^+)$. A further requirement is that the hot gas be electrically neutral, which leads to the condition on the number densities

$$n_{e^-}(\text{total}) = n_{e^-}(\text{free}) + n_{e^+}$$

where "free" refers to those electrons that would normally be associated with nuclei were no pairs created and "total" refers to the total of free plus pair-created electrons. The number density $n_{e^-}(\text{free})$ can be calculated from the density and composition of the material and is a "given." Equation (3.42) for fermion number densities as a function of chemical potential and temperature is true for both electrons and positrons provided that $-\mu(e^-)$ is used for the positrons. This information is sufficient and both n_{e^-} and n_{e^+} may be found although the calculation is not easy. The rate of neutrino emission then follows from application of the Weinberg-Salam-Glashow theory of electro-weak interactions.

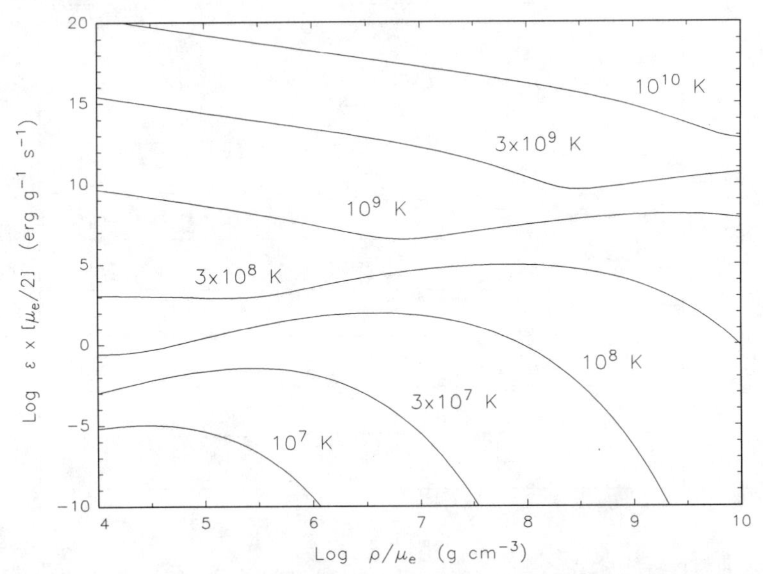

FIGURE 6.10. The combined neutrino loss rates (in erg g^{-1} s^{-1}) for pair annihilation, photo-, and plasma neutrinos versus ρ/μ_e and temperature. Adapted from the calculations of Itoh and collaborators.

Photoneutrinos and Bremsstrahlung Neutrinos

The first of these is the analogue of electron-photon scattering except that instead of a final photon we get a neutrino-antineutrino pair. That is,

$$e^- + \gamma \longrightarrow e^- + \nu_e + \overline{\nu}_e \,. \tag{6.86}$$

The rule seems to be that if you can get an exiting photon then it is also possible to get a ν_e–$\overline{\nu}_e$ pair. Thus ordinary bremsstrahlung, which yields a photon when an electron is scattered off an ion, is a likely candidate. This is an important energy loss mechanism for hot white dwarfs.

Plasma Neutrinos

From elementary physics we know that a free photon cannot create an electron-positron pair because energy and momentum cannot both be conserved in the process. In a very dense plasma, however, electromagnetic waves can be quantized in such a way that they behave like relativistic Bose particles with finite mass, "plasmons," that can decay into either e^-–e^+ or ν_e–$\overline{\nu}_e$ pairs. You might look upon plasmons as heavy photons created especially to cause trouble for some stars.

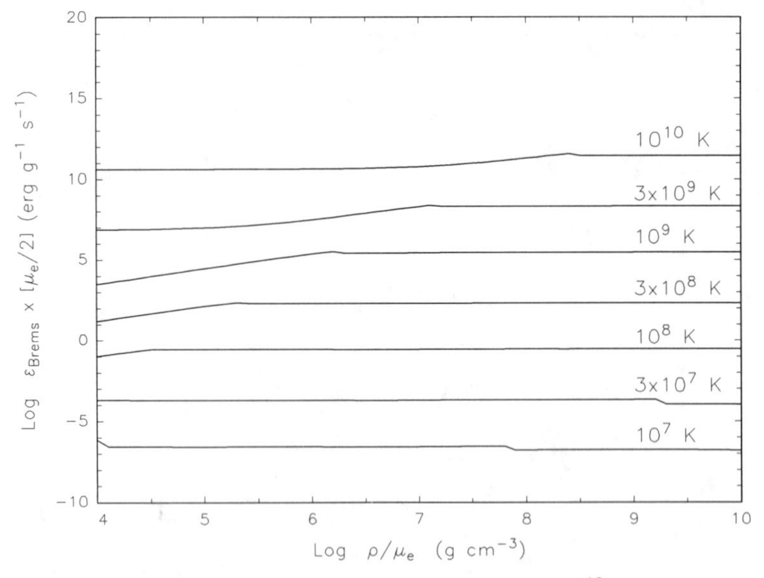

FIGURE 6.11. Bremsstrahlung neutrino loss rates for pure ^{12}C. The kinks in the curves are do to overlap in fitting formulas. Adapted from the calculations of Itoh and collaborators.

Figures 6.10 and 6.11 summarize the neutrino power generated by the above reactions as functions of temperature and density. (Fig. 6.11 is for the case of pure ^{12}C.) In these figures μ_e is the usual mean molecular weight for electrons.[11] As an application of Figure 6.10, consider SN1987A discussed in §2.6.3 and, in particular Figure 2.26, which gave an overview of the evolutionary stages leading to explosion. The next-to-last stage, lasting some two days, consists of the building up of a $1\mathcal{M}_\odot$ iron peak core with a central temperature and density of $T_c \approx 3.7 \times 10^9$ K, $\rho_c \approx 4.9 \times 10^7$ g cm^{-3}. Whether the core is silicon or iron μ_e is still about two. An eyeball estimate from Figure 6.10 of the neutrino power under these conditions is $\varepsilon_\nu \sim 10^{13}$ erg g^{-1} s^{-1}. If the whole core released this specific power (and this is an overestimate), then the total neutrino luminosity is $\mathcal{L}_\nu \sim 5\times10^{12}$ $\mathcal{L}_\odot$ (!) which is not too far distant from the value quoted in Figure 2.26. This extraordinary luminosity loss in neutrinos is primarily due to the pair annihilation process but, in any event, it shows how important these elusive particles can sometimes be for stars.

6.9 Exercises

1. As far as we know, helium is of no use for the metabolism of creatures such as us despite its being the second most abundant element in the universe. But carbon and oxygen are essential. Therefore, let's burn helium (as ^{4}He) and convert it to ^{12}C and ^{16}O. We shall use the triple-α to make ^{12}C, then add on another ^{4}He to make ^{16}O and, further, see how much ^{16}O is destroyed to make ^{20}Ne in one additional reaction. The imagined site for the burning is in the core of an intermediate mass star where $\rho = 10^4$ g cm^{-3} and $T_9 = 0.15$. (These are typical figures for nondegenerate helium-burning in such a star.) We denote the number densities (in cm^{-3}) of the various elements by n_4, n_{12}, n_{16}, and n_{20}, and the corresponding mass fractions by X_4, X_{12}, X_{16}, and X_{20}. The average of the cross section times velocity for the reactions are denoted by: $\langle 3\alpha\rangle$ for the 3-α; $\langle \alpha 12\rangle$ for ^{12}C$(\alpha,\gamma)^{16}$O; $\langle\alpha 16\rangle$ for ^{16}O$(\alpha,\gamma)^{20}$Ne. The $\langle\sigma v\rangle$s are to be deduced from equations (6.78), (6.80), (6.81), and accompanying material.

   (a) Show that the reaction rate equations that govern the creation and destruction of the nuclei are given by

$$\frac{dn_4}{dt} = -\frac{3n_4^3}{6}\langle 3\alpha\rangle - n_4 n_{12}\langle\alpha 12\rangle - n_4 n_{16}\langle\alpha 16\rangle$$

$$\frac{dn_{12}}{dt} = n_4^3\langle 3\alpha\rangle - n_4 n_{12}\langle\alpha 12\rangle$$

[11] The calculations done to construct these figures was based on the work of Naoki Itoh and his collaborators (see references). We used their analytic fits to neutrino rates but they also include tables. If you wish to duplicate the figures using these fits, we warn you that some of their expressions contain obvious errors.

$$\frac{dn_{16}}{dt} = n_4 n_{12} \langle \alpha 12 \rangle - n_4 n_{16} \langle \alpha 16 \rangle$$

$$\frac{dn_{20}}{dt} = n_4 n_{16} \langle \alpha 16 \rangle \, .$$

(b) Now integrate these equations as a function of time where the initial condition is that of pure helium at the temperature and density given above. You are to keep both of the latter fixed for all time. Note that this will involve numerical integration of four simultaneous first-order differential equations. The best way to do this is to first convert the number densities in the rate equations to mass fractions (by $n = \rho X N_{\mathrm{A}}/A$) so that the dependent variables are now the mass fractions. This will be convenient for two reasons. First off, they all range between zero and unity so the scale of the variables is nice. Secondly, we know that

$$X_4 + X_{12} + X_{16} + X_{20} = 1$$

where, initially, $X_4 = 1$ and the others are zero. This provides a conservation check on the computations so that if the sum is *not* one at some stage you are in trouble. Before you integrate the dX_i/dt first try to estimate the kind of time scales you are up against. The 3α reaction basically controls the flow of nuclear processing. Once you run out of αs, you are done. Therefore solve the first rate equation analytically, keeping only the 3α reaction. Then find the time (starting from time zero) when half of the αs are used up using that solution. (The time will be about 10^4 years.) Now you have a rough idea of what time steps to use at the beginning of the time integrations and you can numerically integrate the full set of equations until the αs are essentially used up.

(c) Plot your results for the X_i as functions of time and see how much carbon and oxygen you are left with. You may wish to redo the calculations by changing the problematical $\langle \alpha 12 \rangle$ by a factor of two either way to see what happens.

2. In the normal course of evolution of a massive star, the end products of nuclear burning are elements in the iron region of nucleon number. From our previous discussions, we know this is a disaster because the star continues to contract and heat up. If the temperatures get high enough, the radiation field is capable of initiating photodisintegration reactions and all the iron peak elements end up as a puddle of nucleons. This can happen on such rapid time scales that the abundances of nuclei (as functions of temperature and density) can be calculated approximately as if the gas were in chemical equilibrium.

This sounds like the Saha equation but the folks who do high temperature nuclear astrophysics call it "Nuclear Statistical Equilibrium" or NSE for short. To look at this in a very simplified version consider a gas composed only of ^{56}Ni and ^{4}He where the "chemical reaction" between them is

$$14\,^{4}\mathrm{He} \longleftrightarrow {}^{56}\mathrm{Ni} + Q.$$

You may compute the Q for the reaction from the mass excesses

$$[M - A]c^2(^{4}\mathrm{He}) = 2.42494 \text{ MeV}$$

$$[M - A]c^2(^{56}\mathrm{Ni}) = -53.902 \text{ MeV}.$$

(a) Set up the "Saha equation" for the reaction making believe that you are dealing with atoms and ions and assume that both nuclei are in their ground states. Let the statistical weights be equal to one (and this is all right because the spins of the ground states are zero).

(b) Convert your Saha equation so that the unknowns are mass fractions X_4 and X_{56} where $X_4 + X_{56} = 1$.

(c) Fix the physical density to be $\rho = 10^7$ g cm^{-3} and solve for X_4 and X_{56} for temperatures in the range $4.5 \le T_9 \le 6.5$.

(d) At what temperature is $X_4 = X_{56}$?

(e) And, yes, plot up your results for the Xs versus T_9.

3. We have alluded to the nefarious helium core flash several times, but here is your chance to actually do something about it. Suppose you have a gram of pure helium (as ^{4}He) in the center of a pre–helium flash red supergiant. The density and temperature of the gram are, respectively, $\rho = 2 \times 10^5$ g cm^{-3} and $T = 1.5 \times 10^8$ K. This is hot enough to burn helium by the triple-α reaction—which is the only reaction you will use. The energy generation rate for the reaction is given by equation (6.78).

   You are now to follow the time evolution of the gram as helium-burning proceeds by computing the temperature, $T(t)$, as a function of time. Start the clock running at time zero at the conditions stated. Assume that the density remains constant for all time, that no heat is allowed to leave the gram, and find $T(t)$ until that time when the material begins to become nondegenerate. For this use the nonrelativistic demarcation line $\rho/\mu_e = 6 \times 10^{-9} T^{3/2}$ (see eq. 3.66). Remember to use a specific heat, which is the sum of the electron specific heat (eq. 3.109) and the ideal gas specific heat for pure helium. (Assume that x may be found from $\rho/\mu_e = Bx^3$ of eq. 3.49.)

   Just so this problem doesn't become too difficult, assume that the helium concentration does *not* change with time. Also, plot T versus time in days.

We warn you, this is not an easy problem and you may find yourself in trouble if you are not careful since a good solution requires solving a tough differential equation numerically. You will be able to recognize the flash when it happens because the temperature will suddenly skyrocket after not too many days of burning.

6.10 References and Suggested Readings

§**6.1**: Gravitational Energy Sources

Sections 17.4–17.6 of

▹ Cox, J.P. 1968, *Principles of Stellar Structure*, in two volumes (New York: Gordon and Breach)

contains a fuller discussion of gravitational sources.

§**6.2**: Thermonuclear Energy Sources

Chapters 4 and 5 of

▹ Clayton, D.D. 1968, *Principles of Stellar Evolution and Nucleosynthesis* (New York: McGraw-Hill)

are still the most effective general textbook references for thermonuclear reactions and nucleosynthesis in stars. Some details have changed during the intervening years, but the overall picture he presents is still accurate. Cox (1968), Chapter 17 also contains useful material.

A source for mass excesses (Δ) of nuclei is

▹ Wapstra, A.H., and Bos, K. 1977 *Atomic Data and Nuclear Data Tables*, **19**, 175

or later versions of this series of compilations.

For those of you looking for more information and references on barrier penetration factors as applied to nucelar astrophysics see

▹ Humbler, J., Fowler, W.A., and Zimmerman, B.A. 1987, *Astron. Ap.*, **177**, 317.

The major source for reaction rates (of many kinds) are the compilations and critical reviews of William A. Fowler and his collaborators. The following references should be consulted in sequence of publication because philosophy and nomenclature carry over to the later papers:

▹ Fowler, W.A., Caughlan, G.R., and Zimmerman, B.A. 1967, *Ann. Rev. Astron. Ap.*, **5**, 525

▹ Harris, M.J., Fowler, W.A., Caughlan, G.R., and Zimmerman, B.A. 1983, *Ann. Rev. Astron. Ap.*, **21**, 165

▹ Caughlan, G.R., Fowler, W.A., Harris, M.J., and Zimmerman, B.A. 1985, *Atomic Data Nuc. Data Tables*, **32**, 197

▹ Caughlan, G.R., and Fowler, W.A. 1988, *Atomic Data Nuc. Data Tables*, **40**, 283.

Neutron capture cross sections are reviewed by

▷ Boa, Z.Y., and Käppeler, F. 1987, *Atomic Data Nuc. Data Tables*, **36**, 411.

For a full description of electron capture and other weak reactions in nucleosynthesis see:

▷ Fuller, G.M., Fowler, W.A., and Newman, M.J. 1982, *Ap.J. Suppl.*, **48**, 279

▷ *Ibid* 1985, *Ap.J.*, **293**, 1.

The early developments of the physics of the pp reaction are given in

▷ Bethe, H.A., and Critchfield, C.H. 1938, *Phys. Rev.*, **54**, 248

▷ Bethe, H.A. 1939, *Phys. Rev.*, **55**, p. 103 and 434

▷ Salpeter, E.E. 1952, *Phys. Rev.*, **88**, 547

▷ *Ibid* 1952, *Ap.J.*, **115**, 326.

Recent corrections are due to

▷ Gould, R.J., and Guessoum, N. 1990, *Ap.J.*, **359**, L67

and see

▷ Mampe, M., *et al.* 1989, *Phys. Rev. Letters*, **63**, 593

for an experimentally determined half-life of the neutron.

The description of electron screening used in most modern works is based on the development of

▷ Salpeter, E.E. 1954, *Austrl. J. Phys.*, **7**, 373.

Debye-Hückel theory is discussed by

▷ Landau, L.D., and Lifshitz, E.M. 1958, *Statistical Physics* (London: Pergamon).

§**6.3**: The Proton-Proton Chains

The expression (6.73) for the average Q-value for the pp chains is from

▷ Fowler, W.A., Caughlan, G.R., and Zimmerman, B.A. 1975, *Ann. Rev. Astron. Ap.*, **13**, 69.

An improved, but more complicated, estimate has been given by

▷ Mitalas, R. 1989, *Ap.J.*, **338**, 308.

The papers cited on deuterium and lithium burning and their role in cosmology are

▷ Boesgaard, A.M., and Steigman, G. 1985, *Ann. Rev. Astron. Ap.*, **23** 319

▷ Steigman, G. 1985, in *Nucleosynthesis*, Eds. W. D. Arnett and J. W. Truran (Chicago: University of Chicago Press), p. 48

▷ Yang, J., Turner, M.S., Steigman, G., Schramm, D.N., and Olive, K.A. 1984, *Ap.J.*, **281**, 493.

See also

▷ Deliyannis, C., Demarque, P., Kawaler, S., Krauss, L., and Romanelli, P. 1989, *Phys. Rev. Letters*, **62**, 1583.

§6.4: The Carbon-Nitrogen-Oxygen Cycles

Many of the intricacies of the CNO cycles were worked out by

▹ Caughlan, G.R., and Fowler, W.A. 1962, *Ap.J.*, **136**, 329.

See also

▹ Caughlan, G.R. 1965, *Ap.J.*, **141**, 688.

"Fast" CNO cycles, which require more reactions to treat than are listed in Table 6.2, are reviewed in

▹ Starrfield, S, Sparks, W.M., and Truran, J.W. 1974, *Ap.J. Suppl.*, **28**, 247

in the context of classical novae. The papers referring to CNO abundances in the sun and anomalous abundances in red supergiants are:

▹ Bahcall, J.N., and Ulrich, R.K. 1988, *Rev. Mod. Phys.*, **60**, 297

▹ Iben, I. Jr., and Renzini, A. 1984, *Phys. Reports*, **105**, 329.

§6.5: Helium-Burning Reactions

▹ Fowler, W.A. 1986, in *Highlights of Mod. Ap.*, Eds. S. L. Shapiro and S.A. Teukolosky (New York: Wiley-Interscience), p. 1

is an excellent (and personal) introduction to helium-burning. Landmark papers include

▹ Salpeter, E.E. 1952, *Ap.J.*, **115**, 326

▹ *Ibid* 1953, *Ann. Rev. Nuc. Sci.*, **2**, 41

▹ Hoyle, F. 1954, *Ap.J. Suppl.*, **1**, 121.

Individual rates are given throughout the papers of Fowler and collaborators as listed above. The quote from Fowler (1985) may be found in

▹ Fowler, W.A. 1985, in *Nucleosynthesis*, Eds. W.D. Arnett and J.W. Truran (Chicago: University of Chicago Press) p. 13.

See also

▹ Filippone, B.W. (1986), *Ann. Rev. Nuc. Particle Sci.*, **36**, 717

for further comments on helium-burning.

▹ Buchmann, L. et al. 1993, *Phys. Rev. Letters*, **70**, 726

have obtained new experimental results for part of the S-factor for $^{12}\mathrm{C}(\alpha,\gamma)^{16}\mathrm{O}$, and the nucleosynthetic calculations of

▹ Weaver, T.A., and Woosley, S.E. 1993, *Phys. Reports*, **227**, 1

appear to be consistent with what can be inferred from the experiment.

Screening for the triple-α is discussed in

▹ Fushiki, I., and Lamb, D.Q. 1987, *Ap.J.*, **317**, 368

and references therein. "Pycnonuclear" screening was originally discussed by

▹ Cameron, A.G.W. 1959, *Ap.J.*, **130**, 916

and a newer prespective is offered by

▹ Ichimaru, S., Ogata, S., and Van Horn, H.M. 1992, *Ap.J. Letters*, **401**, L35.

§**6.6**: Carbon and Oxygen Burning

▹ Thielemann, F.-K., and Arnett, W.D. 1985, in *Nucleosynthesis*, Eds. W.D. Arnett and J.W. Truran (Chicago: Univ. of Chicago Press), p. 151,

give an excellent review of the various late thermonuclear burning stages for hydrostatic stars. The rate quoted for the $^{12}C + ^{12}C$ reaction is from

▹ Caughlan, G.R., and Fowler, W.A. 1988, *Atomic Data Nuc. Data Tables*, **40**, 283.

§**6.7**: Silicon "Burning"

The first network calculations for silicon burning were reported in

▹ Truran, J.W., Cameron, A.G.W., and Gilbert, A.A. 1966, *Can. J. Physics*, **44**, 576.

▹ Bodansky, D., Clayton, D.D., and Fowler, W.A. 1968, *Ap.J. Suppl.*, **16**, 299

clarified the quasi-static equilibrium (QSE) nature of this burning and, for a more modern reference with details, see

▹ Woosley, S.E., Arnett, W.D., and Clayton, D.D. 1973, *Ap.J. Suppl.*, **26**, 231.

Another summary of advanced burning, but in the context of the radioactive dating of the elements, may be found in

▹ Cowan, J.J., Thielemann, K.-R., and Truran, J.W. 1991, *Ann. Rev. Astron. Ap.*, **29**, 447.

§**6.8**: Neutrino Emission Mechanisms

Chapter 18 of

▹ Shapiro, S.L., and Teukolsky, S.A. 1983, *White Dwarfs, Neutron Stars, and Black Holes* (New York: Wiley & Sons)

gives an overview of high-energy neutrino emission mechanisms.

▹ Bethe, H.A. 1990, *Rev. Mod. Phys.*, **62**, 901

places some of these in the context of supernova explosions and includes a discussion of electron capture rates.

▹ Chiu H.-Y., 1968, *Stellar Physics* (Waltham: Blaisdell)

discusses weak interactions in considerable detail. He uses an outdated theory for these interactions (not the new unified electro-weak theory) but, for low-energy reactions, his results are perfectly acceptable. The neutrino rates shown in Figures 6.10 and 6.11 are constructed from analytic fitting formulas from the following papers:

▹ Itoh, N., and Kohyama, Y. 1983, *Ap.J.*, **275**, 858

▹ Itoh, N., Matsumoto, N., Seki, M., and Kohyama, Y. 1984, *Ap.J.*, **279**, 413

▹ Itoh, N., Kohyama, Y., Matsumoto, N., and Seki, M. 1984 *Ap.J.*, **280**, 787

- ▷ Itoh, N., Kohyama, Y., Matsumoto, N., and Seki, M. 1984, *Ap.J.*, **285**, 304
- ▷ Munakata, M., Kohyama, Y., and Itoh, N. 1987, *Ap.J.*, **316**, 708
- ▷ Itoh, N., Adachi, T., Nakagawa, M., Kohyama, Y., and Munakata, H. 1989, *Ap.J.*, **339**, 354
- ▷ Itoh, N., Mutoh, H., Hikita, A., and Kohyama, Y. 1992, *Ap.J.*, **395**, 622.

You should also check on errata for some of these papers.

7

Stellar Modeling

To err is human,
but to really foul things up
requires a computer.
— Anonymous

Every novel should have
a beginning, a muddle, and an end.
— Peter De Vries
We are now in the stellar muddling stage.

This chapter will end up having covered a diverse set of topics but all have the same underlying theme: what analytic and numeric techniques are used to model stars? Some of these techniques will yield approximate solutions to the equations of stellar structure, whereas others are designed for the exacting task of comparing model results to real stars. We start with some rather general considerations by reviewing the equations of stellar structure.

7.1 The Equations of Stellar Structure

We shall restrict ourselves, for the moment at least, to discussing what is necessary to model stars in hydrostatic equilibrium and thermal balance while neglecting complicating factors such as nonsphericity, magnetic fields, etc. The assumption of strict equilibrium implies that time-dependent (e.g., evolutionary) processes are ignored for now.

To construct an *ab initio* stellar model we must first specify the total stellar mass and the run of composition as a function of some coordinate such as radius or interior mass. What should come out at the end of the calculation is the run of mass versus radius (or the other way around), and the corresponding local values of pressure, density, temperature, and luminosity. To do this, we need the microscopic constituent physics implied in the following:

$$P = P(\rho, T, \mathbf{X}) \tag{7.1}$$

$$E = E(\rho, T, \mathbf{X}) \tag{7.2}$$

$$\kappa = \kappa(\rho, T, \mathbf{X}) \tag{7.3}$$

$$\varepsilon = \varepsilon(\rho, T, \mathbf{X}) \tag{7.4}$$

and various derivatives of these quantities. Here $\mathbf{X}$ is shorthand for composition (as in a specification of nuclear species). Thus given density, temperature, and composition, the four quantities should be available on demand by the model builder.

The structural and thermal differential relations to be satisfied include:

$$\frac{dP}{dr} = -\frac{G\mathcal{M}_r}{r^2}\rho \quad \text{or} \quad \frac{dP}{d\mathcal{M}_r} = -\frac{G\mathcal{M}_r}{4\pi r^4} \tag{7.5}$$

$$\frac{d\mathcal{M}_r}{dr} = 4\pi r^2 \rho \quad \text{or} \quad \frac{dr}{d\mathcal{M}_r} = \frac{1}{4\pi r^2 \rho} \tag{7.6}$$

$$\frac{d\mathcal{L}_r}{dr} = 4\pi r^2 \varepsilon \rho \quad \text{or} \quad \frac{d\mathcal{L}_r}{d\mathcal{M}_r} = \varepsilon\,. \tag{7.7}$$

The right-hand variants of these equations are set down because, as we shall see, it is often convenient to take $\mathcal{M}_r$ as the independent variable.

Accompanying the above are relations or criteria that establish what the modes of heat transfer are. In the simplest instance of allowing only local adiabatic convection—as in a mixing length theory—these are as follows. (The more general prescription given in Chapter 5 is implemented in the `ZAMS` code discussed in Appendix C.) Compute

$$\nabla_{\text{rad}} = \frac{3}{16\pi ac}\frac{P\kappa}{T^4}\frac{\mathcal{L}_r}{G\mathcal{M}_r} \tag{7.8}$$

and, as was done in Chapter 5, test to see if this exceeds ∇_{ad}. Then set

$$\nabla = \nabla_{\text{rad}} \quad \text{if} \quad \nabla_{\text{rad}} \le \nabla_{\text{ad}} \tag{7.9}$$

for pure diffusive radiative transfer or conduction, or

$$\nabla = \nabla_{\text{ad}} \quad \text{if} \quad \nabla_{\text{rad}} > \nabla_{\text{ad}} \tag{7.10}$$

when adiabatic convection is present locally. The quantity ∇ is given by

$$\nabla = \frac{d\ln T}{d\ln P} = -\frac{r^2 P}{G\mathcal{M}_r \rho}\frac{1}{T}\frac{dT}{dr} = -\frac{4\pi r^4 P}{G\mathcal{M}_r T}\frac{dT}{d\mathcal{M}_r}. \tag{7.11}$$

These computations then establish the local slope of temperature with respect to pressure.

Equations (7.5–7.11), when combined, are equivalent to a fourth-order differential equation in space or mass. If we choose $\mathcal{M}_r$ as the independent variable, then the four boundary conditions required to close the system as applied at the center and surface are:

$$\text{at the center } (\mathcal{M}_r = 0),\ r = \mathcal{L}_r = 0 \tag{7.12}$$

$$\text{and at the surface } (\mathcal{M}_r = \mathcal{M}),\ \rho = T = 0. \tag{7.13}$$

Here "zero" boundary conditions have been applied to the surface for simplicity (see §7.3.2). Note that $\mathcal{M}$ was specified beforehand but other quantities such as total radius, $\mathcal{R}$, and total luminosity, $\mathcal{L}$, must be found as a result of the entire calculation.

In principle we may then solve for the entire structure. However, does a solution always exist, and, if so, is it unique? The answer to both these questions is "no" for many choices of total mass and composition. For the first, you can't make an equilibrium star out of any old thing. The second question is a little more subtle. It turns out that, for some combinations of total mass and composition, multiple solutions to the stellar structure equations are indeed possible. This is not obvious, but a hint is contained, for example, in the observation that the general equation of state for stellar material is exceedingly complicated and a given pressure may be generated at different temperatures and densities by some combination of ion ideal gas, degenerate electrons, radiation, etc., with each of these components having very different thermodynamic properties. The existence of multiple solutions contradicts what was long held to be a "theorem" in stellar astrophysics due to H. Vogt and H.N. Russell. However, the idea of uniqueness is still useful in that among a set of models all having the same mass and run of composition, usually only one seems to correspond to a real star or to have come from some realistic line of stellar evolution. The others are unstable in some fundamental way. Thus, for example, it is now known that it is possible to construct "main sequences" for stars having burnable fuel where the sequence is double valued with respect to mass.[1] As fascinating as these things may be to some theorists, we shall brush them aside and assume that if a stellar model of given mass and run of composition can be constructed, then it is the only one possible.

---

[1] For a review of such problems see Hansen (1978) where this topic and the notion of "secular stability" is discussed.

Before going into how the stellar structure equations are solved in practice, we introduce a simplification which, albeit restrictive, turns out to be of both practical and pedagogical value.

7.2 Polytropic Equations of State and Polytropes

The primary, and classic, reference for the beginning portions of this section is Chandrasekhar (1939). Similar material, although not as exhaustive, may be found in Cox (1968, §23.1), and Kippenhahn and Weigert (1990, §19).

We shall first discuss polytropes in a general way but then interrupt the narrative to consider how these approximations to stellar models and, to some extent, real stars are calculated in practice. This last may seem to take us far afield but, toward the end of the section, we shall return to polytropes for a discussion of how they are used.

7.2.1 General Properties of Polytropes

In previous chapters we encountered equations of state where pressure was only a function of density (and, of course, composition). For example, the equation of state for a completely degenerate, nonrelativistic, electron gas was given by (3.61) as

$$P_{\rm e} = 1.004 \times 10^{13} \left(\frac{\rho}{\mu_{\rm e}}\right)^{5/3} \quad \text{dyne cm}^{-2} \tag{7.14}$$

which is a power law equation of state with $P \propto (\rho/\mu_{\rm e})^{5/3}$. We might then imagine a stellar model composed of a material for which $\mu_{\rm e}$ is a constant throughout and in which *both* the equation of state *and* the actual run of pressure versus density satisfy (7.14). But if this condition is imposed beforehand it might conflict with the complete set of stellar structure equations and a self-consistent model would not be possible. *Polytropes* are pseudo-stellar models for which power law equations of pressure versus density such as (7.14) are assumed *a priori* but where no reference to heat transfer or thermal balance is made. Thus only the hydrostatic and mass equations are used and inconsistencies with respect to the complete set of stellar structure equations are avoided. This may seem to be a high price to pay for consistency, but the resulting polytropic structures have proven to be remarkably useful in the interpretation of many aspects of real stellar structure.

Another motivation for studying polytropes arises from consideration of the structure of certain types of adiabatic convection zones. In a region of efficient convection the actual "del" of (7.11) is given by $\nabla = \nabla_{\rm ad} = 1 - 1/\Gamma_2$ (see eq. 3.88). If Γ_2 is assumed constant, then integrating (7.11) yields

$$P(r) \propto T^{\Gamma_2/(\Gamma_2-1)}(r). \tag{7.15}$$

If, in addition, the gas is an ideal gas with $T \propto P/\rho$, then $P(r) \propto \rho^{\Gamma_2}(r)$ and we have the same situation as above: P obeys a power law relation with respect to density as a function of radius.

In particular, we define a polytropic stellar model to be one in which the pressure is given by

$$P(r) = K\rho^{1+1/n}(r) \tag{7.16}$$

where n, the *polytropic index*, is a constant as is the proportionality constant K.[2] Since the polytrope is to be in hydrostatic equilibrium, then the distribution of pressure and density must be consistent with both the equation of hydrostatic equilibrium and conservation of mass. To best see how this works, divide the hydrostatic equation by ρ, multiply by r^2, and then take the derivative with respect to r of both sides to find

$$\frac{d}{dr}\left(\frac{r^2}{\rho}\frac{dP}{dr}\right) = -G\frac{d\mathcal{M}_r}{dr} = -4\pi G r^2 \rho$$

where the mass equation has been used to obtain the final equality. Rewrite this as

$$\frac{1}{r^2}\frac{d}{dr}\left(\frac{r^2}{\rho}\frac{dP}{dr}\right) = -4\pi G\rho \tag{7.17}$$

which is Poisson's equation. The latter identification is clear if we define the potential Φ such that $g = d\Phi/dr = G\mathcal{M}_r/r^2$, eliminate the pressure derivative (using 7.5), and find $\nabla^2\Phi = 4\pi G\rho$ in spherical coordinates.

We now perform a sequence of transformations with the intent of making (7.17) dimensionless. Define the dimensionless variable θ by

$$\rho(r) = \rho_c\,\theta^n(r) \tag{7.18}$$

where $\rho_c = \rho(r=0)$. The power law for pressure is then

$$P(r) = K\rho_c^{1+1/n}\theta^{n+1} = P_c\,\theta^{1+n}. \tag{7.19}$$

The central pressure, P_c, is clearly equal to

$$P_c = K\rho_c^{1+1/n}. \tag{7.20}$$

Now substitute these into Poisson's equation and find the second-order differential equation for θ

$$\frac{(n+1)P_c}{4\pi G\rho_c^2}\,\frac{1}{r^2}\frac{d}{dr}\left(r^2\frac{d\theta}{dr}\right) = -\theta^n. \tag{7.21}$$

---

[2]Be careful not to confuse this n with the n used as the power law density exponent of opacity.

Finally, introduce the new dimensionless radial coordinate, ξ, by

$$r = r_n \xi \tag{7.22}$$

where the scale length, r_n, is defined as

$$r_n^2 = \frac{(n+1)P_c}{4\pi G \rho_c^2}. \tag{7.23}$$

With this substitution, Poisson's equation becomes

$$\frac{1}{\xi^2}\frac{d}{d\xi}\left(\xi^2 \frac{d\theta}{d\xi}\right) = -\theta^n \tag{7.24}$$

and is now called the *Lane-Emden equation*. Models corresponding to solutions of this equation for a chosen n are called "polytropes of index n" and the solutions themselves are "Lane-Emden solutions" and are denoted by $\theta_n(\xi)$.[3]

Note that if the equation of state for the model material is an ideal gas with $P = \rho N_A kT/\mu$, then some easy manipulations yield

$$P(r) = K' T^{n+1}(r), \quad T(r) = T_c\, \theta(r) \tag{7.25}$$

with

$$K' = \left(\frac{N_A k}{\mu}\right)^{n+1} K^{-n}, \quad T_c = K \rho_c^{1/n} \left(\frac{N_A k}{\mu}\right)^{-1}. \tag{7.26}$$

Thus in a polytrope whose material equation of state is an ideal gas with constant μ, θ measures temperature. Finally, the radial scale factor in this case is

$$r_n^2 = \left(\frac{N_A k}{\mu}\right)^2 \frac{(n+1)T_c^2}{4\pi G P_c} = \frac{(n+1)K\rho_c^{1/n-1}}{4\pi G}. \tag{7.27}$$

To prepare complete polytropic models that might share some resemblance to stars, appropriate boundary conditions must be applied to the Lane-Emden equation. For a complete model, with center at $r=0$ and a surface that has vanishing density, these boundary conditions are as follows. For ρ_c in (7.18) to really be the central density, we require that $\theta(\xi=0)=1$. Furthermore, spherical symmetry at the center (dP/dr vanishing at $r=0$) requires that $\theta' \equiv d\theta/d\xi = 0$ at $\xi = 0$. This last condition pins down the solution at the center so that divergent solutions of the second-order system are suppressed. The regular solutions are called "E-solutions." If the surface is that place where $P=\rho=0$, then we require that the solution θ_n vanish there also. More specifically, the surface is where the *first zero* of θ_n occurs as measured from the center outward. (We do not want the pressure

---

[3]Note that we shall not always append the n subscript to θ_n. The index will usually be obvious from the context.

to vanish both at the "surface" and at some interior point.) We denote the location of the first zero by ξ_1 and it depends on the value of the polytropic index n. To summarize, the boundary conditions for a whole model are:

$$\theta(0) = 1, \quad \theta'(0) = 0 \quad \text{at} \quad \xi = 0 \quad \text{(the center)} \tag{7.28}$$

$$\theta(\xi_1) = 0 \quad \text{at} \quad \xi = \xi_1 \quad \text{(the surface).} \tag{7.29}$$

Since ξ_1 is the location of the surface, then the total (dimensional) radius is at

$$\mathcal{R} = r_n \xi_1 = \left[\frac{(n+1)P_c}{4\pi G \rho_c^2}\right]^{1/2} \xi_1 . \tag{7.30}$$

Thus specifying K, n, and either ρ_c or P_c, yields the radius R.

Analytic E-solutions for θ_n are obtainable for $n = 0$, 1, and 5. Numerical methods must be used to obtain solutions to the Lane-Emden equation for general n.

1. The solution for $n = 0$ is the constant-density sphere discussed in earlier chapters with $\rho(r) = \rho_c$. You may easily verify that

$$\theta_0(\xi) = 1 - \frac{\xi^2}{6}, \quad \text{with} \quad \xi_1 = \sqrt{6} \tag{7.31}$$

   and $P(\xi) = P_c\theta(\xi) = P_c\left[1 - (\xi/\xi_1)^2\right]$. Except that we have not found P_c (which may be found once $\mathcal{M}$ and $\mathcal{R}$ are specified), this is the solution found for the constant-density sphere as given by (1.38). P_c is easily computed using (7.30) with $\xi_1 = \sqrt{6}$ to be $(3/8\pi)(G\mathcal{M}^2/\mathcal{R}^4)$ in accord with (1.39).

2. For $n = 1$, the solution θ_1 is the familiar "sinc" function

$$\theta_1(\xi) = \frac{\sin\xi}{\xi}, \quad \text{with} \quad \xi_1 = \pi. \tag{7.32}$$

   The pressure and density follow from $\rho = \rho_c\theta$ and $P = P_c\theta^2$.

3. The polytrope for $n = 5$ has a finite central density but its radius is unbounded with

$$\theta_5(\xi) = \left[1 + \xi^2/3\right]^{-1/2} \quad \text{and} \quad \xi_1 \to \infty. \tag{7.33}$$

   Despite the infinite radius, this polytrope does has a finite amount of mass associated with it.

Complete and regular solutions with $n > 5$ are also infinite in extent but contain infinite mass. The range of n of interest to us for complete models is then $0 \le n \le 5$.

Given n and K, we can in principle find the dependence of P and ρ on ξ. However, we cannot obtain absolute physical numbers unless $\mathcal{R}$ and either ρ_c or P_c are first specified. This follows from (7.20) and (7.30). The main difficulty is that $\mathcal{R}$ is not known beforehand. But $\mathcal{M}$ is what we wish to specify and this turns out to be enough.

The mass contained in a sphere of radius r is found from (7.6) to be $\mathcal{M}_r = \int_0^r 4\pi r^2 \rho(r)\, dr$. In ξ-space this becomes

$$\mathcal{M}_\xi = 4\pi r_n^3 \rho_c \int_0^\xi \xi^2 \theta^n \, d\xi.$$

The integrand of this expression contains θ^n, but this is just the (negative) of the right-hand side of the Lane-Emden equation (7.24). Therefore insert the left-hand side of (7.24) in place of $-\theta^n$, notice that the factors of ξ^2 cancel and what is left is a perfect differential under the integral. The result is

$$\mathcal{M}_\xi = 4\pi r_n^3 \rho_c \left(-\xi^2 \theta'\right)_\xi \tag{7.34}$$

where $\left(-\xi^2\theta'\right)_\xi$ means "evaluate $(-\xi^2\, d\theta/d\xi)$ at the point ξ." The total mass is given by $\mathcal{M} = \mathcal{M}(\xi_1)$. It should be clear that if $\mathcal{M}$ and $\mathcal{R}$ are specified in physical units, then all else follows. In what comes next, the relations between $\mathcal{M}$, $\mathcal{R}$, etc., are given without derivation. For example,

$$\mathcal{M} = (4\pi)^{-1/2} \left(\frac{n+1}{G}\right)^{3/2} \frac{P_c^{3/2}}{\rho_c^2} \left(-\xi^2\theta'\right)_{\xi_1} \tag{7.35}$$

which, in conjunction with (7.20), gives ρ_c or P_c in terms of $\mathcal{M}$. A little more algebra yields

$$\begin{aligned} P_c &= \frac{1}{4\pi(n+1)\,(\theta')^2_{\xi_1}} \frac{G\mathcal{M}^2}{\mathcal{R}^4} \\ &= \frac{8.952 \times 10^{14}}{(n+1)\,(\theta')^2_{\xi_1}} \left(\frac{\mathcal{M}}{\mathcal{M}_\odot}\right)^2 \left(\frac{\mathcal{R}}{\mathcal{R}_\odot}\right)^{-4} \quad \text{dyne cm}^{-2}. \end{aligned} \tag{7.36}$$

Note that the last result requires n, but not K.

Another result that will prove useful follows from solving for K given n, $\mathcal{M}$, and $\mathcal{R}$:

$$K = \left[\frac{4\pi}{\xi^{n+1}\left(-\theta'\right)^{n-1}}\right]^{1/n}_{\xi_1} \frac{G}{n+1} \mathcal{M}^{1-1/n} \mathcal{R}^{-1+3/n}. \tag{7.37}$$

Note that if $n = 3$, K depends only on $\mathcal{M}$ or, turned around, $\mathcal{M}$ does not depend on $\mathcal{R}$ for any K if $n = 3$.

If the equation of state is that of an ideal gas, then the central temperature is given by

$$\begin{aligned} T_c &= \frac{1}{(n+1)\left(-\xi\theta'\right)_{\xi_1}} \frac{G\mu}{N_{\rm A}k} \frac{\mathcal{M}}{\mathcal{R}} \\ &= \frac{2.293 \times 10^7}{(n+1)\left(-\xi\theta'\right)_{\xi_1}} \mu \left(\frac{\mathcal{M}}{\mathcal{M}_\odot}\right) \left(\frac{\mathcal{R}}{\mathcal{R}_\odot}\right)^{-1} \quad \text{K}. \end{aligned} \tag{7.38}$$

You may easily verify that T_c for the constant-density sphere ($n = 0$) is the same as given by the earlier result (1.53).

A useful quantity that depends only on n is the ratio of central density to mean density. This is given by

$$\frac{\rho_c}{\langle\rho\rangle} = \frac{1}{3}\left(\frac{\xi}{-\theta'}\right)_{\xi_1} . \tag{7.39}$$

Thus the statement will sometimes be made that "this stellar model looks like a polytrope of index so-and-so because its degree of central concentration is such-and-such"; that is, comparison of central to mean density implies an n by way of (7.39). This is often a useful way to look at things—if you know what you're doing.

Finally, it is an easy matter to show that the gravitational potential energy of a polytrope is (see Eq. 1.6 and the discussion preceding that equation for a refresher on Ω)

$$\Omega = -\frac{3}{5-n}\frac{G\mathcal{M}^2}{\mathcal{R}}. \tag{7.40}$$

For the constant-density sphere the coefficient $3/(5-n)$ is just $3/5$ and this is the value quoted for the quantity "q" after (1.7).

Now that some of the formalism is out of the way, what are interesting values for n? The pressure of the completely degenerate but nonrelativistic electron gas goes as $\rho^{5/3}$. Hence, by the definition of the polytropic equation of state (7.16), n for this case is 1.5 (or "a three-halves polytrope"). The density exponent for the fully relativistic case is 4/3 and thus $n = 3$ (or "an n equal three polytrope"). The same indices crop up in other applications as we shall see. For now, recall that $P \propto \rho^{\Gamma_2}$ in an ideal gas convection zone. If no ionization is taking place (almost a contradiction for a real convection zone) then $\Gamma_2 = 5/3$ and $n = 3/2$ again. It will turn out that indices of 1.5 and 3 are the ones usually encountered in simple situations. How unfortunate it then is that neither of these values have analytic E-functions associated with them. Therefore, how *are* these nonanalytic cases computed? The following subsection looks into this question and serves as a brief introduction to how stellar models are computed. After this, we shall use the results from the polytropic calculations.

7.2.2 Numerical Calculation of the Lane-Emden Functions

This section, and others like it, is not intended to be an introduction to numerical analysis, but rather a guide to some techniques used to make stellar models. The subject is obviously important because one practical end of theory is the computation of a number. Get that wrong and you may waste the valuable time of experimentalists and observers.

A primary reference for numerical techniques that work is *Numerical Recipes: The Art of Scientific Computing*, 2d ed., by Press et al. (1992) (we are not biased just because the list of authors is heavily weighted by those practicing the art of astrophysics).

The Lane-Emden equation we wish to solve is

$$\frac{1}{\xi^2}\frac{d}{d\xi}\left(\xi^2\frac{d\theta}{d\xi}\right) = -\theta^n \tag{7.41}$$

with boundary conditions for the complete E-solution, $\theta(0) = 1$, $(d\theta/d\xi) = 0$ at $\xi = 0$, and the vanishing of θ at the surface. There are several ways of obtaining a numerical solution to this problem.

Shooting for a Solution

The first, and most straightforward, method is to treat the system as an initial value problem by starting from the origin ($\xi = 0$), integrating outward, and then stopping when θ goes to zero at the initially undetermined surface. This is a version of the "shooting method" whereby one "shoots" from a starting point and hopes that the shot will end up at the right place. For this we need an "integrator." One of the most useful for this purpose is of the class called "Runge-Kutta" integrators, which are easy to program (even on a programmable handheld calculator—with some patience) and are accurate and stable for simple problems. Note, however, that some problems are not simple and special techniques are required. Runge-Kutta schemes involve evaluating a series of derivatives of the dependent variable, y, at a sequence of points in the interval starting at x in the independent variable and ending at $x+h$. The quantity h is called the "step size." These derivatives are then averaged together in a particular way to eventually find $y(x+h)$. As will be seen below, the solution is "leap-frogged" from x to $x+h$.

The most convenient way to pose the second-order Lane-Emden problem for use in a Runge-Kutta scheme is to cast it in the form of two first-order equations. For notational convenience (and to make what follows better resemble what you will find in *Numerical Recipes*), introduce the new variables $x = \xi$, $y = \theta$, and $z = (d\theta/d\xi) = (dy/dx)$. The Lane-Emden equation now becomes

$$y' = \frac{dy}{dx} = z,$$

$$z' = \frac{dz}{dx} = -y^n - \frac{2}{x}z. \tag{7.42}$$

Suppose we know the values of y and z at some point x_i. Call these values y_i and z_i. If h is some carefully chosen step size, then the goal is to find y_{i+1} and z_{i+1} at $x_{i+1} = x_i + h$. This is, of course, just what is meant by an initial value problem. The particular Runge-Kutta scheme we shall choose to illustrate the technique is the fourth-order Runge-Kutta integrator. (Lower- and higher-order schemes are available.) As promised before, y' and z' are evaluated in a series of steps leading up to x_{i+1} as follows.

Compute the quantities

$$\begin{aligned}
k_1 &= h\,y'(x_i, y_i, z_i) \\
l_1 &= h\,z'(x_i, y_i, z_i) \\
k_2 &= h\,y'(x_i + h/2, y_i + k_1/2, z_i + l_1/2) \\
l_2 &= h\,z'(x_i + h/2, y_i + k_1/2, z_i + l_1/2) \\
k_3 &= h\,y'(x_i + h/2, y_i + k_2/2, z_i + l_2/2) \\
l_3 &= h\,z'(x_i + h/2, y_i + k_2/2, z_i + l_2/2) \\
k_4 &= h\,y'(x_i + h, y_i + k_3, z_i + l_3) \\
l_4 &= h\,z'(x_i + h, y_i + k_3, z_i + l_3).
\end{aligned} \tag{7.43}$$

As you can see, these ks and ls are rough guesses of the changes in the values of the functions y and z at various steps along the way to x_{i+1}. These are then weighted and added to find

$$\begin{aligned}
y_{i+1} &= y_i + \frac{k_1}{6} + \frac{k_2}{3} + \frac{k_3}{3} + \frac{k_4}{6} \\
z_{i+1} &= z_i + \frac{l_1}{6} + \frac{l_2}{3} + \frac{l_3}{3} + \frac{l_4}{6}.
\end{aligned} \tag{7.44}$$

It looks a bit complicated but, once you get in the swing of it, it rolls right along. The error introduced in the integration by discrete steps in this scheme is of the order of the fifth power of the step size h; that is, in moving from x_i to x_{i+1}, y_{i+1} and z_{i+1} are good to $\mathcal{O}(h^4)$. The calculation is a trade-off between the computation time it takes to make many steps of size h and the accuracy desired. Even unlimited time on a computer may not give arbitrary precision, however, for the simple reason that no machine has unlimited precision in the way it represents numbers internally. One therefore always lives with some, albeit small, amount of error. More sophisticated versions of Runge-Kutta (and other) methods are available that automatically adjust step size to maintain some desired, but hopefully reasonable, level of accuracy. One of these is used in the `FORTRAN` programs `ZAMS` and `PULS` discussed in Appendix C.

The polytrope calculation now marches from the origin to the surface. But the origin must be treated with care because the boundary condition

$\theta'(0) = z(0) = 0$ means that equation (7.42) for z' is indeterminate at $x=0$. Since the E-solutions are derived, in part, from the stellar structure equations, we should resolve what to do now before worrying about the same difficulties with more realistic models. The resolution of the problem is to expand $\theta(\xi)$ in the Lane-Emden equation in a series about the origin using the boundary conditions to establish constants in the expansion. This is not particularly difficult to carry out and we quote from Cox (1968, §23.1a):

$$\theta_n(\xi) = 1 - \frac{\xi^2}{6} + \frac{n}{120}\xi^4 - \frac{n(8n-5)}{15120}\xi^6 + \cdots . \tag{7.45}$$

For $\xi \to 0$, find that $z' \to -1/3$, which may be used to start the integration if so desired. A better way is to start at some $0 < \xi \ll 1$, compute y, y', z, and z' from (7.45), and carry on from there. This is a better procedure because there is an irregular solution to the Lane-Emden equation that blows up at the origin, and numerical techniques, no matter how good they are claimed to be, sense this. Therefore, treat the origin as delicately as possible.

The outer surface of the polytrope is reached when $\theta = y$ crosses zero. To resolve just where the first zero lies usually entails adjusting the step size h to be smaller as θ begins to near zero and, perhaps, using some form of interpolation scheme.

Since now you know how to compute polytropes, you may check that Table 7.1 contains the correct values of θ and θ' evaluated at ξ_1, and $\rho_c/\langle\rho\rangle$ for $n = 1.5$, 2, and 3 polytropes. More complete tabulations of the entire functions and other material may be found in Appendix A.5 and Table 23.1 of Cox (1968).

TABLE 7.1. Some parameters for $n = 1.5$, 2, and 3 polytropes

Index n	ξ_1	$\theta'(\xi_1)$	$\rho_c/\langle\rho\rangle$
1.5	3.6538	−0.20330	5.991
2.0	4.3529	−0.12725	11.402
3.0	6.8969	−0.04243	54.183

The Fitting Method

Another method for solving the Lane-Emden equation for E-solutions, and one that is used for real stellar models, is prompted by the following observation. In integrating from the center outward, it is possible that slight errors introduced within the deeper parts of the polytrope will be amplified as the low-density surface is approached. This is in the spirit of having the low-density surface being shaken like the tip of a whip when the heavy handle at the center is moved slightly. The same problems arise in complete stellar models where, because of the exceedingly large contrast between central and surface conditions of, for example, pressure and density, any inaccuracies in numerics near the center are felt manyfold by the time the surface is reached. One method for dealing with this is to integrate starting at both the center *and* the surface and see if the solutions join in some continuous way at a point between the two extremes. This is called the "fitting method" and it is a standard way to construct homogeneous, zero-age main sequence models (as in Appendix C).

The difficulty, as should be clear from what we have seen for polytropes, is that we do not know beforehand exactly where the surface ξ_1 is. Furthermore, since the system is second-order, we require two pieces of information at the surface to start an integration. The first is the requirement that $\theta(\xi)$ must be zero at ξ_1. As a second, we must make a *guess* at the first derivative of θ at ξ_1. Because pressure always decreases outward in a hydrostatic star we at least know that the sign of $\theta'(\xi_1)$ must be negative. Therefore, make reasonable guesses for ξ_1 and $\theta'(\xi_1)$, set $\theta(\xi_1)$ to zero and let the integrator work its way inward from there. At the same time, integrate from the center outward as before. At some interior point in the prospective polytrope (say near $\xi_1/2$) check if both θ and θ' (or the first-order variables y and z) of the inward and outward integrations match at that point—which we now call the "fitting point." They must fit for a complete E-solution because no discontinuities are lurking in the differential equations. If the solutions do not match, then one or both of ξ_1 and $\theta'(\xi_1)$ have been chosen incorrectly. To remedy this situation, the following strategy usually works (see Press et al. 1992, §17.2).

Let x_s (in x-, y-, z-space) be the initial guess at the surface location (i.e., ξ_1). In addition, let x_f be the location of the fitting point and $y_{\rm o}(x_f)$, $z_{\rm o}(x_f)$, $y_{\rm i}(x_f)$, and $z_{\rm i}(x_f)$ be the values of y and z obtained at x_f by means of the outward (subscript "o") and inward (subscript "i") integrations. The situation is pictured in Figure 7.1. Note that in these variables $y(x_s)$ is always zero because it is the value of θ at the surface. What we wish to do is vary x_s and $z_s = z(x_s)$ and see what happens to $y_{\rm i}(x_f)$ and $z_{\rm i}(x_f)$. With this information, we then set up an algorithm to eventually match at the fitting point.

To implement the algorithm, first define $Y(x_s, z_s) = y_{\rm i}(x_f) - y_{\rm o}(x_f)$ and $Z(x_s, z_s) = z_{\rm i}(x_f) - z_{\rm o}(x_f)$. We eventually want $Y = Z = 0$ for continuity of solution. Now change x_s from its initial value to $x_s \to x_s + \delta x_s$, where δx_s is small compared to x_s, while keeping z_s at its original value, and integrate inward. Because x_s has changed, we expect Y and Z to both change. Denote these changes by δY_z and δZ_z where the "z" subscript means "only x_s has changed." Thus $Y \to Y + \delta Y_z$ and $Z \to Z + \delta Z_z$. Similarly, if x_s is fixed at its original value but $z_s \to z_s + \delta z_s$, then $Y \to Y + \delta Y_x$ and $Z \to Z + \delta Z_x$. It should be clear that $\delta Y_z / \delta x_s$ is the difference approximation to the partial

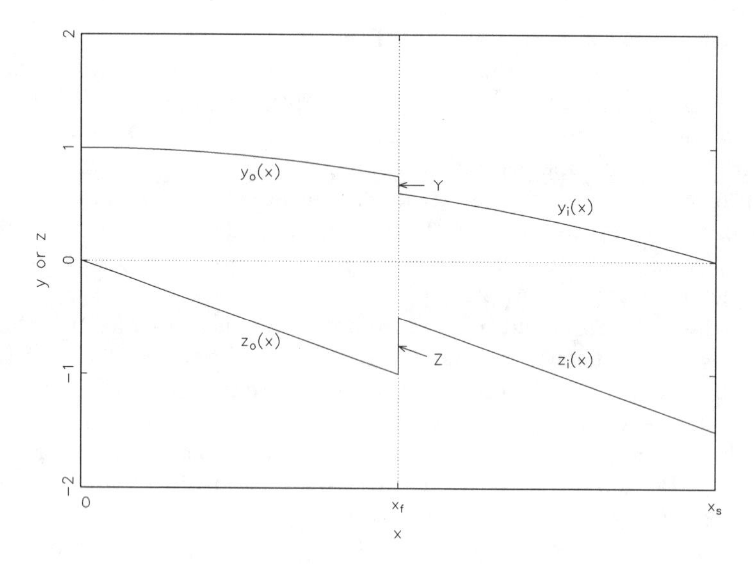

FIGURE 7.1. An illustration of the fitting method for the two variables $y = \theta$ and $z = \theta'$. The solid curves indicate solutions that do not quite satisfy the conditions of continuity. Hence the "jump" in Y and Z at x_f.

derivative of Y with respect to x_s when z_s is kept fixed. In similar fashion, $\delta Y_x/\delta z_s$ represents $(\partial Y/\partial z)_{x_s}$, etc.

The route to an E-solution is to compute the numeric derivatives, as in the above, and then imagine a first-order Taylor expansion of the form

$$\begin{aligned} Y(x_s+\Delta x_s, z_s+\Delta z_s) &= Y(x_s, z_s)+\left(\frac{\delta Y_z}{\delta x_s}\right)\Delta x_s+\left(\frac{\delta Y_x}{\delta z_s}\right)\Delta z_s \\ Z(x_s+\Delta x_s, z_s+\Delta z_s) &= Z(x_s, z_s)+\left(\frac{\delta Z_z}{\delta x_s}\right)\Delta x_s+\left(\frac{\delta Z_x}{\delta z_s}\right)\Delta z_s \end{aligned} \tag{7.46}$$

where the partial derivatives of the expansion have been replaced by their known difference equivalents. The trick is to now find Δx_s and Δz_s such that the left-hand sides of both equations in (7.46) are zero. That is, we wish to find corrections to the initial guesses x_s and z_s such that Y and Z are both zero and thus complete the fitting at x_f. Thus solve the simultaneous, inhomogeneous, but *linear* system

$$\begin{aligned} \left(\frac{\delta Y_z}{\delta x_s}\right)\Delta x_s + \left(\frac{\delta Y_x}{\delta z_s}\right)\Delta z_s &= -Y(x_s, z_s) \\ \left(\frac{\delta Z_z}{\delta x_s}\right)\Delta x_s + \left(\frac{\delta Z_x}{\delta z_s}\right)\Delta z_s &= -Z(x_s, z_s) \end{aligned} \tag{7.47}$$

for the unknowns Δx_s and Δz_s. To first-order, these are the corrections needed to solve for the two roots of the combined equations $Y = 0$ and $Z = 0$.

Because the Lane-Emden equation is nonlinear, however, there is no guarantee that the new values of $x_s \to x_s + \Delta x_s$ and $z_s \to z_s + \Delta z_s$ will satisfy $Y = Z = 0$. (If the equation were linear, you would be done.) This method is therefore iterated with successive corrections until both Δx_s and Δz_s become very small or, better yet, the E-functions approach some preassigned level of continuity at the fitting point. The method we have outlined for converting a nonlinear problem into one that is linear by Taylor expansions, and then solving for some small changes to find roots, is known by many as the "Newton-Raphson method." It comes up so frequently in numerical analysis that it is worth studying carefully—as in *Numerical Recipes* (§9.4) where it is called "Newton's rule."

In the case where the full fourth-order differential system for more realistic stellar models is to be used, the above fitting scheme must be generalized. It is clear that four quantities must be specified: two at the surface and two at the center of the desired model. At the center, two could be chosen out of the three nonzero quantities P_c, ρ_c, and T_c. Given the composition, they are all connected through the equation of state. At the surface, two of $\mathcal{R}$, $\mathcal{L}$, and T_{eff} seem reasonable. Equations (7.5–7.11) are then integrated inward and outward and an attempt is made to match quantities that should be continuous at some interior point specified at some fixed $\mathcal{M}_r$. The variables to be fitted must be four in number and a convenient choice might be r, P, $\mathcal{L}_r$, and T. The same sort of algorithm is then used in which the quartet of surface and center quantities are varied independently and a Newton-Raphson root-finding scheme is employed to calculate corrections. This is more easily said than done but it is conceptually simple and is efficient when done properly. It is a standard method for constructing ZAMS models and it is that used in `ZAMS` in Appendix C.

7.2.3 *The U-V Plane*

The notion of fitting continuous functions in a simplified form of a stellar model raises the question of what quantities in a star (or a model of one) are continuous. The coordinate r must be smooth. So must the pressure be continuous in a hydrostatic star because, otherwise, terrible things would happen to its radial derivative in the equation of motion $\rho\ddot{r} = -(dP/dr) - (G\mathcal{M}_r\rho/r^2)$, which would result in unbounded accelerations. Another continuous quantity is $\mathcal{M}_r$. A discontinuous interior mass would imply unbounded densities in the mass equation. Density itself, however, need not be continuous. (Air resting on lead is fine.) These statements may be recast by introducing two new dimensionless functions that, in effect, summarize the mass and hydrostatic equations. These are

$$\mathrm{U} \equiv \frac{d\ln \mathcal{M}_r}{d\ln r} = \frac{4\pi r^3 \rho}{\mathcal{M}_r} = \frac{3\rho(r)}{\langle \rho(r) \rangle}$$

$$\mathrm{V} \equiv -\frac{d\ln P}{d\ln r} = \frac{G\mathcal{M}_r\rho}{rP} = \frac{r}{\lambda_P}. \tag{7.48}$$

The last part of the equation for U contains $\langle\rho(r)\rangle$, which is the average density interior to the point r. The last equation for V implies that V is the local radius measured in units of the local pressure scale height λ_P. From the preceding remarks it is clear that both U/ρ and V/ρ are continuous for static stars.

We shall only use these functions to illustrate some particular points in stellar structure and modeling but they were of great importance in the earlier history of these subjects. It is still worthwhile to see how U and V were used in the calculations described in Schwarzschild (1958), and Hayashi, Hōshi, and Sugimoto (1962). A useful modern reference is Kippenhahn and Weigert (1990, §23).

We shall need the values of U and V at the center and surface of a typical model for future reference (as in the code `PULS.FOR` in Appendix C). The limit as $r \to 0$ requires that we look at $\mathcal{M}_r$ in that limit. Later on we shall consider in detail how various stellar quantities behave near stellar center but, for now, we state that $\mathcal{M}_r \to (4\pi/3)\rho_c r^3$ for very small r. All that was done here was to realize that ρ has a zero gradient at the center (and, hence, its expansion must be in even powers of r) and to compute the mass in a tiny sphere of radius r of constant density. The surface presents a slightly more difficult problem. We assume here that both the density and pressure go to zero at the surface. Thus the ratio ρ/P in V might seem to be indeterminate but, if the gas is ideal, then ρ/P is just inversely proportional to temperature and, if temperature approaches zero at the surface, then V becomes unbounded. Gathering this together, the boundary conditions on U and V are

$$\begin{aligned} \mathrm{U} &\to 3, \quad \mathrm{V} \to 0, \quad \text{as } r \to 0 \\ \mathrm{U} &\to 0, \quad \mathrm{V} \to \infty \quad \text{as } r \to \mathcal{R}. \end{aligned} \tag{7.49}$$

The U–V variables for polytropes are

$$\begin{aligned} \mathrm{U}(\xi) &= \frac{\xi\theta^n}{(-\theta')} \\ \mathrm{V}(\xi) &= (n+1)\frac{(-\xi\theta')}{\theta}, \end{aligned} \tag{7.50}$$

and an E-solution phrased in terms of these for an $n = 3$ polytrope is illustrated in Figure 7.2. This curve could be reproduced in principle by integrating the Lane-Emden equation from the surface inward starting with the proper values of ξ_1 and $\theta'(\xi_1) \equiv \theta_1'(\mathrm{E})$ and continuing on to the center. It is instructive, however, to consider what happens if a *wrong* value of θ' is chosen at the start. If $\theta'(\xi_1)$ is chosen so that its magnitude is greater than $\theta_1'(\mathrm{E})$ (i.e., $|\theta'(\xi_1)| > |\theta_1'(\mathrm{E})|$), then—and we shall not go through the derivation—$\theta(\xi)$ will first increase as we travel inward but, at some point, it will decrease to zero. That is, the density will go to zero. These are the "F-solutions" (named after R.H. Fowler) or "solutions of the collapsed type" and one of these is denoted by "F" in Figure (7.2).

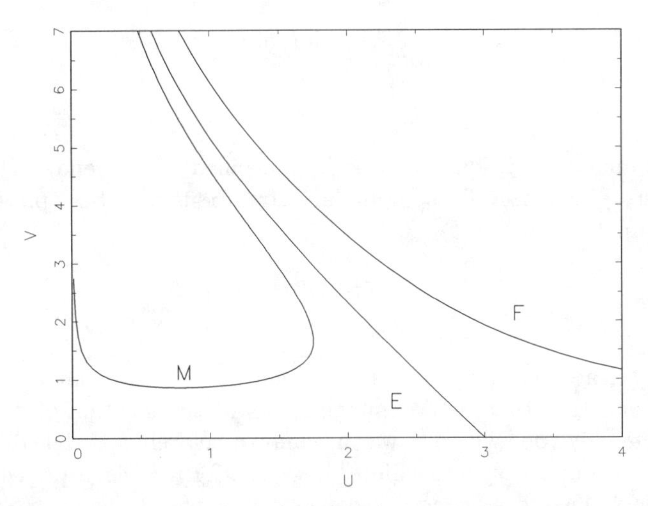

FIGURE 7.2. E–(Lane-Emden), F–(collapsed), and M–(centrally condensed) solutions to the Lane-Emden equation for a polytrope of $n = 3$ (see text). The surface radius ξ_1 was chosen to be its nominal value $\xi_1 = 6.8968486$ while the derivatives of θ at the surface are: $\theta'_1(\mathrm{E}) = -0.0424298$, $\theta'_1(\mathrm{F}) = -0.051$, and $\theta'_1(\mathrm{M}) = -0.038$.

If, on the other hand, $|\theta'(\xi)| < |\theta'_1(\mathrm{E})|$, then $\theta(\xi)$ blows up as the center is approached. These represent irregular solutions to the Lane-Emden equation and are called "centrally condensed" or "M-solutions" (after E.A. Milne); one is shown in the figure.

As they stand, the M– and F– solutions cannot represent complete stellar models because they are unphysical. If, however, polytropic equations of state hold for some portions of stellar interiors, then these solutions are still of some interest. They may, for example, represent pieces of the interiors of stars where other pieces (e.g., E-solutions) are to be tacked on to them in a continuous fashion.

7.2.4 Newton-Raphson or "Henyey" Methods

In integration by the fitting method the idea was to separate the model into two regions and then connect those two regions by continuity. In the Newton-Raphson or Henyey (Henyey, Forbes, and Gould 1964) method, the "integration" of the stellar structure equations is performed over the model as a whole. This is best described by a simple example such as that discussed below.[4] This is a powerful technique (not really originated by astronomers) and is now the standard way to construct evolutionary models.

Consider the second-order system

$$\begin{aligned} \frac{dy}{dx} &= f(x,y,z) \\ \frac{dz}{dx} &= g(x,y,z) \end{aligned} \tag{7.51}$$

with boundary conditions on y and z specified at the endpoints of the interval $x_1 \leq x \leq x_N$. These boundary conditions may be represented by general functions b_1 and b_N such that

$$\begin{aligned} b_1(x_1, y_1, z_1) &= 0 \\ b_N(x_N, y_N, z_N) &= 0 \end{aligned} \tag{7.52}$$

where y_i, z_i are $y(x_i)$, $z(x_i)$. It is assumed that f, g, b_1, and b_N are well behaved and that there is no ambiguity about the location of x_1 and x_N (unlike the polytrope where ξ_1 was not known beforehand). The differential equations are now cast in a "finite difference form" over a predetermined "mesh" in x. That is, we choose a sequence of points $x_1, x_2, \ldots, x_N$ at which y and z are to be evaluated. (This choice must be made with some care for the sake of accuracy.) The question is how do we represent the differential equations. A perfectly respectable way is to replace the differentials with differences and to replace the right-hand sides of (7.51) with an average of the derivatives over the interval covered by the differences. You may think of this as a differential representation of the trapezoidal rule of integration. It is also an example of "implicit" integration to be discussed later. We then have

$$\begin{aligned} \frac{y_{i+1} - y_i}{x_{i+1} - x_n} &= \frac{1}{2}(f_{i+1} + f_i) \\ \frac{z_{i+1} - z_n}{x_{i+1} - x_i} &= \frac{1}{2}(g_{i+1} + g_i) \end{aligned} \tag{7.53}$$

where f_{i+1} and g_{i+1} are shorthand for the functions $f(x_{i+1}, y_{i+1}, z_{i+1})$ and $g(x_{i+1}, y_{i+1}, z_{i+1})$, and $i = 1, 2, \ldots, N-1$. Expressions (7.53) then represent $2N-2$ equations but the two boundary conditions make up the difference so that all $2N$ variables y_i and z_i may, in principle, be found all at once. This is *not* an initial value problem.

To simplify the notation, we assume that the mesh in x is constant so that $x_{i+1} - x_i = \Delta x$ for all i between $i = 1$ and $i = N - 1$. (Real life

[4]An excellent reference for these methods is the classic review by Kippenhahn, Weigert, and Hofmeister (1967) and, in more general language, §17.3 of Press et al. (1992) where they are called "relaxation methods."

is hardly ever so well-behaved that this can be done without some loss of accuracy.) The errors in y and z in this order difference scheme go as $|\Delta x|^3$.

The difficulty with this method is that values of y and z at i and $i+1$ are all mixed up in the difference equations and boundary conditions—but Newton-Raphson comes to the rescue. As in the example of fitting, this means that (7.52–7.53) are to be linearized to find the solution.

Suppose we have some notion of the run of y_i and z_i for all i. These "guesses" do not, in general, satisfy equations (7.52–7.53). We may imagine, however, that there are corrections Δy_i and Δz_i

$$\begin{aligned} y_i &\rightarrow y_i + \Delta y_i \\ z_i &\rightarrow z_i + \Delta z_i \end{aligned} \tag{7.54}$$

that lead to new values of y_i and z_i that might satisfy those equations. We shall now *estimate* the values of Δy_i and Δz_i for all i. If it turns out that these do not quite do the job, then we iterate the following procedure (or "relax" to a solution) until values are found that do.

A first-order estimate of Δy_i and Δz_i is obtained by introducing (7.54) into (7.52–7.53) and expanding those equations to first-order in a Taylor series around the original guesses. For example, the first equation in (7.53) becomes, to first-order in the Δs and after minor arrangement,

$$\begin{aligned} y_{i+1} &- y_i - \frac{\Delta x}{2}\left(f_i + f_{i+1}\right) = \\ &+ \left[\frac{\Delta x}{2}\left(\frac{\partial f}{\partial y}\right)_i + 1\right]\Delta y_i + \left[\frac{\Delta x}{2}\left(\frac{\partial f}{\partial y}\right)_{i+1} - 1\right]\Delta y_{i+1} \\ &+ \left[\frac{\Delta x}{2}\left(\frac{\partial f}{\partial z}\right)_i\right]\Delta z_i + \left[\frac{\Delta x}{2}\left(\frac{\partial f}{\partial z}\right)_{i+1}\right]\Delta z_{i+1} \end{aligned} \tag{7.55}$$

where the partials are taken at constant y or z as the case may be. The corresponding equation for the z-derivative is obtained by replacing ys with zs and fs with gs. Note that the left-hand sides of these equations are zero when the difference equations are satisfied; that is, when the Δys and Δzs go to zero. The linearized forms of the boundary conditions are

$$b_{(1 \text{ or } N)} + \left(\frac{\partial b}{\partial y}\right)_{(1 \text{ or } N)} \Delta y_{(1 \text{ or } N)} + \left(\frac{\partial b}{\partial z}\right)_{(1 \text{ or } N)} \Delta z_{(1 \text{ or } N)} = 0. \tag{7.56}$$

We now arrange all these equations in matrix form so that we shall ultimately solve an equation of the form

$$\mathbf{M} \cdot \mathbf{U} = \mathbf{R} \tag{7.57}$$

where the column vector $\mathbf{U}$ contains the unknown quantities Δy_i and Δz_i, $\mathbf{R}$ is the right-hand side column vector, and $\mathbf{M}$ is the $N \times N$ coefficient matrix. The unknown vector is arranged like so:

$$\mathbf{U} \equiv \left(\Delta y_1, \Delta z_1, \Delta y_2, \Delta z_2, \ldots, \Delta y_N, \Delta z_N\right)^T \tag{7.58}$$

where the superscript T indicates transpose; note the interlacing of the variables. Some of the elements in $\mathbf{R}$ we will denote by

$$\begin{aligned} Y_{i+1/2} &\equiv y_{i+1} - y_i - \frac{\Delta x}{2}(f_{i+1} + f_i) \\ Z_{i+1/2} &\equiv z_{i+1} - z_i - \frac{\Delta x}{2}(f_{i+1} + f_i) \end{aligned} \tag{7.59}$$

and come from terms such as given in (7.55). The "half-step" notation $i+1/2$ is meant to imply that a quantity is evaluated between i and $i+1$. The rest of $\mathbf{R}$ comes from the constant terms in the linearization of the boundary conditions (7.56), which are put in as the first and last elements in $\mathbf{R}$ to give

$$\mathbf{R} = (-b_1, Y_{3/2}, Z_{3/2}, \cdots, Y_{N-1/2}, Z_{N-1/2}, -b_N)^T. \tag{7.60}$$

Parts of the matrix elements in $\mathbf{M}$ come from (7.55) and are denoted by

$$\begin{aligned} A_i &\equiv \frac{\Delta x}{2}\left(\frac{\partial f}{\partial y}\right)_i, \quad C_i \equiv \frac{\Delta x}{2}\left(\frac{\partial g}{\partial y}\right)_i \\ B_i &\equiv \frac{\Delta x}{2}\left(\frac{\partial f}{\partial z}\right)_i, \quad D_i \equiv \frac{\Delta x}{2}\left(\frac{\partial g}{\partial z}\right)_i. \end{aligned} \tag{7.61}$$

Finally (and as is implied by 7.60), the order in which the equations are ranked in the matrix is: (1) boundary condition at x_1; (2) the $2N-2$ difference equations; (3) the boundary condition at x_N. To conserve space we give the following result for $\mathbf{M}$ for a mesh consisting of three points i =1,2,3. Once you construct this for yourself the entire scheme should become clear:

$$\begin{pmatrix} (\partial b/\partial y)_1 & (\partial b/\partial z)_1 & 0 & 0 & 0 & 0 \\ A_1+1 & B_1 & A_2-1 & B_2 & 0 & 0 \\ C_1 & D_1+1 & C_2 & D_2-1 & 0 & 0 \\ 0 & 0 & A_2+1 & B_2 & A_3-1 & B_3 \\ 0 & 0 & C_2 & D_2+1 & C_3 & D_3-1 \\ 0 & 0 & 0 & 0 & (\partial b/\partial y)_3 & (\partial b/\partial z)_3 \end{pmatrix}.$$

You will note the particularly simple structure of this matrix in which no nonzero element is located further than two columns away from the diagonal (as a "band tridiagonal" matrix). This fact makes the problem amenable to several accurate and efficient techniques for solving simultaneous linear equations and you are referred to §17.3 of *Numerical Recipes* for more details.

Once the solution set $\mathbf{U}(\Delta y_i, \Delta z_i)$ is found, then new values of y_i and z_i are immediately obtained by adding Δy_i and Δz_i to the corresponding old guesses. If these new values are sufficiently close to the old values, then the solution has converged and the original difference equations and boundary conditions are presumably satisfied. If not, try again with the new values and iterate until they are. Suffice it to say, the novice (and professional) very often reaches this happy state only after many iterations in the multidimensional solution space of ys and zs. Convergence depends on many factors, the most important being a reasonably good initial guess and that is where experience and intuition count. Initial guesses that bear no resemblance to the desired solution may leave you stranded in a solution space of many dimensions and convergence may not be possible. Even a good guess may not help, however, if the original differential equations are ill-behaved or if the mesh x_i is inappropriate. Another factor often comes in when constructing stellar models where tabulated equations of state, etc., are used. Very often such tables are "noisy" (usually from the introduction of incomplete physics at some temperature or density) and the difference equations see that noise. But, if all goes well, then it goes well indeed and the corrections Δy_i and Δz_i decrease as the square of their absolute values from one iteration to the next and convergence is swift.

With the above as an example, it should be clear how to extend the technique to differential systems of higher than second-order: the bookkeeping gets messier, but the principle remains the same. The stellar problem is fourth-order with two boundary conditions at each end as discussed in §7.1. For stellar evolution off the main sequence, what is usually done is to first construct a ZAMS model using the fitting method and then use that solution as the first guess for a Henyey model.

7.2.5 Eigenvalue Problems and the Henyey Method

The above scheme, as presented, cannot solve the simple problem of an E-solution polytrope. This is because we do not know the radius (ξ_1) beforehand. Thus, on the face of it, a grid in x cannot be established upon which the differences equations and boundary conditions are to be applied. Yet with some minor adjustments, this can be remedied. We note here that the following, and variations thereof, are calculational mainstays in some subareas of stellar astrophysics such as variable star analysis.

Recall (7.42), which is the Lane-Emden equation phrased in terms of $y = \theta$, $z = dy/dx$, with $x = \xi$:

$$\begin{aligned} y' &= \frac{dy}{dx} = z \\ z' &= \frac{dz}{dx} = -y^n - \frac{2}{x}z\,. \end{aligned} \tag{7.62}$$

The boundary conditions are $y = 1$ at $x = 0$, and $y = 0$ at some unknown $x = x_s = \xi_1$; in addition, it is required that $z = dy/dx = 0$ at $x = 0$ to obtain the regular E-solution.

We use the simple trick of rescaling x so that it lies within the closed interval $[0, 1]$ by letting

$$x \to x = \frac{\xi}{\xi_1} = \frac{\xi}{\lambda}. \tag{7.63}$$

Thus given $\xi_1 = \lambda$, the edge of the polytrope is at unity in the new x-coordinate. But, you say, we don't know what this λ is, so it looks as if we haven't gotten anywhere. This is true, so let us find λ as part of the overall problem in a relaxation method.

First transform the Lane-Emden equation into the new x-coordinate:

$$\begin{aligned} y' &= \frac{dy}{dx} = \lambda z \\ z' &= \frac{dz}{dx} = -\lambda y^n - \frac{2}{x} z\,. \end{aligned} \tag{7.64}$$

These equations are in the same form as we had previously but now the parameter λ appears; that is, the functions of (7.51) are now $f = f(x, y, z; \lambda)$ and $g = g(x, y, z; \lambda)$. It is no surprise that an additional variable appears because there are now an overabundance of conditions on the problem. The problem is still second-order but there are two boundary conditions at the center ($y = 1$, $z = 0$), a third at the surface ($y = 0$ at the new $x = 1$), and the system is overdetermined. With λ as an additional degree of freedom, however, the surplus boundary condition is no longer one too many. Here λ is an eigenvalue for the problem in much the same way that eigenvalues appear in other situations in physics. In this situation it yields the radius.

To solve this problem we proceed as in the previous subsection and assume we have a complete run of guesses for y and z over the mesh of x, but we also make a first guess for λ. We then linearize f and g, as before, by letting $y_i \to y_i + \Delta y_i$, etc., but also allow $\lambda \to \lambda + \Delta\lambda$. The scheme is to solve for the $2N + 1$ unknowns y_i, z_i, and λ. There are $2N - 2$ difference equations (as before) but now there are *three* boundary conditions yielding a total of $2N + 1$ equations. It is easy to see (by writing out all the linearized equations) that $\Delta\lambda$ comes in from such terms as the last term in

$$f \to f + \left(\frac{\partial f}{\partial y}\right)_{z_i,\lambda} \Delta y_i + \left(\frac{\partial f}{\partial z}\right)_{y_i,\lambda} \Delta z_i + \left(\frac{\partial f}{\partial \lambda}\right)_{y_i,z_i} \Delta\lambda.$$

Each difference equation thus contributes terms in $\Delta\lambda$ to a matrix algebra problem similar to the one encountered before except that there is now an extra row in the main matrix corresponding to the additional boundary condition and an extra column corresponding to the extra unknown, $\Delta\lambda$. This set of linear equations is somewhat more cumbersome to solve (because of the additional column of nonzero elements) but various techniques can be made to work.

In this specific problem the behavior of the functions near the center must be treated carefully as discussed previously. Either the differential equations are replaced by series solutions up to some point, or the derivatives appearing in the difference equations have to be calculated very accurately. Just how to incorporate series expansions into the Henyey scheme is left as an exercise, but note that all that is required is a redefinition of the boundary conditions so that the proper ones appear at a point slightly removed from the center. Once this has been mastered for polytropes, you will soon find that such eigenvalue problems hold no real terror—maybe.

7.2.6 Dynamic Problems

We can gain more insight into the difficulties of constructing stellar models and evolutionary sequences by examining how rapidly evolving models are handled. As will be pointed out later in this section, some of the techniques used here are also the stock in trade of those who compute models for slowly evolving stars.

By "rapidly" evolving we mean situations where evolutionary processes take place on dynamic time scales such as in supernovae, novae, and most variable stars. Somewhat special techniques are required for this. We shall follow Cox, Brownlee, and Eilers (1966) because it contains a reasonably complete exposition of the methods. You may also wish to consult Chapter 19 of *Numerical Recipes* for more general problems involving the numerical treatment of partial differential equations containing time. This discussion will be restricted to motions that are spherically symmetric.

Consider the equation of motion given earlier in (1.4) but expressed in the Lagrangian form

$$\ddot{r} = -4\pi r^2 \left(\frac{\partial P}{\partial \mathcal{M}_r} \right) - \frac{G\mathcal{M}_r}{r^2}. \tag{7.65}$$

As is usual in a Lagrangian description, the elements of interest in the star are tagged by the mass—in this instance by $\mathcal{M}_r$. To reduce (7.65) to a form suitable for numeric calculation, first divide the star into concentric shells containing masses labeled $\mathcal{M}_{i+1/2}$, where it is understood that $(i + 1/2)$ means that the mass lies between radii r_{i+1} and r_i as illustrated in Figure 7.3. The model center is at r_0. The spherical surfaces corresponding to these various radii will be referred to as "interfaces." The mass interior to the radius at interface r_{i+1} is then

$$\mathcal{M}_r \text{ at } r_{i+1} \rightarrow \sum_{k=0}^{i} \mathcal{M}_{k+1/2}. \tag{7.66}$$

A natural way of expressing the mass equation is to define the density as

$$\rho_{i+1/2} = \frac{\mathcal{M}_{i+1/2}}{(4\pi/3)\left(r_{i+1}^3 - r_i^3\right)} = \frac{1}{V_{i+1/2}} \tag{7.67}$$

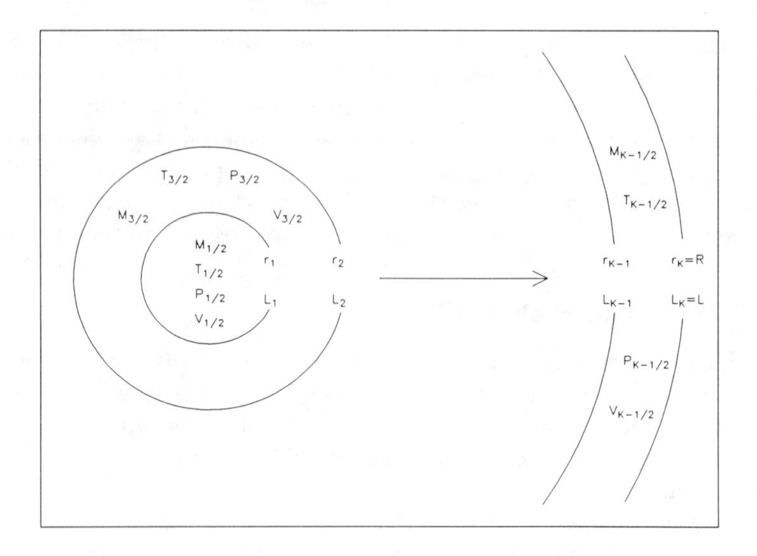

FIGURE 7.3. This figure illustrates the shell and interface partitioning of a stellar model. Some variables are defined within the shells whereas others are defined at the interfaces (see text).

where, for later purposes, we introduce the specific volume, V. The notation implies that the density is defined in shells and not on interfaces. Having made this choice, it is then reasonable to define $T_{i+1/2}$ and $P_{i+1/2}$ as the temperature and pressure associated with the above density so that $P_{i+1/2}$ may be expressed in terms of the corresponding temperature, density, and composition within a shell.

The mass gradient of pressure in (7.65) is simply the difference of the pressures divided by the change in mass across an interface i. The change in mass is constructed as

$$\partial\mathcal{M}_r \text{ at } i = \frac{1}{2}\left[\sum_{k=0}^{i}\mathcal{M}_{k+1/2} - \sum_{k=0}^{i-2}\mathcal{M}_{k+1/2}\right] = \frac{1}{2}\left[\mathcal{M}_{i+1/2} + \mathcal{M}_{i-1/2}\right] \tag{7.68}$$

where $i = 0$ is not needed because the stellar center is never accelerated. Thus (7.65) is transcribed into difference form as

$$\ddot{r}_i = -8\pi r_i^2 \frac{P_{i+1/2} - P_{i-1/2}}{\mathcal{M}_{i+1/2} + \mathcal{M}_{i-1/2}} - \frac{G\sum_{k=0}^{i-1}\mathcal{M}_{k+1/2}}{r_i^2}. \tag{7.69}$$

A technical point now enters. What we are trying to do is to mock up the hydrodynamic time development of a continuous system using discrete time and space steps. The above equation of motion is fine for motions that are subsonic but has real problems if shocks develop.[5] The trouble is that shocks imply some sort of discontinuous behavior in, say, density. In practice, this means that mass shells in a numerical calculation tend to "rattle" around unacceptably near the shock front unless special steps are taken. To alleviate this, an artificial (or numerical, pseudo-, etc.) viscosity is often introduced that smooths out the shock profile and allows the computation to proceed without undue noise. One prescription, and there are quite a few, for an artificial viscosity is (see, for example, Richtmyer and Morton, 1967) $Q_{i+1/2} \sim (\dot{r}_{i+1} - \dot{r}_i)^2 \rho_{i+1/2}$ if a mass shell at $i + 1/2$ is undergoing compression but is set to zero if the shell is expanding. This viscosity is added to the pressure in the difference form of the equation of motion. The net effect is to increase the entropy and spread the shock discontinuity over a few shells with a corresponding loss of resolution in the vicinity of the shock. In the following we shall imagine that some such device is used but we delete specific reference to it.

All the above is imagined to take place at some time t^n. What is required is the status of the system at a new time $t^{n+1} = t^n + \Delta t$, where Δt is a "time step." The method outlined here is called an "explicit" method because the prediction of the system's behavior at t^{n+1} depends only on knowledge of what is happening at t^n. More complicated "implicit" schemes employ an iteration between t^n and t^{n+1} quantities and equations such as was done in the previous section for spatial integrations. The latter schemes have decided advantages, but at the cost of increased complexity. In an explicit method physical quantities march forward in time in a simple way as follows.

Define a time-centered, or average, velocity between t^n and t^{n+1} as

$$\dot{r}_i^{n+1/2} = \dot{r}_i^{n-1/2} + \Delta t^{n+1/2}\, \ddot{r}_i^n \tag{7.70}$$

where $\dot{r}_i^{n-1/2}$ is presumed known from a previous time step and $\Delta t^{n+1/2}$ is the average time step

$$\Delta t^{n+1/2} = \left(t^{n+1} - t^n\right)/2. \tag{7.71}$$

The new radial position is then

$$r_i^{n+1} = r_i^n + \Delta t\, \dot{r}_i^{n+1/2}. \tag{7.72}$$

Since the accelration $\ddot{r}_i^n$ is known from (7.69), all the interface radii may be updated in this fashion. Having these, and assuming that mass is neither lost nor gained within a shell, the densities may similarly be updated by

[5] An excellent pedagogic introduction to shocks is Chapter 1 of *Physics of Shock Waves and High Temperature Hydrodynamic Phenomena* by Zel'dovich and Raizer (1966). That chapter is also available as a separate monograph published in 1968.

application of the mass equation (7.67) using the appropriate new radii. Note that in applying the acceleration equation the pressure external to the last radius at the surface must be specified. The most simple procedure is to set it to zero as if a vacuum were present. (Fancier and more accurate choices are possible.) Thus if the model has K interfaces labeled $i = 1, 2, \cdots, K$ (with the stationary center at $r_0 = 0$), then $P_{K+1/2} = 0$.

The determination of what to use for Δt involves another technical point of some importance. We have not affixed a superscript to this quantity but it should be obvious that its choice depends on how fast the system is changing at the time in question. If the system is evolving rapidly then Δt should be small. A slowly evolving system need not imply that long steps may be taken, however. There is an upper bound on the length of a time step and this is determined by considerations of numerical stability. In an explicit scheme such as this, the time step must be some fraction less than the time it takes for a sound wave to traverse a shell, and all shells must be examined to find this upper bound. (This is known as the Courant condition.) This means that even though the system may be evolving *very* slowly (as in normal stellar evolution on a nuclear time scale), the computation must proceed at a pace comparable to sound travel times across shells—and this is the price paid for computational simplicity. An implicit scheme, which incorporates information about the state of the system at t^{n+1}, is computationally more difficult but is the one used in slow evolution analysis.

To continue further in time requires finding the new accelerations at t^{n+1}, and this requires computation of $P_{i+1/2}$ in (7.69) evaluated at time t^{n+1}, which we call $P^{n+1}_{1+1/2}$. Straight mechanics will not yield this pressure and the remainder of the stellar structure equations must enter. The first is the energy equation (6.2), which, in difference form at t^{n+1}, becomes

$$\begin{aligned} \frac{E^{n+1}_{i+1/2} - E^{n}_{i+1/2}}{\Delta t} = &- \frac{P^{n+1}_{i+1/2} + P^{n}_{i+1/2}}{2} \frac{V^{n+1}_{i+1/2} - V^{n}_{i+1/2}}{\Delta t} \\ &- \frac{\left(\mathcal{L}^{n+1}_{i+1} - \mathcal{L}^{n+1}_{i}\right) + \left(\mathcal{L}^{n}_{i+1} - \mathcal{L}^{n}_{i}\right)}{2\mathcal{M}_{i+1/2}} \\ &+ \frac{\varepsilon^{n+1}_{i+1/2} + \varepsilon^{n}_{i+1/2}}{2}. \end{aligned} \tag{7.73}$$

Note that the luminosity is defined on interfaces whereas ε is defined within a shell (where densities, temperatures, etc., are defined). Also note that time averaging and differencing have been used in accordance with the methods outlined in §7.2.4.

The internal energy, pressure, and energy generation rate in (7.73) are functions of temperature, density, and composition (which, for simplicity, is assumed constant in time here). If only diffusive radiative transfer holds then $\mathcal{L}_r$ is given by (4.25) in Lagrangian form as

$$\mathcal{L}_r = -\frac{(4\pi r^2)^2 ac}{3\kappa}\frac{dT^4}{d\mathcal{M}_r}. \tag{7.74}$$

The mass derivative in (7.74) can be put into difference form in several ways depending on how terms are combined. The simplest, although not necessarily the best, way is to let $r \to r_i$, $\kappa \to \kappa_{i+1/2}$, and let the mass gradient of temperature (to the fourth) be modeled after the mass gradient of pressure in the acceleration equation. In any case, it is easy to see that $\mathcal{L}_i$ will contain temperatures computed on either side of interface i; that is, we need $T_{i\pm 1/2}$. It will also contain reference to $\rho_{i\pm 1/2}$ through the opacity but these are already known (as are the r_i) from the dynamics that got us to time t^{n+1}. The pressure, internal energy, and energy generation rate may also be computed at this time if the temperatures are known. The conclusion is that the temperatures in the K shells are the only unknowns at t^{n+1}. Note, however, that these temperatures are implicitly highly coupled in space in the energy equation because $\mathcal{L}_i$ and $\mathcal{L}_{i+1}$ both appear. Thus the energy equation at shell $i+1/2$ contains $T_{i\pm 1/2}$ and $T_{i+3/2}$ so that three temperatures are associated with the shell at $i+1/2$ (and this will result in a tridiagonal system as indicated below).

To proceed further and find the K temperatures is relatively straightforward and a Newton-Raphson relaxation scheme is the method of choice. The energy equation in the $(i+1/2)$th shell is linearized to yield corrections to guessed temperatures at t^{n+1}, which we denote as $\Delta T_{i+1/2}$. A close inspection of the procedure will reveal that a lot of thermodynamics is required. For example, finding out how the $P\,dV$ term in the energy equation responds to a change in temperature $T_{i+1/2} \to T_{i+1/2} + \Delta T_{i+1/2}$ involves computing the partial of pressure with respect to temperature at constant density. In the end, a set of linear equations in $\Delta T_{i+1/2}$ are obtained of the form

$$\mathbf{M} \cdot \mathbf{\Delta T} = -\mathbf{f}. \tag{7.75}$$

Here $\mathbf{f}_{i+1/2}$ is the differenced energy equation at $i+1/2$ containing temperature guesses, etc., as cast in the form $\mathbf{f} = 0$. If the unknown temperatures are ordered in the column vector $\mathbf{\Delta T}$ as

$$\left(\Delta T_{1/2}, \Delta T_{3/2}, \ldots, \Delta T_{K-3/2}, \Delta T_{K-1/2}\right)^T \tag{7.76}$$

then the matrix $\mathbf{M}$ contains the relevant partials of f with respect to temperature and is tridiagonal in form (zeros everywhere except along the main diagonal and the two diagonals immediately on either side of it). This system is simple to solve (as in *Numerical Recipes*, §2.4). It would be worth your while to try to construct such a linear system of equations on your own if for no other reason than to see how difficult a bookkeeping job is involved.

The sequence in this explicit hydrodynamics scheme is then: compute the accelerations and velocities, update the radii and compute the new densities, and then use the energy equation to find the new temperatures. The transition to a calculation that does not do hydrodynamics but rather insists upon hydrostatic equilibrium is now not too difficult to imagine. The acceleration equation (7.69) is replaced by the hydrostatic equation that is the right-hand side of (7.69) set to zero but, to maintain numerical stability, all the quantities in that equation are time averaged over t^n and t^{n+1} (as in the energy equation). The nasty part of the calculation is that the radii at the future time t^{n+1} must be solved for simultaneously along with all the other variables. This is what makes stellar evolution calculations so difficult. There are many variations on Henyey integrations of the time-dependent stellar problem. Some of these involve choosing clever combinations of variables or rephrasing the structure and evolution equations. For examples of some possibilities see Schwarzschild (1958), Kippenhahn, Weigert, and Hofmeister (1967), and Kippenhahn and Weigert (1990, §11).

If nuclear transformations are present, as they almost always are, changes in abundances must also be accommodated. Changes in mean molecular weight due to ionization usually take place so rapidly that they may be regarded as taking place instantaneously and are therefore incorporated directly into the equation of state. Abundance changes that are very slow compared to other time scales in the system are easy: update the composition after a time step is taken using as simple a difference scheme in time as possible. However, in the most complicated situations, where abundances change rapidly or particular nuclear species must be followed carefully in time for some purpose, then the rate equations for transmutations (or, possibly, an equilibrium version thereof) must be included among the stellar structure equations implicitly. Needless to say, such a full calculation takes its toll of time and patience—but that's the name of the game.

And now, after this long digression, back to polytropes.

7.2.7 *The Eddington Standard Model*

A simple example of the use of polytropes in making a stellar pseudo-model is the "Eddington standard model" in which the energy equation and the equation of diffusive radiative transfer are incorporated in an approximate way.

Recall that the actual run of temperature versus pressure in situations where there is no convection is given by (see Eqs. 7.8–7.11)

$$\nabla = \frac{d\ln T}{d\ln P} = \frac{3}{16\pi ac}\frac{P\kappa}{T^4}\frac{\mathcal{L}_r}{G\mathcal{M}_r}. \tag{7.77}$$

We may also express ∇ in different terms by introducing the radiation pressure $P_{\text{rad}} = aT^4/3$ and unwinding the derivatives in the definition of

∇ to obtain

$$\nabla = \frac{P}{T}\left(\frac{dT/dr}{dP/dr}\right) = \frac{1}{4}\frac{P}{P_{\rm rad}}\frac{dP_{\rm rad}}{dP}. \tag{7.78}$$

Solving for the pressure derivative and combining this with (7.77) then yields

$$\frac{dP_{\rm rad}}{dP} = \frac{\mathcal{L}\kappa}{4\pi cG\mathcal{M}}\frac{\mathcal{L}_r/\mathcal{L}}{\mathcal{M}_r/\mathcal{M}} \tag{7.79}$$

where $\mathcal{L}$ and $\mathcal{M}$ are the total luminosity and mass.

The ratio $(\mathcal{L}_r/\mathcal{L})/(\mathcal{M}_r/\mathcal{M})$ in (7.79) is a normalized average energy generation rate, as may be seen from considering the energy equation in thermal balance. If $(d\mathcal{L}_r/d\mathcal{M}_r) = \varepsilon$ is integrated over $\mathcal{M}_r$, then define

$$\langle\varepsilon(r)\rangle = \frac{\int_0^r \varepsilon\, d\mathcal{M}_r}{\int_0^r d\mathcal{M}_r} = \frac{\mathcal{L}_r}{\mathcal{M}_r} \tag{7.80}$$

with $\langle\varepsilon(\mathcal{R})\rangle = \mathcal{L}/\mathcal{M}$. It is then traditional to introduce $\eta(r)$ as

$$\eta(r) = \frac{\langle\varepsilon(r)\rangle}{\langle\varepsilon(\mathcal{R})\rangle} = \frac{\mathcal{L}_r/\mathcal{L}}{\mathcal{M}_r/\mathcal{M}} \tag{7.81}$$

so that (7.79) becomes

$$\frac{dP_{\rm rad}}{dP} = \frac{\mathcal{L}}{4\pi cG\mathcal{M}}\kappa\eta(r). \tag{7.82}$$

Thus far we have made no other assumptions aside from thermal balance and pure diffusive radiative transfer.

We now formally integrate the last expression from the surface to an interior point r assuming that the surface pressure is zero and find

$$P_{\rm rad} = \frac{\mathcal{L}}{4\pi cG\mathcal{M}}\langle\kappa\eta(r)\rangle\, P(r) \tag{7.83}$$

where the averaged expression is the combination

$$\langle\kappa\eta(r)\rangle = \frac{1}{P(r)}\int_0^{P(r)} \kappa\eta\, dP. \tag{7.84}$$

This is now put in final form by recalling the definition of β, the ratio of ideal gas to total pressure of equation (3.100), so that

$$1 - \beta(r) = \frac{P_{\rm rad}(r)}{P(r)} \tag{7.85}$$

and, thus, after substituting in (7.83),

$$1 - \beta(r) = \frac{\mathcal{L}}{4\pi cG\mathcal{M}}\langle\kappa\eta(r)\rangle. \tag{7.86}$$

To make further progress we must now examine the nature of $\langle \kappa\eta \rangle$. The following will contain the key to the standard model as it was introduced by Eddington (1926) well before stellar processes were completely understood. Yet, as we shall see, this model is not only of historical interest but it also yields insights into how some kinds of stars work.

A reasonable opacity to insert in (7.86) is a combination of electron scattering and Kramers' (see Chapter 3). Thus let

$$\kappa = \kappa_e + \kappa_0 \rho T^{-3.5}. \tag{7.87}$$

Except for the inclusion of an H^- opacity source, this is a good approximation for most main sequence stars. The important thing to note is that this opacity increases outward if ρ does not decrease outward faster than $T^{3.5}$—which it will not in this model.

The quantity $\eta(r)$ is proportional to the average energy generation rate and, if we restrict ourselves to main sequence–like objects, it should decrease outward fairly rapidly to unity because of the relatively high positive temperature exponent of ε. Thus the product $\kappa\eta$ should not vary as strongly with position as does either of its components. The crucial assumption in the standard model is that $\kappa\eta$ varies so weakly with position that it may be taken as a constant throughout the star. This means that the right-hand side of (7.85) is constant and so is β.

The constancy of β may be translated into a temperature versus density relation as follows. If we assume that the pressure is made up of the sum of ideal gas plus radiation pressure only, then

$$P_{\text{rad}} = \frac{1-\beta}{\beta} P_{\text{gas}} = \frac{1-\beta}{\beta} \frac{N_A k}{\mu} \rho T = \frac{1}{3} a T^4 \tag{7.88}$$

from previous work. Solving for temperature then yields

$$T(r) = \left(\frac{3 N_A k}{a\mu} \frac{1-\beta}{\beta} \right)^{1/3} \rho^{1/3}(r). \tag{7.89}$$

Note that this relation is similar to that found using the virial theorem approach (of 1.33) but for the constant multiplying $\rho^{1/3}$. What is more important, however, is that this is not a relation between virial average quantities but rather gives the run of T versus ρ through the star.

To proceed, we now use

$$P = \frac{P_{\text{gas}}}{\beta} = \frac{N_A k}{\mu} \frac{\rho T}{\beta} \tag{7.90}$$

to find the pressure-density relation

$$P(r) = \left[\left(\frac{N_A k}{\mu} \right)^4 \frac{3}{a} \frac{1-\beta}{\beta^4} \right]^{1/3} \rho^{4/3}(r). \tag{7.91}$$

If we restrict ourselves to situations where μ is constant (homogeneous, constant state of ionization, etc.), then the term within the brackets is a constant (because β is) and what we are left with is a polytropic equation of the form $P \sim \rho^{4/3}$. The exponent on the density immediately tells us that we are dealing with a polytrope of index $n = 3$ and that the coefficient K of (7.16) is

$$K = \left[\left(\frac{N_{\rm A} k}{\mu} \right)^4 \frac{3}{a} \frac{1-\beta}{\beta^4} \right]^{1/3} . \tag{7.92}$$

On the other hand, K is also given by (7.37) which, for $n = 3$, is

$$K = \frac{(4\pi)^{1/3}}{4} \frac{G\mathcal{M}^{2/3}}{\left[\xi^4 \left(-\theta_3' \right)^2 \right]_{\xi_1}^{1/3}} . \tag{7.93}$$

We now equate the two expressions for K, substitute the relevant polytropic quantities from Table 7.1, and find

$$\frac{1-\beta}{\beta^4} = 0.002996\, \mu^4 \left(\frac{\mathcal{M}}{\mathcal{M}_\odot} \right)^2 . \tag{7.94}$$

To find the temperature, combine (7.89) and (7.94):

$$T(r) = 4.62 \times 10^6 \beta\mu \left(\frac{\mathcal{M}}{\mathcal{M}_\odot} \right)^{2/3} \rho^{1/3}(r). \tag{7.95}$$

Note that β and $\mathcal{M}$ in (7.95) are not independent but are connected through relation (7.94) so that things are not as simple as they may look. Table 7.2 summarizes this connection. The general trend is now clear and agrees with what we know about the zero-age main sequence: more massive stars have higher temperatures and radiation pressure plays a greater role (lower βs).

TABLE 7.2. Eddington standard model

$\mu^2 \mathcal{M}/\mathcal{M}_\odot$	β
1.0	0.9970
2.0	0.9885
5.0	0.9412
10.0	0.8463
50.0	0.5066

What information does the standard model *not* provide? First of all, it does not yield absolute numbers for temperatures, densities, and pressures unless we specify both the mass and radius of the configuration. For example, (7.95) yields the run of temperature versus density but where is the normalization of density to be found? Equation (7.39) contains the central density but the *average* density is required and, hence, the mass and radius. The reason for this need is that, although we have shown the standard model to be a polytrope of index three using the presumed constancy of $\langle \kappa\eta \rangle$, we have not then gone back and really solved the energy and heat transport equations. In this sense, the analysis is incomplete. You may look into this further (*q.v.*, Chandrasekhar 1939, Chapter VI) but that will lead us far afield (and the standard model can only be pushed so far). However, we shall now assume that both the mass and total radius of a model are specified and see how well the standard model does with that information.

Given the mass and radius, and thus average density, the central density for an $n = 3$ polytrope is $\rho_c = 54.18\langle\rho\rangle$ from Table 7.1. The run of T versus ρ then follows from (7.95). We can then compare this result to that from a ZAMS model with the same mass and total radius. Figure 7.4 shows the run of temperature versus density for a ZAMS solar model with $X = 0.743$, $Y = 0.237$, and $Z = 0.02$ (yielding $\mu \approx 0.6$ with complete ionization). The average density is $\langle\rho\rangle = 2.023$ g cm^{-3} and central density $\rho_c = 82.49$ g cm^{-3}. A standard model with this average density has $\rho_c = 109.6$ g cm^{-3}; we use this to compute $T(r)$ versus $\rho(r)$ as shown in the figure. The standard model does remarkably well compared to the ZAMS sun model in the inner regions where density and temperature are high but does not fare so well in the outer regions. These outer regions starting at $\rho \approx 0.1$ g cm^{-3} behave more like an $n = 3/2$ polytrope in this instance because of convection but this convective region constitutes only about 0.6% of the mass of the star. Most of the star, in mass units, has a structure close to that of an $n = 3$ polytrope. Note that if the interior pressure is controlled by an ideal gas, then $P \propto T^4$, as can be inferred from (7.91) and (7.95).

7.2.8 Applications to Zero Temperature White Dwarfs

We are sure that it has occurred to you that polytropes also represent excellent approximations to zero temperature white dwarfs in either the limits of completely nonrelativistic or relativistic degeneracy (that is how we started our discussion). Therefore, recall that the nonrelativistic equation of state is $P_e = 1.004 \times 10^{13} \left(\rho/\mu_e\right)^{5/3}$ dyne cm^{-2} from either (3.61) or (7.14). This is a polytropic equation of state and, hence, we know the regular solution is a polytrope of index $n = 3/2$. From (7.16) we may readily identify the equation of state coefficient K and equate it to the polytropic K given by (7.37), use Table 7.1, and find

$$K = \frac{1.004 \times 10^{13}}{{\mu_e}^{5/3}} = 2.477 \times 10^{14} \left(\frac{\mathcal{M}}{\mathcal{M}_\odot}\right)^{1/3} \left(\frac{\mathcal{R}}{\mathcal{R}_\odot}\right). \tag{7.96}$$

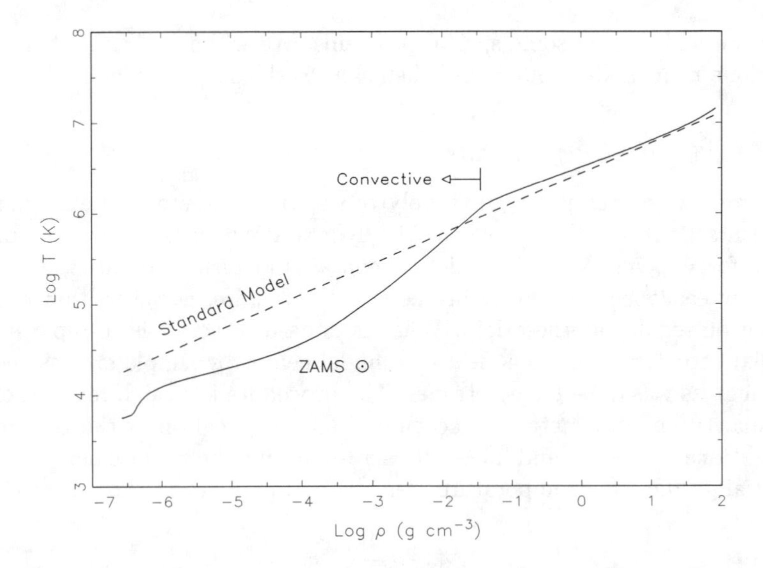

FIGURE 7.4. The solid line is the temperature-density relation through a full-blown model of the zero-age homogeneous main sequence sun with a total radius of 6.168×10^{10} cm or $0.886\ \mathcal{R}_\odot$. The dashed line shows the standard model result where the same total radius has been assumed.

This yields the mass-radius relation

$$\frac{\mathcal{M}}{\mathcal{M}_\odot} = 2.08 \times 10^{-6} \left(\frac{2}{\mu_e}\right)^5 \left(\frac{\mathcal{R}}{\mathcal{R}_\odot}\right)^{-3} \tag{7.97}$$

which is close to what we found in the virial estimate of (3.59).

The corresponding completely relativistic result is found using the equation of state (3.62), $P_e = 1.243 \times 10^{15} (\rho/\mu_e)^{4/3}$ g cm^{-3}, and by realizing that we are dealing with an $n = 3$ polytrope; namely,

$$K = \frac{1.243 \times 10^{15}}{{\mu_e}^{4/3}} = 3.841 \times 10^{14} \left(\frac{\mathcal{M}}{\mathcal{M}_\odot}\right)^{2/3} \tag{7.98}$$

or

$$\frac{\mathcal{M}_\infty}{\mathcal{M}_\odot} = 1.456 \left(\frac{2}{\mu_e}\right)^2 \tag{7.99}$$

This is the Chandrasekhar limiting mass discussed in §3.52 (and see 3.63). Real white dwarfs will be discussed in Chapter 9.

7.3 The Approach to Real Models

This section will deal with some aspects of modeling real stars. We have already introduced sketches of various numerical modeling techniques and now we will discuss some special problems. We will start with what is to be done near model centers and then skip to the surface layers.

7.3.1 Central Expansions

As we saw in treating regular polytropes, the center of a stellar model presents some peculiar problems. The hydrostatic equation and the expressions for $\nabla_{\rm rad}$ and ∇ (of 7.8 and 7.11) contain indeterminate ratios at $r = 0$, as may easily be verified by inspection. We require, however, that everything be regular at the origin. What is done in practice is to replace the stellar structure equations near the model center by simple expansions in r, much as was done for polytropes. The procedure is straightforward once regularity and symmetry are accounted for. Application of the boundary conditions $\mathcal{M}_r \to 0$ and $\mathcal{L}_r \to 0$ as $r \to 0$ and the requirement of zero spatial gradients of temperature, density, and pressure at the center yield

$$\mathcal{M}_r \to \frac{4\pi}{3}\rho_c r^3 \tag{7.100}$$

$$P(r) \to P_c - \frac{2\pi}{3}G\rho_c^2 r^2 \tag{7.101}$$

$$\mathcal{L}_r \to \frac{4\pi}{3}\rho_c\varepsilon_c r^3 . \tag{7.102}$$

The subscript "c" implies the central value of a quantity. The behavior of temperature near center depends on the mode of energy transport and is

$$T(r) \to T_c - \frac{1}{8ac}\frac{\kappa_c\rho_c^2\varepsilon_c}{T_c^3} r^2 \tag{7.103}$$

if there is no convection, and

$$T(r) \to T_c - \frac{2\pi}{3}G\,\nabla_{\rm ad,c}\frac{\rho_c^2 T_c}{P_c} r^2 \tag{7.104}$$

if adiabatic convection is present (so that $\nabla = \nabla_{\rm ad}$). An easy way to verify the temperature relations is to eliminate r^2 between $P(r)$ and $T(r)$ (either version) and construct a numerical logarithmic derivative. Note that (7.103) may be replaced with (7.104) if $\nabla_{\rm ad}$ is replaced by $\nabla_{\rm rad}$ in the latter equation.

How to treat the surface is more difficult.

7.3.2 The Radiative Stellar Envelope

The term "envelope" is used in various ways in stellar physics and you will see it used many ways here also. Examples of its use are as follows. When describing the structure of red supergiants, which have a dense degenerate core surrounded by an immense diffuse region (both of which may contain substantial mass), the latter region is often referred to as the envelope. On the other hand, the "envelope" of a white dwarf is considered to be the very thin (in mass and depth) nondegenerate region overlying the massive degenerate core. For the purposes of this section, an envelope consists of the portion of a star that starts at the surface, continues inward, but contains negligible mass, has no thermonuclear or gravitational energy sources, and is in hydrostatic equilibrium. This implies, among other things, that the luminosity in the envelope is fixed at its surface value with $\mathcal{L}_r = \mathcal{L}$ and that $\mathcal{M}_r$ is essentially just the total mass $\mathcal{M}$. As we shall see, these requirements allow for considerable simplification in determining the structure of the envelope. Before rushing madly ahead with equations, however, we have to consider more closely what is meant by the stellar "surface."

Photospheric or Zero Boundary Conditions?

In previous chapters we have had to apply boundary conditions at the stellar surface for various reasons. What exactly is the stellar surface? That question is not that easily answered, even for the sun. We know that the sun has a photosphere (and that, for us, is the surface from which the visible light emanates), a transition region, a chromosphere, a corona, and a mild stellar wind. Depending on your viewpoint, the sun either has a well-defined surface or it may continue out well into the solar system. For us, the surface of a star will be somewhere near the gaseous photosphere. Anything outside that fuzzy surface will be assumed to have no effect on what happens inside the star. (Exceptions of course have to be made for accreting systems or those stars with rapid mass loss.) In any case, the "zero boundary conditions" of temperature, pressure, etc., used thus far really do not apply in any real sense although we have used them with some success. But, if successful, they must be reasonably good approximations in some situations. So then, what are these situations and how good is the approximation and, also to the point, how do we describe a photosphere?

To discuss what happens at a stellar photosphere we return to some elementary stellar atmospheres theory. Recall from Chapter 4 that the relation between the frequency-dependent flux and the Planck function is given by (from 4.20)

$$\mathcal{F}_\nu = -\frac{4\pi}{3}\frac{1}{\kappa_\nu \rho}\frac{\partial B_\nu}{\partial r} = \frac{\mathcal{L}_\nu}{4\pi r^2} \tag{7.105}$$

at large optical depths. This also defines the frequency-dependent luminosity, $\mathcal{L}_\nu$. At the same level of approximation it is easy to show that the frequency-dependent radiation pressure is given by

$$P_{\mathrm{rad},\nu} = \frac{4\pi}{3c}B_\nu\,. \tag{7.106}$$

This is consistent with the statement that $P_{\mathrm{rad},\nu} = U_\nu/3$ (with U_ν being the radiation energy density) because $U_\nu = 4\pi B_\nu / c$ in LTE (see Eqs. 3.18–3.20 and 4.5–4.6).

Putting this together yields

$$\frac{c}{\rho} \frac{\partial P_{\mathrm{rad},\nu}}{\partial r} = -\frac{\kappa_\nu \mathcal{L}_\nu}{4\pi r^2}. \tag{7.107}$$

This is now integrated over frequency so that

$$\frac{dP_{\mathrm{rad}}}{dr} = -\frac{\kappa \rho \mathcal{L}}{4\pi r^2 c} \tag{7.108}$$

where κ is defined as

$$\kappa = \frac{1}{\mathcal{L}} \int_0^\infty \kappa_\nu \mathcal{L}_\nu \, d\nu \tag{7.109}$$

and it is *not* the Rosseland mean opacity defined previously by (4.22). (Later we shall use this opacity but make believe it is Rosseland as one of a series of approximations.)

Because at this juncture we are only interested in the radiation properties of the stellar material very near the photosphere, we simplify (7.108) by replacing r on the right-hand side with $\mathcal{R}$, which is defined as the radius at which we find the photosphere. We also define the effective temperature, T_{eff}, as the temperature that satisfies the relation

$$\mathcal{L} = 4\pi\sigma\mathcal{R}^2 T_{\mathrm{eff}}^4 \tag{7.110}$$

in accord with usual nomenclature and (2.1). Thus T_{eff} is the temperature the photosphere (at $\mathcal{R}$) would have if that surface radiated as a black body. Note that of the three quantities in (7.110) only $\mathcal{L}$ is directly observable; $\mathcal{R}$ and T_{eff} may both turn out to be convenient fictions. This is because the term "visible surface" is really a spectrum-dependent statement (photons of one frequency may emerge to final visibility from different depths compared to other photons) and, in any case, no star really emits radiation into space as a pure black body. Thus consider the concepts of stellar radius and effective temperature with caution.

One final point before we proceed. Recall from our discussions in Chapter 4 that we had defined the "true surface" as that level in the star where there was no incoming radiation. This was set at optical depth $\tau = 0$ (see Eqs. 4.8, 4.9, and discussion). What we will have to determine is how that level is related to the level at $\mathcal{R}$ or how the photosphere differs, if at all, from the true surface?

With the above in mind, now integrate (7.108) with $r = \mathcal{R}$ from $\tau = 0$ to some arbitrary depth τ and find

$$P_{\mathrm{rad}} = -\int_{\text{true surface}}^{\text{arbitrary point}} \frac{\mathcal{L}}{4\pi\mathcal{R}^2 c} \kappa\rho \, dr = \int_0^\tau \frac{\mathcal{L}}{4\pi\mathcal{R}^2 c} \, d\tau \tag{7.111}$$

or

$$P_{\rm rad}(\tau) = \frac{\mathcal{L}}{4\pi\mathcal{R}^2 c}\tau + P_{\rm rad}(\tau\!=\!0) = \frac{\sigma T_{\rm eff}^4}{c}\tau + P_{\rm rad}(\tau\!=\!0). \tag{7.112}$$

We now have to determine what the radiation pressure at the true surface is. A general expression for radiation pressure may be constructed by considering the momentum transferred by radiation across an imaginary surface at some position. If we realize that $I(\theta)/c$ is that flux (energy flux/c), then

$$P_{\rm rad} = \frac{2\pi}{c}\int_0^\pi I(\theta)\cos^2\theta\,\sin\theta\,d\theta \tag{7.113}$$

by arguments similar to those used in deriving the total flux (see 4.3). The additional factor of $\cos\theta$ comes about because we require a projection of the momentum to the radial direction. To make further progress, $I(\theta)$ must be specified. There are several strategies possible here but the most straightforward is to invoke a version of the Eddington approximation. (For a much fuller discussion of the virtues and faults of this approximation, see Mihalas 1978.) A primary consequence of this approximation is that the radiation pressure is given everywhere, *except at* $\tau=0$, by its LTE value $P_{\rm rad} = aT^4/3$. This is the same result as would be obtained were $I(\theta)$ isotropic with $I = (\sigma/\pi)T^4$ (from 4.6). Assume, therefore, that $I(\theta)$ is isotropic everywhere with this value except at the true surface. At $\tau = 0$ we shall compromise and let $I(\theta)$ be isotropic for all outgoing angles but set it to zero for $\pi \geq \theta > \pi/2$. Thus no radiation enters the true surface from the outside. Equation (7.113) then yields

$$P_{\rm rad}(\tau\!=\!0) = \frac{2\pi}{3c}I(\tau\!=\!0). \tag{7.114}$$

We now find $I(\tau=0)$ by computing the flux at zero optical depth and assuming, as a further minor approximation, that the position of the true surface is at $\mathcal{R}$ (and remember that a relatively large change in optical depth does not mean a correspondingly large change in radius). Using expression (4.3) for the flux we have

$$\mathcal{L} = 4\pi\mathcal{R}^2\,2\pi\int_0^{\pi/2} I(\tau\!=\!0)\cos\theta\sin\theta\,d\theta = 4\pi\mathcal{R}^2\pi I(\tau\!=\!0). \tag{7.115}$$

Use this to eliminate $I(\tau=0)$ in (7.114) and find

$$P_{\rm rad}(\tau\!=\!0) = \frac{2}{3c}\frac{\mathcal{L}}{4\pi\mathcal{R}^2} = \frac{2}{3c}\sigma T_{\rm eff}^4. \tag{7.116}$$

The complete expression for the radiation pressure at depth is then

$$P_{\rm rad}(\tau) = \frac{1}{3}aT^4(\tau) = \frac{\sigma}{c}(\tau + 2/3)T_{\rm eff}^4. \tag{7.117}$$

From this also obtain the run of temperature in the very outermost layers:

$$T^4(\tau) = \frac{1}{2} T_{\text{eff}}^4 \left(1 + \frac{3}{2}\tau\right) \tag{7.118}$$

after recalling that $a = 4\sigma/c$. Thus in these approximations the photosphere lies at the optical depth $\tau_{\text{p}} = 2/3$ where $T(\tau_{\text{p}}) = T_{\text{eff}}$ (and "p" stands for "photosphere"). Note also that the temperature is nonzero even at the surface, where it has the value $2^{-1/4} T_{\text{eff}}$, and not zero as assumed for zero boundary conditions.

To find the run of total pressure in the outer layers requires solving the hydrostatic equilibrium equation

$$\frac{dP}{dr} = -g\rho. \tag{7.119}$$

If mass and radius are regarded as fixed in the local gravity, then g is a constant, with $g_s = G\mathcal{M}/\mathcal{R}^2$, and the hydrostatic equation can immediately be integrated from the true surface down to some optical depth to yield

$$P(\tau) = g_s \int_0^\tau \frac{d\tau}{\kappa}. \tag{7.120}$$

What we want is the pressure at the photosphere, which is now known to lie at $\tau = 2/3$ (or nearby, depending on how the previous analysis is done in detail). To again make matters simple, consider the case where opacity is constant (as a version of the "grey" atmosphere) and equal to its value at the photosphere. Denote that opacity by κ_{p}. Equation (7.120) can then be integrated and becomes

$$P(\tau_{\text{p}}) = \frac{2}{3}\frac{g_s}{\kappa_{\text{p}}} + P(\tau{=}0). \tag{7.121}$$

If the material gas contributes little or nothing to the total pressure at the true surface (as seems reasonable because nothing should act there to reverse the flow of radiation outward), then setting $P(\tau{=}0) = P_{\text{rad}}(\tau{=}0)$ yields

$$P(\tau_{\text{p}}) = \frac{2}{3}\frac{g_s}{\kappa_{\text{p}}} \left(1 + \frac{\kappa_{\text{p}}\mathcal{L}}{4\pi c G\mathcal{M}}\right) \tag{7.122}$$

after a little algebra and the use of (7.116).

For most stars the last factor is the parentheses is small with

$$\frac{\kappa_{\text{p}}\mathcal{L}}{4\pi c G\mathcal{M}} = 7.8 \times 10^{-5}\, \kappa_{\text{p}} \left(\frac{\mathcal{L}}{\mathcal{L}_\odot}\right) \left(\frac{\mathcal{M}}{\mathcal{M}_\odot}\right)^{-1} \tag{7.123}$$

and it can almost always be ignored. For some very massive and luminous stars, however, it cannot ignored as the following argument shows. Near the true surface where radiation pressure dominates, the hydrostatic equation is as given above but with dP/dr replaced by $dP_{\rm rad}/dr$. If the luminosity is very high and the radiation field very intense, we can imagine that the force due to radiation pressure might overwhelm the local gravitational force. This situation can be written as

$$-\frac{dP_{\rm rad}}{dr} > g_s \rho. \tag{7.124}$$

If we further make the (actually contradictory) assumption that radiative diffusion is still responsible for energy transport, then a slightly rewritten form of the transport equation (from 4.26) is

$$\mathcal{L} = -\frac{4\pi \mathcal{R}^2 c}{\kappa_{\rm p}\rho}\frac{dP_{\rm rad}}{dr}. \tag{7.125}$$

Eliminating the pressure gradient between the two equations then yields an estimate of how large the luminosity must be so that radiative forces exceed gravitational forces. That limiting luminosity, called the *Eddington critical luminosity* or *Eddington limit*, is

$$\mathcal{L}_{\rm Edd} = \frac{4\pi c G \mathcal{M}}{\kappa_{\rm p}} \tag{7.126}$$

and this overall combination is exactly that which appears as the second term of (7.122). If that term exceeds unity then the Eddington limit has been exceeded. It should be obvious that this subject is intimately connected with mass loss (for more on the implications, see Shapiro and Teukolsky 1983). As a practical matter, the opacity usually used in (7.126) is electron scattering because high luminosities usually imply high temperatures. With a hydrogen mass fraction of $X = 0.7$ and $\kappa_{\rm e} = .34$ cm^2 g^{-1} used for the photospheric opacity, the Eddington limit is

$$\left(\frac{\mathcal{L}_{\rm Edd}}{\mathcal{L}_\odot}\right) \approx 3.5 \times 10^4 \left(\frac{\mathcal{M}}{\mathcal{M}_\odot}\right). \tag{7.127}$$

It is to be understood that if the luminosity approaches (within a couple of tens of percent) of this number, then a simple static stellar atmosphere will not adequately describe what is going on; the dynamics of momentum and energy transfer between the radiation field and matter must be done correctly and this is difficult.

If the Eddington term is neglected then the photospheric pressure is given by

$$P(\tau_{\rm p}) \approx \frac{2}{3}\frac{g_s}{\kappa_{\rm p}}. \tag{7.128}$$

This may now be used to find the density at the photosphere. If the gas is assumed to be composed only of the sum of ideal gas plus radiation, we set

(7.128) equal to that sum and find

$$\frac{1}{3}aT_{\text{eff}}^4 + \frac{N_{\text{A}}k}{\mu}\rho_p T_{\text{eff}} = \frac{2}{3}\frac{g_s}{\kappa_0 \rho_p^n T_{\text{eff}}^{-s}} \tag{7.129}$$

where the power law expression $\kappa_0 \rho_{\text{p}}^n T_{\text{eff}}^{-s}$ has replaced the opacity. If n and s are known and T_{eff} is fixed, then ρ_{p} may be found using some iterative method.[6] This implies, incidentally, that one must have some idea of photospheric conditions beforehand in order that the *kind* of opacity and its exponents be known. In the simple case, where it is assumed that radiation pressure is unimportant,

$$\rho_{\text{p}}^{n+1} \approx \frac{2}{3}\frac{g_s}{\kappa_0}\frac{\mu}{N_{\text{A}}k}T_{\text{eff}}^{s-1}. \tag{7.130}$$

We know typical ranges for T_{eff} and gravity so we can easily find out what kinds of numbers are associated with photospheric densities. For example, if the gravity and T_{eff} are chosen as solar ($g_s \approx 2.7 \times 10^4$ cm s^{-2} and $T_{\text{eff}} \approx 5770$ K), and the opacity is pure electron scattering ($n = s = 0$), then (7.130) yields $\rho_{\text{p}} \sim 10^{-7}$ g cm^{-3}. Using the same conditions but with the more realistic H^- opacity (see 4.38) gives $\sim 10^{-6}$ g cm^{-3}, which is essentially the same number at our level of approximation. In any case, photospheric densities are far smaller than those deeper down.

Note in all the above that it has been assumed that convection plays no role in heat transport between the true and photospheric surfaces. This is consistent with our notion of a radiating, static, visible surface, and we shall continue to think of the photosphere in those terms. However, even in the sun the effects of underlying convection may easily be seen in the form of cells, granulation, etc., so that if the photospheric regions are to be modeled correctly, much care must be taken (do not forget magnetic fields, and so on). We shall not go to such extremes, but we will find that convection can extend right up to the base of the photosphere.

When making stellar models in practice, things can get complicated. What is done is to construct a "grid" of realistic stellar atmospheres where each model atmosphere in the grid is labeled by, for example, a different combination of effective temperature and surface gravity. If, during the course of some sort of iterative procedure used in making a complete stellar model, a set of boundary conditions is required at the photosphere, then interpolation is done in the grid to yield these boundary conditions for a given effective temperature and gravity. A description of one strategy for such an interpolation is given in Kippenhahn, Weigert, and Hofmeister (1967, §IV).

We may now go on to see what to do about the stellar envelope and to find out when the application of photospheric conditions is important.

---

[6]This is the method used in the **ZAMS** code in Appendix C.

The Radiative Stellar Envelope

Here we compute the structure of envelopes in which convection is negligible or nonexistent. We still assume that $\mathcal{L}_r = \mathcal{L}$ and $\mathcal{M}_r$ is essentially the same as the total mass $\mathcal{M}$. The primary reference here is Cox (1968, chap. 20).

If convection is neglected, then ∇ is equal to ∇_{rad} and

$$\nabla = \frac{d\ln T}{d\ln P} = \frac{3}{16\pi acG}\frac{P\kappa}{T^4}\frac{\mathcal{L}}{\mathcal{M}} \tag{7.131}$$

where we have set $\mathcal{M}_r = \mathcal{M}$ and $\mathcal{L}_r = \mathcal{L}$ and

$$\frac{3\mathcal{L}}{16\pi acG\mathcal{M}} = 7.59 \times 10^9 \left(\frac{\mathcal{M}}{\mathcal{M}_\odot}\right)^{-1}\left(\frac{\mathcal{L}}{\mathcal{L}_\odot}\right). \tag{7.132}$$

We shall assume for convenience that the pressure is due solely to an ideal gas with no radiation pressure. Once the envelope solution is obtained, this assumption can be examined and corrections made. The opacity in (7.131) is then written in the interpolation form $\kappa = \kappa_0\rho^n T^{-s}$ which, for an ideal gas, becomes

$$\kappa = \kappa_g P^n T^{-n-s} \tag{7.133}$$

where

$$\kappa_g = \kappa_0\left(\frac{\mu}{N_{\text{A}}k}\right)^n. \tag{7.134}$$

With this substitution the differential equation (7.131) now contains only P and T as the active variables and may be rewritten

$$P^n\,dP = \frac{16\pi acG\mathcal{M}}{3\kappa_g\mathcal{L}}T^{n+s+3}\,dT. \tag{7.135}$$

If T_0 and P_0 refer to some upper reference level in the envelope (such as the photosphere) with $P(r) \geq P_0$ and $T(r) \geq T_0$, then an easy integration of (7.135) yields

$$P^{n+1} = \frac{n+1}{n+s+4}\frac{16\pi acG\mathcal{M}}{3\kappa_g\mathcal{L}}T^{n+s+4}\left[\frac{1-(T_0/T)^{n+s+4}}{1-(P_0/P)^{n+1}}\right] \tag{7.136}$$

for $n+s+4 \neq 0$. If this sum of exponents does equal zero through some unlikely numerical accident, then the solution differs from the above but is still easy to find.

We now examine some likely combination of exponents n and s to see what (7.136) implies for the run of pressure versus temperature in the envelope. Note first that if $n+s+4$ and $n+1$ are both positive then the terms $[T_0/T(r)]^{n+s+4}$ and $[P_0/P(r)]^{n+1}$ get small rapidly as we go to deep depths in the envelope. Thus for $(n+s+4)$, $(n+1) > 0$ and $T(r) >> T_0$, $P(r) >> P_0$

$$P^{n+1} \rightarrow \frac{n+1}{n+s+4}\frac{16\pi acG\mathcal{M}}{3\kappa_g\mathcal{L}}T^{n+s+4}. \tag{7.137}$$

If T_0 and P_0 refer to photospheric values then, in this instance, the solution for pressure versus temperature deep in the envelope below the photosphere converges rapidly to a common solution independent of photospheric conditions. Thus, as far as the interior structure is concerned, we could just as well have used zero boundary conditions for pressure and temperature. Examples include Kramers' opacity ($n = 1$, $s = 3.5$) and electron scattering ($n = s = 0$).

An important counterexample to the above is where the envelope opacity is due to H^- as in the estimate given by (4.38). Here $n = 1/2$, $s = -9$ with $n + s + 4 = -4.5$ and photospheric boundary conditions have a strong influence on the underlying layers. It is also true that in cool stars where H^- opacity is important, the underlying layers are convective and the above analysis does not apply. We shall return to this shortly.

If we assume (7.137) holds, then the logarithmic run of temperature versus pressure at depth is

$$\nabla(r) \rightarrow \frac{n+1}{n+s+4} = \frac{1}{1+n_{\text{eff}}} \tag{7.138}$$

where $n_{\text{eff}} = (s+3)/(n+1)$ is the "effective polytropic index." The reason for this name is as follows. Equation (7.137) also yields

$$P = K' T^{1+n_{\text{eff}}} \tag{7.139}$$

where it is assumed that we may use zero boundary conditions on T and P, and where K' may readily be established from (7.137). If the gas is ideal and of constant composition, with $P \propto \rho T/\mu$, then $P \propto \rho^{1+1/n_{\text{eff}}}$ and the structure is polytropic (as in 7.20). Recalling our discussion of polytropes and ideal gases, the coefficient K' is the same as that identified in (7.25) and, from the present analysis, is equal to

$$K' = \left[\frac{1}{1+n_{\text{eff}}} \frac{16\pi acG\mathcal{M}}{3\kappa_0 \mathcal{L}} \left(\frac{N_{\text{A}}k}{\mu}\right)^n\right]^{1/(n+1)} \tag{7.140}$$

from which we may calculate the polytropic constant K by way of (7.26). A practical difficulty in carrying out the analysis further in terms of polytrope language is that the solution (7.137) must eventually be joined to the rest of the star, and the present analysis cannot do that. Suffice it to say that you might well imagine how this might be done (with appropriate conditions of continuity, etc.). For now we remark that the polytropic-like solution in the envelope corresponding to $\theta(\xi)$ of the previous sections need not be of the complete E-solution variety and may be of F- or M-solution character.

The constancy of ∇ in (7.138) implies that the combination $P\kappa/T^4$ is a constant by virtue of (7.131). Specifically, if Kramers' opacity holds in the envelope, then (with $n = 1$, $s = 3.5$, and $n_{\text{eff}} = 3.25$) the solution for P versus T is

$$P(r) = \left[\frac{1}{4.25} \frac{16\pi acG\mathcal{M}}{3\kappa_0 \mathcal{L}} \frac{N_{\text{A}}k}{\mu}\right]^{1/2} T^{4.25}(r) \tag{7.141}$$

and $\nabla = 0.2353$. If no ionization processes are in progress and $\Gamma_2 = 5/3$, then $\nabla_{\text{ad}} = 1 - 1/\Gamma_2 = 0.4$ and thus $\nabla < \nabla_{\text{ad}}$, which implies no convection; the envelope is radiative as originally assumed. For electron scattering with $n = s = 0$, $n_{\text{eff}} = 3$ (an "$n = 3$" polytrope again!), $\nabla = 0.25$ and the same conclusion holds. Note, however, these results are only a rough guide and must be applied with caution; ionization processes and convection, albeit almost negligible, occur in the outer layers of nearly all stars and a complete and accurate integration including all effects is necessary in modeling real stars. The present analysis remains a rough guide but, on the other hand, and even in practice, constant luminosity and negligible envelope mass are often used to simplify envelope integrations.

The Radiative Temperature Structure

If the envelope is radiative with n_{eff} constant and certain restrictions given below apply, then temperature as a function of radius may be found. We assume that n and s are such that zero boundary conditions are adequate (as discussed above). This means that we know $\nabla = 1/(1 + n_{\text{eff}})$. Thus rewrite the equation of hydrostatic equilibrium in the form

$$\frac{dP}{dr} = \frac{P}{\nabla}\frac{1}{T}\frac{dT}{dr} = -\frac{G\mathcal{M}}{r^2}\rho \tag{7.142}$$

where $\mathcal{M}_r = \mathcal{M}$ is still assumed. If the gas is ideal and $P = \rho N_A kT/\mu$ is used to replace the pressure in the middle term of (7.142), we find

$$(n_{\text{eff}} + 1)\frac{dT}{dr} = -\frac{G\mathcal{M}\mu}{N_A k}\frac{1}{r^2}. \tag{7.143}$$

This is then integrated to yield T(r) as

$$\begin{aligned} T(r) &= \frac{1}{1+n_{\text{eff}}}\frac{G\mathcal{M}\mu}{N_A k}\left(\frac{1}{r} - \frac{1}{\mathcal{R}}\right) \\ &= \frac{2.293 \times 10^7}{1+n_{\text{eff}}}\mu\left(\frac{\mathcal{M}}{\mathcal{M}_\odot}\right)\left(\frac{\mathcal{R}}{\mathcal{R}_\odot}\right)^{-1}\left(\frac{1}{x} - 1\right)\ \text{K} \end{aligned} \tag{7.144}$$

where $x = r/\mathcal{R}$. Note that if Kramers' opacity could be used everywhere in the solar envelope (so that $n_{\text{eff}} = 3.25$) and the composition were ionized Pop I ($\mu \approx 0.6$), then the temperature at a level only 1% below the surface ($x = 0.99$) would be about 33,000 K as compared to T_{eff} of 5770 K. In other words, large positive values of the temperature and density exponents of opacity imply a rapid increase of temperature inward and the outer boundary conditions matter little.

We may also find the envelope mass using the above. If the gas is ideal $P = \rho N_A kT/\mu$ may be equated to $P = K'T^{1+n_{\text{eff}}}$ of (7.139) and the density is $\rho(r) = K'\mu T^{n_{\text{eff}}}(r)/N_A k$. But since $T(r)$ is given by (7.144), we then have $\rho(r)$. The latter is integrated (after weighting by $4\pi r^2$) from some envelope level r to $\mathcal{R}$ to give the mass, $\mathcal{M} - \mathcal{M}_r$, above that level. As an exercise for you, consider electron scattering opacity (κ_e, $n = s = 0$, $n_{\text{eff}} = 3$), use (7.140) to compute K', and show that

$$1 - \frac{\mathcal{M}_r}{\mathcal{M}} = \frac{\pi^2 ac}{12\kappa_e}\left(\frac{\mu G}{N_A k}\right)^4 \frac{\mathcal{M}^3}{\mathcal{L}} \times \left[\frac{x^3}{3} - \frac{3}{2}x^2 + 3x - \frac{11}{6} - \ln x\right]. \tag{7.145}$$

If x is very close to unity then the term in the brackets is approximately $(1-x)^4/4$. For a solar mass and luminosity, equation (7.145) implies, for example, that traversing 15% of the total radius inward from the surface uses up only a little less than 1% of the mass. This confirms our assumption that $\mathcal{M}_r \approx \mathcal{M}$ through the envelope.

7.3.3 Completely Convective Stars

We know from previous considerations that cool stars tend to have convective envelopes. In this section we shall carry this to the extreme and discuss some of the properties of *fully* convective stars and how such objects come to be. The analysis may become algebraically tedious in spots, but the result will bear directly on pre–main sequence evolution.

Consider a cool star whose surface layer opacity is dominated by H^- which is, from (4.38),

$$\kappa_{H^-} \approx 2.5 \times 10^{-31}\left(\frac{Z}{0.02}\right)\rho^{1/2}T^9 \quad \text{cm}^2\ \text{g}^{-1} \tag{7.146}$$

for hydrogen mass fractions X around 0.70. We already know that the exponents for this opacity ($n = 1/2$, $s = -9$) spell trouble for a simple envelope analysis because the outer boundary conditions are felt deep down into the envelope. Thus, from now on, we use photospheric boundary conditions with temperature $T_p = T_{\text{eff}}$ and pressure $P_p = 2g_s/3\kappa_p$ as derived earlier. Now to find the structure of the envelope.

Consider the pressure-temperature relation (7.136) and transform it into an equation for ∇ as a function of temperature. You may easily verify that the result is

$$\nabla = \frac{1}{1+n_{\text{eff}}} + \left(\frac{T_{\text{eff}}}{T}\right)^{n+s+4}\left[\nabla_p - \frac{1}{1+n_{\text{eff}}}\right] \tag{7.147}$$

with $n_{\text{eff}} = (s+3)/(n+1)$ as given by (7.138). The "p" means photospheric, ∇_p is ∇ evaluated at the photosphere, and (see 7.131)

$$\nabla_p = \frac{3\kappa_0\mathcal{L}}{16\pi acG\mathcal{M}}\left(\frac{\mu}{N_A k}\right)^n \frac{P_p^{n+1}}{T_{\text{eff}}^{n+s+4}} = \frac{3\mathcal{L}}{16\pi acG\mathcal{M}}\frac{P_p\kappa_p}{T_{\text{eff}}^4}. \tag{7.148}$$

At the photosphere $P_p = 2g_s/3\kappa_p$, $g_s = G\mathcal{M}/\mathcal{R}^2$, and, of course, $\mathcal{L} = 4\pi\mathcal{R}^2\sigma T_{\text{eff}}^4$. Inserting this information into (7.148) yields $\nabla_p = 1/8$. The run of ∇ below the photosphere is obtained from (7.147) which, for H^- opacity (with $n_{\text{eff}} = -4$), is

$$\nabla(r) = -\frac{1}{3} + \frac{11}{24}\left[\frac{T_{\text{eff}}}{T(r)}\right]^{-9/2} . \tag{7.149}$$

Note that since temperature increases with depth, so does ∇. The implication of this observation is that at some depth ∇ must eventually become larger than the thermodynamic derivative ∇_{ad}. If, as an approximation, we assume that ∇_{ad} is given by its ideal gas value $\nabla_{\text{ad}} = 0.4$ (with no ionization taking place), then we can estimate where (in temperature) ∇ is equal to, and thereafter would exceed, ∇_{ad}. Thus we can set $\nabla(r)$ of (7.149) equal to 0.4 and solve for $T(r)$ at this critical depth—and we shall do this in just a bit. For now, observe that if $\nabla > \nabla_{\text{ad}}$, then the stellar material becomes convective and, for simplicity, we assume that the convection is adiabatic. Thus at depths deeper than the critical depth, $\nabla(r) = \nabla_{\text{ad}} = 0.4$. As remarked upon at the end of §7.2.1, this behavior of $\nabla(r)$, along with the ideal gas assumption, implies a polytrope of index 3/2 and

$$P = K'T^{5/2} \tag{7.150}$$

gives the run of pressure with temperature (see 7.25). The picture is then that of a photosphere from which escapes the visible radiation, underlain by a radiative layer of depth to be determined (and probably a shallow layer at that), and, under that, convection. This represents the outer layers of the sun as we know it.

In writing down (7.150) we note the following. In the extreme case where convection continues down to the stellar center, the constant K' cannot be arbitrary because (7.150), as a complete polytrope, must have solutions corresponding to a complete model with appropriate central boundary conditions. In other words, given $\mathcal{M}$ and $\mathcal{R}$, K' must satisfy the combination of relations (7.37) and (7.26) for K' and K given earlier for ideal gas polytropes. One way to approach this is to recast pressure and temperature in the dimensionless variables discussed by Schwarzschild (1958, §13) where

$$p = \frac{4\pi}{G}\frac{\mathcal{R}^4}{\mathcal{M}^2}P \tag{7.151}$$

$$t = \frac{N_A k}{G}\frac{\mathcal{R}}{\mu\mathcal{M}}T. \tag{7.152}$$

Equation (7.150) then becomes

$$p = E_0 t^{5/2} \tag{7.153}$$

with

$$E_0 = K' 4\pi \left(\frac{\mu}{N_A k}\right)^{5/2} G^{3/2}\mathcal{M}^{1/2}\mathcal{R}^{3/2}. \tag{7.154}$$

But for an ideal gas, $n = 3/2$, E-solution polytrope, K' is given by (7.26) as

$$K'_{n=3/2} = \left(\frac{N_A k}{\mu}\right)^{5/2} K_{n=3/2}^{-3/2}. \tag{7.155}$$

After substituting for K of (7.37) this becomes

$$K'_{n=3/2} = \frac{2.5^{3/2}}{4\pi}\left[\xi_{3/2}^{5/2}\left(-\theta'_{3/2}\right)^{1/2}\right]_{\xi_1}\left(\frac{N_A k}{\mu}\right)^{5/2}\frac{1}{G^{3/2}\mathcal{M}^{1/2}\mathcal{R}^{3/2}}. \tag{7.156}$$

Putting this together we find the surprising result that E_0 does not depend on any of the physical parameters of the model (mass, radius, composition) but rather contains only the surface values of the polytropic variables and is the constant

$$E_0 = \left(\frac{-125}{8}\xi_{3/2}^5\theta'_{3/2}\right)_{\xi_1}^{1/2} = 45.48 \tag{7.157}$$

using the results from Table 7.1.

After these introductory remarks we may now compute some of the parameters of a completely convective star. In particular, we shall seek a relation between mass, luminosity, effective temperature, and composition. This will take a few steps. We first need the temperature and density at that level in the star below the photosphere where $\nabla = \nabla_{ad} = 0.4$. This may be found from (7.147) using $\nabla_p = 1/8$ and the exponents $n = 1/2$ and $s = -9$ from the estimate for the H^- opacity. You may readily check that the temperature at that level, denoted by T_f, is given by $(T_f/T_{\text{eff}}) = (8/5)^{2/9} \approx 1.11$; that is, T_f is a mere 11% higher than T_{eff}. The implication is that convection starts just below the photosphere. The pressure, P_f, at the top of the convective interior is found by rewriting (7.136) in the form

$$\left(\frac{P}{P_p}\right)^{n+1} = 1 + \frac{1}{1+n_{\text{eff}}}\frac{1}{\nabla_p}\left[\left(\frac{T}{T_{\text{eff}}}\right)^{n+s+4} - 1\right] \tag{7.158}$$

which, for the case in question, yields $P_f = 2^{2/3}P_p$.

We now apply (7.150) in the form $P_f = K'T_f^{5/2}$. The polytropic parameter K' is obtained from the combination of equations (7.154) to (7.157) and is

$$K' = \frac{3.564\times10^{-4}E_0}{\mu^{2.5}}\left(\frac{\mathcal{M}}{\mathcal{M}_\odot}\right)^{-1/2}\left(\frac{\mathcal{R}}{\mathcal{R}_\odot}\right)^{-3/2} \tag{7.159}$$

or, in functional dependence, $K' = K'(\mathcal{M}, \mathcal{R}, \mu)$.[7] To express K' in terms of T_{eff} and $\mathcal{L}$, use $\mathcal{L} = 4\pi\sigma\mathcal{R}^2T_{\text{eff}}^4$ and find $K' = K'(\mathcal{M}, T_{\text{eff}}, \mathcal{L}, \mu)$. Now

[7]From now on, we shall write down little in the way of explicit formulas, but will wait until the very end to give the answer. You are advised to work out all the tedious numbers as we go along.

for $P_f = 2^{2/3} P_\mathrm{p}$. The photospheric pressure is $2g_s/3\kappa_\mathrm{p}$, which is a function of $\mathcal{M}$ and $\mathcal{R}$ (from g_s) and T_eff and photospheric density (from $\kappa_\mathrm{p} = \kappa_0 \rho_\mathrm{p}^n T_\mathrm{eff}^{-s}$). The density is eliminated using the ideal gas equation of state to yield

$$P_\mathrm{p} = \left(\frac{2}{3} \frac{G\mathcal{M}}{\kappa_0 \mathcal{R}^2} \right)^{1/(n+1)} \left(\frac{\mu}{N_\mathrm{A} k} \right)^{-n/(n+1)} T_\mathrm{eff}^{(n+s)/(n+1)} \tag{7.160}$$

which has the dependence $P_\mathrm{p} = P_\mathrm{p}(\mathcal{M}, \mathcal{R}, \mu, T_\mathrm{eff})$. Again get rid of $\mathcal{R}$ and find $P_\mathrm{p} = P_\mathrm{p}(\mathcal{M}, T_\mathrm{eff}, \mathcal{L}, \mu)$. It should now be clear that the polytropic equation $P_f = K' T_f^{5/2}$ becomes a power law equation containing only the variables $\mathcal{M}$, T_eff, $\mathcal{L}$, κ_0, and μ (and, in the general case, n and s). For our estimate of H^- opacity, this relation becomes

$$T_\mathrm{eff} \approx 2600 \mu^{13/51} \left(\frac{\mathcal{M}}{\mathcal{M}_\odot} \right)^{7/51} \left(\frac{\mathcal{L}}{\mathcal{L}_\odot} \right)^{1/102} \ \mathrm{K}. \tag{7.161}$$

The strange exponents appearing here are a good indication of how messy this calculation is. Also, as will be pointed out later, the temperature coefficient of 2600 K is too low in this simple calculation. It should be more like 4000 K. In any case, this relation shows up as a set of nearly vertical lines in the Hertzsprung-Russell diagram with $\mathcal{M}$ being the parameter labeling the lines and where T_eff is virtually independent of $\mathcal{L}$ for a given $\mathcal{M}$. Thus completely convective stars (with a radiative photosphere) of a given mass and (uniform) composition in hydrostatic equilibrium lie at nearly constant low effective temperature independent of luminosity. Or, phrased another way, the effective temperatures of such stars are nearly independent of how the energy is generated. The next section discusses to what kinds of stars these models correspond.

A Question of Entropy

To interpret the above, we return to a comment made in Chapter 5 about the role of entropy in convective stars (and we shall follow the excellent historical review of Stahler 1988 in much of what follows). For infinitesimal and reversible changes, the first and second laws state that

$$T\,dS = dE + P\,d\left(\frac{1}{\rho}\right) = dE - \frac{P}{\rho^2}\,d\rho \tag{7.162}$$

where E is in erg g^{-1} and the entropy, S, has the units erg g^{-1} K^{-1}. We wish to transform this to something more familiar in the stellar context. To do so, express E and ρ in terms of P and T; that is, let $E = E(P,T)$ and $\rho = \rho(P,T)$. The differentials are then expanded out into partials with respect to P and T. These partials are then transformed using standard thermodynamic rules (as in Landau and Lifshitz 1958, §16, and Cox 1968,

§9.14) and the relation

$$c_P \nabla_{\rm ad} = \frac{P}{\rho T} \frac{\chi_T}{\chi_\rho} \tag{7.163}$$

(from equation 3.93) is applied to finally arrive at

$$\frac{dS}{dr} = c_P \left(\nabla - \nabla_{\rm ad} \right) \frac{d \ln P}{dr}. \tag{7.164}$$

The key here is the presence of $\nabla - \nabla_{\rm ad}$. Since hydrostatic equilibrium requires that $d \ln P/dr \leq 0$ everywhere, then the following are true:

1. If the star is locally radiative with $\nabla < \nabla_{\rm ad}$, then $dS/dr > 0$ and entropy increases outward at that location.

2. If $\nabla > \nabla_{\rm ad}$ so that the star is convectively unstable, then $dS/dr < 0$ and entropy decreases outward. In the special case of very efficient adiabatic convection, ∇ exceeds $\nabla_{\rm ad}$ by so little that $\nabla - \nabla_{\rm ad} = 0^+$ and we may effectively set $\nabla = \nabla_{\rm ad}$. If we restrict ourselves to this situation, then S is effectively constant through a convective region.

Combinations of these statements are shown in Figure 7.5 for three stars; one is completely convective, the second entirely radiative, and the last has a radiative interior but convective envelope where, if you look closely, the convective stars have thin radiative photospheres.

Having settled on the above behavior for the run of entropy, we now determine some of the thermodynamics of entropy for an ideal gas. First rewrite (7.162) as

$$T\,dS = \left(\frac{\partial E}{\partial T} \right)_\rho dT + \left(\frac{\partial E}{\partial \rho} \right)_T d\rho - \frac{P}{\rho^2}\, d\rho \tag{7.165}$$

where for a constant-composition ideal monatomic gas (assumed not to be in the process of ionization) $E = 3N_{\rm A}kT/2\mu$ erg g^{-1} and $P = \rho N_{\rm A}kT/\mu$. Performing the indicated operations yields

$$T \frac{dS}{dr} = \frac{N_{\rm A}k}{\mu} \frac{d \ln \left(T^{3/2}/\rho \right)}{dr} \tag{7.166}$$

which, to within an additive constant after integration, becomes

$$S(r) = \frac{N_{\rm A}k}{\mu} \ln \left[T^{3/2}(r)/\rho(r) \right]. \tag{7.167}$$

This may be recast into two other convenient forms by using the equation of state to eliminate either temperature or density. Thus

$$S(r) = \frac{N_{\rm A}k}{\mu} \ln \left[T^{5/2}(r)/P(r) \right] \tag{7.168}$$

$$S(r) = \frac{3}{2} \frac{N_{\rm A}k}{\mu} \ln \left[P(r)/\rho^{5/3}(r) \right]. \tag{7.169}$$

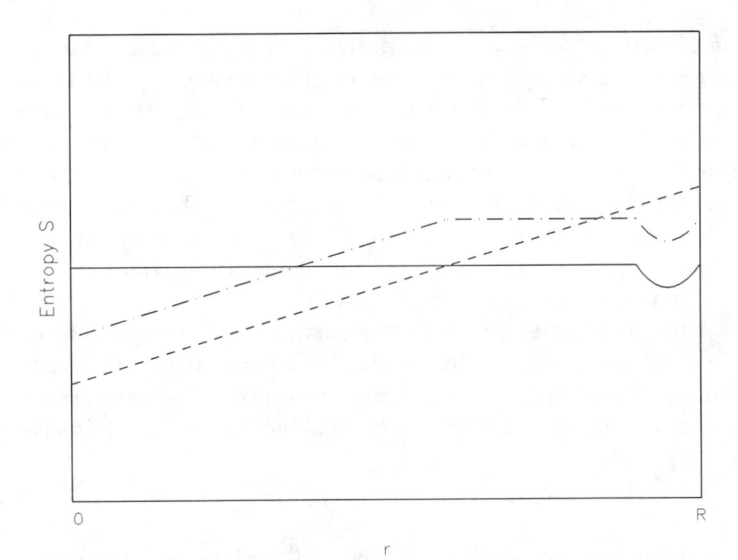

FIGURE 7.5. The run of entropy with radius for a completely convective star with very efficient convection (solid line), a star that is radiative throughout (dashed line), and one with a radiative core and convective envelope (dash-dot). Note that the convective stars have thin radiative layers near their photospheres under which is a transition layer of inefficient convection.

Note that various constants have been absorbed into the constant terms implicit in the right-hand sides of these expressions.

The first application of the last two equations is evident if $\nabla = \nabla_{\rm ad}$ in an ideal gas where $\Gamma_2 = 5/3$; namely, these conditions imply that $P \propto \rho^{5/3}$ and $P \propto T^{5/2}$ so that adiabatic convection means constancy of entropy as in Figure 7.5. A second observation allows us to estimate the relation between entropy and, say, total radius in a star. If the star is in hydrostatic equilibrium, then, from dimensional considerations, pressure is approximately $P \approx G\mathcal{M}^2/\mathcal{R}^4$ and $\rho \approx \mathcal{M}/\mathcal{R}^3$. Equation (7.169) then yields the total entropy $S_{\rm tot} \approx \mathcal{M}(3N_{\rm A}k/2\mu)\ln\left(G\mathcal{M}^{1/3}\mathcal{R}\right)$, or

$$\mathcal{R} \propto \exp\left(\frac{2\mu}{3N_{\rm A}k}\frac{S_{\rm tot}}{\mathcal{M}}\right) . \tag{7.170}$$

This states that the total radius is a sensitive increasing function of entropy for fixed mass and (uniform) composition.[8] Thus if Figure 7.5 represents

[8]Another way to look at this is to consider gravitationally contracting stars with no thermonuclear energy generation. If processes are reversible, then the energy equation (6.2) implies $dQ/dt = T\,dS/dt = -\partial\mathcal{L}_r/\partial\mathcal{M}_r$. Contraction means that $d\mathcal{R}/dt < 0$ and $\partial\mathcal{L}_r/\partial\mathcal{M}_r < 0$ so that entropy decreases with time along with radius.

the run of entropy for stars of fixed mass, composition, *and* radius, then the integrals over mass of the entropy must be very nearly the same. This is why, for example, the entropy in the inner regions of the radiative star is shown to be lower than for the convective star, and the reverse is true in the outer layers. The intermediate case in the figure shows what happens if just the outer layers are convective. As an afternote, the standard model is completely radiative according to the criteria given above (as it should be for consistency). Namely, $\rho \propto T^3$ for this model (from 7.95) so that $S(r)$ increases outward by application of (7.167).

We are now in a better position to understand the result given by (7.161). Equation (7.168) may be evaluated for the entropy at the photosphere by substituting T_{eff} and the photospheric pressure for the temperature and pressure in the logarithm. For H^- opacity, the entropy is easily shown to be

$$S_{\text{p}} = \frac{N_{\text{A}}k}{\mu} \ln\left(3T_{\text{eff}}^{11.5}\rho_{\text{p}}^{0.5}/2\kappa_0 g_s\right). \tag{7.171}$$

Thus as T_{eff} is reduced, so is S_{p}. However, T_{eff} cannot fall too low for two reasons. Too low an effective temperature means that the photosphere can become optically thin (as opacity decreases) and this violates our notion of a photosphere. In addition, the entropy cannot drop below the interior value in this simple picture because otherwise the very outer layers will have decreasing entropy implying convection—and this is a contradiction. There is then a minimum to S_{p} and a corresponding minimum for T_{eff}.

Application to Pre–Main Sequence Evolution

A direct application of this discussion is to the evolution of pre–main sequence stars. If we suppose that they have no interior thermonuclear energy sources (although burning of deuterium, which may have been present in the protostar nebula, may play a role), then contraction from a protostellar cloud will eventually yield high luminosities at large radii, and large luminosities usually require convection. If accretion of matter onto the forming star may be neglected (and this is not really true), the object follows a path on the Hertzsprung-Russell diagram along a path of effective temperature given by (7.161). At any given stage its effective temperature cannot fail below that value because of the arguments given above. Were this not true, then the star would enter into the "forbidden region" on the diagram at lower effective temperatures as shown long ago by Hayashi (1961). These paths are appropriately known as "Hayashi tracks" and are those taken by the T Tauri stars (see below). As the star continues to contract, however, its luminosity may decrease to the point where the deep interior ultimately becomes radiative and the foregoing discussion does not apply. To see what happens then, assume that the deep interior opacity is Kramers' and apply homology arguments. If the interior gas is ideal and there is still no thermonuclear energy generation, then a simple exercise yields the mass-luminosity–effective temperature relation (show this because it's easy stuff from Chapter 1)

$$\mathcal{L} \propto \mathcal{M}^{22/5} T_{\rm eff}^{4/5} \,. \tag{7.172}$$

The implication is that when the luminosity of the contracting star falls below a critical value, evolution proceeds along a track given by (7.172) to higher effective temperatures until, finally, interior temperatures reach the point of hydrogen ignition and the main sequence stage of evolution begins. Note that if the mass is *too* low, then the track given by (7.172) may lie below the intersection of the main sequence and the Hayashi track. If so, then hydrogen burning will commence at that intersection but the star will remain completely convection on the main sequence. If the star is even less massive than this, then hydrogen burning may never begin and the result is a brown dwarf. From such calculations, the minimum mass of a hydrogen-burning main sequence star is estimated to be close to 0.1 $\mathcal{M}_\odot$. Less massive objects result in brown dwarfs (see §2.2).

The essentials of the above results were summarized in Figure 2.2 where Pre–main sequence evolutionary tracks were shown on the Hertzsprung-Russell diagram for various mass stars. The direction of evolution is down the Hayashi track to lower luminosities until the equivalent of relation (7.172) is reached and then follows the march to the main sequence. Also shown in Figure 2.2 are the locations of observed T Tauri stars, which are now believed to be stars in the process of contracting to the main sequence. The heavy line that defines an upper envelope to where these stars appear implies that we have not told the whole story; namely, why are these T Tauri stars not seen above this "birthline" (Stahler 1988)? A major reason is that we have neglected the actual hydrodynamical processes of star formation from interstellar clouds. Among these processes is accretion of gas onto the forming star. This provides a high luminosity at the accretion surface but this is obscured by dust and gas. It is only after the accretion ends that the star is fully revealed below the birthline.

7.4 References and Suggested Readings

§7.1: The Equations of Stellar Structure

The subject of multiple solutions to the stellar structure equations has been reviewed by

▹ Hansen, C.J. 1978, *Ann. Rev. Astron. Ap.*, **16**, 15

in the context of secular stability.

§7.2: Polytropic Equations of State and Polytropes

The material on polytropes in this chapter is standard. We recommend

▹ Chandrasekhar, S. 1939, *Introduction to the Study of Stellar Structure* (New York: Dover)

for more details than you might ever want. We warn you that the mathematics can get a bit rough at times. Other references include §23.1 of

▷ Cox, J.P. 1968, *Principles of Stellar Structure* (New York: Gordon and Breach)

and Chapter 19 of

▷ Kippenhahn, R., and Weigert, A. 1990, *Stellar Structure and Evolution* (Berlin: Springer-Verlag).

Note that the symbols used for the polytropic variables in the last reference are not standard.

We strongly recommend that you have

▷ Press, W.H., Teukolsky, S.A., Flannery, W.T., Vetterling, W.T. 1992, *Numerical Recipes, The Art of Scientific Computing*, 2d ed. (Cambridge: Cambridge University Press)

on your bookshelf. A calculation poorly done is an abomination in this day of the high-speed computer (microcomputers included). Note that the earliest editions of *Numerical Recipes* have several mistakes in the `FORTRAN` computer programs accompanying the text. These have hopefully all been corrected. These programs are also available on floppy disks (as are "`C`" versions).

Again we recommend

▷ Hayashi, C., Hōshi, R., and Sugimoto, D. 1962, *Prog. Theo. Phys. Suppl.* of Japan, No. 22.

Their use of the U-V plane is a virtuoso performance. Earlier examples are discussed in the text by

▷ Schwarzschild, M. 1958, *Structure and Evolution of the Stars* (Princeton: Princeton University Press).

The most complete discussion of model making is still to be found in

▷ Kippenhahn, R., Weigert, A., and Hofmeister, E. 1967, *Meth. Comp. Phys.*, **7**, 53.

The original "Henyey method" appears in

▷ Henyey, L.G., Forbes, J.E., and Gould, N.L. 1964, *Ap.J*, **139**, 306.

The original LANL (then called LASL) method for one-dimensional hydrodynamics is discussed in

▷ Cox, A.N., Brownlee, R.R., and Eilers, D.D. 1966, *Ap.J.*, **144**, 1024.

There are newer techniques and some of these are reviewed in Press et al. (1992).

The most lucid introduction to shock phenomena we know of is the first chapter of

▷ Zel'dovich, Ya.B., and Raizer, Yu.P. 1966, *Physics of Shock Waves and High Temperature Hydrodynamic Phenomena*, in two volumes, Eds. W.D. Hayes and R.F. Probstein (New York: Academic Press).

These two volumes contain much of interest for the astrophysicist. Chapter 1 has been reprinted separately in 1966 as *Elements of Gasdynamics and the Classical Theory of Shock Waves* (also from Academic Press). There are more recent texts containing numerical methods for dealing with shock waves but you cannot go wrong with

▹ Richtmyer, R.D., and Morton, K.W. 1967, *Difference Methods for Initial Value Problems* 2d ed. (New York: WileyInterscience.)

Sir Arthur Eddington's name appears in many astronomical contexts during the earlier years of the 20th century ranging from the gravitational bending of light to stellar interiors and variable stars. If you wish to find out how well science can be explained read through

▹ Eddington, A.S. 1926, *Internal Constitution of the Stars* (Cambridge: Cambridge University Press).

It is also available in a 1959 Dover edition.

The original work on constructing zero temperature white dwarfs appears in

▹ Chandrasekhar, S. 1939, *Introduction to the Study of Stellar Structure* (New York: Dover)

where the notion of polytropes is extended to include the combination of nonrelativistic and relativistic degenerate equations of state.

§7.3: The Approach to Real Models

The primary reference for stellar atmospheres is

▹ Mihalas, D. 1978, *Stellar Atmospheres*, 2nd ed. (San Francisco: Freeman).

Various applications of the Eddington limit are discussed in

▹ Shapiro, S.L., and Teukolsky, S.A. 1983, *Black Holes, White Dwarfs, and Neutron Stars* (New York: Wiley Interscience).

The use of interpolation among atmospheres in making stellar models is reviewed in

▹ Kippenhahn, R., Weigert, A., and Hofmeister, E. 1967, *Meth. Comp. Phys.*, **7**, 53.

Chapter 20 of

▹ Cox, J.P. 1968, *Principles of Stellar Structure* (New York: Gordon and Breach)

has a more complete discussion of envelope construction than we have attempted. Included are fairly realistic convective envelopes and the use of the dimensionless Schwarzschild variables.

As in Chapter 2 we recommend the review by

▹ Stahler, S.W. 1988, *Pub. Astron. Soc. Pacific.*, **100**, 1474

for a discussion of pre–main sequence evolution. See also the original paper of

▹ Hayashi, C. 1961, *Pub. Astron. Soc. Japan*, **13**, 450

for Hayashi tracks. For some thermodynamic conversions see

▹ Landau, L.D., and Lifshitz, E.M. 1958, *Statistical Physics* (London: Pergamon).

8

Structure and Evolution of the Sun

Truly the light is sweet,
and a pleasant thing it is
for the eyes to behold the sun.
— Ecclesiastes 11:7

On the other hand,

I hate the beach. I hate the sun.
I'm pale and I'm red-headed.
I don't tan—I stroke.
— Woody Allen (*Play it again Sam*, 1972)

This chapter will discuss what is known about the interior of the present-day sun, as *the* prototype star, and how it has evolved to this state. We shall emphasize not only what strides have been made but also what uncertainties remain. Note that our proximity to the sun and the level of detail visible from our viewpoint on earth means that we see far more than we completely understand at present. Observations over the past several decades have revealed a rich variety of surface phenomena associated with magnetic fields and their almost cyclical behavior; magnetic fields, like convection, are difficult to include in stellar models. The general pattern of surface rotation has the equator rotating faster than higher latitude regions and interactions between this differential rotation, modulations in the magnetic field, and subsurface convection, remain particularly thorny subjects and active areas of research. Neutrinos have been observed emanating from the deep interior but at a rate less than that predicted from standard models. The sun is a variable star—albeit variable on only a low-amplitude scale—and we have to see what this can tell us.

So while the sun is our best observed star, the uncertainties listed above and phenomena we do not fully understand are multiplied many-fold when extrapolated to other far more distant objects. In light of this we shall discuss some of these issues here.

8.1 The Sun as the Prototype Star

The mass of the sun is determined from measurements of the dynamics of planets and natural or artificial satellites in the solar system and it is known far more accurately than for any other stellar object, with the possible exception of some binary pulsars. A currently accepted Figure is $\mathcal{M}_\odot = (1.9891 \pm 0.0004) \times 10^{33}$ g (derived from Cohen and Taylor 1987). The solar luminosity is known to almost the same precision although variations of up to 0.5% have been reported. These may be due to variations during its magnetic cycle or other short-term effects such as solar flares and the like. A more probable variation due to the solar magnetic cycle is only about 0.07% around a mean value for the solar flux of $(1.368 \pm 0.001) \times 10^6$ erg cm^{-2} s^{-1} at one Astronomical Unit (Willson et al. 1986, and see Newkirk 1983). This yields $\mathcal{L}_\odot = (3.847 \pm 0.003) \times 10^{33}$ erg s^{-1}. Detection of a secular change in luminosity due to evolution is not possible at this time. The solar radius appears to be as stable in size as the luminosity and a value of 6.96×10^{10} cm is currently accepted for the radius at the optical photosphere.

The composition of the sun is not directly observable except at the photosphere, and even there it requires theoretical interpretation of spectral features or solar wind abundances with attendant uncertainties. If no mixing has occurred to change the surface composition of the sun during its evolution, then it is the same as that of the material from which the sun was formed in the first place (with the possible exception of some very reactive nuclei such as deuterium and lithium). It appears that the ratio of the mass fraction of heavy elements (Z, in the nomenclature of Chapter 1) to hydrogen (X) is $Z/X = 0.02739$–0.02765 to quote from Aller (1986) (the lower figure) and Grevesse (1984) (the upper figure). Not so well determined are the individual values of either X or Y (the helium mass fraction) but it is known that $Y \approx 0.25$. One way to establish the value of Y relies on constructing evolutionary sequences for the sun and then matching the present-day luminosity and radius to an estimated solar age (see below). Given Z/X, the value of Y that gives the best match is then the adopted mass fraction of helium. Note, however, that the validity of this procedure is no better than the input physics characterizing the opacity, nuclear reaction rates, and other processes such as convection. Granting this caveat, Bahcall and Ulrich (1988), for example, find $Y = 0.271$ using $Z/X = 0.02765$. This is then one estimate for the primordial helium composition of the sun. Since $X+Y+Z=1$, this yields $X=0.7094$ and $Z=0.0196$ where the number of significant digits quoted will most probably turn out to be illusory.

The individual "mix" of heavy elements is also important because of its effect on opacities and, to some extent, the operation of the CNO cycles (see Chapter 5). Given in Table 8.1 are the Grevesse (1984) fractional abundances of the most abundant heavy elements, where Z_i is the nuclear charge for an element and n_i is the number density (in cm^{-3}) normalized to a total of unity ($\sum_i n_i = 1$). (Note that these are not isotopic abundances.) We see that carbon and oxygen are, by far, the most abundant elements at the sun's surface after hydrogen and helium.

TABLE 8.1. Solar heavy element abundances

	Z_i	n_i		Z_i	n_i		Z_i	n_i
C	6	0.29661	Al	13	0.00179	Ca	20	0.00139
N	7	0.05918	Si	14	0.02149	Ti	22	0.00006
O	8	0.49226	P	15	0.00017	Cr	24	0.00028
Ne	10	0.06056	S	16	0.00982	Mn	25	0.00017
Na	11	0.00129	Cl	17	0.00019	Fe	26	0.02833
Mg	12	0.02302	Ar	18	0.00230	Ni	28	0.00108

The age of the sun can only be established from radiometric dating of terrestrial rocks (plus some geological estimates of melting and cooling times), lunar material, and meteorites. The time of condensation of solar matter is placed at somewhat less than five billion years ago. If the sun was formed on the main sequence before the planets and other material, then a best estimate for the present age of the sun is $(4.5 \pm 0.1) \times 10^9$ years. Stricter error bars are quoted by some—as in Guenther (1989)—but, at least at present, a possible error of 10^8 years is not at all serious for evolutionary studies.

The significance of the sun for stellar structure and evolution is clear: it is the primary proving ground because we know its mass, luminosity, radius, effective temperature, initial composition, and age. If model-building procedures fail to reproduce these solar properties then all other studies using these procedures are in grave danger of being just plain wrong.

8.2 From the ZAMS to the Present

This section will review the structure and evolution of the sun from the zero-age main sequence (ZAMS) to the present in light of material developed in earlier chapters. After this is completed we shall go on to discuss how well our models compare to the real sun when it is looked at closely.

8.2.1 The Sun on the ZAMS

Constructing a consistent homogeneous ZAMS model for the sun is like doing stellar evolution in reverse; the model must be such that when evolved forward in time it yields the present-day sun at its present age. The givens for our ZAMS sun are its mass and composition (see §§2.2 and 7.1). Rotation and magnetic fields are assumed to have no effect on the evolution. Furthermore, mass loss is almost universally neglected for the main sequence stage of solar evolution because the present mass loss rate of $\dot{\mathcal{M}} \approx 10^{-14}$ $\mathcal{M}_\odot$ yr^{-1} (Cassinelli and MacGregor 1986) amounts to a small fraction of the total mass even when integrated over the main sequence lifetime of 10^{10} years. If these effects are ignored, then current practice recognizes the following parameters which, aside from some practical limits, are varied in solar ZAMS modeling until 4.5×10^9 years of evolution yields the present-day sun. These are:

1. The helium content of the surface layers is not precisely determined by observation but it must be around $Y \approx 0.25$. The exact value (as reflecting the original solar content) affects the structure through the mean molecular weight of the mixture, the opacity, and the behavior of convection zones. If, as discussed previously, the metal content relative to hydrogen is fixed by observation, then Y is varied as a parameter affecting the overall composition.
2. The method of treating convective transport must be decided at the outset. As we discussed in Chapter 5, the majority of model builders choose some version of the mixing length theory for this purpose because of computational simplicity. The results to be reported here do this also. There are, however, undetermined parameters in any version of the MLT. Foremost among these is the mixing length, ℓ. If the pressure scale height, λ_P, is taken as a measure of the mixing length, then the mixing length parameter, $\alpha = \ell/\lambda_P$, of (5.69) is the usual free parameter. From experience with both general and solar modeling, α should be close to unity. Note, however, that a single value of α is used everywhere and for all time. There is no guarantee that this is reasonable even in the context of the MLT.

The procedure is then to choose various combinations of Y and α in different ZAMS models, evolve them to the solar age, and settle on the combination that gives the correct radius and luminosity.

The results quoted here are either abstracted from the Yale models of Guenther, Jaffe, and Demarque (1989) and Guenther et al. (1992) or from calculations using similar codes and input physics. These models are representative of "standard models," which use just the sort of physics and techniques outlined in this text. For example, the Yale models use the Anders and Grevesse (1989) mix of metals with Los Alamos opacities and auxiliary tables (§4.6) and an Eddington gray atmosphere (§7.3.2). The latter is computationally efficient and perfectly adequate for general studies although, for some purposes (e.g., solar oscillations), a real atmosphere should be used. To reproduce the present-day sun, a helium mass fraction of $Y = 0.288$ and mixing length to pressure scale height of $\alpha = 1.2$ are required. The version of the MLT used in these calculations is that reviewed in our §5.1.7 and, more fully, in Cox (1968, Chapter 14). An elapsed age of 4.5×10^9 years for the present-day sun is assumed.

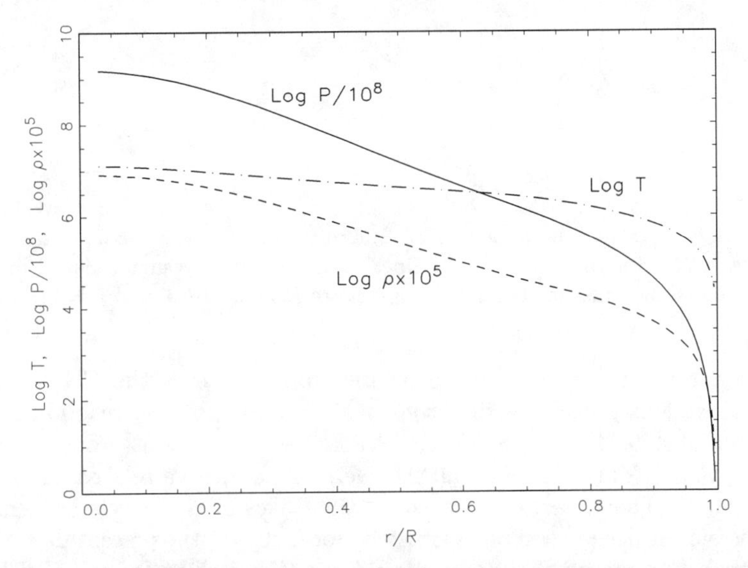

FIGURE 8.1. Shown are the runs of pressure, temperature, and density for a model of the zero-age sun. Note that the pressure has been multiplied by 10^{-8} and the density by 10^5. The abscissa is the relative radius $r/\mathcal{R}$ where $\mathcal{R} = 0.886\mathcal{R}_\odot$.

Figures 8.1 and 8.2 show some results for the ZAMS sun. The runs of pressure, temperature, and density versus radius in Figure 8.1 are smooth and show the rapid decrease in these variables to the surface. We have already remarked (in §7.2.7) that the Eddington standard model reproduces the behavior of these variables through most of the model remarkably well given, say, the average density.

The solid lines in Figure 8.2 illustrate the run of relative luminosity $\mathcal{L}_r/\mathcal{L}$ (with $\mathcal{L} = 0.725\mathcal{L}_\odot$) and $\mathcal{M}_r/\mathcal{M}_\odot$ versus relative radius for the ZAMS sun. (The dotted lines refer to the present–day sun). The total radius is $\mathcal{R} = 0.886\mathcal{R}_\odot$. We note immediately that the sun on the ZAMS, at 4.5×10^9 years in the past, was some 12% smaller and 25% less luminous than it is now. The luminosity shows the characteristic rapid rise at small radius and this is associated with the temperature sensitivity of the nuclear

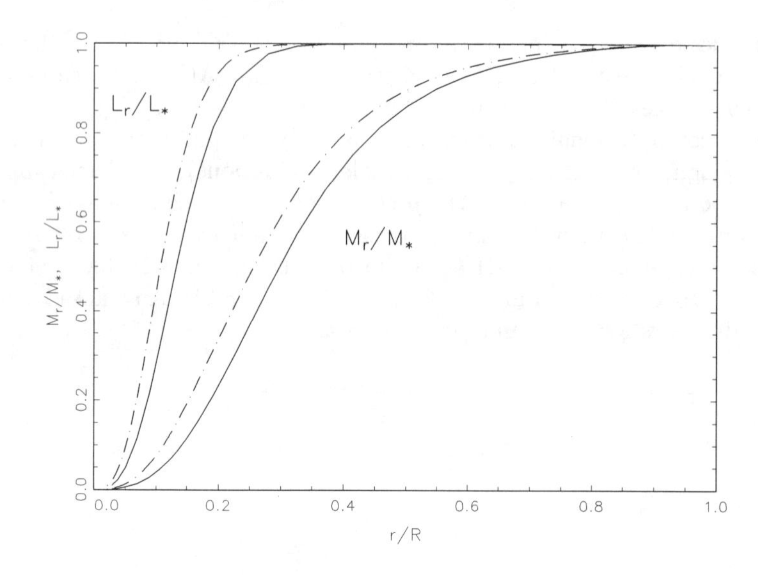

FIGURE 8.2. Relative luminosity, $\mathcal{L}_r/\mathcal{L}_\star$, and mass, $\mathcal{M}_r/\mathcal{M}_\star$, are plotted versus relative radius for the ZAMS (solid lines) and present–day suns (dashed lines). Total radius and mass for the ZAMS model are $\mathcal{R} = 0.886\mathcal{R}_\odot$ and $\mathcal{L} = 0.725\mathcal{L}_\odot$.

reaction rate. The proton-proton chains dominate over the CNO cycles and, as you may easily verify from our estimates for the reaction rates in equations (6.74) and (6.75), $\varepsilon_{\rm CNO}/\varepsilon$ is less than 1% at model center. Radiative diffusion transports all the energy flux out to a relative radius of $r/\mathcal{R} \approx 0.73$. Thereafter, efficient convection takes over to levels just below the visible surface (as will be illustrated more fully for the present-day sun). Although this means that convection is most important for some 30% of the total radius, the corresponding figure in mass is only about 3% because of the greatly lower densities in the outer layers.

How does this structure change as nuclear transmutations take place and the sun evolves from the zero age main sequence?

8.2.2 Evolution from the ZAMS

The most obvious changes in the sun as it evolves from the ZAMS to its present state are increases in both radius and luminosity. This is a characteristic of all standard calculations. To estimate the magnitude of the increase in luminosity, for example, we shall use simple dimensional arguments plus the fact that hydrogen burning to helium means an increase in mean molecular weight. In doing so, we follow the discussion of Endal (1981). Almost all of what we need can be found in the first chapter.

Recall the virial theorem analysis (§1.3.4), which yielded the relation

$$T \propto \mu \mathcal{M}^{2/3} \rho^{1/3} \tag{8.1}$$

for an ideal gas star in hydrostatic equilibrium. (Radiation pressure contributes much less than 1% to the solar pressure even at the center and may be safely ignored.) Here, as usual, μ is the mean molecular weight. If, furthermore, we assume that radiative diffusion controls the energy flow, then

$$\mathcal{L} \propto \frac{\mathcal{R}T^4}{\kappa\rho} \tag{8.2}$$

using (1.57). If Kramers' is the dominant opacity with $\kappa = \kappa_0 \rho T^{-3.5}$, then elimination of T and application of the mass equation $\mathcal{R} \propto (\mathcal{M}/\rho)^{1/3}$ yields

$$\mathcal{L} \propto \frac{\mathcal{M}^{5.33} \rho^{0.117} \mu^{7.5}}{\kappa_0}. \tag{8.3}$$

This expression can be used to calculate an estimate for the change in luminosity with time as composition changes.

Because of the small exponent of density (and, from knowing beforehand how relatively little the radius will change) we neglect the term $\rho^{0.117}$. From estimates of κ_0 (as in §4.4.3) we know that κ_0 does not vary strongly with either X or Y and so we neglect it also. Equation (8.3) is then rewritten in time-dependent form relating changes in $\mathcal{L}$ and μ from time $t = 0$ to some arbitrary time t:

$$\frac{\mathcal{L}(t)}{\mathcal{L}(0)} = \left[\frac{\mu(t)}{\mu(0)}\right]^{7.5}. \tag{8.4}$$

The task is now to find how μ varies with time.

If the bulk of the stellar interior is assumed to be completely ionized and the metal content is small compared to hydrogen and helium, then equation (1.52) is appropriate with

$$\mu(t) = \frac{4}{3 + 5X(t)}. \tag{8.5}$$

But we know something about how X changes with time. Because hydrogen burning releases approximately $Q = 6 \times 10^{18}$ ergs for every gram of hydrogen converted to helium, the instantaneous rate of change of a spatially averaged X is

$$\frac{dX(t)}{dt} = -\frac{\mathcal{L}(t)}{\mathcal{M}Q}. \tag{8.6}$$

Taking the time derivative of μ then yields

$$\frac{d\mu(t)}{dt} = -\frac{5}{4}\mu^2(t)\frac{dX}{dt} = \frac{5}{4}\mu^2(t)\frac{\mathcal{L}(t)}{\mathcal{M}Q}. \tag{8.7}$$

We now differentiate (8.4), substitute (8.7), get rid of $\mu(t)$ using (8.4), and find

$$\frac{d\mathcal{L}(t)}{dt} = \frac{75}{8}\frac{\mu(0)}{\mathcal{M}Q}\frac{\mathcal{L}^{1+17/15}(t)}{\mathcal{L}^{-1+17/15}(0)} \tag{8.8}$$

with solution

$$\mathcal{L}(t) = \mathcal{L}(0)\left[1 - \frac{85}{8}\frac{\mu(0)\mathcal{L}(0)}{\mathcal{M}Q}t\right]^{-15/17} . \tag{8.9}$$

Putting in numbers by expressing luminosities in units of $\mathcal{L}_\odot$ and introducing the present solar age $t_\odot$ of 4.5×10^9 years and letting $\mu(0) \approx 0.6$ then gives

$$\frac{\mathcal{L}(t)}{\mathcal{L}_\odot} \approx \frac{\mathcal{L}(0)}{\mathcal{L}_\odot}\left[1 - 0.3\frac{\mathcal{L}(0)}{\mathcal{L}_\odot}\frac{t}{t_\odot}\right]^{-15/17} . \tag{8.10}$$

If $t = t_\odot$ (i.e., $\mathcal{L}(t) = \mathcal{L}_\odot$), then the luminosity on the ZAMS must have been $\mathcal{L}(0) \approx 0.79\mathcal{L}_\odot$ from the solution of (8.10). The models we have quoted give an agreeably close value of 0.73.

The above result is interesting from not only our stellar evolution perspective, but it bears on how life must have evolved on earth. The earliest microorganisms appear in the fossil record about 3.5×10^9 years ago. At that time, by application of the above with an adjustment given by the evolutionary models, the sun was nearly 25% less luminous than now. Because descendants of some of those same microorganisms are alive today in essentially unchanged form, there must be some explanation for how the earlier life forms could have survived and propagated with a significantly lower solar constant. The answer probably lies in the evolution of the earth's atmosphere which, as fascinating as that topic may be, we shall have to pass by.

Another way to look at the above is to consider what happens as the mean molecular weight increases with time in the hydrogen-burning core of a hydrostatic star. If $P \propto \rho T/\mu$, then an increase in μ without a corresponding increase in the product ρT would be accompanied by a decrease in pressure. But since the core must still support the unchanged mass layers above it, this situation would lead to an imbalance of forces and hydrostatic equilibrium would be impossible to maintain. The result would then be a compression of the core with a corresponding increase in density. This process would take place very rapidly compared to nuclear time scales because we know that the dynamic readjustment time of §1.3.3, $t_{\rm dyn}$, is only about an hour for the sun. The conclusion is that ρT must increase. In particular, ρ must increase but so should T by virtue of the virial result (8.1). An increase in T then implies an increase in the energy generation rate (to the fourth power of T for the proton-proton reaction) and thus the overall rate of power output increases also.

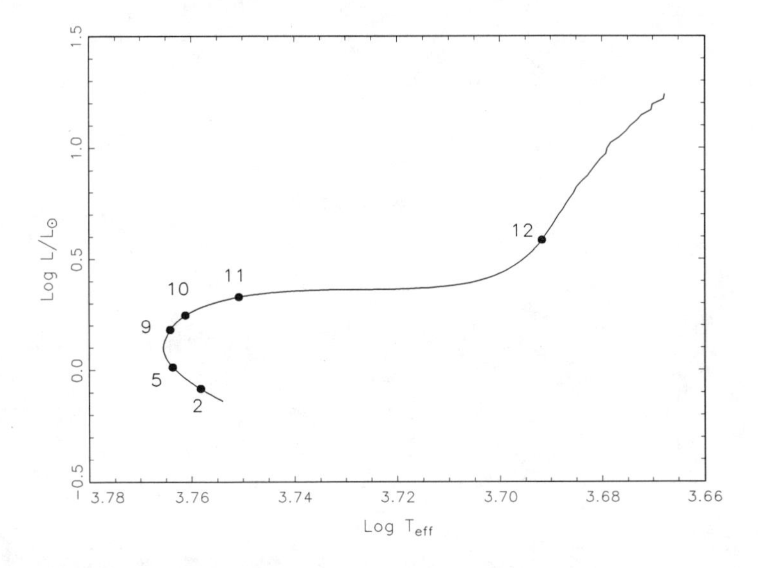

FIGURE 8.3. A solar model evolution track in the Hertzsprung–Russell diagram from the ZAMS to the red giant stage. Elapsed evolutionary time from the ZAMS is indicated by the filled circles where the units are Gyr.

8.2.3 The Present-Day Sun

Figure 8.3 shows the evolutionary track of a model of the sun on an H-R diagram from the ZAMS to the point where it is clearly a red giant. Elapsed time is indicated by the labeled circles in Gyr (10^9 years). The present-day sun, even after some 4.5 Gyr of evolution, is still very close in $\mathcal{L}$, $\mathcal{R}$, and T_{eff} to its original ZAMS position. The inner core, however, has changed substantially.

Figure 8.2 showed the current run of relative luminosity and mass as a function of relative radius (as dotted lines). The figure shows that $\mathcal{L}_r/\mathcal{L}_\odot$ and $\mathcal{M}_r/\mathcal{M}_\odot$ for the present-day sun rise more steeply than for the ZAMS sun. In the case of the luminosity, the reason for this is that increased central temperatures have intensified the energy production and, from the energy equation (1.54), the luminosity gradient must steepen accordingly. (The contribution from gravitational energy sources due to contraction is always very small during the initial stages of evolution off the main sequence.) Contraction and the implied increase in density account for the steeper gradient in mass.

The run of pressure, temperature, and density versus relative radius at an elapsed age of 4.5 Gyr is very similar to that of Figure 8.1 for the ZAMS except that the central values are now increased to $T_c = 1.53 \times 10^7$ K, $P_c = 2.26 \times 10^{17}$ dyne cm^{-2}, and $\rho_c = 146$ g cm^{-3} as a consequence of the contraction of the inner regions. The burning of hydrogen to helium

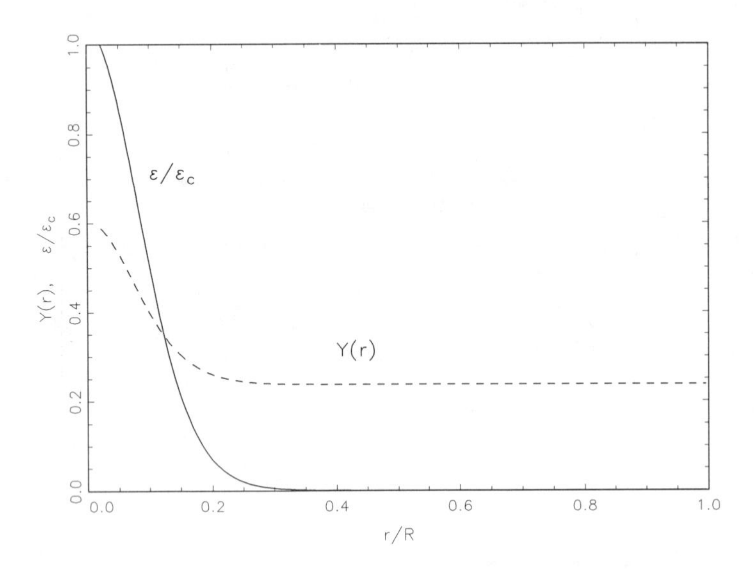

FIGURE 8.4. Because of hydrogen-burning in the radiative solar core, the mass fraction, Y, of helium in the present-day sun has increased while the mixture becomes less hydrogen-rich. Also shown is the energy generation rate $\varepsilon(r)$ as compared to its central value ε_c.

in the still radiative core of the sun has depleted the former and this is shown in Figure 8.4. No longer are the mass fraction profiles, $X(r)$ and $Y(r)$, flat—as was initially assumed for the ZAMS—and Y_c has increased by nearly a factor of two, which is made up for by a corresponding decrease in X_c. Also shown in the figure is the ratio of the energy generation rate $\varepsilon(r)$ to its central value $\varepsilon_c = 16.2$ erg g^{-1} s^{-1} (an increase of 17% from the ZAMS). The contribution from the CNO cycles is now 7% and will continue to rise as central temperatures get hotter because of the high temperature sensitivity of the CNO cycles.

The convection zone of the sun is moderately extensive (but not compared to lower-mass main sequence stars) and occupies the outer 30% of the radius (but only 2% of the mass). At these high levels in the star the total luminosity is constant because little or no energy is being generated there. Thus Figure 8.5, which shows the run of $\mathcal{L}_r/\mathcal{L}$ and convective to total luminosity ($\mathcal{L}_{r,\text{conv}}/\mathcal{L}_{\text{tot}}$), indicates that a major fraction of the solar energy flux is carried by convection before the luminous power is finally radiated to space at the photosphere. The detailed model results show that the convection is nearly adiabatic through almost all of the convection zone except for the very bottom and top of the zone. Phrased another way, the gradients ∇ and ∇_{ad} of Chapter 5 are very nearly equal.

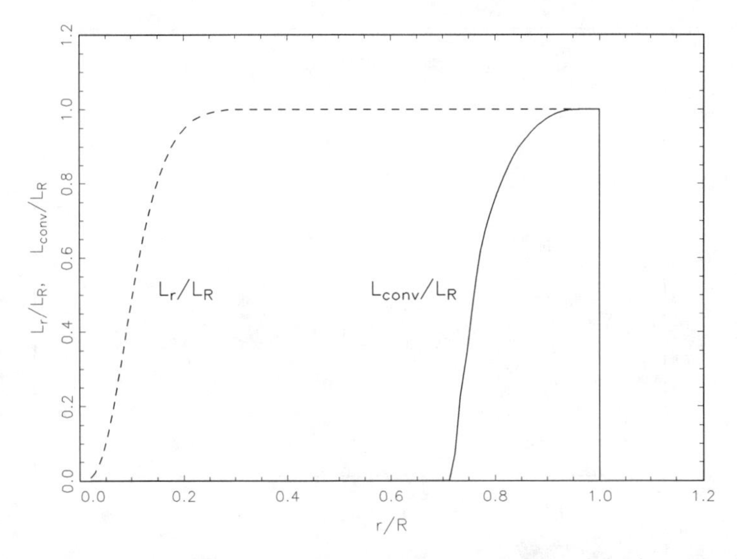

FIGURE 8.5. The ratios $\mathcal{L}_{\text{tot}}/\mathcal{L}_R$ (dashed line) and $\mathcal{L}_{\text{conv}}/\mathcal{L}_R$ (solid line) are shown as a function of relative radius in the present-day sun. $\mathcal{L}_R$ is equal to the luminosity at the surface. The convective luminosity drops to zero just as the photosphere is reached at $r = R$.

Another way to show the extent of the convection zone is to examine the square of the Brunt-Väisälä frequency

$$N^2 = g\left[\frac{1}{\Gamma_1 P}\frac{dP}{dr} - \frac{1}{\rho}\frac{d\rho}{dr}\right] = -\frac{\chi_T}{\chi_\rho}\left(\nabla - \nabla_{\text{ad}}\right)\frac{g}{\lambda_{\text{P}}}. \tag{8.11}$$

which was originally given as equation (5.35). Remember that in a radiative zone N is the frequency of oscillation for a fluid blob that has been displaced from its equilibrium position when buoyancy is the restoring force (§5.1.4). In convectively unstable regions, where $\nabla > \nabla_{\text{ad}}$, N^2 is negative and $1/|N|$ measures the e-folding time for increases in velocity and temperature perturbations. All the information necessary to construct N^2 as a function of radius may be found from the solar model and this is shown in Figure 8.6. Here, the abscissa is $\log\left(1 - r/\mathcal{R}_\odot\right)$, which is a scale that heavily emphasizes the outer regions. Note that the stellar center is now at the right end of the figure. The ordinate is in the units of frequency^2 (Hz^2) to accommodate the other variable in the figure (the "Lamb frequency"), which will be used later when discussing solar oscillations, as will N^2. With this abscissa the bottom of the solar convection zone is at $\log\left(1 - r/\mathcal{R}_\odot\right) \approx - - 0.55$ while the top is at a value of approximately -3.7. The latter corresponds to a depth of only a couple of hundred kilometers below the photosphere—the exact value will depend on just how the atmosphere is constructed—and is located at a height where temperatures are sufficiently low that hydrogen recombination has finally been completed (as discussed in §3.4).

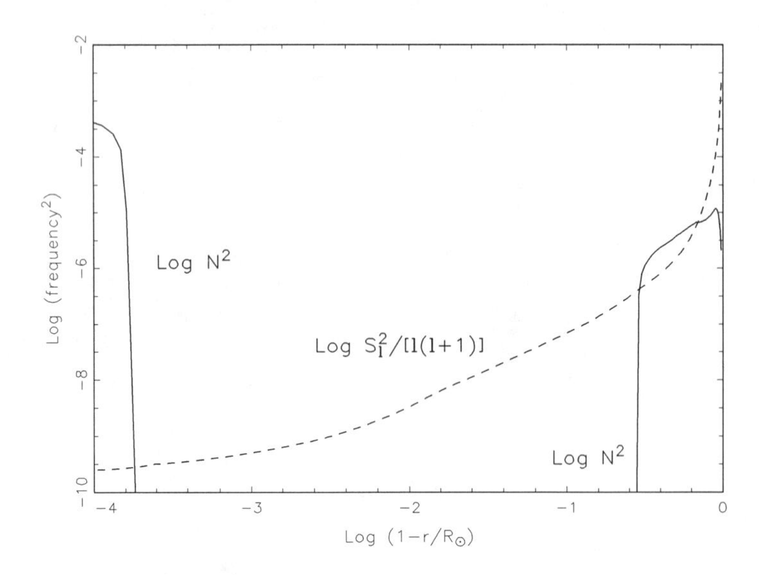

FIGURE 8.6. Shown are N^2 and $S_\ell^2/[\ell(\ell+1)]$ in a standard solar model. The abscissa is $\log(1 - r/\mathcal{R}_\odot)$, which places the solar center at the right-hand edge of the figure. S_ℓ^2 is a quantity to be discussed in Chapter 10.

From all external appearances, the standard model reproduces the sun as we see it. It has been constructed so that the age, surface composition, luminosity, radius, and effective temperature match the object in our daytime sky. This is a significant achievement but, aside from the apparent consistency of thinking that we have the inside right because the outside looks right, are there other observations that probe beneath the visible surface that can reinforce our optimism or poke holes in it?

8.3 The Solar Neutrino "Problem"

Since 1968, in the Homestake Mine, Kellogg, South Dakota, at a depth of nearly 1500 meters, Raymond Davis Jr. and his collaborators have been detecting electron neutrinos emitted from deep within the sun. Using some 600 tons of the cleaning fluid compound perchloroethylene (C_2Cl_4)—and a great deal of ingenuity—they count the rate at which the radioactive isotope ^{37}Ar (half-life of 35 days) is produced by the reaction

$$\nu_e + {}^{37}\mathrm{Cl} \longrightarrow \mathrm{e}^- + {}^{37}\mathrm{Ar}. \tag{8.12}$$

This requires an incoming neutrino with an energy exceeding 0.81 MeV for the reaction to proceed. (For a review of the history and results of this ^{37}Cl experiment see Rowley et al 1985, and references therein.) Besides a small amount accounted for from extra-solar sources, these incoming neutrinos are produced from hydrogen-burning reactions in the sun. Ordinary material is remarkably transparent to neutrinos whose typical absorption cross sections are in the 10^{-44} cm^2 range for energies of order MeV. Thus, using this figure, a typical mean free path for neutrino absorption within the sun is $\lambda = 1/\sigma\langle n\rangle \sim 10^9 \mathcal{R}_\odot$ (give or take a couple orders of magnitude) where $\langle n\rangle$ is the average number density of particles. Therefore, any neutrinos produced in the sun escape easily.[1]

The particular reactions responsible for producing neutrinos in the solar interior may be inferred (aside from two reactions to be given shortly) from Tables 6.1 and 6.2 listing the pp chains and CNO cycles. These are

$$\begin{aligned}
{}^1\mathrm{H} + {}^1\mathrm{H} &\longrightarrow {}^2\mathrm{H} + \mathrm{e}^+ + \nu_\mathrm{e} \\
{}^7\mathrm{Be} + \mathrm{e}^- &\longrightarrow {}^7\mathrm{Li} + \nu_\mathrm{e}(+\gamma) \\
{}^8\mathrm{B} &\longrightarrow {}^8\mathrm{Be} + \mathrm{e}^+ + \nu_\mathrm{e} \\
&\cdots \\
{}^{13}\mathrm{N} &\longrightarrow {}^{13}\mathrm{C} + \mathrm{e}^+ + \nu_\mathrm{e} \\
{}^{15}\mathrm{O} &\longrightarrow {}^{15}\mathrm{N} + \mathrm{e}^+ + \nu_\mathrm{e} \\
{}^{17}\mathrm{F} &\longrightarrow {}^{17}\mathrm{O} + \mathrm{e}^+ + \nu_\mathrm{e}.
\end{aligned}$$

The first reaction is the pp reaction from the PP–I chain and the next two are from the PP–II and PP–III chains. The neutrinos from the pp reaction are emitted in a continuum up to an endpoint energy of 0.42 MeV. These cannot be detected by the Davis experiment because the endpoint energy is less than the threshold energy for the neutrino capture on ^{37}Cl. The decay of ^{8}B also yields a continuum of neutrinos with energies up to 15 MeV and these are accessible to the experiment. The ^{7}Be reaction is actually two reactions and the final neutrino energy depends on the final nuclear state of ^{7}Li. The result is a monoenergetic neutrino at an energy of either 0.862 MeV or 0.384 MeV. The first decay is more probable and occurs 90% of the time. The three CNO reactions yield continuum neutrinos with endpoint energies of 1.20, 1.73, and 1.74 MeV respectively. Two additional pp–chain reactions—which we have not discussed—also yield neutrinos. These are the three-body reactions "pep" [^{1}H(pe$^-$, ν_e)^{2}H] and "hep" [^{3}He(pe$^-$, ν_e)^{4}He]. These rare reactions contribute almost nothing to energy generation in the sun but pep does emit line neutrinos at 1.44 MeV while hep emits in a continuum up to 18.8 MeV.

We now examine what are the predicted fluxes of these neutrinos from a standard solar model and what should be the rate of detection by the chlorine experiment. It should be clear that the rate of neutrino emission by the above individual reactions must depend sensitively on temperature

---

[1]The best overview of the neutrino problem is to be found in Bahcall (1989). We strongly suggest you look through that material if you want a complete picture up to 1989. New experimental results seem to appear almost every month and anything we may add to Bahcall's discussion is bound to be out of date. So keep your eye on the recent literature.

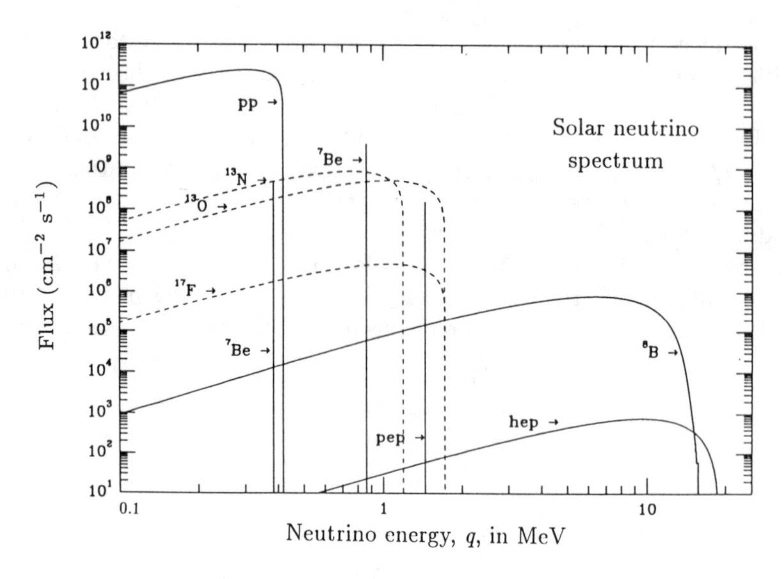

FIGURE 8.7. The energy spectrum of neutrinos predicted from a standard solar model. The solid (dashed) lines are pp–chain (CNO) reaction neutrinos. The units for continuum neutrinos are cm^{-2} s^{-1} MeV^{-1} and, for line neutrinos, cm^{-2} s^{-1}. From Bahcall (1989).

because temperature primarily determines the relative competition between PP–I through PP–III chains and the CNO cycles. Thus neutrino-detecting experiments potentially offer a unique probe into the solar interior and that is why so much effort has been expended in designing such experiments.

Figure 8.7 shows the predicted neutrino spectrum from a standard solar model (from Bahcall 1989, §1.4, see also Bahcall and Ulrich 1988, Bahcall et al. 1988). The units of the flux are cm^{-2} s^{-1} MeV^{-1} for continuum sources and cm^{-2} s^{-1} for line sources. All fluxes are calibrated so that they are those that should be seen at one astronomical unit from the sun. The flux from the more energetic ^{7}Be neutrino, which ^{37}Cl can see, is about 10^{10} neutrinos cm^{-2} s^{-1}. To find how many absorption reactions would take place per second per target on targets with absorption cross section $\sigma \sim 10^{-44}$ cm^2 we multiply σ by the flux to find 10^{-35} captures s^{-1} per target. This estimate is not entirely correct but the final result is not too far from what is observed by the Davis experiment. The convenient unit used in this business is the "solar neutrino unit," or SNU (pronounced "snoo"), and it is defined as 1 SNU $= 10^{-36}$ captures s^{-1} target^{-1}. We shall use these SNU units from now on.

The solar neutrino "problem" is the following. The neutrino flux observed at the Homestake Mine is 2.07 ± 0.3 (1σ error) SNU over the period 1970–1988 (see the above references). The flux predicted from various versions of standard models is 7.9 ± 2.4 SNU (an "effective" 3σ error) where errors cover estimates of independent uncertainties in nuclear reaction rates, opacities, model-building techniques, idiosyncrasies of various researchers, and practically all the other items discussed thus far in this text (see Bahcall 1989). Of the 7.9 SNU, 6.1 SNU is contributed by the ^{8}B decay and 1.1 SNU by the ^{7}Be electron capture. The theoretical prediction is well outside the experimental error bars.

There are presently three other neutrino experiments in operation. The Kamiokande II natural–water experiment detects Cherenkov radiation from electrons scattered by neutrinos (with a 7.3 MeV threshold) whereas SAGE (Soviet–American Gallium Experiment) and GALLEX both rely on the reaction $^{71}\mathrm{Ga}(\nu_e, e^-)^{71}\mathrm{Ge}$ with a threshold of only 0.23 MeV, which is less than that needed to detect pp reaction neutrinos. Thus far *all* of these experiments detect neutrinos but at rates less than are predicted for the particular reactions involved.[2]

This has been the situation from about 1969 until the present writing (late 1992) and a consensus solution is not in sight. On one hand, we seem to have a good grasp on what makes the sun work but, when looked at closely using one probe that senses the deep interior, something is greatly amiss. No reasonable tinkering with standard models seems capable of removing the neutrino discrepancy.

There is insufficient space here to discuss at any length "nonstandard" solar models that try to address the neutrino problem. These include, for example, models that have inhomogeneous outer layers. If there were some way to cause elements to diffuse and separate in the sun during some earlier stage of evolution, then we may be fooled by the presently observed solar surface composition into thinking we know what the interior composition was when the sun formed. There are several mechanisms capable of causing elemental segregation and numerical experiments using these can "solve" the neutrino problem. There are two difficulties with this and other solutions however: the prescriptions tend to be *ad hoc* and are underconstrained leaving no observational way to test them. Many nonstandard models do run afoul of the observed solar oscillation spectrum, which we shall discuss in Chapter 10.

Elementary particle physics offers a solution that leaves the standard model intact. In some grand unified theories ("GUTs"), the electron neutrino is not massless—and this is not ruled out by experiment—but the mass must be small. If so, it is possible that an electron neutrino may be converted into a muon neutrino under the proper circumstances (or the other way around). Muon neutrinos, on the other hand, are not detectable by the chlorine and other experiments and, hence, even though the sun produced electron neutrinos at exactly the rate predicted by the standard

[2]See, for example, the articles: *Physics Today*, Oct 1990, **43**, No. 10, p. 17; *Science News*, June 1992, **141**, p. 388; *Physics Today*, August 1992, **45**, No. 8, p. 17.

model, some fraction of them would not be detected at the earth. This process of "neutrino oscillation" may be enhanced when neutrinos pass through a material medium (the "MSW" effect—see Bahcall et al. 1988, Bahcall and Bethe 1990). It is believed by many that this is the most promising line of inquiry to follow and may yet turn out to do the trick. Thus it may be that the solar neutrino problem tells us more about physical nature at its most fundamental level than it does about stellar astrophysics.

8.4 The Role of Rotation in Evolution

It is fortunate for makers of solar models that the sun rotates slowly. Were it to rotate rapidly, at speeds close to breaking up by centrifugal forces, the solar test bed for stellar evolution would almost be uncomputable using present technology. On the other hand, the sun does rotate and the effects are observable.

Rapidly rotating stars are not unusual but, on the main sequence at least, almost all of these are massive and bright. Less massive main sequence stars tend to be slow rotators. The observational evidence for this is presented in Figure 8.8 (from Kawaler 1987), which shows the average equatorial rotation velocities versus mass of a large sample of main sequence stars compiled by Fukuda (1982). The crosses denote complete samples, whereas data indicated by circles exclude Am stars, which tend to rotate anomalously slowly, and Be stars, which rotate rapidly but show peculiar emission features. (The Am stars are of spectral class A but with peculiar abundances and they are almost exclusively members of binary systems.) The nearly horizontal line labeled $\langle v \rangle = v_{\text{crit}}$ is the locus where stars of equatorial velocity v are at breakup; that is, where surface gravitational and centrifugal forces are equal with $v_{\text{crit}} = \sqrt{G\mathcal{M}/\mathcal{R}}$. It is evident that the more massive stars are rapid rotators with equatorial velocities only about a factor of two less than breakup.

Once masses drop below about $1.5\mathcal{M}_\odot$ ($\log \mathcal{M}/\mathcal{M}_\odot \approx 0.2$), however, average rotation velocities decrease precipitously. This mass corresponds to around spectral type F0 ($\mathcal{L} \approx 6\mathcal{L}_\odot$) on the main sequence and to the mass (and T_{eff}) below which stars have significant envelope convection (see Fig. 5.3). Thus it appears that convective envelopes and slow rotation are connected in some way.

Rotational angular momentum can be carried away from a star by a stellar wind, but how much angular momentum is lost depends on the distance at which the wind decouples from the stellar interior in terms of rotation. For a simple wind, this decoupling occurs near the photosphere. However, if the wind couples to the stellar magnetic field, it can be forced to corotate with the star well beyond the photosphere. The magnetic field is rooted in the stellar interior and as the corotating wind moves beyond the photosphere it gains angular momentum at the expense of the interior.

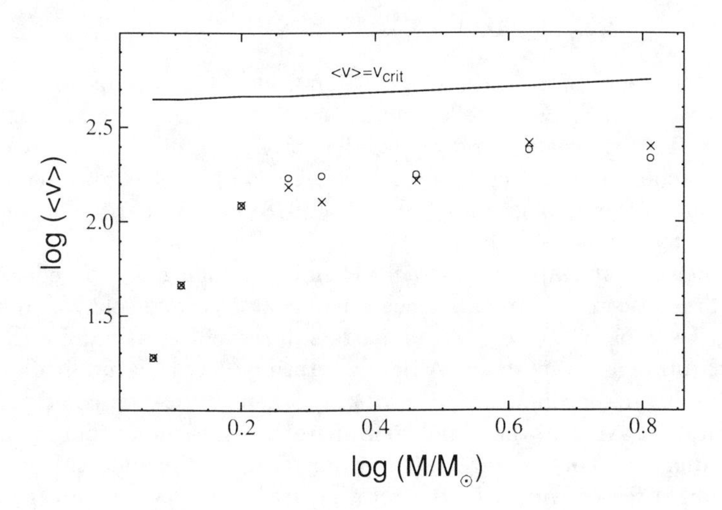

FIGURE 8.8. The observed average equatorial rotation velocity $\langle v \rangle$ is plotted as a function of mass for main sequence stars. Note the clear change in the behavior of $\langle v \rangle$ at $\log(\mathcal{M}/\mathcal{M}_\odot) \approx 0.2$ ($\mathcal{M} \approx 1.5\mathcal{M}_\odot$). Reproduced, with permission, from Kawaler (1987).

This gain of angular momentum by the wind is proportional to the square of the distance above the photosphere. Thus magnetic fields can greatly amplify the angular momentum loss a star experiences when coupled with a wind.

Since stars later than F0 have significant surface convection zones that can drive stellar winds and, through dynamo action, produce and modulate internal magnetic fields, these stars will experience much more angular momentum loss than their higher mass cousins. Support for this idea comes from the observation that the more massive main sequence stars generally have feeble surface convection zones and weak magnetic fields. Even though some have strong winds, most rotate near the average value of 1/3 v_{crit}. For more on the theory of magnetic braking, see the reviews by Mestel (1984), Collier Cameron et al. (1991), and Kawaler (1990). Observations of young stars that lend support to the theory are reviewed in Stauffer and Hartmann (1986) and Stauffer (1991).

The other side of this picture implies that the sun initially formed rotating much more rapidly than it does today. Could earlier rapid rotation have influenced evolution? Before we discuss this, it is worthwhile pointing out some of the subtle implications of rotation for stellar structure even when the rotation is slow.

8.4.1 von Zeipel's Paradox

Stellar rotation, of even the simplest kind, introduces complexities in the construction of realistic stellar models that are beyond our present capabilities. This is because rotation is inherently three dimensional. We shall only touch on one aspect of the subject. For an excellent introduction we recommend the monograph by Tassoul (1978) and, in particular for what follows, his Chapters 7 and 8.

Suppose we attempt to construct a chemically homogeneous stellar model in hydrostatic and thermal balance where heat is carried solely by diffusive radiative transfer. To complicate matters, however, let us require that the model rotate as a rigid body. A rigidly rotating body is naturally described in cylindrical coordinates $\mathbf{r} = \mathbf{r}(\varpi, \varphi, z)$ where the z-axis coincides with the rotation axis and the radial coordinate ϖ is measured from the z-axis. The equation of hydrostatic equilibrium must now include the effects of centrifugal forces which, in the rotating frame of the star, are given by the term $\rho\Omega^2\varpi\, \boldsymbol{e_\varpi}$ where $\boldsymbol{e_\varpi}$ is the radial unit vector. The potential corresponding to this force is $\Phi_{\text{cent}} = -\Omega^2\varpi^2/2$. The total potential, centrifugal plus gravitational is then

$$\Phi_{\text{eff}}(\varpi, z) = \Phi_{\text{grav}}(\varpi, z) - \frac{\Omega^2\varpi^2}{2}. \tag{8.13}$$

Note that we have assumed azimuthal symmetry (no φ dependence) as seems reasonable. The equation of hydrostatic equilibrium then becomes

$$\frac{1}{\rho}\nabla P = -\nabla\Phi_{\text{eff}} = -\nabla\Phi_{\text{grav}} + \Omega^2\varpi\, \boldsymbol{e_\varpi} = \mathbf{g}_{\text{eff}}. \tag{8.14}$$

The acceleration $\mathbf{g}_{\text{eff}}$ is the local effective gravity, which now includes centrifugal effects. In addition to the above we also need Poisson's equation, which is still given by

$$\nabla^2\Phi_{\text{grav}} = 4\pi G\rho \tag{8.15}$$

(as in §7.2.1). Note that only Φ_{grav} appears here.

The first application of these equations comes about from considering "level" surfaces on which Φ_{eff} is constant. The gradient $\nabla\Phi_{\text{eff}}$ evaluated on such a level surface is, of course, perpendicular to that surface. If $d\mathbf{r}$ is an infinitesimal unit of length lying tangent to the surface then, by usual arguments in the vector calculus, $d\mathbf{r} \bullet \nabla\Phi_{\text{eff}} = d\Phi_{\text{eff}} = 0$ (as it must). But this implies, by dotting $\mathbf{r}$ into (8.14), that $dP = 0$ on a level surface or, equivalently, pressure is a constant on such a surface. Thus level surfaces are also "isobaric" surfaces with $P = P(\Phi_{\text{eff}})$. This may be reversed to read $\Phi_{\text{eff}} = \Phi_{\text{eff}}(P)$ from which follows (from 8.14) $\rho^{-1} = d\Phi_{\text{eff}}/dP$. Density is then also constant on a level surface (making it an "isopycnic" surface).

If the equation of state of our chemically homogeneous model is that of an ideal gas with $P = \rho N_{\text{A}}kT/\mu$, then $T = T(\Phi_{\text{eff}})$. Thus far, the only quantity in sight that is *not* a constant on a level surface is $\mathbf{g}_{\text{eff}}$. It is normal to the surface but, because the level surfaces need not be the same distance apart in (ϖ, φ, z)-space, it may vary in magnitude over the surface. (Note that if Ω is zero, then the level surfaces are concentric spheres and we regain the constancy of gravity for fixed radius.)

The second constraint we have placed on the model is that of thermal balance. That is, we require $d\mathcal{L}_r/dr = 4\pi r^2 \rho\varepsilon$ of (1.54). A more general way of putting this is in terms of the vector flux $\boldsymbol{\mathcal{F}}$. Thus write

$$\nabla \bullet \boldsymbol{\mathcal{F}} = \rho\varepsilon. \tag{8.16}$$

The requirement of radiative diffusive transfer specifies how the flux is transported through a level surface so that, with a slight rearrangement of gradients,

$$\boldsymbol{\mathcal{F}} = -\frac{4ac}{3}\frac{T^3}{\kappa\rho}\frac{dT}{d\Phi_{\text{eff}}}\nabla\Phi_{\text{eff}}. \tag{8.17}$$

Note that T is regarded as a function of Φ_{eff} as above. Furthermore, ε and κ are also functions of Φ_{eff} because they contain only $\rho(\Phi_{\text{eff}})$ and temperature. Thus, and still following Tassoul (1978), we write

$$\boldsymbol{\mathcal{F}} = f(\Phi_{\text{eff}})\,\nabla\Phi_{\text{eff}} = -f(\Phi_{\text{eff}})\mathbf{g}_{\text{eff}} \tag{8.18}$$

where the function $f(\Phi_{\text{eff}})$ takes care of the Φ_{eff}–dependent terms in (8.17). Now take the divergence of this expression and use the fact that ∇f and $\nabla\Phi_{\text{eff}}$ are perpendicular to the level surface to find

$$\nabla \bullet \boldsymbol{\mathcal{F}} = \frac{df}{d\Phi_{\text{eff}}}\left(\frac{d\Phi_{\text{eff}}}{dn}\right)^2 + f\nabla^2\Phi_{\text{eff}} = \rho\varepsilon. \tag{8.19}$$

Here $\mathbf{n}$ is an outward unit normal to the level surface.

A nearly final result is obtained by realizing that $d\Phi_{\text{eff}}/dn$ is the same as $|\mathbf{g}_{\text{eff}}|$ so that, using Poisson's equation and doing some rearranging, you should find

$$\frac{df}{d\Phi_{\text{eff}}}\,\mathbf{g}_{\text{eff}}^2 + f(\Phi_{\text{eff}})\left[4\pi G\rho - 2\Omega^2\right] = \rho\varepsilon. \tag{8.20}$$

In order to satisfy all the constraints set on the model, this equation must be satisfied everywhere. At first glance, nothing seems peculiar because it is just the divergence of the radiative flux set equal to the energy generation rate (per unit volume) for thermal balance. However, if the structure is such that any two level surfaces are not spaced everywhere equidistantly apart (as they are if there were no rotation), then there is trouble. To see this, note that $\rho\varepsilon$ and the terms within the brackets depend only on Φ_{eff} *but* $\mathbf{g}_{\text{eff}}$ is *not* constant on arbitrary level surfaces. Therefore the coefficient $(df/d\Phi_{\text{eff}})$ must be zero because everything else is a constant. Thus drop this term from (8.20) and what remains is a relation between variables on a level surface and, in particular, a requirement on ε. Solving for ε yields

$$\varepsilon \propto \left[1 - \frac{\Omega^2}{2\pi G\rho}\right]. \tag{8.21}$$

But how can this be? The energy generation rate cannot be determined by how the star rotates!

The difficulty is that we have overly constrained the problem. This is von Zeipel's (1924) "paradox": namely, a uniformly rotating star (and the situation is more general than this) cannot be in steady-state radiative thermal equilibrium. Something must give.

The solution lies in relaxing the constraints. We refer you to Tassoul (1978, Chapter 8), where this is discussed. Briefly, it appears that either the angular rotation frequency Ω must depend on both ϖ and z, or fluid motions (e.g., "meridional" circulation) must take part in the transfer of heat. Consideration of these topics would take us into fascinating, very difficult, and not fully resolved territory. Our stance here is to back off and assume that rotation is sufficiently slow that—as a good approximation—many such effects can safely be ignored. But we do so at our peril with the realization that our description of the interiors of many stars is incomplete.

8.4.2 Rotational Mixing of Stellar Interiors

From the discussion above, we see that rotation should have little direct effect on overall the evolutionary changes in temperature, density, and pressure within the sun during the course of its evolution. However, some observed properties of the sun—and, in particular, the elemental abundance of rare species in the photosphere—may be telling us a significant story about the internal rotation of the sun and how it has changed with time. For example, the element ^{7}Li has an abundance in the solar photosphere that is a factor of 200 smaller than the abundance of ^{7}Li found in meteorites, terrestrial rocks, and younger stars. (See §6.3.1.) This would be an easy thing to explain if this ^{7}Li was destroyed by nuclear reactions within the solar convection zone. If so, then convective mixing would dilute the lithium abundance of the entire convection zone and lead to a depleted surface value compared to the primordial abundance. However, standard solar models indicate that the base of the solar convection zone never gets quite hot enough to burn ^{7}Li very much; the standard solar model depletes lithium to about 1/3 of its initial value. If the solar convection zone was deeper by about half a pressure scale height, then the observed level of ^{7}Li destruction could occur, but to do this would result in a solar model that is at odds with observations of other younger stars. Rotation provides us with a way out of this dilemma.

Recall that in our discussion of convection convective material undergoes presumably turbulent mixing. Therefore, if such a convection zone is rotating, then convective mixing should result in angular momentum exchange, leaving the zone rotating essentially as a solid body. Radiatively stable material, on the other hand, can support a gradient in angular velocity with depth. Now consider the young sun, which started its life as a fully convective pre–main sequence star. As such, it initially rotated as a solid body, with a much larger angular momentum than today. As it settled onto the main sequence, the interior had become radiative while the envelope remained convective (see the section on the ZAMS sun). During contraction, the early sun became more centrally concentrated. If specific angular momentum was conserved (that is, each mass shell had an angular momentum that did not change with time by transfer to other mass shells) then this central concentration resulted in a spin–up of the solar core, with decreasing angular velocity $\Omega(r)$ from the center outward.

Recall also that for stars with surface convection zones, angular momentum can be lost from the surface by a magnetized stellar wind. The surface convection zone, which should have continued rotating as a solid body, thus experienced a continual loss of angular momentum to space. Therefore, in addition to a smooth gradient in $\Omega(r)$ in the radiative interior, the surface spun down quickly and *somewhat* independently of the interior. At the base of the convection zone, then, a discontinuity in $\Omega(r)$ may have developed. Numerous analytic and laboratory studies of rotating fluids have shown that steep gradients of Ω can be hydrodynamically unstable (see, e.g., Zahn 1987). Thus these instabilities will trigger mixing of stellar material and its angular momentum to reduce the gradients in $\Omega(r)$ to a state of (at least) marginal stability. Such redistribution of angular momentum can occur on either short time scales (i.e., the free-fall time) or longer time scales (i.e., the thermal time scale), but it is hard to see how angular momentum redistribution can be completely avoided. Because the principal generator of shear is the braking of the convective envelope, the base of the envelope will be where the angular momentum redistribution and resulting mixing of material will be most noticeable.

Rotation therefore could result in mixing of material near the base of the solar convection zone and the solar lithium depletion may be a signature of angular momentum redistribution. It should be clear that a discussion of computations of solar evolution that includes such mixing is beyond the scope of this text; it involves simultaneous solutions of the usual equations of stellar evolution and, in addition, treatment of hydrodynamic instabilities on many different time scales (and in three dimensions!). Some computational work in this area, with simplifying assumptions about distribution of angular momentum in latitude and longitude, has addressed the problem of rotational mixing. Comparison of these model results with observations of the lithium abundance seen in young stars and very old stars, as well as the sun, has met with remarkable success. (See, for example, Pinsonneault 1988 and the review by Sofia, Pinsonneault, and Deliyannis 1991 and references therein)

8.5 References and Suggested Readings

§**8.1**: The Sun as the Prototype Star

Every few years there is a review article that lists presently accepted values of fundamental constants. Our reference is to

▹ Cohen, E.R., and Taylor, B.N. 1987, *Rev. Mod. Phys.*, **59**, 1151.

Our sources for variations in solar luminosity are

▹ Newkirk, G. Jr. 1983, *Ann. Rev. Astron. Ap.*, **21**, 429

▹ Willson, R.C., Hudson, H.S., Frohlich, C., and Brusa, R.W. 1986, *Science*, **234**, 1114.

The details of quoted solar surface abundances are bound to change through the years. We quote from the spectroscopic work of

▹ Aller, L.H. 1986, in *Spectroscopy of Astrophysical Plasmas*, Eds. A. Dalgarno and D. Layzer (Cambridge, U.K.: Cambridge University Press), p. 89

▹ Grevesse, N. 1984, *Phys. Scr.*, **T8**, 49

and the helioseismic analysis of

▹ Bahcall, J.N., and Ulrich, R.K. 1988, *Rev. Mod. Phys.*, **60**, 297.

The age of the sun is probably not known to within better then 10^8 years because of uncertainties of phasing between solar, planetary, and minor body formation within the solar system. For comments see

▹ Guenther, D.B. 1989, *Ap.J.*, **339**, 1156.

§**8.2**: From the ZAMS to the Present

Unlike luminous stars, the sun does not have much of a wind. The mass loss rate we quote is from

▹ Cassinelli, J.P., and MacGregor, K.B. 1986, in *Physics of the Sun* Vol III, Ed. P.A. Sturrock (Boston: Reidel), p. 47.

"Standard" models have become more standard through the years as model makers have tended to agree upon the best strategies and physics for making models. The models of

▹ Guenther, D.B., Jaffe, A., and Demarque, P. 1989, *Ap.J.*, **345**, 1022

▹ Guenther, D.B, Demarque, P., Kim, Y.–C., and Pinsonneault, M.H. 1992, *Ap.J.*, **387**, 372

are representative. (This does not say that everyone gets the same results, however!) The 1989 Yale models use the mix of

▹ Anders, E., and Grevesse, N. 1989, *Geochim. Cosmochin. Acta*, **53**, 197.

As usual, we recommend

▹ Cox, J.P. 1968, *Principles of Stellar Structure*, in two volumes (New York: Gordon and Breach)

for some in-depth prespectives on stellar structure.

It is always gratifying to use a simple calculation to help make sense out of all the numbers pouring out of a computer. The virial estimate for luminosity change of the evolving sun is from

▹ Endal, A.S. 1981, in *Variations of the Solar Constant*, NASA Conf. Pub. 2191, 175.

§8.3: The Solar Neutrino "Problem"

The observational results of the Homestake Mine experiment are reviewed in

▹ Rowley, J.K., Cleveland, B.T., Davis, R. Jr. 1985, in *Solar Neutrinos and Neutrino Astronomy*, Eds. M.L. Cherry, W.A. Fowler, and K. Lande (AIP: New York), pp. 1–21.

A good overall review of neutrinos, solar and otherwise, may be found in

▹ Bahcall, J.N. 1989, *Neutrino Astrophysics* (Cambridge: Cambridge University Press).

For more on solar neutrinos see also

▹ Bahcall, J.N., Huebner, W.F., Lubow, S.H., Parker, P.D., and Ulrich, R.K. 1982, *Rev. Mod. Phys.*, **54**, 767

▹ Bahcall, J.N., and Ulrich, R.K. 1988, *Rev. Mod. Phys.*, **60**, 297

▹ Bahcall, J.N., Davis, R. Jr., and Wolfenstein, L. 1988, *Nature*, **334**, 487.

Solutions of the solar neutrino problem involving fundamental particle physics are discussed in

▹ Bahcall, J.N., and Bethe, H. 1990, *Phys. Rev. Lett.*, **65**, 2233.

§8.4: The Role of Rotation in Evolution

The best textbook on the subject of stellar rotation is due to

▹ Tassoul, J.-L. 1978, *Theory of Rotating Stars* (Princeton: Princeton University Press)

and we recommend it for your general bookshelf.

Figure 8.8 is taken from

▹ Kawaler, S.D. 1987, *P.A.S.P.*, **99**, 1322

using material from

▹ Fukuda, I. 1982, *P.A.S.P.*, **94**, 271.

Papers concerning the role of magnetic breaking and rotation include

▹ Mestel, L. 1984, in *3d Cambridge Workshop on Cool Stars, Stellar Systems, and the Sun*, Eds. S. Baliunas and L. Hartmann, (New York: Springer), p. 49

▹ Stauffer, J.R., and Hartmann, L. 1986, *P.A.S.P.*, **98**, 1233

▹ Kawaler, S.D. 1990, in *Angular Momentum and Mass Loss for Hot Stars*, Eds. L.A. Willson and R. Stallio, (Dordrecht: Kluwer), p. 55

▹ Collier Cameron, A., Li, J. and Mestel, L. 1991, in *Angular Momentum Evolution of Young Stars*, Eds. S. Catalano and J. R. Stauffer, (Dordrecht: Kluwer), p. 297

▹ Stauffer, J.R. 1991, in *Angular Momentum Evolution of Young Stars*, Eds. S. Catalano and J.R. Stauffer, (Kluwer: Dordrecht), p. 117.

The original von Zeipel's "paradox" is discussed in

▹ von Zeipel, H. 1924, *M.N.R.A.S. London*, **84**, 665, 684.

General questions of instabilities induced by steep rotation gradients is discussed by

▹ Zahn, J.–P. 1987, in *The Internal Solar Angular Velocity*, Eds. B. Durney and S. Sofia (Dordrecht: Reidel).

Some papers concerning the sun, rotation, and the lithium problem include

▹ Pinsonneault, M. 1988, *Evolutionary Models of the Rotating Sun and Implications for Other Low Mass Stars*, Ph.D. Dissertation, Yale University

▹ Sofia, S., Kawaler, S., Larson, R., and Pinsonneault, M. 1991, in *The Solar Interior and Atmosphere*, Eds. A.N. Cox, W.C. Livingston, and M.S. Matthews (Tucson: University of Arizona), p. 140

▹ Sofia, S., Pinsonneault, M., and Deliyannis, C. 1991, in *Angular Momentum Evolution of Young Stars*, Eds. S. Catalano and J. R. Stauffer (Dordrecht: Kluwer), p. 333.

9

Structure and Evolution of White Dwarfs

Any fool can make a white dwarf.
— Icko Iben Jr. (1985)

We have already discussed the evolutionary stages leading to the white dwarfs (§2.4.1) and described their internal structure as being determined by the combination of high gravities and an electron degenerate equation of state. This chapter will elaborate on their structure, evolution, and importance as the endpoint of evolution for most stars. There is no one text that deals solely with these objects but, for further reading, we suggest Shapiro and Teukolsky (1983), Liebert (1980), Iben and Tutukov (1984), Tassoul *et al.* (1990), D'Antona and Mazzitelli (1990), and Wood (1992).

9.1 Observed Properties of White Dwarfs

In some respects white dwarfs form a remarkably homogeneous class of star. Figure 9.1 shows a color-magnitude diagram where a good sample of these stars are plotted. They form a well-defined sequence to luminosities down to around $3 \times 10^{-5}\ \mathcal{L}_\odot$, below which, as far as can be determined, we do not find cooler objects (Liebert et al. 1988). The tight correlation of luminosity with effective temperature (i.e., M_v with $B - V$) immediately demonstrates that their radii are all very nearly the same with $\mathcal{R} \approx 0.01\,\mathcal{R}_\odot \approx 7 \times 10^8$ cm. Spectroscopic observations coupled with theoretical stellar atmosphere calculations have determined that their surface gravities are near $\log g \approx 8.2$ ($g \approx 1.5 \times 10^8$ cm s^{-2}) which, considering the radii, yields masses of $\mathcal{M} \sim 0.6\,\mathcal{M}_\odot$. Spectroscopic results for individual single (nonbinary) stars of the most common types (DA and DB, as discussed later) firm this up further and indicate that an average mass is $0.6\ \mathcal{M}_\odot$ with a surprisingly low dispersion of only around $0.1\,\mathcal{M}_\odot$ about this figure.[1]

White dwarfs in binary systems have a wider range of masses determined from reliable binary orbit solutions; for example, the mass of Sirius B is $1.053 \pm 0.028\ \mathcal{M}_\odot$ while 40 Eri B is below the single-star mean with $\mathcal{M} \approx 0.43\ \mathcal{M}_\odot$ (as reviewed in Liebert 1980).

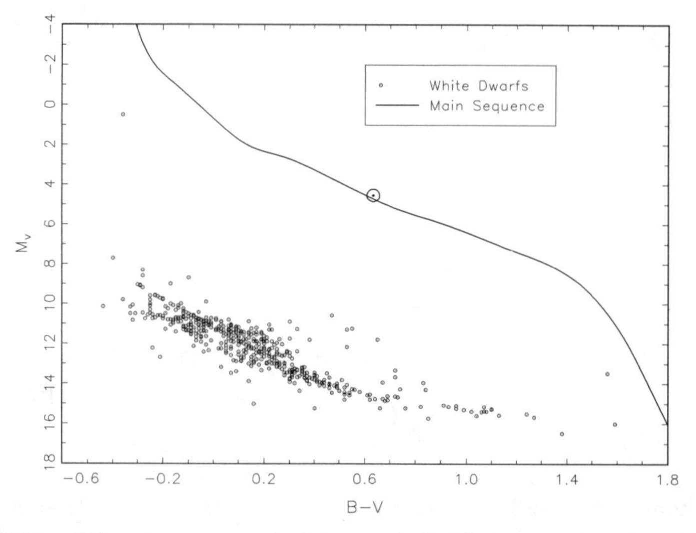

FIGURE 9.1. The color-magnitude (observer's H–R) diagram for white dwarfs in terms of M_v and $B - V$. Data from McCook and Sion (1987). The solid line is the main sequence with the sun indicated by the $\odot$.

Spectroscopic observations also reveal that the atmospheric composition of white dwarfs may differ wildly from one to the next. Most common are those whose surfaces consist almost entirely of hydrogen with contamination by other elements exceeding, in some instances, no more than one part in a million by number of atoms. These are the DA white dwarfs and they comprise some 80% of all white dwarfs, although the exact percentage does depend on effective temperature class. Next most common are the DB white dwarfs with helium atmospheres, which make up almost 20%

[1]The realization that most single white dwarfs have nearly the same mass is a relatively recent development. Since they have evolved from stars with different initial masses this uniformity must be telling us something about how mass is lost in the AGB stage. However, there are a small number of single objects whose masses lie in the high-mass tail of the distribution. Relevant papers and reviews are Weidemann and Koester (1984), Oke et al. (1984), Weidemann (1990), Bergeron et al. (1992), Trimble (1991, 1992).

of the total. The remainder consists of stars with hybrid atmospheres or those with peculiar abundances. The most commonly used spectroscopic classification scheme is summarized in Table 9.1 (adapted from Sion et al. 1983, and see Jaschek and Jaschek 1987, Chapter 15). Note that there is evidence that the surface abundance, and therefore spectral classification, for a given white dwarf may change as it evolves. As new data appear, we also expect the classification scheme to evolve with time.

TABLE 9.1. White dwarf spectroscopic classification scheme

Spectral Type	Characteristics
DA	Balmer Lines only: no He I or metals present
DB	He I lines: no H or metals present
DC	Continuous spectrum with no readily apparent lines
DO	He II strong: He I or H present
DZ	Metal lines only: no H or He
DQ	Carbon features of any kind

Composite classifications are also possible, such as "DBA" for a white dwarf with He I and H lines. A "V" suffix is used to denote variable white dwarfs; e.g., DAV. Adapted from Sion *et al.* (1983).

Effective temperatures for white dwarfs range from well over 100,000 K to as low as about 4000 K. The majority of known white dwarfs have temperatures higher than the sun and hence the "white" in their name. As will soon be apparent, the sequence in Figure 9.1 can best be explained by cooling where, as time progresses, hot white dwarfs gradually evolve to lower temperatures along the sequence.

White dwarfs are observed to rotate but with periods usually longer than a few hours (Greenstein and Peterson 1973, Pilachowski and Milkey 1987, and Koester and Herrero 1988). This is a remarkable observation in itself because if, for example, we were to let the sun evolve to the white dwarf stage without losing either mass (an unlikely assumption) or angular momentum (equally unlikely), then the resultant carbon-oxygen object, with a radius of 5.6×10^8 cm (see eq. 3.64), would have a solid body rotation period of only about 2.5 minutes. Angular momentum loss must therefore be a common feature of stellar evolution.

Many white dwarfs are variable stars (and probably the most common overtly variable stars in the universe) and a small number also have the strongest magnetic fields known for "normal" stars (of up to 10^8 gauss and we exclude pulsars here for which there is only indirect evidence for even stronger fields). The magnetic white dwarfs will be the subject of a §9.4 and we reserve §10.5 for a discussion of the seismology of variable white dwarfs.

9.2 White Dwarf Evolution

We have yet to establish that evolutionary models of white dwarfs actually do reproduce the observed objects but, if our earlier ideas are correct, then the interior should be largely electron degenerate. On the other hand, the very surface cannot be degenerate because white dwarfs are observed to have high effective temperatures. The surface layers should therefore be nondegenerate and this means very different equations of state and opacity sources. We shall see that energy is transported rapidly through the degenerate interior but has to diffuse gradually through the nondegenerate envelope. Thus the cooling of white dwarfs involves the properties of matter under a wider range of conditions than most other problems in physics or astrophysics. However, this degenerate core–nondegenerate envelope picture results in an elegant simplification that permits us to construct a very simple model for how white dwarfs evolve.

9.2.1 Simple Cooling Curves

The white dwarf model we shall now consider has the following elements. Imagine that the core of the star, which comprises nearly all of the mass and radius, is degenerate. Overlying the core is a thin envelope of nondegenerate material and the transition between degeneracy and nondegeneracy is assumed to take place abruptly at a radius $r_{\rm tr}$. If the electrons are nonrelativistic at $r_{\rm tr}$, then the relation between density and temperature there is given by (3.66), $\rho_{\rm tr} \approx 6 \times 10^{-9} \mu_e T_{\rm tr}^{3/2}$. In the electron-degenerate core interior to $r_{\rm tr}$, electron conduction is very efficient at transporting heat (according to the arguments of §4.5) and only a mild temperature gradient is required to drive the flux. Thus, for simplicity, we assume that the core is isothermal with temperature $T_{\rm core} = T_{\rm tr}$.[2]

To determine $r_{\rm tr}$ we need to be more specific about the model. If the envelope does not support convection (and this is *not* true for many white dwarfs), then the envelope approximations of §7.3.2 should describe the run of pressure versus temperature and density. For zero boundary conditions equation (7.139) states that $P = K'T^{1+n_{\rm eff}}$ where K' is given by (7.140) and $n_{\rm eff}$ is $(s+3)/(n+1)$. The exponents n and s are those in $\kappa = \kappa_0 \rho^n T^{-s}$ of (1.59). We now use this information to establish a relation between $T_{\rm tr}$ (and hence $T_{\rm core}$), luminosity, and mass.

The pressure must be continuous across $r_{\rm tr}$. Above $r_{\rm tr}$ the gas is nonde-

---

[2]This assumption of isothermality also requires that there be no strong sources or sinks of energy in the core such as might be associated with nuclear burning, gravitational contraction, or neutrinos.

generate and we use the ideal gas law $P_{\rm tr} = \rho_{\rm tr} N_{\rm A} k T_{\rm tr}/\mu$. We then have

$$P_{\rm tr} = K' T_{\rm tr}^{1+n_{\rm eff}} = \rho_{\rm tr} \frac{N_{\rm A} k}{\mu} T_{\rm tr} = 6 \times 10^{-9} \mu_{\rm e} \frac{N_{\rm A} k}{\mu} T_{\rm tr}^{1+3/2} \tag{9.1}$$

where the transition relation between density and temperature (3.66) has been used to eliminate density in the third term. The coefficient K' contains $\mathcal{L}$, $\mathcal{M}$, μ, and the opacity. At this point we have to decide what is the dominant opacity source, which also means specifying the composition. To make matters as simple as possible, suppose the white dwarf is composed entirely of elements with atomic masses greater than ^{4}He. In fact, we expect the cores of most white dwarfs to be composed of some combination of ionized carbon and oxygen. This is because ^{12}C and ^{16}O, both with $\mu_{\rm e} = 2$ (§1.4.1), are the products of helium burning and, for single white dwarfs of average mass 0.6 $\mathcal{M}_\odot$, this is as far as core evolution has gone. Choosing the same composition for the surface layers is not so good but, as we shall see, the final result we obtain shall be quite reasonable. For the opacity source we choose bound-free Kramers' which, from (4.36), is approximated by $\kappa_{\rm b-f} \approx 4 \times 10^{25} \rho T^{-3.5}$ cm^2 g^{-1}. This analytic opacity, while crude, still gives the flavor of what goes on, so we use it here with no further apologies. There is no way to get what we want without a little judicious fudging.

The model we are setting up is that of a highly conducting core surrounded by a thin insulating blanket. Heat flows easily out of the core but must work its way out through the envelope. When we discuss cooling, it is the envelope that controls the rate of cooling and it is thus here that the choice of opacity is critical. Applying (7.140) for K' yields

$$K' \approx 8.1 \times 10^{-15} \mu^{-1/2} \left[\frac{\mathcal{M}/\mathcal{M}_\odot}{\mathcal{L}/\mathcal{L}_\odot} \right]^{1/2} \tag{9.2}$$

where we have used $\kappa_{\rm b-f}$ and $n_{\rm eff} = 3.25$. Setting $\mu_{\rm e} = 2$ and solving the combination of (9.1) and (9.2) for luminosity gives us the relation

$$\frac{\mathcal{L}}{\mathcal{L}_\odot} \approx 6.6 \times 10^{-29} \mu \frac{\mathcal{M}}{\mathcal{M}_\odot} T_{\rm tr}^{7/2}. \tag{9.3}$$

Note that for $\mu = 12$ (^{12}C) and $\mathcal{M} = 0.6\mathcal{M}_\odot$, $\mathcal{L}/\mathcal{L}_\odot = 100$ (10^{-4}), $T_{\rm tr} = T_{\rm core}$ is about 2.4×10^8 K (4.6×10^6 K). The high temperature at the high luminosity rules out the presence of either hydrogen or helium in the core; either would have been burned under these conditions (see Cox 1968, §25.5).

We can now estimate the thickness of the surface layer by using (7.144), which gives temperature as a function of $r/\mathcal{R}$ for a thin radiative envelope. You may easily check that for $\mathcal{L}/\mathcal{L}_\odot = 10^{-4}$, $r_{\rm tr}/\mathcal{R} \approx 2 \times 10^{-2}$, which means that the nondegenerate envelope can indeed be thin.

The next step is to find how the white dwarf cools. We still assume that there are no internal energy sources such as nuclear burning or gravitational contraction. The last means that the total radius remains roughly constant with time. This approximation becomes better and better as the star cools and the only energy source is the internal heat of the star, which gradually leaks out through the blanketing envelope.[3] These are the essentials of the Mestel (1952) cooling theory for white dwarfs which, except for some refinements, has stood the test of time.

To apply this theory first recall that the specific heat of a nonzero temperature degenerate gas is controlled by the ions. From (3.112), $c_{V_\rho} = 1.247 \times 10^8/\mu_I$ erg g^{-1} K^{-1}. Since temperature is constant in the core and the core takes up essentially all the stellar mass, then the rate at which the ions release heat on cooling is

$$\mathcal{L} = -\frac{dE_{\text{ions}}}{dt} = -c_{V_\rho}\mathcal{M}\frac{dT_{\text{tr}}}{dt}. \tag{9.4}$$

To find how luminosity changes with time, differentiate (9.3) with respect to time, use (9.3) to get rid of T_{tr}, and then use (9.4) to get rid of the temperature derivative. The resulting differential equation, which should now contain only the dependent variable $\mathcal{L}(t)$, is then integrated to obtain a "cooling time"

$$t_{\text{cool}} = 6.3 \times 10^6 \left(\frac{A}{12}\right)^{-1} \left(\frac{\mathcal{M}}{\mathcal{M}_\odot}\right)^{\frac{5}{7}} \left(\frac{\mu}{2}\right)^{-\frac{2}{7}} \left[\left(\frac{\mathcal{L}}{\mathcal{L}_\odot}\right)^{-\frac{5}{7}} - \left(\frac{\mathcal{L}_0}{\mathcal{L}_\odot}\right)^{-\frac{5}{7}}\right] \tag{9.5}$$

where t_{cool} is in years. Here A is the mean atomic weight of the nuclei in amu and $\mathcal{L}_0$ is the luminosity at the start of cooling. This equation has the property that after long elapsed times the second term in the brackets becomes negligible. We therefore drop it for simplicity and, at the same time, change the leading coefficient to match more accurate results for cooling white dwarfs calculated by Iben and Tutukov (1984) (and see Iben and Laughlin 1989). The final result is

$$t_{\text{cool}} = 8.8 \times 10^6 \left(\frac{A}{12}\right)^{-1} \left(\frac{\mathcal{M}}{\mathcal{M}_\odot}\right)^{5/7} \left(\frac{\mu}{2}\right)^{-2/7} \left(\frac{\mathcal{L}}{\mathcal{L}_\odot}\right)^{-5/7} \text{yr}. \tag{9.6}$$

Figure 9.2 shows cooling curves for pure carbon white dwarfs derived from evolutionary calculations. Were these results pure Mestel (1952) cooling, we would only have straight lines in this log–log plot as shown by the dotted line in the figure for $0.6\mathcal{M}_\odot$ as derived from (9.6). The very close,

[3]Most stars heat up when they lose energy (see the virial theorem results of §1.3.2), which makes for a self-regulating mechanism even though it sounds peculiar. Thus it is that white dwarfs are odd because they follow what seems like the more reasonable path and cool when they lose energy. The reason has partially to do with the very low specific heats of degenerate electrons versus ions and the strong influence of density on the internal energy of the electrons (§3.7.1). See the discussion in Cox (1968, §25.3b).

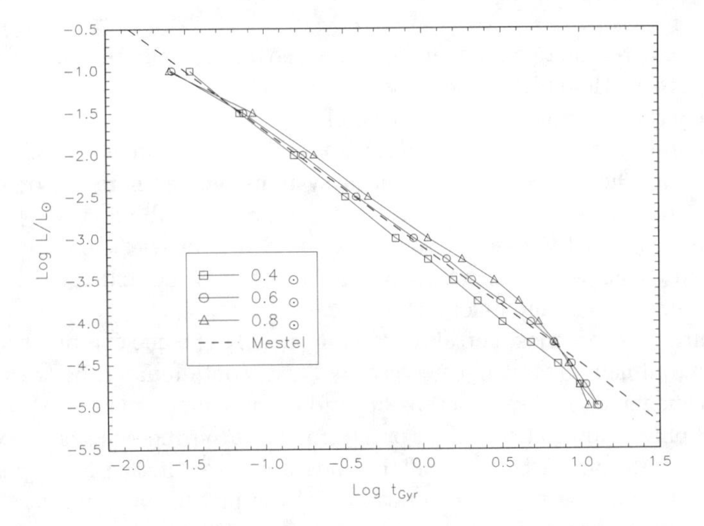

FIGURE 9.2. Cooling curves for pure carbon white dwarf models as adapted from Winget et al. (1987). The dotted line is a Mestel (1952) cooling curve for a $0.6\mathcal{M}_\odot$ carbon white dwarf using equation (9.6).

but perhaps fortuitous, match indicates that we have captured the essentials of the physics of white dwarf cooling. It is evident that the hottest, and therefore most luminous white dwarfs, cool the fastest. At the cool end, we see by plugging $\mathcal{L} = 10^{-4.5}\mathcal{L}_\odot$ into (9.6) that the cooling time for the coolest white dwarfs is approximately 10^{10} years. This is the main point we wish to make here, and one to be elaborated upon later: the coolest white dwarfs have ages we associate with the oldest stars.

9.2.2 Realistic Evolutionary Calculations

The ideal calculation of evolving white dwarfs has yet to be realized. In addition to physical processes that are difficult to model in the white dwarfs themselves, the starting conditions require knowledge of how white dwarfs are formed. This means that the origin of white dwarfs as planetary nebula nuclei ("PNNs") must be understood to provide realistic starting models. Unfortunately, the process of planetary nebula formation is still somewhat of a mystery. White dwarfs can play a critical role in helping explain that stage because whatever PNN models are made must eventually reproduce the observed statistics of those stars.

What is usually done in white dwarf evolutionary studies is to either start off with a completely *ab initio* model of a very hot white dwarf where the initial structure (run of composition, $T(r)$, etc.) is specified beforehand or, with the hope of realism, to start a model from the main sequence and continue on to the white dwarf stage. The latter requires evolving the model into the asymptotic giant branch phase and then lifting off the outer layers in some reasonable manner to expose the underlying pre–white dwarf object. Both methods have their champions.

It is generally assumed that the cores of most white dwarfs are composed of a carbon-oxygen mixture. Exceptions to this may occur for special objects such as the white dwarfs in nova systems where there is not only evidence that these are more massive than average (about $1\,\mathcal{M}_\odot$) but, from observations of the ejecta, some may be rich in heavier elements such as those in the nuclear mass range Na to Al (e.g., Truran and Livio 1989.) We shall ignore such anomalous objects.[4]

The surface layers most certainly consist initially of some combination of hydrogen and helium (but not necessarily both) contaminated by traces of heavier elements. The exact details depend on how much mass was lost in the AGB phase, how much nuclear processing has occurred, whether mixing has taken place, and whether stellar winds are active in the PNN phase. Determination of the surface composition of the just-formed and very hot object is difficult because temperatures exceeding 100,000 K put interesting details of the spectrum in the hard UV and soft X-ray. However, based on the observation that cooler white dwarfs show either nearly pure hydrogen (DA) or helium (DB) in their spectra, two different classes of model have been examined in detail. The first assumes that all hydrogen has been lost. These evolve into DB white dwarfs. The second has a hydrogen surface layer above a layer of helium. Whether the DB-like object has trace amounts of hydrogen (which can later "float" to the surface to convert the star to a DA) is a matter of controversy at the present time. Similarly, the thickness of the hydrogen layer in the DA objects is also in doubt. If thin enough, convection at later stages could convert the DA to a spectroscopic DB. We shall avoid such unresolved issues here because, for the most part,they are refinements. However, the total mass of hydrogen cannot exceed $10^{-4}\mathcal{M}_\odot$ because, if it did, nuclear burning would occur. Similarly, the helium layer mass should not exceed about $10^{-2}\,\mathcal{M}_\odot$.

Whether DA or DB, the newly formed objects have some common characteristics. The surface layers are still hot enough that shell CNO hydrogen or helium burning may still be taking place. These energy sources can, for the initial stages of evolution, provide the dominant energy source. There are also energy losses due to neutrino emission. Because of the extreme conditions in the deep interior, various processes come into play that produce neutrinos that easily pass through the star and carry away energy. We shall not discuss these processes here (see §6.8) but in very hot white dwarfs they are efficient in cooling the interior and may cause a central temperature inversion. Under the latter circumstance heat flows inward and some

---

[4]There may be a selection effect operating here because the more massive objects in close binary systems are expected to erupt more often and more brightly than their less massive counterparts. A paucity of ejecta for the less massive objects could also hide their true composition.

is eventually given to the neutrinos. The luminosity loss due to neutrinos can rival or exceed optical luminosity of the entire star (see below).

Several new pieces of physics enter as the white dwarf cools. We have already mentioned solid-state effects in §3.6 and these can radically alter the equation of state. Because these are phase changes (e.g., crystallization), the thermal properties of the medium are also effected. Thus, for example, if the stellar material crystallizes, then latent heat is liberated, which can slow the cooling of the star. All these effects are incorporated into modern models (as in Iben and Tutukov 1984; Lamb and Van Horn 1975; Koester and Schönberner 1986; Tassoul, Fontaine, and Winget 1990).

In the simple radiative model of the preceding section we ignored convection completely. This is a serious omission. As the white dwarf cools, surface convection zones grow and die out, thus affecting the rate of cooling in these important outer layers (see, e.g., Fontaine and Van Horn 1976, Tassoul, Fontaine, and Winget 1990). As part of the modeling of these convection zones we also require adequate and consistent equations of state for the envelope. These can be very difficult to compute because of nonideal effects for the multicomponent gases involved. One tabulation of such an equation of state may be found in Fontaine, Graboske, and Van Horn (1977).

Mixing of elements by convection is also possible, which can change the photospheric abundances and confuse the issue of spectral classification. We again remind you of the theoretical difficulties in describing stellar convection. Thus far, the MLT (see Chapter 5) is used for these evolutionary studies with its attendant problems.

Now how do we account for the near purity of elements in the atmospheres of DA and DB white dwarfs? This is not only an important observational issue but it also impacts on the evolution: hydrogen and helium are different. Even at the trace element level it has important consequences because, for example, opacities are strongly effected by even trace amounts of heavy elements. The prime cause of this purity is "gravitational settling." This term, although it is frequently used in the literature, is somewhat of a misnomer, although it is convenient to picture light elements floating and heavy elements sinking. It is true that the gravitational field is ultimately responsible for separation of heavy from light, but the immediate cause is the presence of pressure gradients and the resulting imbalance of forces on ions. The derivation of the rate of separation is beyond what we shall do here but it contains some very interesting physics. A classical derivation is contained in Chapman and Cowling (1960).

Countering the effects of gravitational settling is the normal process of diffusion whereby gradients in composition force elements to diffuse and thus reduce the gradients. In addition, "radiative levitation" can cause elements to rise by means of radiative forces acting on specific trace atoms through bound-bound and bound-free transitions. The net effect of these diffusive processes is very complex but the bottom line is that evolutionary time scales in high-gravity white dwarfs are amply long for separation to become complete (although some elements in the deep core may resist). Other consequences of separation and diffusion for white dwarfs are discussed in Fontaine and Michaud (1979), Michaud and Fontaine (1984), and Iben and MacDonald (1985). We also remark that these processes are of considerable interest for other stars as in the peculiar abundance spectral class A stars (Michaud 1970).

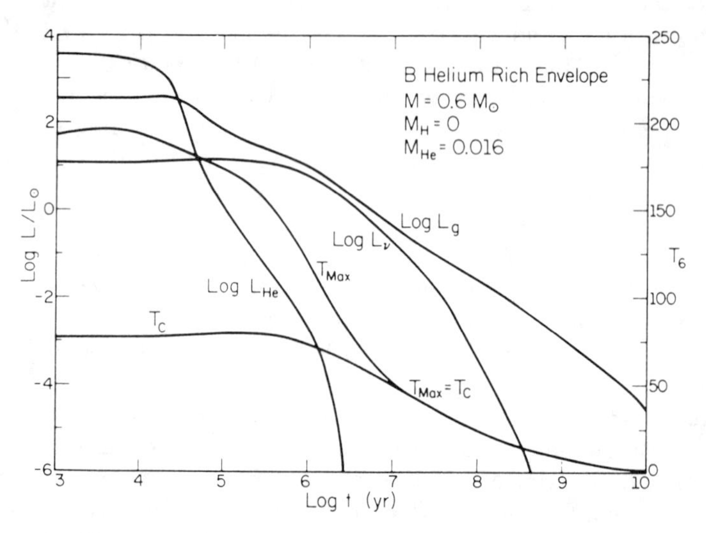

FIGURE 9.3. Shown as a function of time are luminosities during the evolution of a $0.6\,\mathcal{M}_\odot$ helium atmosphere white dwarf. Maximum and central temperatures are also indicated. Reproduced, with permission, from Iben and Tutukov (1984).

Finally, we show some evolutionary results from the work of Iben and Tutukov (1984). Their methods parallel those of other authors and are representative of modern efforts. The model is that of a $0.6\,\mathcal{M}_\odot$ DB helium atmosphere white dwarf and the mass of the helium layer is $0.016\,\mathcal{M}_\odot$. Figure 9.3 shows the time evolution of various components of luminosity and temperature. The quantity $\mathcal{L}_\text{g}$ is the total of the luminosity released from internal thermal and gravitational potential energies. The luminosity generated by helium shell burning is denoted by $\mathcal{L}_\text{He}$ and neutrino losses are represented by $\mathcal{L}_\nu$. Also shown is the maximum temperature in the model (T_max) which, because of neutrino losses, is not necessarily located at the center of the model where the temperature is T_c.

The total photon luminosity for the same DB model is shown in Figure 9.4 as "Model B" (as are the results for a DA white dwarf sequence). Also shown are the late evolutionary effects of liquefaction and crystallization. The legends involving $\dot{\mathcal{M}}_\text{acc}$ indicate the luminosity released by gravitational potential energy if mass is accreted onto the stellar surface. All these effects are only noticeable after the total luminosity falls below

$10^{-2}\,\mathcal{L}_\odot$. You may wish to compare this figure to the dotted line in Figure 9.2, which shows the simple analytic cooling curve.

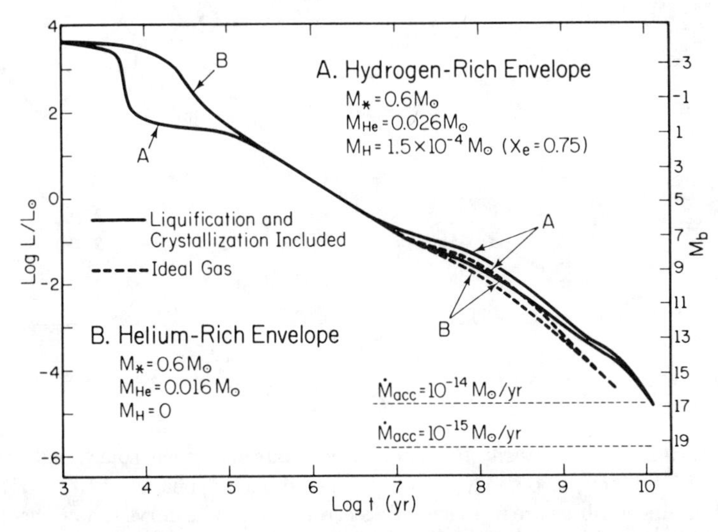

FIGURE 9.4. Luminosity versus time for two evolutionary sequences. Model "B" is the same as in Figure 9.3. Model "A" is a DA star with parameters given in the figure. Reproduced, with permission, from Iben and Tutukov (1984).

Use has recently been made of realistic white dwarf cooling calculations to estimate the ages of the oldest stars in the solar neighborhood of our galactic disk. White dwarfs comprise about 30% of all such stars. By matching observed luminosity functions of white dwarfs (number of stars per unit luminosity or magnitude per pc^3) to calculations of mass-dependent cooling ages coupled with estimates of main sequence progenitor ages and lifetimes and birth rates, the final result gives a maximum age of solar neighborhood stars of several billion years. A complete discussion here would be too lengthy but the ingredients of the problem include almost all facets of the evolution of low-mass stars. See the previous references to Winget et al. (1987), Iben and Laughlin (1989) and Wood (1992) as recommended reading.

A key factor in the calibration of this dating technique is the question of what is the luminosity function for white dwarfs. We indicated earlier that those with luminosities less than $\mathcal{L} = 10^{-4.5}\,\mathcal{L}_\odot$ are rare or even absent (see Figure 9.1 and discussion). But the least luminous and coolest are also the oldest and these are probable candidates for the oldest stars. Shown in Figure 9.5 is a luminosity function from Liebert, Dahn, and Monet (1988) as used by Wood (1992). The vital thing to note is the rapid decline in the space density of white dwarfs with luminosities near $\mathcal{L} \approx 10^{-4.5}\,\mathcal{L}_\odot$. Even though the statistics are very poor at this time (the last bin contains only two stars) the overall effect is most certainly real. The problem is that the

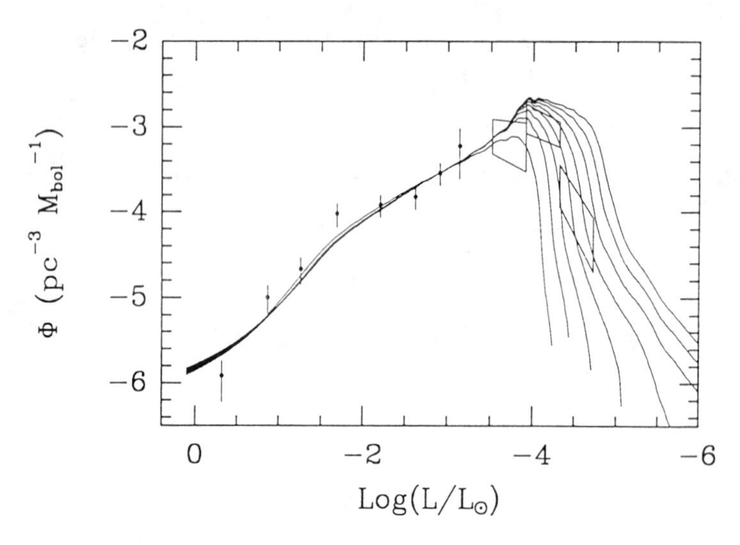

FIGURE 9.5. The luminosity function for white dwarfs in the solar neighborhood in units of number per pc^3–M_{bol}. Note the rapid decline near $\mathcal{L} \approx 10^{-4.5}\,\mathcal{L}_\odot$. The lowest luminosity bins are represented as error boxes. The curves show theoretical results for model populations with ages from 5 Gyr to 15 Gyr. Reproduced, with permission, from Wood (1992).

cooling age of these coolest objects is very sensitive to luminosity (and, to a lesser extent, white dwarf mass) and this can change the results for the oldest stellar ages. Be that as it may, the method is in principle simple and has considerable promise. This is also illustrated in Figure 9.5, where the observed luminosity function is fit by theoretical calculations for a sample of stars that was originally formed on the main sequence around 10^{10} years ago. This particular result is from Wood (1992). Other possibilities, and other ages, are given in Iben and Laughlin (1989).

Since the coolest white dwarfs are only detectable in the immediate neighborhood of the sun, this result strictly applies only to the solar neighborhood. There is good reason to believe, however, that this result is valid for most or all of the galactic disk. A major challenge for the future is to observe dim white dwarfs in the galactic halo and thus establish a relative timing of halo versus disk formation.

Note that the cooling curve for models of white dwarfs, when applied to the observed population, is calibrated only roughly using the slope of the luminosity function above the low luminosity cutoff. There is hope to provide a precise calibration of cooling times as a function of effective temperature if we could watch individual white dwarfs cool. Such a remarkable observation can indeed be made by careful analysis of pulsating white dwarf stars, as discussed further in §10.5.

9.3 The Magnetic White Dwarfs

There are two primary ways of detecting magnetic fields in white dwarfs. The first, and most sensitive, is by measuring linear and quadratic Zeeman effects in spectral lines if those lines can be recognized in the strong field objects. This technique requires strong lines but these are not always to be found and, even if they are present, the inherent dimness of white dwarfs often defeats the observer. (A check on some results may be made by observing gravitational redshifts of the lines.) The second method depends on the detection of continuum circular polarization and is especially useful when magnetic field strengths are high. At the present time, the lower limit for detectable fields is about 10^5 Gauss except under unusually favorable circumstances, and that's 100,000 times bigger than the sun's average magnetic field!

The compilation of magnetic field strengths by Angel et al. (1981) lists measurements for over 100 white dwarfs. The results fall into three categories: (1) upper limits of a few thousand Gauss (these are the relatively rare cases where the observing conditions are favorable); (2) possible detections at around 10^5 Gauss (where, in most cases, the errors bars on the measurements preclude a firm determination at that level); and (3) fields clearly in excess of 10^6 Gauss (=1 MG). Thus it appears that white dwarfs either have "weak" fields or very strong fields. The number of these strong-field white dwarfs is, however, very small and they are now referred to as the "magnetic white dwarfs." Schmidt (1988) lists the 24 known magnetic white dwarfs. The inferred strengths of the polar field (assuming dipole geometry) range from about 2 MG up to the strongest at perhaps over 500 MG for the star PG1031+234. (For a recent review of the magnetic objects, see Chanmugam 1992.)

The polarization in this last object is modulated with a period of three and a half hours due to rotation and this allows the surface of star to be "scanned" as a function of time. From this, the geometry of the magnetic field may be inferred, and this is shown in Figure 9.6 (from Schmidt et al. 1986). There appears to be a global field that is dipolar in nature but the axis of the field is inclined away from the rotation axis (an "oblique" rotator). In addition there is a magnetic "spot" on the surface whose central field may approach 1000 MG (Latter et al. 1987)!

Are strong (or weak) fields in white dwarfs a surprise or not? The origin and evolution of magnetic fields in stars is still not at all well understood. Some fields may be the remains of interstellar medium magnetic fields that were trapped during the process of star formation. It is more likely, however, that fields are produced *in situ* by dynamo action caused by the interaction of convection and rotation (see, for example, the review by Parker 1977). Important as it is, we shall not attempt to develop this topic. We can, on the other hand, do some simple calculations to see if we should be surprised at the strength of these fields.

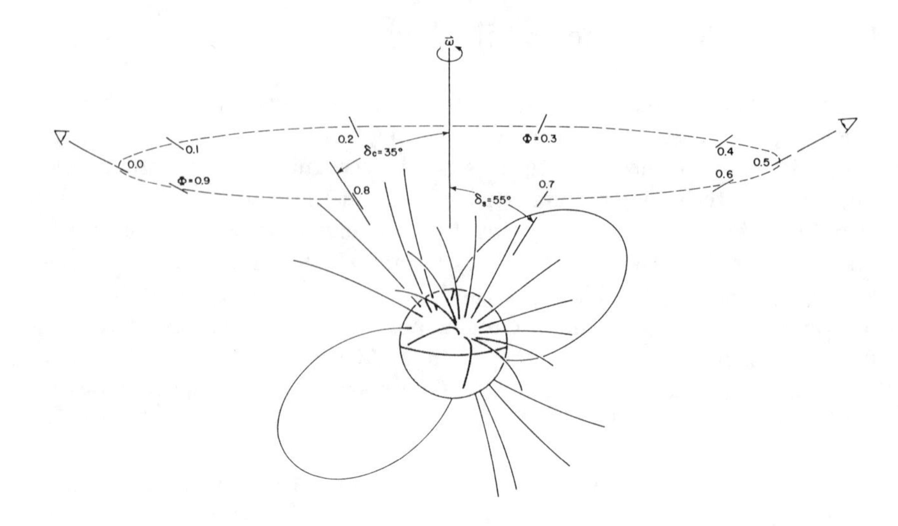

FIGURE 9.6. The inferred field geometry of the magnetic white dwarf PG1031+234. A dipole-like field that is not aligned with the rotation axis is accompanied by a magnetic "spot." Fields on the surface of this star may approach 10^9 Gauss. Reproduced, with permission, from Schmidt et al. (1986).

The average surface magnetic field of the sun (as a typical white dwarf progenitor star) is of the order of one Gauss. Suppose this is representative of the interior down through the convection zone (although one Gauss is probably too low). If we were to now suddenly turn off the mechanism for producing this field, how long would the field persist? The time development of a magnetic field is given by (see Jackson 1975, §10.3)

$$\frac{\partial \mathbf{B}}{\partial t} = \boldsymbol{\nabla}\times(\mathbf{v}\times\mathbf{B}) + \frac{c^2}{4\pi\sigma}\nabla^2\mathbf{B}. \tag{9.7}$$

The units used here are Gaussian (see the Appendix of Jackson 1975) and there is no clear agreement within the astronomical community on a common set of units. The quantity σ is the conductivity and has the units s^{-1}. If we ignore the first term on the right-hand side, then this is a diffusion equation, meaning that the magnetic field will decay away unless replenished. A characteristic diffusion time is

$$\tau = \frac{4\pi\sigma L^2}{c^2} \tag{9.8}$$

where the length L is a measure of the spatial variation of the field. For the remainder of the discussion this will be set equal to the stellar radius $\mathcal{R}$.

An estimate for the conductivity of the ideal ionized gas is given by Spitzer (1962, §5.4) which is, after changing to our units,

$$\sigma = \gamma_e \frac{2(2kT)^{3/2}}{\pi^{3/2} m_e Z e^2 \ln \Lambda}. \tag{9.9}$$

The quantity γ_e depends on the ionic nuclear charge, Z (for complete ionization), and ranges from unity for very large Z down to 0.582 for hydrogen. Our equation (5.30) gives an expression for Λ in the hydrogen case. If we assume that the sun is pure hydrogen, then its conductivity is

$$\sigma(\mathrm{H}) \approx 1.4 \times 10^8 \frac{T^{3/2}}{\ln \Lambda}. \tag{9.10}$$

For a typical virial estimate temperature of $T \approx 2 \times 10^6$ K from (1.33) and average density of $\langle \rho \rangle = 1.4$ g cm^{-3}, the solar magnetic field decay time is $\tau \sim 7 \times 10^{18}$ s or 2×10^{11} years. This is an overestimate (the time should be more like $\sim 10^{10}$ years), but the end result is that the field should persist at reasonable strength through the main sequence stage. If this field is not lost in later stages (and planetary nebula ejection might do just this), then the final white dwarf should retain the remnants of the field.

To estimate what the field strength in the just-formed white dwarf might be, we return to the original equation for the magnetic field evolution. Since the conductivity seems to be large (and, as we shall show, will remain so), we neglect the diffusion term; that is, we go to the infinite conductivity limit. In this limit the field lines are frozen into the plasma and the magnetic flux, Φ_{mag}, is conserved (in a Lagrangian sense—see Jackson 1975, §6.1, §10.3). This may be put crudely as "$\Phi_{\mathrm{mag}} \sim B\mathcal{R}^2$ remains constant as radius changes." Thus if the sun has a roughly 1 G field, then the just-formed white dwarf (with a radius of a little over $0.01\,\mathcal{R}_\odot$) should have a field of the order 10^4 G. Pushing all other uncertainties aside, this means that typical white dwarfs should have weak fields. Where do the magnetic objects come from?

A best guess is that the progenitors of the magnetic white dwarfs are the Ap stars. (See, for example, the review by Angel 1978 and, for the Ap stars, the monograph by Jaschek and Jaschek 1987.) These stars have anomalously intense magnetic fields that may be as high as 4000 G and their population statistics are consistent with the number of magnetic white dwarfs versus weak field objects. And, not so incidentally, there is a subclass called the "rapidly oscillating Ap stars," which are nonradial variables in which the magnetic field is aligned obliquely to the rotation axis. (For reviews see Kurtz 1986, and §9 of Unno et al. 1989.)

Once having been formed with strong magnetic fields, these objects can retain their fields for long times. To substantiate this we now allow for diffusion and consider the finite conductivities of a nonrelativistically degenerate pure carbon plasma given by Wendell, Van Horn, and Sargent

(1987):

$$\sigma = 10^9 \frac{T^2}{\rho \kappa_{\text{cond}}} \quad \text{s}^{-1}. \tag{9.11}$$

Here κ_{cond} is the conductive opacity for which we gave an estimate in (4.45). Inserting that estimate for pure carbon we find that

$$\sigma \approx 2 \times 10^{15} \rho \quad \text{s}^{-1}. \tag{9.12}$$

For an average white dwarf density of 5×10^5 g cm^{-3} this yields a decay time of 2×10^{11} years. This is an overestimate (as in the case of the sun) but the detailed calculations of Wendell et al. (1987) indicate that the decay times for simple dipole fields are always longer than the evolutionary time. More complicated fields with higher multipole moments tend to decay faster, implying that after long times the geometry of the fields should simplify.

Finally, are strong magnetic fields a factor in evolution? A rough way to gauge their importance is to compare the magnetic field pressure $P_{\text{mag}} = B^2/8\pi$ to the gas pressure. Wendell et al. (1987) point out that the central values of the fields in their models are some ten times the surface values. For a surface field of $B = 10^9$ G, as an upper limit thus far, this implies a central magnetic pressure of $P_{\text{mag}} \approx 4 \times 10^{18}$ dyne cm^{-2}. But this is far smaller than the hydrostatic estimate $P \sim G\mathcal{M}^2/\mathcal{R}^4 \sim 4 \times 10^{23}$ dyne cm^{-2} required for equilibrium in a typical white dwarf. They also suggest that $P_{\text{mag}} \ll P$ holds for all times because flux conservation ($B\mathcal{R}^2$ a constant) implies $P_{\text{mag}} \sim \mathcal{R}^{-4}$, which has the same dependence on $\mathcal{R}$ as the hydrostatic pressure. The conclusion is that we may safely neglect the effects of magnetic fields as far as deep interior evolutionary calculations are concerned. The same may not be true for regions of the star near the surface because the pressures are relatively low there. At the very least, opacities are affected by fields because of their effects on atomic energy levels, and this, after all, is how the fields may be detected in the first place.

9.4 References and Suggested Readings

§9.1: Observed Properties of White Dwarfs

There are several texts, papers, and reviews worth reading. Among these are Chapters 3 and 4 of

- ▹ Shapiro, S.L., and Teukolsky, S.A. 1983, *Black Holes, White Dwarfs, and Neutron Stars* (New York: Wiley & Sons)

Chapter 25 of

- ▹ Cox, J.P. 1968, *Principles of Stellar Structure*, in two volumes (New York: Gordon and Breach)

and Chapter 35 of

- ▹ Kippenhahn, R., and Weigert, A. 1990, *Stellar Structure and Evolution* (Berlin: Springer–Verlag).

Important recent papers and reviews include

▷ Liebert, J. 1980, *Ann. Rev. Astron. Ap.*, **18**, 363

▷ Iben, I., Jr., and Tutukov, A.V. 1984, *Ap.J.*, **282**, 615

▷ Tassoul, M., Fontaine, G., and Winget, D.E. 1990, *Ap.J. Suppl*, **72**, 335

▷ D'Antona, F., and Mazzitelli I. 1990, *Ann. Rev. Astron. Ap.*, **28**, 139

▷ Wood, M.A. 1992, *Ap.J.*, **386**, 539.

Masses of single white dwarfs are discussed in

▷ Weidemann, V., and Koester, D. 1984, *Astron. Ap.*, **132**, 195

▷ Oke, J.B., Weidemann, V., and Koester, D. 1984, *Ap.J.*, **281**, 276

▷ Weidemann, V. 1990, *Ann. Rev. Astron. Ap.*, **28**, 103

▷ Bergeron, P., Saffer, R.A., and Liebert, J. 1992, *Ap.J.*, **394**, 228

▷ Trimble, V. 1991, *P.A.S.P.*, **104**, 1, §10.5

▷ Trimble, V. 1992, *P.A.S.P.*, **105**, 1, §13.6.

The classification scheme outlined in Table 9.1 is relatively new and was first proposed by a group of astronomers who have had much to do with establishing the properties of white dwarfs. See

▷ Sion, E.M., Greenstein, J.L., Landstreet, J.D., Liebert, J., Shipman, H.L., and Wegner, G. 1983, *Ap.J.*, **269**, 253.

Chapter 15 of

▷ Jaschek, C., and Jaschek, M. 1987, *The Classification of Stars* (Cambridge: Cambridge University Press)

also contains a lot of useful observational material.

The M_V versus $B - V$ data used to construct Figure 9.1 was taken from

▷ McCook, G.P., and Sion, E.M. 1987, *Ap.J. Suppl.*, **65**, 603.

If you look carefully at this figure you will detect some odd linear trends: some of the data as presented are not entirely unbiased. The observed lower limit on white dwarf luminosities is discussed by

▷ Liebert, J., Dahn, C.C., and Monet, D.G. 1988, *Ap.J.*, **332**, 891.

Detecting rotation in white dwarfs is a difficult enterprise as you may see by reading

▷ Greenstein, J.L., and Peterson, D.M. 1973, *Astron. Ap.*, **25**, 29

▷ Pilachowski, C.A., and Milkey, R.W. 1987, *Pub. Astron. Soc. Pac.*, **99**, 836

▷ Koester, D., and Herrero, A. 1988, *Ap.J.*, **332**, 910.

§**9.2**: White Dwarf Evolution

The essentials of white dwarf cooling were spelled out over forty years ago in the classic paper by

▷ Mestel, L. 1952, *M.N.R.A.S.*, **112**, 583.

The cooling time of equation (9.6) is derived from

▹ Iben, I. Jr., and Tutukov, A.V. 1984, *Ap.J.*, **282**, 615

and see

▹ Iben, I. Jr., and Laughlin, G. 1989, *Ap.J.*, **341**, 312

for more material. The cooling curves of Figure 9.2, on the other hand, are from

▹ Winget, D.E., Hansen, C.J., Liebert, J., Van Horn, H.M., Fontaine, G., Nather, R.E., Kepler, S.O., and Lamb, D.Q. 1987, *Ap.J.*, **315**, L77.

The white dwarfs found in nova systems are definitely unlike ordinary single objects. See, for example,

▹ Truran, J.W., and Livio, M. 1989, in IAU Colloquium No. 114 *White Dwarfs*, Ed. G. Wegner (Berlin: Springer), p. 498.

A selection of evolutionary calculations for white dwarfs includes

▹ Lamb, D.Q., and Van Horn, H.M. 1975, *Ap.J.*, **200**, 306

▹ Iben, I. Jr., and Tutukov, A.V. 1984, *Ap.J.*, **282**, 615

▹ Koester, D., and Schönberner, D. 1986, *Astron. Ap.*, **154**, 125

▹ Tassoul, M., Fontaine, G., and Winget, D.E. 1990, *Ap.J. Suppl*, **72**, 335.

An older, but still useful, study of the effects of convection is the work of

▹ Fontaine, G., and Van Horn, H.M. 1976, *Ap.J. Suppl.*, **31**, 467.

Nonideal effects in multicomponent mixtures is discussed in

▹ Fontaine, G., Graboske, H.C. Jr., and Van Horn, H.M. 1977, *Ap.J. Suppl.*, **35**, 293.

The diffusion of heavy versus light elements is discussed in

▹ Chapman, S., and Cowling, T.G. 1960, *The Mathematical Theory of Non-Uniform Gases* (Cambridge: Cambridge University Press).

Astrophysical applications, for both white dwarfs and other stars, may be found in

▹ Michaud, G. 1970, *Ap.J.*, **160**, 641

▹ Fontaine, G., and Michaud, G. 1979, *Ap.J.*, **231**, 826

▹ Michaud, G., and Fontaine, G. 1984, *Ap.J.*, **283**, 787

▹ Iben, I. Jr., and MacDonald, J. 1985, *Ap.J.*, **296**, 540.

§**9.3**: The Magnetic White Dwarfs

A series of paper and reviews concerning the magnetic white dwarfs includes

▹ Angel, J.R.P., Borra, E.F., and Landstreet, J.D. 1981, *Ap.J. Suppl.*, **45**, 457

▹ Schmidt, G.D., West, S.C., Liebert, J., Green, R.F., and Stockman, H.S. 1986, *Ap.J.*, **309**, 218

▷ Schmidt, G.D. 1988, in IAU Colloquium No. 95, *Second Conference on Faint Blue Stars*, Eds. A.G.D. Philip, P.S. Hayes, and J. Liebert (Schenectady: L. Davis Press), p. 377

▷ Latter, W.B., Schmidt, G.D., and Green, R.F. 1987, *Ap.J.*, **320**, 308

▷ Chanmugam, G. 1992, *Ann. Rev. Astron. Ap.*, **30**, 143

Generation of magnetic fields in stars is discussed in

▷ Parker, E.N. 1977, *Ann. Rev. Astron. Ap.*, **15**, 45.

Other references pertinent to our discussion include

▷ Wendell, C.E., Van Horn, H.M., and Sargent, D. 1987, *Ap.J.*, **313**, 284

▷ Jackson, J.D. 1975, *Classical Electrodynamics*, 2nd ed. (New York: Wiley & Sons)

▷ Spitzer, L. 1962, *Physics of Fully Ionized Gases* (New York: Interscience)

▷ Angel, J.R.P. 1978, *Ann. Rev. Astron. Ap.*, **16**, 487

▷ Kurtz, D.W. 1990, *Ann. Rev. Astron. Ap.*, **28**, 607

▷ Unno, W., Osaki, Y., Ando, H., Saio, H., and Shibahashi, H. 1989, *Nonradial Oscillations of Stars* (Tokyo: University of Tokyo Press)

▷ Jaschek, C. , and Jaschek, M. 1987, *The Classification of Stars* (Cambridge: Cambridge University Press).

10
Asteroseismology

Shake, rattle, and roll.
Title and most of the lyrics of a popular song.
— Bill Haley & the Comets (1954)

With few exceptions we have treated stars as hydrostatic objects in which gravitational and pressure gradient forces are everywhere in balance. The exceptions have been supernovae, novae, and intrinsically variable stars, in order of decreasing violence of fluid motions. None of these objects are in hydrostatic equilibrium because force balance is not maintained and local accelerations cause the stellar fluid to move. This chapter shall consider only variable stars where mass motions involve relatively modest excursions of material and overall kinetic energies of motion are small compared to, for example, gravitational binding energies. The title of the chapter is nearly self-explanatory: we shall use observed surface luminosity, radius, and color variations to probe the interior in much the same spirit as in terrestrial seismology. This is not a new subject but it has blossomed in recent years to include stars ranging from the sun to classical Cepheid variables and white dwarfs, as a representative few of many objects. Even this short list makes it clear that the subject spans all the phases of stellar evolution.

The plan of the chapter is as follows. We first treat small-amplitude motions that are strictly periodic and radially symmetric. If the motions are strictly periodic then the time-averaged energy content of the star remains constant, which is the same as saying that the oscillations are *adiabatic*. This is, of course, an approximation to a real situation where energy redistribution within the star takes place over time scales that are very long compared to a period of oscillation (or pulsation, or one of a few other terms used to describe the variability). We shall then introduce nonadiabatic effects to briefly explore the causes of variability. Finally, we shall see what happens when the motions are not radially symmetric. Since there are three excellent texts in the literature (Cox 1980, and Unno *et al.* 1979, 1989) our treatment of these topics will sometimes be quick and dirty. Examples of applications to real stars will be given where appropriate with particular emphasis put on the sun and variable white dwarfs.

10.1 Adiabatic Radial Pulsations

We have frequently emphasized the point that a star is an object whose structure is primarily determined by mechanics. To understand this more clearly in the present context recall, from Chapter 1, that dynamic times ($t_{\rm dyn}$) are usually short compared to times characteristic of internal energy exchange within a star (e.g., $t_{\rm KH}$). This is not strictly true for all stars, or even the outer portions of most stars, but it forms the basis of the "adiabatic approximation" for the study of stellar pulsation. In this approximation it is assumed that *all* heat exchange mechanisms may be ignored so that the system is purely mechanical. The problem then reduces to a rather complicated exercise equivalent to studying the normal modes of a coupled system of pendulums and springs or, more appropriately, the behavior of sound waves confined in a box. The adiabatic approximation is remarkably useful in variable star theory because not only does it greatly simplify the analysis but it also yields accurate models of dynamic response for most stars. The penalty paid is severe, however, because it cannot tell us what causes *real* stars to pulsate. In this section we shall restrict the discussion to radially symmetric motions. This means that the star is always radially symmetric and all effects due to rotation, magnetic fields, etc., may be safely ignored.

Since heat transfer is ignored in the adiabatic approximation, we can completely describe the mechanical structure with only the mass and force equations

$$\frac{\partial \mathcal{M}_r}{\partial r} = 4\pi r^2 \rho \tag{10.1}$$

$$\ddot{r} = -4\pi r^2 \left(\frac{\partial P}{\partial \mathcal{M}_r}\right) - \frac{G\mathcal{M}_r}{r^2} \tag{10.2}$$

where we explicitly introduce partial derivatives to make sure derivatives with respect to time appear only where appropriate. If the star were purely static, then $\ddot{r}$ would be zero everywhere. Imagine that this indeed is initially the case but, by some means, the star is forced to depart from this initial hydrostatic equilibrium state in a radially symmetric, but otherwise arbitrary, manner. Furthermore, and to make the problem tractable, suppose that any departures from the static state are small in the following sense. Let a zero subscript on radius (r_0) or density (ρ_0) denote the local values of these quantities in the static state at some given mass level $\mathcal{M}_r$. As the motion commences both radius and density will, in general, depart from their static values at that same mass level and be functions of time and the particular mass level in question. This constitutes a Lagrangian description of the motion because we follow a particular mass level on which, we can imagine, all particles are painted red for identification to distinguish them from particles at other mass levels. We now describe the motion by letting

$$r(t,\mathcal{M}_r) = r_0(\mathcal{M}_r)\left[1+\delta r(t,\mathcal{M}_r)/r_0(\mathcal{M}_r)\right], \tag{10.3}$$

$$\rho(t,\mathcal{M}_r) = \rho_0(\mathcal{M}_r)\left[1+\delta \rho(t,\mathcal{M}_r)/\rho_0(\mathcal{M}_r)\right] \tag{10.4}$$

where δr and $\delta\rho$ are the *Lagrangian perturbations* of density and radius. These last two quantities are used to describe the motion through time at a given mass level. The requirement that departures from the static state be small is $|\delta r/r_0| \ll 1$ and $|\delta\rho/\rho_0| \ll 1$.

We now linearize the mass and force equations by replacing the position (radius) and density at a mass level by the perturbed values of (10.3) and (10.4) and, in the result, keeping only those terms that are of first or lower order in $\delta r/r_0$ and $\delta\rho/\rho_0$. (Recall that we did the same sort of thing back in §1.1 in examining a variational principle.) To see how this goes, first consider the mass equation

$$\frac{\partial \mathcal{M}_r}{\partial\left[r_0(1+\delta r/r_0)\right]} = 4\pi\left[r_0(1+\delta r/r_0)\right]^2\left[\rho_0(1+\delta\rho/\rho_0)\right]. \tag{10.5}$$

Now carry through the derivative in the denominator of the left-hand side and expand out the products on the right. The first operation yields a new denominator $(1+\delta r/r_0)\,\partial r_0 + r_0\,\partial(\delta r/r_0)$. The derivative ∂r_0 is then factored out so that the overall left-hand side contains the factor $\partial\mathcal{M}_r/\partial r_0$. The small terms remaining in the denominator are then brought up using a binomial expansion to first order to yield

$$\frac{\partial \mathcal{M}_r}{\partial r_0}\left[1-\frac{\delta r}{r_0}-r_0\frac{\partial\left(\delta r/r_0\right)}{\partial r_0}\right].$$

The right-hand side of the mass equation is simpler because all we need do is expand out the factors to first order to obtain

$$4\pi r_0^2\rho_0\left(1+2\frac{\delta r}{r_0}+\frac{\delta\rho}{\rho_0}\right).$$

When the two sides of the linearized mass equation are set equal we find that the result contains the zero-order equation

$$\frac{\partial\mathcal{M}_r}{\partial r_0} = 4\pi r_0^2\rho_0,$$

which is the mass equation for the unperturbed configuration. Since this is automatically satisfied, we take advantage of the equality and subtract this from the linearized equation. (This is a typical result of linearization about an equilibrium state.) After some easy rearrangement we find the following relation between the Lagrangian perturbations that must be satisfied in order that mass conservation be met for the configuration as it evolves in time:

$$\frac{\delta\rho}{\rho_0} = -3\,\frac{\delta r}{r_0} - r_0\,\frac{\partial\left(\delta r/r_0\right)}{\partial r_0}. \tag{10.6}$$

Note that part of this equation is familiar because, if we ignore the derivative term, it is merely the logarithmic form of the homology relation between density and radius given by (1.62) of §1.6.

The force equation is relatively straightforward and we leave it as an exercise to show that its linearization yields[1]

$$\rho_0 r_0\,\frac{d^2\delta r/r_0}{dt^2} = \rho_0 r_0\left(\frac{\ddot{\delta r}}{r_0}\right) = -\left(4\,\frac{\delta r}{r_0} + \frac{\delta P}{P_0}\right)\frac{\partial P_0}{\partial r_0} - P_0\frac{\partial\,\delta P/P_0}{\partial r_0}. \tag{10.7}$$

Implicit in the derivation of this equation are the conditions $\ddot{r}_0 = 0$ and $\dot{r}_0 = 0$, which must apply since the reference state is completely static.

At this point in the analysis we take the usual path in perturbation theory and assume that *all* perturbations prefixed by δs may be decomposed into Fourier components with the time element represented by exponentials. Thus, for example, introduce the space component of relative fluid displacement, $\zeta(r_0)$, by

$$\frac{\delta r}{r_0}(t, r_0) = \zeta(r_0)\,e^{i\sigma t} \tag{10.8}$$

where the exponential takes over the duties of describing the time evolution of displacement and $\zeta(r_0)$, which depends only on r_0 (i.e., the mass level), can be considered as the shape of the displacement at zero time. Note that both σ and $\zeta(r_0)$ can be complex. The left-hand side of the force equation now becomes $-\rho_0 r_0 \sigma^2 \zeta(r_0)\,e^{i\sigma t}$.

It should be clear that we are now in trouble because the two linearized equations for force and mass contain the three variables $\zeta(r_0)$, and the space parts of the pressure and density perturbations. This comes about because we have neglected the energetics of the real system and so our description is incomplete. To turn this into a purely mechanical problem we now couple $\delta\rho$ and δP in the adiabatic approximation by recalling the Lagrangian relation between changes in pressure to changes in density given by (3.87)

$$\Gamma_1 = \left(\frac{\partial \ln P}{\partial \ln \rho}\right)_{\rm ad}. \tag{10.9}$$

[1]You may wish to look through Cox (1980, §5.3) for a discussion about how δs and total and partial time derivatives work with one another.

Since this is shorthand for $P \propto \rho^{\Gamma_1}$ and δ is a Lagrangian differential operator, we merely take logarithmic δ-derivatives to find

$$\frac{\delta P}{P_0} = \Gamma_1 \frac{\delta\rho}{\rho_0}. \tag{10.10}$$

This relation takes the place of any energy and heat transfer equations that would normally appear and we now have as many variables as equations.

There are several paths we could take now but we choose the following: (1) make sure all perturbations are replaced by their spatial Fourier components with common factors of $e^{i\sigma t}$ cancelled; (2) replace all occurrences of $\delta\rho$ by δP using the adiabatic condition; (3) rearrange the two linearized equations so that space derivatives appear on the left-hand side; (4) replace partial derivatives by total space derivatives (with the understanding that all variables depend only on r_0); (5) delete all reference to zero subscripts because all that really appears are perturbations and quantities from the static configuration. The result is

$$\frac{d\zeta}{dr} = -\frac{1}{r}\left(3\zeta + \frac{1}{\Gamma_1}\frac{\delta P}{P}\right), \tag{10.11}$$

$$\frac{d\left(\delta P/P\right)}{dr} = -\frac{d\ln P}{dr}\left(4\zeta + \frac{\sigma^2 r^3}{G\mathcal{M}_r}\zeta + \frac{\delta P}{P}\right) \tag{10.12}$$

where the factor $r^3/G\mathcal{M}_r$ appears as a result of using the hydrostatic equation to get rid of some terms containing dP/dr (which you may wish to retain rather than introducing $\mathcal{M}_r$).

We now have a set of coupled first-order differential equations but we need boundary conditions. The first of these is simple because we require that δr be zero at the center ($r = 0$). To see how this comes about consider a particle of infinitesimal extent at the very center of the equilibrium star. There is no place for this particle to move to ($\delta r \neq 0$) without violating the condition of radial symmetry. Physical regularity of the solutions also requires that both ζ and $d\zeta/dr$ be finite at the center. The only way to arrange for all this to be true is to have the term in parenthesis on the right-hand side of (10.11) vanish at stellar center. This yields the first boundary condition

$$3\zeta + \frac{1}{\Gamma_1}\frac{\delta P}{P} = 0, \quad \text{at } r = 0. \tag{10.13}$$

The second boundary condition is applied at the surface. For our purposes it is adequate to assume zero boundary condition for the static model star (as in §7.1). Specifically, we assume $P \to 0$ as $r \to \mathcal{R}$. More complicated conditions are possible—such as for a photospheric surface—but they add nothing of real importance for our discussion. The first thing to realize is that the leading coefficient of the right-hand side of the linearized force equation is just $1/\lambda_P$ where λ_P is the pressure scale height. This latter quantity rapidly goes to zero as the surface is approached so that in order for the relative pressure perturbation, $\delta P/P$, to remain finite we must have

$$4\zeta + \frac{\sigma^2 \mathcal{R}^3}{G\mathcal{M}}\zeta + \frac{\delta P}{P} = 0, \quad \text{at } r = \mathcal{R}. \tag{10.14}$$

Though not immediately evident, this condition is equivalent to requiring that all interior disturbances be reflected at the surface (as it itself moves) back into the interior; that is, no pulsation energy is lost from the star because all is reflected back inward from the surface.

Thus far all looks well. We have an equal number of differential equations and boundary conditions. But, all the equations derived thus far are linear and homogeneous in ζ and $\delta P/P$ so the question remains as to how these quantities are to be normalized. As it stands, any scaling is permitted for either perturbation at some unspecified point in the star and the overall solution may be as small or large as we like. To pin things down we must choose a nonzero normalization. This is completely arbitrary but certain choices are preferred (and differ among different investigators). We choose

$$\zeta = \frac{\delta r}{r} = 1, \quad \text{at } r = \mathcal{R}. \tag{10.15}$$

We now realize that this places an additional constraint on the problem and, in effect, we have exceeded the permissible number of boundary conditions. The way out of this apparent dilemma is to recognize that the (perhaps complex) frequency σ has not been specified. In fact, it can only take on a value (or values) such that all boundary conditions are satisfied including the normalization condition. (Note that σ cannot depend on the normalization condition because the latter just scales the solutions.) Thus σ or, more properly, σ^2—because only that quantity appears in our equations—is an *eigenvalue* and the corresponding perturbations are *eigenfunctions* for that particular σ^2. We now discuss the properties of the eigenvalues for this adiabatic problem and this will involve a little mathematics.

10.1.1 The Linear Adiabatic Wave Equation

First, we leave it to you to collapse the two first-order differential equations for ζ and $\delta P/P$ down into one second-order equation in ζ. (This involves differentiating 10.11 and then eliminating any reference to $\delta P/P$ or its derivative by using 10.11 and 10.12.) The result is

$$\mathbf{L}(\zeta) \equiv -\frac{1}{\rho r^4}\frac{d}{dr}\left(\Gamma_1 P r^4 \frac{d\zeta}{dr}\right) - \frac{1}{r\rho}\left\{\frac{d}{dr}\left[(3\Gamma_1 - 4)\, P\right]\right\}\zeta = \sigma^2\zeta. \tag{10.16}$$

Here $\mathbf{L}$ is a second-order differential operator that is shorthand for the middle part of the whole equation where, in this case, ζ is the operand. We can write the above in simple form as $\mathbf{L}(\zeta) = \sigma^2\zeta$. It may not look like it at first sight but, with some hindsight, this equation is a wave equation and, in this context, is called the *linear adiabatic wave equation* or LAWE.[2]

This all may look pretty formidable but there are redeeming features. (You should consult any decent text on mathematical physics—such as Arfken 1985—for what follows if you wish to do serious work with the theory of pulsating stars.) All the quantities in $\mathbf{L}$ are well behaved and $\mathbf{L}$ is a Sturm–Liouville operator. Furthermore, we can symbolically integrate over the star and show (subject to our boundary conditions and other constraints—see Cox 1980, §8.8) that

$$\int_0^{\mathcal{M}} \zeta^*(\mathbf{L}\zeta)\, r^2\, d\mathcal{M}_r = \int_0^{\mathcal{M}} \zeta(\mathbf{L}\zeta)^*\, r^2\, d\mathcal{M}_r \tag{10.17}$$

where ζ^* is the complex conjugate of ζ. Now we may as well be doing quantum mechanics because this equality means that the Sturm–Liouville operator $\mathbf{L}$ is self-adjoint (or Hermitian) and the following statements about σ^2 and its eigenfunctions are true:

1. All eigenvalues σ^2 of the system are real, as are the corresponding eigenfunctions. There are then two possibilities. If $\sigma^2 > 0$ then σ is purely real and the complete eigenfunction $\zeta(r)\, e^{i\sigma t}$ is oscillatory in time by virtue of the temporal factor $e^{i\sigma t}$. Otherwise, if $\sigma^2 < 0$, then σ is pure imaginary and the perturbations grow or decay exponentially with time. We shall only concern ourselves with the first possibility in practical situations. Thus if $\sigma^2 > 0$, then σ is the angular frequency of the oscillation with corresponding period $\Pi = 2\pi/\sigma$.
2. There exists a minimum value for σ^2 which, were we doing quantum mechanics, would correspond to the ground state.
3. If ζ_j and ζ_k are two eigenfunction solutions for eigenvalues σ_j^2 and σ_k^2, then

$$\int_0^{\mathcal{M}} \zeta_j^* \zeta_k\, r^2\, d\mathcal{M}_r = 0 \quad \text{if } j \neq k. \tag{10.18}$$

   The eigenfunctions are then orthogonal in this sense.

What we then have for $\sigma^2 > 0$ are standing waves such that the star passes through the equilibrium state twice each period.

[2]Some investigators refer to more general versions of this equation as the LAWE. Cox (1980), for example, includes nonradial motions in his formulation.

10.1.2 Some Examples

To get an idea of what is going on here we first consider the (unrealistic) case where both ζ and Γ_1 are supposed constant throughout the star. (Were such a situation possible it would correspond to homologous motions.) The LAWE then reduces to

$$-\frac{1}{r\rho}\left(3\Gamma_1 - 4\right)\frac{dP}{dr}\zeta = \sigma^2\zeta. \tag{10.19}$$

In the simple case of the constant-density model $[\rho(r) = \langle\rho\rangle]$ we replace $-(1/\rho r)\, dP/dr$ by $G\mathcal{M}_r/r^3$, which becomes $4\pi G\langle\rho\rangle/3$. The result is

$$\left(3\Gamma_1 - 4\right)\frac{4\pi G}{3}\langle\rho\rangle = \sigma^2. \tag{10.20}$$

If $\Gamma_1 > 4/3$, then σ is real and the corresponding period is

$$\Pi = \frac{2\pi}{\sigma} = \frac{2\pi}{\sqrt{\left(3\Gamma_1 - 4\right)\langle\rho\rangle\, 4\pi G/3}}. \tag{10.21}$$

This is just the "period-mean density" relation derived in §1.3.5 but it is now clear how Γ_1 enters. If, on the other hand, $\Gamma_1 < 4/3$ we know enough to expect trouble. Here σ is imaginary and the e-folding time for either growth or decay of the motions is

$$\tau = \frac{1}{|\sigma|} = \frac{1}{\sqrt{|3\Gamma_1 - 4|\langle\rho\rangle\, 4\pi G/3}}. \tag{10.22}$$

This is the free-fall time (corrected for various factors), $t_{\rm dyn}$, discussed in §1.3.3.

More realistic examples are periodic oscillations in polytropes. Recall from §7.2.1 that the dependent variable for polytropes is θ, which is related to the pressure by $P = P_c\theta^{1+n}$ where n is the polytropic index. The density is given by $\rho = \rho_c\theta^n$ and the independent variable is ξ, which is proportional to radius. We introduce these variables into the two differential equations for adiabatic radial pulsations (10.11) and (10.12), use various relations derived in §7.2.1, and find

$$\frac{d\zeta}{d\xi} = -\frac{1}{\xi}\left[3\zeta + \frac{1}{\Gamma_1}\frac{\delta P}{P}\right] \tag{10.23}$$

$$\frac{d\left(\delta P/P\right)}{d\xi} = -(1+n)\frac{\theta'}{\theta}\left[4\zeta + \omega^2\frac{\xi/\theta'}{(\xi/\theta')_1}\zeta + \frac{\delta P}{P}\right] \tag{10.24}$$

where ω^2 is the dimensionless frequency (squared)

$$\omega^2 = \frac{\mathcal{R}^3}{G\mathcal{M}}\sigma^2. \tag{10.25}$$

The prime on θ denotes $d\theta/d\xi$ and the subscript 1 in the middle term of (10.24) means that the term is to be evaluated at the surface of the polytrope where $\xi = \xi_1$. Both $\theta(\xi)$ and $\theta'(\xi)$ are known from the equilibrium polytrope solution. We could have phrased the above in terms of $\delta\xi/\xi$ and $\delta\theta/\theta$ but keeping the more physical variables $\zeta = \delta r/r$ and $\delta P/P$ is preferred.

As an example, consider the oscillations of an $n = 2$ polytrope with Γ_1 of 5/3.[3] The insert in Figure 10.1 shows $\theta(\xi)$ versus ξ for that index polytrope. The main curves show three solutions for ζ, each corresponding to a different eigenvalue ω^2. The curves are labeled "fundamental" for the smallest $\omega^2(= 4.001)$, "first overtone" for the next largest ($\omega^2 = 13.34$), and "second overtone" for the third largest ($\omega^2 = 26.58$). The nomenclature for these modes of oscillation agrees with that used in acoustics. We could also correctly have called the fundamental the "first harmonic," and the first overtone the "second harmonic," and so on. But beware, astronomers are not consistent (nor always correct) in their use of these terms and you will often see the first overtone given the name "first harmonic."

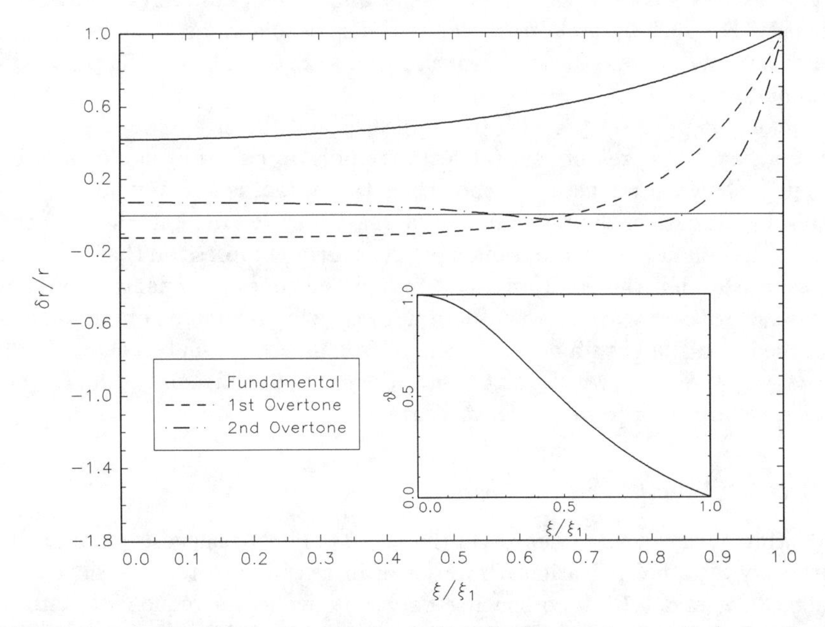

FIGURE 10.1. Shown are the relative radial displacements $\zeta = \delta r/r$ for three modes in an $n = 2$ polytrope. The abscissa is the polytropic variable ξ given in units of its value at the surface (or, in other words, $r/\mathcal{R}$). Also shown in the insert is the run of θ with ξ for the equilibrium polytrope.

[3]With some care you can do this yourself as a computing exercise. See, for example, §38.3 in Kippenhahn and Weigert (1990).

You should note that the fundamental has no nodes (crossings of the $\zeta = 0$ axis) for $\delta r/r$ whereas each successive overtone has one more node than the preceding. Thus there is a one-to-one correspondence of complexity in eigenfunction with ordering of eigenvalue and we might as well be discussing vibrating strings, the hydrogen atom, or almost any other wave phenomenon. The fluid motions of the fundamental mode are not homologous because ζ is not a constant but they are still simple since the star expands and contracts almost uniformly and in phase.[4] (Some refer to this as a "breathing mode.") As a rough rule of thumb, the contrast between the central value of ζ to its surface value (here normalized to one) is $\zeta(\mathcal{R})/\zeta(r=0) \sim \rho_c/\langle\rho\rangle$. In our example, the degree of central concentration for the $n = 2$ polytrope is $\rho_c/\langle\rho\rangle = 11.4$ and the corresponding ratio of the displacements is a little over three. The motion of the fundamental mode, and even more so the overtones, has been likened to a whip whose heavy handle moves only slightly in contrast to the wild excursions of the whip's end. An $n = 2$ polytrope is only mildly centrally condensed but, on the other hand, a highly evolved star, such as a classical Cepheid variable, has a core of very high density and a comparatively low average density. For such variable stars the fluid displacements near the center are miniscule and, in practice, the central regions are often ignored in pulsation calculations.

Before we press on, we wish to emphasize that the eigenfunctions shown in Figure 10.1 represent real (at least for polytropes) fluid motions. Each mode is shown at an instant of time when the surface is at maximum expansion. In the case of the fundamental, a quarter cycle later (at $t = \Pi/4$) the "star" has already undergone enough subsequent compression that it is just passing through the equilibrium state and the time-dependent eigenfunction is zero everywhere. In another quarter cycle maximum compression is reached and the eigenfunction is the mirror image of what is shown with $\delta r/r$ negative everywhere. Expansion then starts again and, a half cycle later, we are back from where we started.

10.1.3 Asymptotic Analysis

A listing of the eigenfrequencies of the $n = 2$ polytrope suggests yet another property of radial pulsations. In order of increasing overtone we have $\omega =$ 2.00, 3.65, and 5.16. The spacing between successive frequencies here is certainly not constant but it does hover around 1.6. Were we to compute

[4]The quantity Q, defined as $Q = \Pi\sqrt{\langle\rho\rangle/\langle\rho_\odot\rangle}$ is equal to 0.058 days for the fundamental mode of an $n = 2$, $\Gamma_1 = 5/3$ oscillating polytrope. The corresponding value for Q in the standard model ($n = 3$, see §7.2.7) is 0.038 days. This last value was used in the remarks made after equation (1.37) for the period-mean density relation. Q depends on the central concentration of the model with the largest value belonging to a constant-density ($n = 0$) model.

successive overtones to high radial order (in terms of number of nodes) we would find that the spacing in frequency between modes would approach a constant. This is a common characteristic of mechanical systems and it follows from an "asymptotic analysis" of the wave equation. To show how this works we first convert the LAWE (10.16) into something that looks more like a wave equation by the substitution (see Tassoul and Tassoul 1968)

$$w(r) = r^2 (\Gamma_1 P)^{1/2} \zeta(r). \tag{10.26}$$

A bit of manipulation then yields the new wave equation in $w(r)$

$$\frac{d^2 w}{dr^2} + \left[\frac{\sigma^2 \rho}{\Gamma_1 P} - \phi(r)\right] w = 0 \tag{10.27}$$

where $\phi(r)$ is the not very edifying function

$$\phi(r) = \frac{2}{r^2} + \frac{2}{\Gamma_1 P r}\frac{d(\Gamma_1 P)}{dr} - \left[\frac{1}{2\Gamma_1 P}\frac{d(\Gamma_1 P)}{dr}\right]^2$$
$$+\frac{1}{2\Gamma_1 P}\frac{d^2(\Gamma_1 P)}{dr^2} - \frac{1}{\Gamma_1 P r}\frac{d}{dr}\left[(3\Gamma_1 - 4) P\right]. \tag{10.28}$$

The first term in the brackets of the new wave equation is simpler than it appears because the combination $\Gamma_1 P/\rho$ is just the square of the local sound speed v_s^2 (as in eq. 1.35). Thus

$$\frac{d^2 w}{dr^2} + \left[\frac{\sigma^2}{v_s^2} - \phi\right] w = 0. \tag{10.29}$$

Now assume that the wave function $w(r)$ may be represented as

$$w(r) \propto e^{i k_r r} \tag{10.30}$$

where k_r is a wave number which, in general, is a function of r. Because we shall eventually consider only large wave numbers, the terms on the right-hand side of (10.29) can be taken as roughly constant over a wavelength $\lambda_r = 2\pi/k_r$. Hence radial derivatives of k_r shall be assumed to be negligible; that is, k_r is *nearly* constant. The wave equation in disguised form then becomes

$$k_r^2 = \frac{\sigma^2}{v_s^2} - \phi \tag{10.31}$$

which is a dispersion relation for k_r. Thus if $\sigma^2 > v_s^2\phi$ then the argument of the exponential in w of (10.30) is purely imaginary and w is sinusoidal. This is characteristic of *propagating* or standing waves. However, if $\sigma^2 < v_s^2\phi$ then the solution for w contains exponentially decreasing or increasing components and the solution is said to be *evanescent*. The quantum mechanical analogue here is a particle in a potential well. If the particle's energy (which might be proportional to some σ^2) is greater than that of the bottom of the well, then the eigenfunction is sinusoidal within the well. The likelihood is small, however, of finding the particle in regions where its energy is less than the local potential because the solutions decay exponentially in those regions.

We now take the road into "asymptopia" and imagine that $\sigma^2/v_s^2 \gg \phi$ and $k_r r \gg 1$ within some interval of radius, say, $a \leq r \leq b$. This means that there are many wavelengths packed into that interval and the "particle" is high above the bottom of the potential well. We idealize the situation further by supposing that the potential well at a and b is so steep that the wave function is contained solely within those endpoints. Under these conditions the eigenfunction is trapped within the well and a mode, which consists of a standing wave for the mechanical system, must have an integer or half-integer number of wavelengths within $[a, b]$. To measure the number of wavelengths we must integrate k_r over the interval because k_r is still a function of r. (The simple expression $k_r[b - a]$ is not correct.) Thus our requirement for a true mode is, as you may verify with a little thought

$$\int_a^b k_r \, dr = (n+1)\pi \tag{10.32}$$

where n is the number of nodes the eigenfunction has within the interval not including the endpoints.[5] Thus $n = 0$ corresponds to a half-wave and the fundamental mode, and $n = 1$ is the first overtone with one complete wavelength packed into $[a, b]$, and so on. We are, of course, not in this low-overtone domain, but this is the general scheme.

Since ϕ is assumed to be small then $k_r = \sigma/v_s$ which gives us

$$\sigma = (n+1)\pi \left[\int_a^b \frac{dr}{v_s}\right]^{-1} . \tag{10.33}$$

If we define the constant frequency $\sigma_0 = \pi / \int_a^b v_s^{-1} \, dr$, then the asymptotic behavior for σ is

$$\sigma = (n+1)\sigma_0 \tag{10.34}$$

so that the eigenfrequencies σ are equally spaced from one mode to the next with spacing σ_0. To date there is only one star for which we can actually observe high-order radial modes and this is the sun, although such pulsations may exist for some white dwarfs (Kawaler 1993). Whether of high overtone or not, these modes are often called radial pressure modes

[5] We use n here to count the number of nodes and to denote mode order. This is the usual nomenclature used by those doing solar seismology. The white dwarf seismology community, on the other hand, tends to use k for the same purposes. Our apologies to our colleagues among the white dwarfers but k is too easily confused with wave number.

(p-modes), or radial acoustic modes. The reason for this designation should be obvious from §1.3.5 where we considered the sound travel time across a star. Sound waves are propagating pressure disturbances and the preceding discussion containing ratios of dr/v_s, which measure travel times for sound waves, is the same thing.

Adiabatic radial pulsations in stars are relatively well understood and modern calculations do a good job in representing these purely mechanical motions. This is a desirable state of affairs because most classes of intrinsically variable stars are radial pulsators (as discussed in §2.6.1). However, "intrinsically" implies that it is something inside the star that causes the pulsation. What we have done thus far cannot tell us what that something is because the purely adiabatic periodic motions we have considered have no beginning, end, or apparent cause—which means we have to relax the constraint of adiabaticity.

• Appendix C discusses an adiabatic radial pulsation code (written in `FORTRAN`) called `PULS.FOR` that uses output from `ZAMS`. (The code will also compute nonradial modes.) You may wish to compute a series of harmonics for some ZAMS model and investigate the ordering of eigenvalues and the structure of the corresponding eigenfunctions. •

10.2 Nonadiabatic Radial Motions

By nonadiabatic we mean that heat may be exchanged between moving fluid elements of the star. The take-off point for the analysis is the energy equation given in time-dependent form. A good deal of the work was already done in Chapter 6, where gravitational energy sources were discussed. We now combine equations (6.4) and (6.5) from that chapter to obtain an expression for the mass gradient of luminosity that implicitly contains gravitational terms. Thus

$$\frac{\partial \mathcal{L}_r}{\partial \mathcal{M}_r} = \varepsilon - \frac{P}{\rho\,(\Gamma_3 - 1)}\left[\frac{\partial \ln P}{\partial t} - \Gamma_1 \frac{\partial \ln \rho}{\partial t}\right] \tag{10.35}$$

where ε is the thermonuclear energy generation rate. What we wish to do, for reasons to be made clear later, is to eventually replace the time derivative of pressure with one of temperature. This requires some thermodynamics, and the first step is to use (3.91) to convert the multiplier of the brackets to

$$\frac{P}{\rho\,(\Gamma_3 - 1)} = \frac{c_{V_\rho} T}{\chi_T}.$$

The energy equation then becomes, after minor rearrangement,

$$\frac{\partial \ln P}{\partial t} = \Gamma_1 \frac{\partial \ln \rho}{\partial t} + \frac{\chi_T}{c_{V_\rho} T}\left[\varepsilon - \frac{\partial \mathcal{L}_r}{\partial \mathcal{M}_r}\right]. \tag{10.36}$$

(Note that the adiabatic case is regained if the last term is always zero for thermal balance.) Now use the two middle terms of (3.94) to get rid of Γ_1. It is then relatively easy to manipulate χ_T and χ_ρ and ratios of partial derivatives to obtain the desired result

$$\frac{\partial \ln T}{\partial t} = (\Gamma_3 - 1)\frac{\partial \ln \rho}{\partial t} + \frac{1}{c_{V_\rho} T}\left[\varepsilon - \frac{\partial \mathcal{L}_r}{\partial \mathcal{M}_r}\right]. \tag{10.37}$$

We now linearize the energy equation with the replacements

$$\begin{aligned} T &\longrightarrow T_0 + \delta T \\ \rho &\longrightarrow \rho_0 + \delta\rho \\ \varepsilon &\longrightarrow \varepsilon_0 + \delta\varepsilon \\ \mathcal{L}_r &\longrightarrow \mathcal{L}_{r,0} + \delta\mathcal{L}_r \end{aligned}$$

where, as usual, zero subscripts refer to the equilibrium state. (You may easily confirm later that we need not vary c_{V_ρ} or Γ_3 because the resulting variations will not appear in our final expressions.) If we insist that the equilibrium state is in both hydrostatic equilibrium and thermal balance (with $\varepsilon_0 = \partial\mathcal{L}_{r,0}/\partial\mathcal{M}_r$ and zero time derivatives of T_0 and ρ_0) then, after dropping zero subscripts as before,

$$\frac{\partial(\delta T/T)}{\partial t} = (\Gamma_3 - 1)\frac{\partial(\delta\rho/\rho)}{\partial t} + \frac{1}{c_{V_\rho} T}\left[\delta\varepsilon - \frac{\partial\,\delta\mathcal{L}_r}{\partial\mathcal{M}_r}\right]. \tag{10.38}$$

Finally let all perturbations vary as $e^{i\omega t}$ to find our final form of the linearized energy equation

$$\delta\varepsilon - \frac{\partial\,\delta\mathcal{L}_r}{\partial\mathcal{M}_r} = i\omega\, c_{V_\rho} T\left[\frac{\delta T}{T} - (\Gamma_3 - 1)\frac{\delta\rho}{\rho}\right]. \tag{10.39}$$

The δs again refer only to the space parts of the perturbations. (The perhaps complex frequency ω used here is not the dimensionless one introduced for polytropes.) Note, first off, that this equation contains the imaginary unit $i = \sqrt{-1}$ and we suspect, correctly, that the nonadiabatic problem results in eigenvalues and eigenfunctions that are complex. The solutions then automatically contain either exponentially growing ("driving" or "unstable") or decaying ("damping" or "stable") properties. An intrinsically variable star is one in which nonadiabatic effects drive the star to pulsational instability.

To go on and discuss the full nonadiabatic problem is beyond the scope of this text. You are invited to peruse Cox (1980) to appreciate the difficulty of the subject. We shall, however, give some relevant pointers to what is going on by first considering an analysis due to Sir Arthur S. Eddington (1926) that is reviewed in Cox (1980, §9.4) and Clayton (1968, §6–10). His approximation hinges on the assumption that the motions in a variable star are *almost* adiabatic in that the time scale for growth of an instability is long compared to the period of an oscillation. That is, over the bulk of a star the mechanics (operating on short time scales) dominate over the slower effects of heat exchange. This is paramount to saying that if we observe a variable star at the beginning of one cycle of oscillation (whenever that may be), then, if we wait for one mechanically determined period, the star will return *almost* precisely back to that beginning state. In fact, to the lowest order of approximation, we shall assume that initial and final states are thermodynamically identical. What we shall look for are higher-order effects that give us a sense of how big is the "almost" in the above.

Consider then a shell of mass $\Delta\mathcal{M}$ at some radius in a spherical star. If the physical properties in this shell vary over some portion of a pulsation cycle we can use the first law of thermodynamics to describe the relation between the work done (dW) by the shell on its surroundings, the internal energy (dE) gained by the shell, and the heat added to the shell (dQ):

$$dQ = dE + dW. \tag{10.40}$$

For a complete cycle this becomes

$$\oint dQ = \oint dE + \oint dW \tag{10.41}$$

where $\oint$ means that we compute the integral over only one cycle. If the shell of mass truly returns to its initial thermodynamic state over one cycle then the whole process is reversible and $\oint dE$ vanishes because E is a state variable. We are left with $W = \oint dQ$ as the work done by the shell on its surroundings in one cycle. To proceed further, the change in entropy for this reversible system is given by $dS = dQ/T$. Suppose we start the temperature at T_0 in the initial state. After some time has elapsed, the temperature of the shell will have changed to, say, $T = T_0 + \delta T$. To first-order the entropy change between the two states will be

$$dS = \frac{dQ}{T} \approx \frac{dQ}{T_0} - \frac{\delta T}{T^2}\, dQ \tag{10.42}$$

where dQ is the heat added over that time. Over one reversible cycle, however, $\oint dS = 0$ because S is also a state variable. This leaves us with

$$\oint \frac{dQ}{T} = \oint \frac{dQ}{T_0} - \oint \frac{\delta T}{T_0^2}\, dQ = 0. \tag{10.43}$$

We may pull the constant T_0 from out of the first integral in the middle term (leaving just $\oint dQ$) and substitute into the expression for W to find, to first-order,

$$W = \oint \frac{\delta T}{T}\, dQ. \tag{10.44}$$

This result looks peculiar because $\delta T/T$ now appears in the integrand for W. What has happened here is that W would be precisely zero over a cycle if the star really returned to its initial state. What we have picked up by our argument is that small piece of W that may differ from zero.

Finally, we consider all mass shells in the star by integrating over mass to find the total work done

$$W_{\text{tot}} = \int_{\mathcal{M}} \oint \frac{\delta T}{T} \, dQ \, d\mathcal{M}_r. \tag{10.45}$$

This is interpreted as follows. If $W_{\text{tot}} > 0$ then the pulsating star has done work on itself over one cycle (by means still to be explored) and any initial perturbation will increase. Note that some mass elements may contribute negatively to the whole but the overall effect is positive. In this case we say that the star is driving the pulsations and is unstable. If we take the integral apart, we see that a positive contribution comes about if heat is absorbed ($dQ > 0$) when temperature is on the increase ($\delta T > 0$) but heat must be emitted when temperatures are on the decline. In this sense a variable star is a self-contained heat engine. An ordinary stable star cannot accomplish self-driven motion because were it to try, so to speak, perturbations would be damped out because the preponderance of perturbed mass elements would lose heat to their surroundings as temperatures increase and gain heat as they cool.

We now associate dQ with $(\delta\varepsilon - \partial\,\delta\mathcal{L}_r/\partial\mathcal{M}_r)\,dt$ (as in 10.38) because the latter is the heat added to (or subtracted from) the mass shell over a time dt. Thus W_{tot} becomes

$$W_{\text{tot}} = \int_{\mathcal{M}} \oint \frac{\delta T}{T} \left[\delta\varepsilon - \frac{\partial\,\delta\mathcal{L}_r}{\partial\mathcal{M}_r}\right] dt \, d\mathcal{M}_r. \tag{10.46}$$

This tells us something about the reality of eigenvalues and eigenfunctions because the term in brackets sits at the left-hand side of equation (10.39). Suppose that ω, δT, and $\delta\rho$ are all real. This implies that $[\delta\varepsilon - \partial\delta\mathcal{L}_r/\partial\mathcal{M}_r]$ is pure imaginary and we are led to a contradiction as seen from the following. One of the integrals for W_{tot} is over time, so we must include the factor $e^{i\omega t}$ in all variations. Since the spatial part of $dQ\,dt$—which is the right-hand side of (10.39) multiplied by dt—contains i it must be 90° out of phase with $\delta T\, e^{i\omega t}$. If you sketch the time behavior of the integrand of W_{tot} on the complex plane and then transfer the real part of the integrand to a real-time axis (to get physical results) you will immediately see that the result is a purely periodic curve over one period. That is, W_{tot} must be zero and the system is strictly conservative. Thus, in order for stars to be self-excited, the frequency ω and associated quantities must be complex so that the resultant phasing in W_{tot} yields a nonzero result.

10.2.1 The Quasi-Adiabatic Approximation

Another way to approach the question of instability is to derive a differential equation for $\delta r(t, \mathcal{M}_r)/r$ in nonadiabatic form. The result will be an expanded version of the LAWE which, when used in a useful approximation, will give estimates of the imaginary part of ω and hence the e-folding times for driving or damping.

We first differentiate the linearized force equation (10.7) with respect to time. The result will contain a term like $\partial\left[\delta P/P_0\right]/\partial t$, which can be eliminated using a linearized verison of the energy equation (10.36). (This is where we depart from the adiabatic analysis.) Terms in $\partial\left[\delta\rho/\rho_0\right]/\partial t$ are dealt with by using the mass equation (10.6). A fair amount of algebraic simplification (see Cox 1980, §7.7) and introduction of $\delta r/r = \zeta\, e^{i\omega t}$ then yields

$$i\omega\mathbf{L}(\zeta) - i\omega^3\zeta = -\frac{1}{r\rho}\frac{d}{dr}\left[\rho\left(\Gamma_3 - 1\right)\left(\delta\varepsilon - \frac{\partial\,\delta\mathcal{L}_r}{\partial\mathcal{M}_r}\right)\right] \tag{10.47}$$

where $\mathbf{L}$ is the linear operator of the LAWE (10.16). We use this equation as follows. Suppose we separate ω into two parts so that

$$\omega = \sigma + i\kappa_{\text{qa}} \tag{10.48}$$

where both σ and κ_{qa} are real. If our suspicions are correct, then σ (as the frequency of pulsation) should be determined primarily by adiabatic and mechanical processes and thus, as a presumably good approximation, we set it equal to the adiabatic frequency. The quantity κ_{qa} will then measure the rate of driving or damping because $e^{i\omega t} = e^{i\sigma t - \kappa_{\text{qa}} t}$; that is, $-\kappa_{\text{qa}}$ is an inverse e-folding time with $\kappa_{\text{qa}} < 0$ (> 0) implying driving (damping). We expect $\sigma \gg |\kappa_{\text{qa}}|$ because of the assumed rapidity of mechanical versus thermal effects.

The next step is essentially the equivalent of first-order perturbation theory in quantum mechanics. If we have a process that dominates the behavior of a system—such as in the mechanical oscillations of a star—but this is modified by a weaker process (the nonadiabatic effects), then the shift in the energy of the system (the shift in eigenvalue) may be estimated by using only the eigenfunctions of the mechanical system in an integral method without actually solving the complete problem. It goes like this.

We assume that any eigenfunctions required in the application of (10.47) are those obtained from an adiabatic calculation. Thus we replace any occurrence of ζ by ζ_{ad} where the subscript "ad" implies adiabatic. Now multiply (10.47) on the left by ζ_{ad} and integrate the result over $r^2\, d\mathcal{M}_r$ through the entire mass. This yields

$$i\omega\int_{\mathcal{M}} \zeta_{\text{ad}}\mathbf{L}\zeta_{\text{ad}}\, r^2\, d\mathcal{M}_r - i\omega^3\int_{\mathcal{M}} \zeta_{\text{ad}}\zeta_{\text{ad}}\, r^2\, d\mathcal{M}_r =$$

$$
\begin{aligned}
& - \int_{\mathcal{M}} \frac{\zeta_{\rm ad}}{r\rho} \frac{d}{dr} \left[\rho \left(\Gamma_3 - 1 \right) \left(\delta\varepsilon - \frac{\partial \delta \mathcal{L}_r}{\partial \mathcal{M}_r} \right)_{\rm ad} \right] r^2 \, d\mathcal{M}_r \qquad (10.49) \\
\equiv \quad & -C_{\rm qa}
\end{aligned}
$$

and defines C_{qa} as the integral on the second line. The "qa" subscript that appears here and in $\kappa_{\rm qa}$ means "quasi-adiabatic" because some of what we are doing is adiabatic, whereas other parts contain nonadiabatic elements.[6] Since σ is large compared to $|\kappa_{\rm qa}|$, the $i\omega^3$ in the second term becomes approximately equal to $\sigma^2 \left(i\sigma - 3\kappa_{\rm qa} \right)$ after small terms are dropped. The factor $i\omega$ is just $i\omega = i\sigma - \kappa_{\rm qa}$. We now recall from (10.17) that the integrated LAWE is, in our new notation for the adiabatic eigenfunction,

$$
\int_{\mathcal{M}} \zeta_{\rm ad} \mathbf{L} \zeta_{\rm ad} \, r^2 \, d\mathcal{M}_r = \sigma^2 \int_{\mathcal{M}} \zeta_{\rm ad} \zeta_{\rm ad} \, r^2 \, d\mathcal{M}_r \qquad (10.50)
$$

which allows us to get rid of the first integral in (10.49). We shall assume that the star is not dynamically unstable so that σ^2 is positive. Collecting all terms and solving for $\kappa_{\rm qa}$ gives

$$
\kappa_{\rm qa} = - \frac{C_{\rm qa}}{2\sigma^2 \int_{\mathcal{M}} \zeta_{\rm ad}^2 \, r^2 \, d\mathcal{M}_r} = - \frac{C_{\rm qa}}{2\sigma^2 J} \qquad (10.51)
$$

where J, the *oscillatory moment of inertia*, is $J = \int_{\mathcal{M}} \zeta_{\rm ad}^2 r^2 \, d\mathcal{M}_r$. (The quantity J often appears in pulsation theory. The "moment of inertia" part of its name comes about because $\int r^2 \, d\mathcal{M}_r$ is the moment of inertia for the star.) Since both σ^2 and J are positive, the sign of the growth rate $\kappa_{\rm qa}$ depends only on the sign of $C_{\rm qa}$. We now show how $\kappa_{\rm qa}$ and $W_{\rm tot}$ of (10.46) are related.

First change the integration variable in $C_{\rm qa}$ to r rather than $\mathcal{M}_r$ by using the mass equation $d\mathcal{M}_r = 4\pi r^2 \rho \, dr$. An integration by parts in (10.49) yields the constant term

$$
4\pi r^3 \zeta_{\rm ad} \rho \left(\Gamma_3 - 1 \right) \left(\delta\varepsilon - \frac{\partial \delta \mathcal{L}_r}{\partial \mathcal{M}_r} \right) \Bigg|_0^{\mathcal{R}}
$$

which vanishes at both limits if the density at the outer surface is taken to be zero. To deal with the remaining integral, note that $d \left(4\pi r^3 \zeta_{\rm ad} \right) / dr = 4\pi r^2 \left(3\zeta_{\rm ad} + r \, d\zeta_{\rm ad}/dr \right)$. But, from the linearized mass equation (10.6),

$$
\left(\frac{\delta \rho}{\rho} \right)_{\rm ad} = -3\zeta_{\rm ad} - r \frac{d\zeta_{\rm ad}}{dr} \qquad (10.52)
$$

which we use to get rid of all reference to $\zeta_{\rm ad}$ in favor of the adiabatic density variation. The next to last step is to realize that adiabatic density versus

[6]The particular version of the quasi-adiabatic derivation given here is due to H.M. Van Horn from correspondence dating back to the middle 1970s.

temperature perturbations are connected through Γ_3 by (3.89) which, in this context, reads

$$\left(\frac{\delta\rho}{\rho}\right)_{\rm ad}(\Gamma_3 - 1) = \left(\frac{\delta T}{T}\right)_{\rm ad}. \tag{10.53}$$

Finally, after converting back to a mass integration, we arrive at an expression for the quasi-adiabatic growth rate

$$\kappa_{\rm qa} = -\frac{\int_{\mathcal{M}} (\delta T/T)_{\rm ad} \left(\delta\varepsilon - \partial\,\delta\mathcal{L}_r/\partial\mathcal{M}_r\right) d\mathcal{M}_r}{2\sigma^2 J}. \tag{10.54}$$

Elements of this result should be suspiciously familiar because the numerator looks like the mass integral part of $W_{\rm tot}$. Note that the two integrals in $W_{\rm tot}$ give the work done on the star by itself over one period. If we divide that by the total kinetic energy of oscillation, then the resulting ratio gives us a rate of change increase (or decrease) of pulsation energy per period. The kinetic energy of oscillation is derived from

$$\int_{\mathcal{M}} \frac{1}{2} (\delta\dot{r})^2 \, d\mathcal{M}_r$$

which is $mv^2/2$ in disguise. But a time derivative of δr yields $i\sigma\delta r$ or $i\sigma r\zeta_{\rm ad}$ in the adiabatic case. Therefore the kinetic energy of oscillation is just $\sigma^2 J/2$. The last step is to recognize that $\kappa_{\rm qa}$ measures the e-folding time for a change in amplitude of a perturbation, whereas energies go as the square of amplitudes. To first-order this introduces a factor of 1/2 into $W_{\rm tot}/\left(\sigma^2 J/2\right)$. The two formulations ($W_{\rm tot}$ versus $\kappa_{\rm qa}$) tell us the same thing if we accept the validity of the quasi-adiabatic approximation.

How may we use all this? Consider first the role of $\delta\varepsilon$ in equation (10.54) for $\kappa_{\rm qa}$ and ignore for now the presence of $(\partial\delta\mathcal{L}_r/\partial\mathcal{M}_r)$. As we have often done before, first write ε in the power law form $\varepsilon \propto \rho^\lambda T^\nu$. (See, e.g., 1.56 or 6.48.) Then, treating the perturbation operator as a differential, we find

$$\left(\frac{\delta\varepsilon}{\varepsilon}\right)_{\rm ad} = \lambda\left(\frac{\delta\rho}{\rho}\right)_{\rm ad} + \nu\left(\frac{\delta T}{T}\right)_{\rm ad} \tag{10.55}$$

where the adiabatic subscript is appended to remind us—for the last time—that all the eigenfunctions are derived from an adiabatic calculation. We shall leave the subscript off from now on, but only for clarity. To get everything in terms of $\delta T/T$ we use the adiabatic relation (10.53), which introduces $(\Gamma_3 - 1)$ so that the integrand in $\kappa_{\rm qa}$ is proportional to $(\delta T/T)^2$. Except for some very unusual circumstances not often met in stars, λ and ν (and certainly $[\Gamma_3 - 1]$) are all positive. In other words, the integrand of $\kappa_{\rm qa}$ is always positive and, hence, $\kappa_{\rm qa}$ is negative—implying instability. Thus the effect of thermonuclear reactions in stars is to push them toward instability. This should be obvious because ε increases and adds heat to a compressing element and this is our criterion for driving. The fact of the matter is, however, that no intrinsically variable star has been unambiguously shown to be unstable due to this effect (although very massive upper main sequence stars may feel the effects of this mechanism). So, what physical mechanisms are present in stable stars that override thermonuclear destabilization, and what causes variable stars to pulsate? To answer this we have to look into the mass gradient term of the luminosity perturbation in (10.54).

10.2.2 The κ– and γ–Mechanisms

Thus far we have not considered how heat is transported. Since there are still outstanding problems associated with the interaction of convection and pulsation, we shall consider only heat transport by radiative diffusion and how it is modulated by pulsation.

We first linearize the diffusion equation, say in the form given by (4.25). An easy exercise yields the relative variation of radiative luminosity,

$$\frac{\delta \mathcal{L}_r}{\mathcal{L}_r} = 4\zeta - \frac{\delta \kappa}{\kappa} + 4\frac{\delta T}{T} + \frac{1}{dT/dr}\frac{d}{dr}\left(\frac{\delta T}{T}\right) \tag{10.56}$$

where κ is the opacity. This expression is the same as given by Cox (1980, eq. 7.11a). For our analysis we shall ignore the derivative term and, in the spirit of the quasi-adiabatic calculation, assume that the variations are adiabatic. Then, using: (1) the power law expression for the opacity, $\kappa \propto \rho^n T^{-s}$ (as in equation 1.59); (2) the linearized mass equation (10.6) without the derivative term; (3) the adiabatic relation between $\delta\rho/\rho$ and $\delta T/T$ (10.53), we find

$$\frac{\delta \mathcal{L}_r}{\mathcal{L}_r} \approx -\left(\frac{4/3+n}{\Gamma_3 - 1}\right)\frac{\delta T}{T} + (s+4)\frac{\delta T}{T}. \tag{10.57}$$

This equation will now be used in the quasi-adiabatic expression (10.54) which, without the ε term, is

$$\kappa_{\mathrm{qa}} = \frac{\int_{\mathcal{M}} (\delta T/T)\,(\partial\, \delta\mathcal{L}_r / \partial \mathcal{M}_r)\, d\mathcal{M}_r}{2\sigma^2 J}. \tag{10.58}$$

Recall that a positive value of κ_{qa} means that the star is stable. We can get that positive value if, through most of the star, the mass gradient of $\delta\mathcal{L}_r$ is positive upon compression (i.e., when $\delta T > 0$).

An example is a region within a star where Kramers' opacity operates with $n = 1$ and $s = 3.5$. From our discussion of opacities in §4.4 *et seq.*, this opacity dominates where ionization is not that active (as in the downward slope portions of Fig. 4.2). Thus also set $\Gamma_3 = 5/3$ assuming a nonionizing ideal gas. Putting these numbers into (10.57) yields $\delta\mathcal{L}_r/\mathcal{L}_r = 4(\delta T/T)$.

The mass derivative is then

$$\frac{\partial\,\delta\mathcal{L}_r}{\partial\mathcal{M}_r} \approx 4\mathcal{L}_r\,\frac{\partial(\delta T/T)}{\partial\mathcal{M}_r} \tag{10.59}$$

if the gradient of $\mathcal{L}_r$ can be neglected, as it can where there is no energy generation going on ($\partial\mathcal{L}_r/\partial\mathcal{M}_r = 0$). In the simple case of a fundamental mode the mass gradient of δT is positive upon adiabatic compression because the absolute values of the variations in radius or density are small near the center and increase outward. Equation (10.59) then implies that the mass gradient of $\delta\mathcal{L}_r$ is also positive on compression in this case of Kramers' opacity. Hence, in the circumstance we describe, the quasi-adiabatic approximation states that the particular region (or entire star) is stable because $\kappa_{\rm qa}$ of (10.58) is positive. What is happening here is consistent with our earlier analysis. If $\partial\delta\mathcal{L}_r/\partial\mathcal{M}_r > 0$ upon compression, then more radiative power leaks out of a mass shell than is entering it. Thus the mass shell loses energy when compressed and this was our criteria for stability. You may turn the argument around by considering expansion and come to the same conclusion about the stabilizing nature of the process. The root cause of the stabilization is that Kramers' opacity decreases upon compression with its accompanying increase of temperature: the material becomes leakier to radiation.

If, on the other hand, we are in an active ionization zone, then κ either may not decrease nearly as rapidly as in the above case—which means s is considerably less than 3.5—or may actually increase with temperature—as in the positive slope portions of Figure 4.2. Furthermore, ionization means that Γ_3 is less than 5/3. The net combination of these two circumstances can force the factors in front of δT in (10.57) to cause the mass gradient of $\delta\mathcal{L}_r$ to be *negative* when δT is positive. This situation is destabilizing: a mass element *gains* heat upon compression. In the extreme form of this argument, the destabilization due to opacity comes about because increases in opacity due to increases in temperature tend to dam up the normal flow of radiation from out of an ionizing mass element. The mass element, in effect, thus heats up relative to its surroundings. Destabilization in this instance is due to the "κ–effect."

It may be the case that the exponent s in the opacity law differs only slightly from a Kramers'–like law in an ionization zone but $\Gamma_3 - 1$ is still small. Such is the case for second helium ionization, which takes place at about 4×10^4 K, where the last electron of helium is being removed and recombined. (Fig. 4.2 shows the seemingly slight effect on opacity of second helium ionization for a typical Pop I mixture.) The effect on Γ_3 alone may be enough to cause instability when (10.57) is applied. What happens here is that the work of compression goes partially into ionization and temperatures do not rise as much if ionization is not taking place. Thus an ionization region tends to be somewhat cooler than the surrounding non-ionizing regions upon compression and heat tends to flow into the ionizing region. This part of the destabilization process is called the γ–effect. Note that in most instances the κ– and γ–effects go hand in hand.

But stars generally have zones that are ionizing and others that are not. Which win out? Another related question is what regions are the most important in establishing the mechanical response of a star. This last may be answered by rearranging the LAWE in its integrated form (10.50) to read

$$\sigma^2 = \frac{\int_{\mathcal{M}} \zeta \mathbf{L} \zeta \, r^2 \, d\mathcal{M}_r}{\int_{\mathcal{M}} \zeta^2 \, r^2 \, d\mathcal{M}_r}. \tag{10.60}$$

All this says is that the eigenvalue is equal to a ratio of integrals containing its eigenfunctions. It is then reasonable to suppose that the integrand of the numerator on the right-hand side of this equation is a "weight function" that tells us where it is in the star that most of the action occurs that determines σ^2 (see Cox 1980, §8.13). This was investigated long ago by Epstein (1950). He specifically computed the weight functions for the fundamental modes of centrally condensed stellar models—such as those expected for evolved classical Cepheid variables—and found that the weight function reached a strong peak at $r/\mathcal{R} \approx 0.75$. If this radius does correspond to the region where we expect the pulsation properties of a centrally condensed model to be primarily determined, then what are the thermal properties of that region with respect to ionization?

We can easily compute the temperature of a star at $r/\mathcal{R} = 0.75$ if we assume that a simple envelope integration is justified. The temperature is then given as a function of radius by (7.144). If, in that expression, we use $\beta = 1$, $\mu = 0.6$, and $n_{\text{eff}} \approx 3.25$ (i.e., for Kramers' opacity), then

$$T(r/\mathcal{R} = 0.75) \approx 1.1 \times 10^6 \, \frac{\mathcal{M}/\mathcal{M}\odot}{\mathcal{R}/\mathcal{R}\odot} \text{ K}. \tag{10.61}$$

A classical Cepheid variable has a mass of $8\,\mathcal{M}\odot$ and radius $90\,\mathcal{R}\odot$. This combination yields $T(r/\mathcal{R} = 0.75)$ of about 10^5 K, which corresponds roughly to the second ionization zone of helium. (See exercise 1 of Chapter 3 and the opacity plot for pure helium in Fig. 4.3.) Detailed nonadiabatic studies of classical Cepheid models have indeed confirmed that it is this ionization zone that is responsible for their variability. Other regions of the star may try to stabilize but this zone ultimately wins out.

10.2.3 Nonadiabaticity and the Classical Cepheid Instability Strip

Thus far we have relied on the quasi-adiabatic approximation as a diagnostic tool. This is not always wise because some regions in variable stars react very nonadiabatically to any perturbation. Our measure of how good we expected the quasi-adiabatic approximation to be was based on the ratio of thermal to dynamic time scales. If that ratio is large then motions are close to adiabatic. The dynamic time scale, $t_{\rm dyn}$, for a variable star is determined by the period, Π, of oscillation for the whole star and, for the fundamental mode, it may be approximated by the period-mean density relation. The thermal time scale for a particular region in a star, on the other hand, is a measure of how long it takes heat to be transported through the region. We can estimate this time scale for a shell of mass content $\Delta\mathcal{M}$ by dividing its total heat content by the luminosity that must pass through it. (This is similar to what we did in deriving the Kelvin-Helmholtz time scale of §1.3.2.) As an approximation to the heat content we use $c_{V_\rho} T\,\Delta\mathcal{M}$ where c_{V_ρ} and T are suitable averages within the shell. Thus the thermal time scale, $t_{\rm th}$, is

$$t_{\rm th} \approx \frac{c_{V_\rho} T\,\Delta\mathcal{M}}{\mathcal{L}}. \tag{10.62}$$

The ratio we use to estimate how adiabatically or nonadiabatically the shell responds to perturbations is $t_{\rm th}/t_{\rm dyn}$ with

$$\Phi(\Delta\mathcal{M}) \equiv \frac{t_{\rm th}}{t_{\rm dyn}} = \frac{c_{V_\rho} T\,\Delta\mathcal{M}}{\Pi\mathcal{L}}. \tag{10.63}$$

If Φ is large then the motion should be nearly adiabatic and the quasi-adiabatic approximation should be close to the truth. If, however, Φ happens to be small then the thermal time scale is short compared to the period and a full nonadiabatic treatment is warranted. We are *not* going to show you how to do a nonadiabatic analysis here but rather we shall extract from such analyses some properties of the classical Cepheid instability strip discussed in §2.6.1.

The place we expect Φ to be small is in the outer envelope of a star where temperatures are low and the heat content is correspondingly small. We therefore define the "transition temperature," $T_{\rm TR}$, as the temperature at that point in the envelope where Φ is unity. Interior to that point motions are quasi–adiabatic, whereas to the exterior and up to the surface all must be treated nonadiabatically. If $\Delta\mathcal{M}$ is now the mass of the envelope above this transition level then $T_{\rm TR}$ is given by

$$\Phi(\Delta\mathcal{M}) = \frac{c_{V_\rho} T_{\rm TR}\,\Delta\mathcal{M}}{\Pi\mathcal{L}} = 1 \tag{10.64}$$

where our "suitable average" for T is $T_{\rm TR}$ itself.

We shall not go into the details here but the region of maximum driving in a classical Cepheid variable coincides with the location of the transition temperature which, from our previous discussion, should then also be the temperature for second helium ionization. (Arguments for this and for what follows are discussed in Cox 1968, §27.7 and Cox 1980, §10.1.) Given this, we can then estimate the slope of the Cepheid strip in the $\mathcal{L}$–$T_{\rm eff}$ Hertzsprung-Russell diagram. To start, we set $\Phi(\Delta\mathcal{M})$ to unity as in the above, assume c_{V_ρ} is roughly constant, and delete reference to $T_{\rm TR}$ because it too is constant (around 4×10^4 K). What is left is a relation between $\Delta\mathcal{M}$, $\mathcal{L}$, and Π which, ignoring constants, is $\Delta\mathcal{M} \propto \Pi\mathcal{L}$. If the mass shell $\Delta\mathcal{M}$ is assumed to be thin, then $\Delta\mathcal{M}$ can be eliminated by using an estimate of the pressure at the transition point needed to support the overlying mass $\Delta\mathcal{M}$; that is, $P \approx G\mathcal{M}\,\Delta\mathcal{M}/\mathcal{R}^4$. Now equate this to the envelope solution for pressure from (7.139) $P = K'T^{1+n_{\rm eff}}$ where K' is given by (7.140). The temperature T here is just a constant ($T_{\rm TR}$); note that K' depends only weakly on $\mathcal{L}$ and $\mathcal{M}$ so that we can ignore it also. (For $n = 1$, $K' \sim \sqrt{\mathcal{M}/\mathcal{L}}$. Other reasonable choices for n will not change our overall conclusion.) Thus $\Delta\mathcal{M} \propto \mathcal{R}^4/\mathcal{M}$. The period-mean density relation gives us the estimate $\Pi \propto \mathcal{M}^{-1/2}\mathcal{R}^{3/2}$ so that our condition for $\Phi = 1$ becomes, after some simple algebra,

$$\mathcal{L} \propto \frac{\mathcal{R}^{5/2}}{\mathcal{M}^{1/2}}.$$

To get rid of the mass, recall that the luminosity of classical Cepheid variables depends on mass through (2.27), $\mathcal{L} \propto \mathcal{M}^4$. This implies that the factor $\mathcal{M}^{1/2}$ in the above brings in $\mathcal{L}^{1/8}$, which is a mild enough dependence that we ignore it. Finally, substitute $\mathcal{L} \propto \mathcal{R}^2 T_{\rm eff}^4$, get rid of $\mathcal{R}$, and after clearing away the debris find

$$\mathcal{L} \propto T_{\rm eff}^{20} \tag{10.65}$$

along the instability strip. You can play with various powers in this "derivation" but, in the end, the clear result is that the Cepheid strip should be very steep in the H–R diagram and, as is the same thing, at nearly constant effective temperature. Figure 2.20, which shows the observed strip, confirms our theoretical result. The strip leans only slightly to lower effective temperatures as luminosity increases.

Nonadiabatic calculations can nicely predict the blue (hot) edge of the instability strip but the red edge is another matter. On comparing models of Cepheids with the observed red edge, it is found that vigorous envelope convection begins near the low-temperature side of the strip. (See Fig. 5.3 for a sketch of where we expect efficient envelope convection on the H–R diagram.) Thus, in one way or the other, it is thought that convection must inhibit instability. Exactly how it does so is still not clear despite some heroic attempts (as partially reviewed by Cox 1980, §19.3, and see particularly Toomre 1982). The trouble is that convection is difficult enough without having to couple it to pulsation.

A Footnote on Nonlinear Modeling

There have also been many successful studies of radially variable stars using one-dimensional hydrodynamics. The methods used are like those briefly discussed in §7.2.6. One particularly satisfying result that illustrates instability is described in Cox (1968, §27.8). A model classical Cepheid envelope is constructed in hydrostatic equilibrium with parameters (total mass, luminosity, etc.) in the range expected for a variable. The hydrodynamics are then turned on and the model is followed in time. By virtue of numerical noise in the initial model (no computer or calculation is perfect) the "Cepheid" quivers and this quivering is amplified as nonadiabatic effects begin to be felt. After an elapsed time corresponding to about 400 periods (in this particular calculation) maximum nonlinear amplitudes are reached and the model sitting in the computer acts like a real Cepheid you might observe in the sky. This type of work is very different from what we have described for linear theory. There are no imposed infinitesimal bounds on amplitudes and the dynamic model finds its own final pulsating state. The modeling of what processes (such as atmospheric shocks) are responsible for limiting the amplitude is still an active field of study.

Finally, we mention the work of J.R. Buchler and his collaborators. (For a review see Buchler 1990.) Since we are dealing with very complex nonlinear systems we can expect some surprises. One of the most interesting fields of endeavor in the physical sciences in recent years is that of nonlinear dynamics. It has been shown that even some seemingly simple systems are prone to behavior that is almost counterintuitive. Under certain conditions, whether in the laboratory or the computer, these systems allow for regular limit cycles (what we have been discussing), irregular pulses and period doubling, and, in some instances, chaotic behavior. The minor adjustment of a laboratory or computer parameter can lead the system from one of these states to another with astonishing rapidity. So it may be with stars. Not all variable stars behave as nicely as we have led you to believe in this chapter. If you review the variables listed in Table 2.4, some are indicated as being "irregular." It may well be that such variables, and others, are in this murky land of near-chaos. For a taste of what is being done in this field you may want to look through articles such as those by Kovács and Buchler (1988a, 1988b) where theoretical nonlinear calculations are described for Type II Cepheids and RR Lyrae variables. They can do strange things on the computer.

10.3 An Introduction to Nonradial Oscillations

What we shall develop here are the tools necessary to describe how a star may oscillate in modes that do *not* preserve radial symmetry. These are called *nonradial* modes. The prime references here are Cox (1980), Unno et al. (1979, 1989), and Ledoux and Walraven (1958).

10.3.1 Linearization of the Hydrodynamic Equations

The path we shall take differs somewhat from the preceding discussion of radial pulsations. Instead of starting with the standard stellar structure equations, we shall delve into simple fluid mechanics although, for the sake of simplicity, adiabatic motions will still be assumed. In this vein we neglect all dissipative effects (such as viscosity) and assume that the stellar fluid cannot support shear stresses. (What happens when shear is introduced—as would be the case for the earth—will be briefly touched upon later.) The equations required to describe how the fluid behaves dynamically are: Poisson's equation for the gravitational potential Φ (as introduced after equation 7.17), the equation of continuity, and the equation of motion. In that order, these equations are (*cf.* Landau and Lifshitz 1959, §15)

$$\nabla^2\Phi = 4\pi G\rho \tag{10.66}$$

$$\frac{\partial\rho}{\partial t} + \boldsymbol{\nabla}\bullet(\rho\mathbf{v}) = 0 \tag{10.67}$$

$$\rho\left(\frac{\partial}{\partial t} + \mathbf{v}\bullet\boldsymbol{\nabla}\right)\mathbf{v} = -\boldsymbol{\nabla}P - \rho\boldsymbol{\nabla}\Phi\,. \tag{10.68}$$

Here $\mathbf{v} = \mathbf{v}(\mathbf{r}, t)$ is the fluid velocity and Φ is the gravitational potential, which is related to the local (vector) gravity by $\mathbf{g} = -\boldsymbol{\nabla}\Phi$. (The scalar gravity g we have been using in this text is then the negative of the radial component of $\mathbf{g}$.) As phrased, these provide a *Eulerian* description of the motion wherein we place ourselves at a particular location, $\mathbf{r}$, in the star and watch what happens to $\mathbf{v}(\mathbf{r}, t)$, $\rho(\mathbf{r}, t)$, etc., as functions of time. In a nonrotating hydrostatic star, $\mathbf{v}$ is zero everywhere. (We ignore fluid motions associated with convection.)

Observe first that we know the values of all the above physical variables in the unperturbed star as solely a function of $r = |\mathbf{r}|$. Now imagine that each fluid element in the star is displaced from its equilibrium position at $\mathbf{r}$ by an arbitrary, but infinitesimal, vector distance $\boldsymbol{\xi}(\mathbf{r}, t)$. (The radial component of this quantity previously was called δr.) Remember that this kind of displacement—which takes an identifiable parcel of fluid and moves it somewhere else—is a Lagrangian displacement. If $\mathbf{v} = 0$, then the Eulerian and Lagrangian perturbations of $\mathbf{v}$, denoted respectively by $\mathbf{v}'$ and $\delta\mathbf{v}$, are equal and are given by

$$\mathbf{v}' = \delta\mathbf{v} = \frac{d\boldsymbol{\xi}}{dt} \tag{10.69}$$

where d/dt is the Stokes derivative

$$\frac{d}{dt} = \frac{\partial}{\partial t} + \mathbf{v}\bullet\boldsymbol{\nabla}\,. \tag{10.70}$$

(For a complete derivation of these statements see Cox 1980, §5.3. The Stoke's derivative was also used in §5.1.2 when discussing convection and it appears in eq. 10.68.) We shall continue to use $'$ and δ to denote the Eulerian and Lagrangian perturbations of quantities.

As the fluid is displaced, all other physical variables are perturbed accordingly. Thus, for example, the pressure $P(\mathbf{r})$, which was originally associated with the fluid parcel at $\mathbf{r}$, becomes $P(\mathbf{r}) + \delta P(\mathbf{r}, t)$ when the parcel is moved to $\mathbf{r} + \boldsymbol{\xi}(\mathbf{r}, t)$. The same statement applies to the density and its perturbation $\delta\rho(\mathbf{r}, t)$. Note again that these are Lagrangian displacements and thus require the δ operator.

If the motion is adiabatic then the relation between δP and $\delta\rho$ is the same as that used for radial oscillations:

$$\frac{\delta P}{P} = \Gamma_1 \frac{\delta\rho}{\rho}. \tag{10.71}$$

Note that we cannot use a like relation for the Eulerian perturbations $P'(\mathbf{r}, t)$ and $\rho'(\mathbf{r})$ because these perturbations are used to find the new pressures and densities at a *given* point $\mathbf{r}$ without saying where the fluid came from. The Eulerian and Lagrangian variations are connected, however, by a relation derived in any book on hydrodynamics (or see Cox 1980, §5.3); namely, and using density as an example,

$$\delta\rho = \rho' + \boldsymbol{\xi} \bullet \boldsymbol{\nabla}\rho. \tag{10.72}$$

Such relations will be used extensively later.

The analysis proceeds by replacing P, ρ, Φ, and $\mathbf{v}$, by $P+P'$, $\rho+\rho'$, $\Phi+\Phi'$, and $\mathbf{v}'$ in equations (10.66–10.68), multiplying everything out, and keeping terms to only first order in the perturbations. We are again performing a linear analysis of the system.[7] Thus, for example, the force equation becomes

$$\rho \frac{\partial^2 \boldsymbol{\xi}}{\partial t^2} = -\boldsymbol{\nabla}P - \rho\boldsymbol{\nabla}\Phi - \boldsymbol{\nabla}P' - \rho\boldsymbol{\nabla}\Phi' - \rho'\boldsymbol{\nabla}\Phi. \tag{10.73}$$

The first two terms on the right-hand side cancel because

$$-\boldsymbol{\nabla}P - \rho\boldsymbol{\nabla}\Phi = 0 \tag{10.74}$$

is the equation of hydrostatic equilibrium for the unperturbed star. What is left is an equation that contains only the perturbed quantities as first-order variables. Similarly, the continuity and Poisson equations become

$$\rho' + \boldsymbol{\nabla} \bullet (\rho\boldsymbol{\xi}) = 0 \tag{10.75}$$

$$\nabla^2\Phi' = 4\pi G\rho'. \tag{10.76}$$

In setting down the linearized form of the continuity equation we have given what results after an integration over time and the removal of a constant of integration by insisting that $\rho' = 0$ when $\boldsymbol{\xi} = 0$.

Even though we have linearized the equations, the above set of partial differential equations is still daunting because the system is second-order in time and fourth order in space. To reduce the system further requires

[7]Note that we use Eulerian perturbations here but we could have used Lagrangian forms. It is a matter of taste and tradition but see Pesnell (1990).

a bit more work. The aim will be to convert what we have to *ordinary* differential equations. We first assume, as was done for radial oscillations, that all the variations may be Fourier analyzed with $\boldsymbol{\xi}$, P', ρ', and Φ' being proportional to $e^{i\sigma t}$ where σ is an angular frequency. Thus, for example,

$$\boldsymbol{\xi}(\mathbf{r},t) = \boldsymbol{\xi}(\mathbf{r})\, e^{i\sigma t}. \tag{10.77}$$

With this substitution the time variable is separated out and all variations become functions solely of the radius vector $\mathbf{r}$.

The second step is to completely ignore the variation in gravitational potential, Φ'. This step, called the "Cowling approximation" (Cowling 1941), is remarkably good, provided that little mass is thrown around during the motion of the fluid. We cannot justify it here (see the above references) but it introduces only minor errors in many cases of practical interest and it reduces our labors by nearly a factor of two. With this in mind, we proceed.

First observe that the continuity equation, (10.75), expands out to

$$\frac{\rho'}{\rho} = -\nabla \bullet \boldsymbol{\xi} - \frac{\boldsymbol{\xi} \bullet \nabla\rho}{\rho} \tag{10.78}$$

with minor rearrangement and where all perturbations are only functions of $\mathbf{r}$. The last term here, however, is just the last term of (10.72) divided by the density. Thus these two equations yield the following for the Lagrangian variation of density:

$$\frac{\delta\rho}{\rho} = -\nabla \bullet \boldsymbol{\xi}. \tag{10.79}$$

(A little thought about the meaning of $\nabla \bullet \boldsymbol{\xi}$, which is sometimes called the dilatation, would have yielded this immediately.) This result also relates the relative Lagrangian variation of pressure to $\nabla \bullet \boldsymbol{\xi}$ through the adiabatic condition (10.71).

We now expand (10.73) in vector components dealing with the radial component first; that is,

$$\sigma^2 \xi_r = \frac{1}{\rho}\frac{\partial P'}{\partial r} - \frac{\rho'}{\rho^2}\frac{dP}{dr} \tag{10.80}$$

where $\nabla\Phi$ has been replaced by $-\nabla P/\rho$ using the hydrostatic condition for the unperturbed star and ξ_r is the radial component of $\boldsymbol{\xi}(\mathbf{r})$. Note that P is a function of r only and thus we do not need partials for its gradient. Now, for reasons to be made apparent later, the term containing P' is manipulated so that the radial derivative acts on P'/ρ instead and, as you may easily check, (10.80) becomes

$$\sigma^2 \xi_r = \frac{P'}{\rho^2}\frac{d\rho}{dr} + \frac{\partial}{\partial r}\left(\frac{P'}{\rho}\right) - \frac{\rho'}{\rho^2}\frac{dP}{dr}. \tag{10.81}$$

The next devious steps are aimed at converting this into final form. We keep the second term on the right-hand side as it stands but manipulate the sum of the first and third terms so that P' and ρ' are replaced by their Lagrangian forms δP and $\delta\rho$. Use (10.72) to do this. Then use the adiabatic condition (10.71) to replace δP by $\delta\rho$. Finally, get rid of $\delta\rho$ in favor of $\nabla \bullet \boldsymbol{\xi}$ with the help of (10.79). After the smoke clears we find

$$\sigma^2 \xi_r = \frac{\partial}{\partial r}\left(\frac{P'}{\rho}\right) - A\frac{\Gamma_1 P}{\rho}\nabla \bullet \boldsymbol{\xi} \tag{10.82}$$

for the radial equation where $A(r)$ is the Schwarzschild discriminant of (5.36), which played an important role in convection (and see Eq. 10.92). Note that $\Gamma_1 P/\rho$ is the square of the local sound speed. The combined presence of v_s and A will mean that nonradial oscillations come in two distinct flavors and not just acoustic waves.

The equations for the other two components of $\boldsymbol{\xi}$ are straightforward once we define the rest of our coordinate system. Since the equilibrium star is spherical we shall use spherical coordinates, (r, θ, φ), and see how well they work. The gradient terms are easy and, remembering that ρ depends only on r, we find

$$\sigma^2 \xi_\theta(r,\theta,\varphi) = \frac{\partial}{\partial\theta}\left[\frac{1}{r}\frac{P'}{\rho}(\mathbf{r})\right] \tag{10.83}$$

$$\sigma^2 \xi_\varphi(r,\theta,\varphi) = \frac{1}{\sin\theta}\frac{\partial}{\partial\varphi}\left[\frac{1}{r}\frac{P'}{\rho}(\mathbf{r})\right]. \tag{10.84}$$

The final stage for the separation of variables is now set.

10.3.2 Separation of the Pulsation Equations

The three equations (10.82–10.84) contain the four unknowns ξ_r, ξ_θ, ξ_φ, and P'/ρ. We shall now show that there are really only *two* independent unknowns and that the remaining two degrees of freedom collapse down into a well-known function from mathematical physics. Our method for demonstrating this is to give the answer and then see if it works.

We propose the following solution for $\boldsymbol{\xi}(\mathbf{r})$ and $P'(\mathbf{r})/\rho$:

$$\begin{aligned}\boldsymbol{\xi}(r,\theta,\varphi) &= \xi_r(r,\theta,\varphi)\,\mathbf{e_r} + \xi_\theta(r,\theta,\varphi)\,\mathbf{e_\theta} + \xi_\varphi(r,\theta,\varphi)\,\mathbf{e_\varphi} \\ &= \left[\xi_r(r)\,\mathbf{e_r} + \xi_t(r)\,\mathbf{e_\theta}\frac{\partial}{\partial\theta} + \xi_t(r)\,\mathbf{e_\varphi}\frac{1}{\sin\theta}\frac{\partial}{\partial\varphi}\right] Y_{\ell m}(\theta,\varphi)\end{aligned} \tag{10.85}$$

and

$$\frac{P'(\mathbf{r})}{\rho} = \frac{P'(r)}{\rho} Y_{\ell m}(\theta,\varphi). \tag{10.86}$$

Here the $\mathbf{e_i}$ are the dimensionless unit vectors in spherical coordinates, and $\xi_t(r)$ and $P'(r)/\rho$ are new functions of r only, which are related to each other by

$$\xi_t(r) = \frac{1}{\sigma^2}\frac{1}{r}\frac{P'(r)}{\rho}. \tag{10.87}$$

The function $\xi_t(r)$ effectively replaces P'/ρ and the θ and φ components of $\boldsymbol{\xi}$ and it will be referred to as the "transverse displacement." Finally, the angle-dependent function $Y_{\ell m}(\theta, \varphi)$ is the *spherical harmonic* (or surface harmonic) of combined indices ℓ and m. This function, which does the angular separation for us, arises frequently in physics (such as in the hydrogen atom and applications in electricity and magnetism) and is the regular solution of the second-order partial differential equation

$$\frac{1}{\sin\theta}\frac{\partial}{\partial\theta}\left(\sin\theta\,\frac{\partial Y_{\ell m}}{\partial\theta}\right)+\frac{1}{\sin^2\theta}\,\frac{\partial^2 Y_{\ell m}}{\partial\varphi^2}+\ell(\ell+1)Y_{\ell m}=0. \qquad (10.88)$$

Here ℓ must be zero or a positive integer and m can only take on the integer values $-\ell$, $-\ell+1$, $\cdots$, 0, $\cdots$, $\ell-1$, ℓ. Thus for a given value of ℓ there are only $2\ell+1$ permitted values of m. The separation given above gives rise to "spheroidal modes." Another separation is possible, resulting in "toroidal modes," but we shall ignore them. Note that $\ell = 0$ is a special case because $Y_{00}(\theta, \varphi)$ is a constant and, more to the point, it does not depend on either θ or φ. In other words, solutions for $\ell = 0$ depend only on r and thus correspond to radial modes. Since we have already discussed them we shall concentrate only on solutions for which $\ell > 0$. Some of the properties of the $Y_{\ell m}$ will be discussed in a bit but the important point for now is the following.

If the angular components of equation (10.85) and equation (10.87) are introduced into the angular components of the force equation (10.83) and (10.84), then the derivatives of $Y_{\ell m}$ cancel out of both sides of the resulting equations. For example, the left-hand side of (10.83) is (using 10.85)

$$\sigma^2\xi_\theta(r,\theta,\varphi)=\sigma^2\xi_t(r)\frac{\partial Y_{\ell m}}{\partial\theta}$$

while the right-hand side becomes (using 10.86–10.87)

$$\frac{\partial}{\partial\theta}\left[\frac{1}{r}\frac{P'(\mathbf{r})}{\rho}\right]=\sigma^2\frac{\partial\xi_t(r)Y_{\ell m}}{\partial\theta}=\sigma^2\xi_\theta(r,\theta,\phi).$$

Thus the θ component of the force equation is satisfied with our choice of ξ_t and separation of variables. You may easily show that the φ component is also consistent but aside from consistency it yields no further information. We need the radial component for this.

The difficult term in the radial equation (10.82) is the divergence of $\boldsymbol{\xi}(r, \theta, \varphi)$. Written out in full it is

$$\begin{aligned}\nabla\cdot\boldsymbol{\xi} &= \frac{1}{r^2}\frac{d}{dr}\left(r^2\xi_r\right)\\ &+\frac{1}{r\sin\theta}\frac{\partial}{\partial\theta}\left(\xi_\theta\sin\theta\right)+\frac{1}{r\sin\theta}\frac{\partial}{\partial\varphi}\left(\xi_\varphi\right).\end{aligned} \qquad (10.89)$$

But if (10.85) is inserted here, then you may easily verify that you regain the first two terms of the differential equation for $Y_{\ell m}$ (eq. 10.88) so that

$$\nabla \bullet \boldsymbol{\xi} = \frac{1}{r^2}\frac{d}{dr}\left(r^2 \xi_r\right) Y_{\ell m} - \frac{\ell(\ell+1)}{r}\, \xi_t\, Y_{\ell m} \tag{10.90}$$

where ξ_r now depends only on r. On the other hand, $\nabla \bullet \boldsymbol{\xi}$ is, from (10.79), the same as $-\delta\rho/\rho$, which can be written (from parts of the derivation leading to 10.82 and 10.87) as

$$\frac{\delta \rho}{\rho} = \frac{\rho}{\Gamma_1 P}\left(\sigma^2 r\, \xi_t - g\, \xi_r\right) \tag{10.91}$$

where the equation of hydrostatic equilibrium has been used to replace the pressure derivative by $-\rho g$ and common factors of $Y_{\ell m}$ have been eliminated. The net result of these manipulations is that we obtain a first-order ordinary differential equation after equating the divergence and $-\delta\rho/\rho$. The final form of this will be given shortly after the following important frequencies are defined.

The first of these frequencies is the Brunt-Väisälä frequency N given by (5.36) and (5.38) as

$$N^2 = -Ag = -g\left[\frac{d\ln\rho}{dr} - \frac{1}{\Gamma_1}\frac{d\ln P}{dr}\right]. \tag{10.92}$$

Recall that N, in the simplest interpretation, is the frequency of oscillation associated with a perturbed parcel of fluid in a convectively stable medium ($N^2 > 0$).

The second frequency is the *Lamb frequency*, S_ℓ, defined by

$$S_\ell^2 = \frac{\ell(\ell+1)}{r^2}\frac{\Gamma_1 P}{\rho} = \frac{\ell(\ell+1)}{r^2}\, v_s^2. \tag{10.93}$$

In addition, we introduce the *transverse wave number*, k_t (with units cm^{-1}),

$$k_t^2 = \frac{\ell(\ell+1)}{r^2} = \frac{S_\ell^2}{v_s^2}. \tag{10.94}$$

If we relate a transverse wavelength $\lambda_t = 2\pi/k_t$ to k_t then S_ℓ^{-1} is the time it takes a sound wave to travel the distance $\lambda_t/2\pi$.

The differential equation resulting from equating $\nabla \bullet \boldsymbol{\xi}$ and $-\delta\rho/\rho$ now becomes, after some easy algebra and using (10.90–10.91),

$$r\frac{d\xi_r}{dr} = \left[\frac{k_t^2 g r}{S_\ell^2} - 2\right]\xi_r + r^2 k_t^2\left[1 - \frac{\sigma^2}{S_\ell^2}\right]\xi_t. \tag{10.95}$$

A second differential equation is also gotten from the radial force equation (10.82), the definition of ξ_t from (10.87), and using $\delta\rho/\rho$ of (10.91) instead of $\nabla \bullet \boldsymbol{\xi}$ in (10.90). You may verify that the result is

$$r\frac{d\xi_t}{dr} = \left[1 - \frac{N^2}{\sigma^2}\right]\xi_r + \left[\frac{r}{g}N^2 - 1\right]\xi_t. \tag{10.96}$$

We can't apologize for the algebra you have had to go through to reach this point, but we're almost done.

Equations (10.95) and (10.96) taken together constitute a second-order ordinary differential equation. If we had kept in the potential field variations the system would have ended up as fourth order with the additional variables Φ' and its radial derivative. Except for its use in precise applications, the second-order Cowling approximation captures the essence of the low-amplitude behavior of stars.

The boundary conditions on our set of two equations derive from the behavior of the oscillating star at the surface and the center. We refer you to Cox (1980, §17.6) and Unno et al. (1979, §13.1 and §17.1) for the major details of how these are derived. The central boundary condition depends on how various hydrostatic quantities such as A and P vary with radius near the center and the insistence that ξ_r and ξ_t be finite there. These regular solutions go as $(\xi_r, \xi_t) \propto r^{\ell-1}$ for small r and the relation between them is

$$\xi_r(r) = \ell\, \xi_t(r) \quad \text{as } r \to 0. \tag{10.97}$$

The condition at the surface depends on the atmospheric conditions of the static star. In the simplest instance of zero boundary conditions (as discussed in §7.3.2), the perturbations must be such that the surface pressure remains zero. This is the same as requiring that δP be zero at the surface. To express this as a relation between ξ_r and ξ_t consider (10.91), which can be rewritten as

$$\frac{\delta P}{P} = \frac{\rho}{P}\left[\sigma^2 r\, \xi_t - g\, \xi_r\right].$$

Just under the surface the ratio $\delta P/P$ should be finite from physical considerations. As we approach the surface this should remain true even as P goes to zero. However, the factor ρ/P outside the expression in brackets in the above tends to infinity because, in the case of an ideal gas, for example, it is inversely proportional to temperature and T goes to zero at the surface (as in 7.144). Thus for the relative pressure perturbation to remain finite at the surface we require that

$$\xi_t(\mathcal{R}) = \frac{g_s \xi_r(\mathcal{R})}{\sigma^2 \mathcal{R}} \tag{10.98}$$

where g_s is the surface gravity. As in the case of radial oscillations this gives complete reflection and standing waves. We remark here that under some conditions perfect reflection is not possible for real atmospheres and pulsation energy may escape through the surface causing heating of circumstellar material.

We now have two ordinary differential equations and two independent boundary conditions. But, as for radial oscillations, our equations and boundary conditions are linear and homogeneous and we thus have to fix a normalization. Again the choice of normalization is arbitrary and we choose

$\xi_r(\mathcal{R}) = 1$. The system is now overdetermined and σ^2 is an eigenvalue and the perturbations $\xi_r(r)$ and $\xi_t(r)$ are the eigenfunctions.

At this juncture we note an important mathematical difference between radial and nonradial oscillations. Recall that the LAWE for radial oscillations was Sturm–Liouville, which led to a nice ordering of the eigenfrequencies and eigenfunctions with a definite lower bound for σ^2. Such is not the case here even in the Cowling approximation although the system is still self-adjoint. This means that even though the eigenvalues, σ^2, are still real, there is no guarantee that nonradial modes are ordered in any simple way and, in particular, there may be no lower bound on σ^2. We still retain, however, orthogonality of eigenfunctions (see Unno et al. 1989, §14.2 for a derivation of self-adjointness).

We now return briefly to the properties of the $Y_{\ell m}(\theta, \varphi)$. These angular functions are given by

$$Y_{\ell m}(\theta, \varphi) = \sqrt{\frac{2\ell+1}{4\pi}\frac{(\ell-m)!}{(\ell+m)!}}\, P_\ell^m(\cos\theta)\, e^{im\varphi} \tag{10.99}$$

where the $P_\ell^m(\cos\theta)$ are the associated Legendre polynomials generated by

$$P_\ell^m(x) = \frac{(-1)^m}{2^\ell \ell!}\left(1-x^2\right)^{m/2} \frac{d^{\ell+m}}{dx^{\ell+m}}\left(x^2-1\right)^\ell . \tag{10.100}$$

Here x denotes $\cos\theta$.[8] As noted before, the restrictions on ℓ and m for these functions are $\ell = 0, 1, \ldots$ (an integer), and m is an integer with $|m| \le \ell$ for reasons of regularity and single-valuedness of solution. You may wish to play with these functions, but Figure 10.2 shows what they look like on the surface of a sphere where light areas correspond to positive values of the real part of $Y_{\ell m}$ and dark areas to negative values. The symmetry axis defining $\theta = 0$ is almost vertical in the figure; it is actually tilted towards you by 10°. Modes with $m = 0$ are called zonal modes while those with $|m| = \ell$ resemble the segments of an orange and are called sectoral modes. Tesseral modes are those of intermediate type.

Although the figure gives some idea of what is happening on spherical surfaces, the actual motion of the fluid is more complicated. The eigenfunctions have nodal lines on the surface of a sphere at any radius and instant of time but there may also be nodes at different radial positions. This is very difficult to picture. In addition, the oscillatory time dependence means that fluid sloshes back and forth periodically. As a simple example, consider the sectoral mode $\ell = m = 1$. This has one angular node passing through the poles, but this node moves in a retrograde azimuthal direction

[8]Depending on the author and use, you may see other factors of $(-1)^m$ appearing in these formulas. These constitute different phase conventions but do not change the physics. We use the convention of Jackson (1975), §3.5. Arfken (1985) and other texts use other choices of phase.

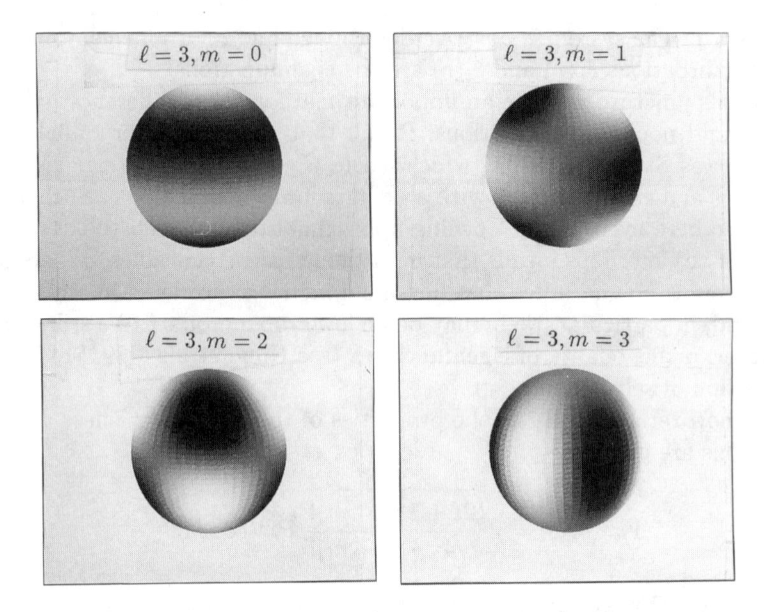

FIGURE 10.2. Shown are the patterns of light and dark on the surface of a sphere corresponding to positive and negative values of the real part of the $Y_{\ell m}$. This figure gives illustrations for (clockwise from the upper left) Y_{30} (zonal), Y_{31} (tesseral), Y_{32} (tesseral), and Y_{33} (sectoral).

(to smaller φ) because of the factor $e^{i(\sigma t + m\varphi)}$ coming from the time dependence and the exponential in $Y_{\ell m}$. That is, lines of constant phase are such that $d\varphi/dt = -\sigma/m$ and, for m positive, we have a running retrograde wave. The zonal ($m = 0$) case is easier to visualize because we can imagine the light and dark portions of a surface (as in the figure) alternating periodically in brightness with a period $\Pi = 2\pi/\sigma$ for σ^2 positive. For one view of nonradial motions see the vector displacement fields shown in Smeyers (1967) for massive upper main sequence stars.

10.3.3 Properties of the Solutions

A great deal can be learned about the solutions to the ordinary differential equations for ξ_r and ξ_t of (10.95) and (10.96) by performing a local analysis of the system. We assume that ξ_r and ξ_t vary much more rapidly in space than do the other physical variables appearing in those equations—such as N^2—so that those variables can be considered constant over some limited range of radius. To quantify this, assume that both ξ_r and ξ_t vary spatially as $e^{ik_r r}$ where the wave number k_r is very large compared to r. Thus both eigenfunctions have many wiggles over a short span of space. Inserting this complex exponential into the differential equations then yields a homogeneous set of algebraic equations in ξ_r and ξ_t whose coefficient determinant must be zero in order to obtain nontrivial solutions. Keeping terms dominant in k_r then yields the dispersion relation

$$k_r^2 = \frac{k_t^2}{\sigma^2 S_\ell^2}\left(\sigma^2 - N^2\right)\left(\sigma^2 - S_\ell^2\right) \tag{10.101}$$

where, as before, we assume σ^2 is positive. The implications of this are:

1. If σ^2 is greater or less than *both* of N^2 and S_ℓ^2, then $k_r^2 > 0$ and sinusoidal propagating solutions are present because $e^{ik_r r}$ reduces to sines and cosines.
2. If σ^2 lies between N^2 and S_ℓ^2, then k_r^2 is negative and solutions show exponential, or evanescent, behavior.

Thus N^2 and S_ℓ^2 are critical frequencies for wave propagation.

The dispersion relation (10.101) may be used to solve for σ^2 in two limits for propagating waves. To facilitate this we also define the total wave number, K, by $K^2 = k_r^2 + k_t^2$ (see Unno et al. 1979, §14). This gives more of the flavor of a wave that can travel in a combination of radial and transverse directions. The understanding is that K is large for a local analysis. Then, if σ^2 is much greater than both N^2 and S_ℓ^2, and $|N^2|$ is smaller than $\S_\ell^2$ (which is usually the case), the "large" root of (10.101) is

$$\sigma_p^2 \approx \frac{K^2}{k_t^2}S_\ell^2 = (k_r^2 + k_t^2)v_s^2, \quad (\sigma^2 \gg N^2, S_\ell^2). \tag{10.102}$$

The subscript "p" has been appended to σ^2 to denote "pressure" because only the sound speed enters. These are *pressure* or *acoustic* modes but we shall often refer to them as "p-modes." You should note here the resemblance to radial modes where ℓ is zero. In that case k_t is zero and we regain (10.31 with $\phi = 0$) derived from our earlier asymptotic analysis.

The small root follows if σ^2 is much less than N^2 and S_ℓ^2 and is given by

$$\sigma_g^2 \approx \frac{k_t^2}{k_r^2 + k_t^2}N^2, \quad (\sigma^2 \ll N^2, S_\ell^2). \tag{10.103}$$

These are *gravity* or "g-modes," so-called because buoyancy in the gravitational field is the restoring force. Note that if N^2 is negative, implying convection, then σ_g is pure imaginary and the perturbation either grows or decays exponentially in time. (These correspond to g^-–modes whereas those associated with $N^2 > 0$ are called g^+–modes.) This is the pulsation analogue to our discussion of convective time scales in §5.1.4.

Thus, in summary, p-modes constitute the high-frequency end of the nonradial oscillation spectrum whereas g-modes are of low frequency.

If each mode in a spectrum is orthogonal with respect to the others, then the eigenfunctions corresponding to each eigenvalue σ^2 must differ in important respects. Following our local analysis as an approximation to what happens, k_r and ℓ must measure this difference. Since k_r is a wavenumber, the corresponding local wavelength is $\lambda_r = 2\pi/k_r$. The total number of nodes (denoted by n) in either eigenfunction is then $n \approx 2\int_0^{\mathcal{R}} dr/\lambda_r$ where the "2" counts the two nodes per wavelength. Thus $n \approx \int_0^{\mathcal{R}} k_r\, dr/\pi$. If (10.102) is integrated such that the integral of k_r appears by itself and if ℓ is small so that k_t^2 may be neglected (for simplicity), we again obtain the estimate

$$\sigma_p \approx n\pi \left[\int_0^{\mathcal{R}} \frac{dr}{v_s}\right]^{-1} . \tag{10.104}$$

Thus for large n the p-mode frequencies are equispaced. (For more exact treatments using JWKB methods see Unno et al. 1979, §15, or Tassoul 1980. Our estimates hide many sins.) Note that the frequency spacing depends only on the run of the sound speed which, for an ideal gas, depends primarily on temperature. Thus, in stars such as the sun, p-modes effectively sample the temperature structure.

A corresponding estimate for the *periods* of g-modes is

$$\Pi_g = \frac{2\pi}{\sigma_g} \approx n\,\frac{2\pi^2}{\left[\ell(\ell+1)\right]^{1/2}}\left[\int_0^{\mathcal{R}} \frac{N}{r}\, dr\right]^{-1} . \tag{10.105}$$

Here it is the period that is equally spaced in n (a fact to be taken advantage of when discussing variable white dwarfs) and it depends sensitively on ℓ. Also, the frequencies (periods) decrease (increase) with n, in direct contrast to the p–modes. We shall give a numerical example of the relative ordering of nonradial modes in §10.4 (see Fig. 10.4).

The same limits on σ^2 relative to N^2 and S_ℓ^2 also yield the following rough estimates for the ratio of radial to tangential eigenfunctions when used in the differential equations (10.95) and (10.96):

$$\left|\frac{\xi_r}{\xi_t}\right| \sim \begin{cases} rk_r & \text{p-modes} \\ \ell(\ell+1)/rk_r & \text{g-modes.} \end{cases} \tag{10.106}$$

For large radial wavenumber ($rk_r \gg 1$) the fluid motions for p-modes are primarily radial, whereas they are primarily transverse for g-modes.

Mode Classification

We have seen that the character of a particular mode depends on n, ℓ, and the relative amplitudes of the radial and tangential displacements. In addition, p-modes are of higher frequency than g-modes. The frequency of a given mode is denoted by $\sigma_{n\ell}$ where it is understood that two different modes (p– and g–) may exist for a given combination of n and ℓ. How does the azimuthal "quantum" number m enter this picture? If the unperturbed star is spherically symmetric, then the eigenvalue (the frequency) is independent of m even though the eigenfunction must depend on m through the appearance of $e^{im\varphi}$ in $Y_{\ell m}$. This is true because there is no preferred axis of symmetry in a spherically symmetric system. We may arbitrarily choose such an axis—and this establishes the pole for measuring the co-latitude angle θ—but, since φ enters only as a phase factor in $e^{im\varphi}$ and we may choose any great polar circle to start measuring φ, it cannot enter in the final analysis for the eigenvalue. (The same is true for the isolated hydrogen atom where $Y_{\ell m}$ also appears in the eigenfunction: the energy eigenvalue does not depend on m.) Another way to look at this is to review our derivation of the pulsation equations to ordinary differential form where m did not appear. If, on the other hand, there were effects that destroyed spherical symmetry—such as rotation or magnetic fields—then m would play a role and we would have to include that in the specification of σ.

We have left out quite a few things in this introduction to nonradial adiabatic stellar oscillations. If you peruse the advanced literature, you will discover that complicated evolved stars have modes that look like p-modes in some portions of the interior but g-modes elsewhere. As interesting as the authors find these niceties, they would take us far afield here. We shall, however, discuss applications of this theory to the sun and variable white dwarfs, which represent the further extremes in stellar evolution.

• For those of you who would welcome the challenge of computing your own nonradial eigenfrequencies and eigenfunctions for ZAMS models (and we suggest trying a ZAMS sun), we discuss an adiabatic Cowling code in the program `PULS.FOR` in Appendix C.[9] •

10.4 Helioseismology

The sun, and perhaps all stars at some level, is variable but you need keen instruments to detect its variability. Leighton, Noyes, and Simon (1962) first observed a five-minute correlation in velocity-induced Doppler shifts of absorption lines formed at the solar surface. These were (and are) interpreted as vertical oscillations of large patches of fluid having velocities of 1 km s^{-1} or less with a coherence time of around five minutes. It was not until nearly ten years later that Ulrich (1970) and Leibacher and Stein (1971) independently suggested that what was being observed was the result of global oscillations with periods of around five minutes wherein the sun acted as a resonant cavity for acoustic waves propagating through its

[9]`PULS.FOR` uses the dimensionless "Dziembowski (1971) variables," which have certain computational advantages. In the Cowling approximation these are $y_1(x) = \xi_r(r)/r$ and $y_2(x) = \xi_t(r)\sigma^2/g$. The independent variable that replaces r in that code is $x = \ln(r/P)$. This has the advantage of "stretching" the computational mesh to give better resolution at the center and surface. (See Osaki and Hansen 1973.)

interior. It is now well established that the majority of these waves are nonradial acoustic modes.

We shall not discuss here how these waves are excited but it is not by the same mechanism that is responsible for the variability of other overtly variable stars. The best current model involves the convection zone of the sun wherein the noise generated by turbulence effectively causes the whole sun to quiver in response.

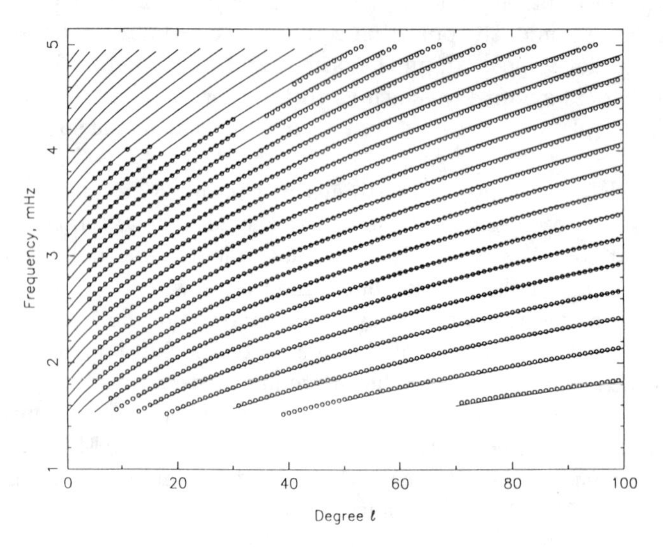

FIGURE 10.3. Measured solar p-mode frequencies, $f_{n\ell}$, for $\ell < 100$ are shown by the circles. Theoretical results are indicated by the solid lines where individually computed frequencies for fixed n have been made continuous. The lowest "ridge line" has $n = 2$ and n increases by one for each higher ridge. Note that the frequency scale is in mHz so that a typical frequency of $f = 3$ mHz corresponds to a period of $\Pi = 1/f \approx 5$ minutes. (From Libbrecht 1988 based on data from Duvall et al. 1988.) Reprinted by permission of Kluwer Academic Publishers.

10.4.1 Probing the Interior of the Sun

Nonradial modes have now been observed by several methods and it is known that they consist of the incoherent superposition of millions of acoustic p-modes. The current literature on the subject, both observational and theoretical, is expanding at a rate comparable to the number of modes and the results described here will surely soon be outdated. For the present, however, we recommend the review articles by Christensen-Dalsgaard *et al.* (1985), Leibacher et al. (1985), Wentzel (1991), Libbrecht and Woodard (1991), Brown et al. (1986), Toomre (1986), and Libbrecht (1988). The first four are in a semipopular vein and are excellent introductions. You should also peruse the conference proceedings literature for reports of the annual meetings of the "Global Oscillation Network Group" (GONG). This consortium of astronomers is dedicated to setting up a global network of stations to observe oscillations of the sun.[10]

To give a picture of a large piece of the observational and theoretical results, we first present Figure 10.3 (from Libbrecht 1988), which shows measured p-mode frequencies, $f_{n\ell}$ (in mHz with $2\pi f_{n\ell} = \sigma_{n\ell}$), versus ℓ. The individual points are observed modes (from Duvall et al. 1988) where n, the mode order, is as indicated in the figure caption. Such measurements are extraordinarily difficult and involve two-dimensional Doppler imaging of the solar surface and other techniques. The solid lines are from theoretical nonradial pulsation calculations for a standard model of the sun by Christensen-Dalsgaard (1985). These calculations were done using essentially the same theory as discussed above but included the gravitational perturbations we left out (plus a few other bells and whistles).

These modes are readily identifiable as p-modes by several means. The way we shall go about it is to first examine the structure of a ZAMS sun model using the code `ZAMS.FOR` discussed in Appendix C. (And you can do this yourself.) A ZAMS sun is perfectly adequate for our purposes because we believe that the regions in the sun most readily identified with p-modes have not changed that much over the years. The key element to look for is the behavior of the two critical frequencies S_ℓ^2 and N^2 that were part of our local analysis leading to (10.101). Shown in Figure 10.4 is S_ℓ^2 (for $\ell = 2$) and N^2 versus radius for a ZAMS sun with composition $X=0.74$ and $Y=0.24$. The ordinate has the units of the square of angular frequency (rad^2 s^{-2} in base 10 logarithm). Such a figure is called a "propagation diagram."

The behavior of S_ℓ^2 is understood as follows. As $r \to 0$, $S_\ell^2 = \ell(\ell+1)v_s^2/r^2$ approaches positive infinity because of the factor r^{-2}. Near the surface it approaches zero (negative infinity in the log) because the temperature of the ideal gas effectively goes to zero and thus does the sound speed. On the other hand, N^2 goes to zero at the center because it contains the factor g. It then increases to a maximum, tails off, and then drops to negative values starting at $r/\mathcal{R} \approx 0.83$. This last precipitous drop signals the onset of vigorous envelope convection.[11] You cannot see it in this figure but N^2 then rises to finite positive values as the photosphere is reached.

The next step in interpretation is to recall that p-modes are supposed to propagate where σ^2 is greater than both S_ℓ^2 and N^2. What we have done

[10]There are six of these automated stations planned. The two-dimensional imaging of velocity fields should yield an enormous amount of data. You can find a brief but informative description of the system on page 85 of Wentzel (1991). For example, when the system is running at full tilt it should generate over 10^{12} bytes of information over a three-year period equivalent to the total text contained in a decent-size public library!

[11]The base of the convection zone for the present-day sun is at $r/\mathcal{R} \approx 0.73$, which can be inferred from Figure 8.6. Differences in composition, choice of mixing length, and evolutionary effects account for the `ZAMS` result.

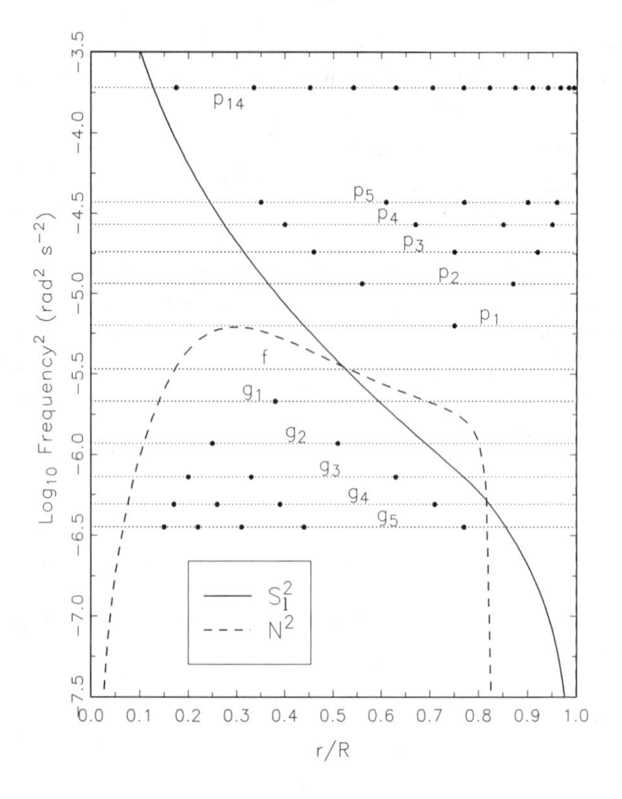

FIGURE 10.4. Shown is a propagation diagram for a model of the ZAMS sun. See the text for details of how this figure was constructed and for its significance.

is to compute p-mode eigenfrequencies and eigenfunctions of this ZAMS model using the code `PULS.FOR` discussed in Appendix C. The results of this calculation are summarized in Figure 10.4. The horizontal dotted lines give the location in frequency squared (σ^2) of six p-modes of angular order $\ell = 2$. The one labeled "p_1" is the simplest in that it has only one node in the radial displacement ξ_r. The location of that node with respect to $r/\mathcal{R}$ is indicated by the large dot on the dotted line at $r/\mathcal{R} \approx 0.75$. The next higher frequency mode is the p_2 mode with two nodes indicated for ξ_r, and so on.

We note that the complexity of these p-modes increases regularly with frequency (as it does for radial modes) and all the large dots on the figure are confined to the region where σ^2 is greater than both S_ℓ^2 and N^2. This is as it should be because the dots represent nodes where the eigenfunctions behave like sinusoids and the waves are propagating. Outside this region there are no nodes, which implies that the modes are evanescent. Finally, a p_{14} mode with 14 nodes is also shown (leaving out modes p_{6-13} for clarity).

The frequency of this mode is $\sigma = 1.386 \times 10^{-2}$ rad s^{-1} ($f = 2.206 \times 10^{-3}$ Hz), which yields a period, Π, of 7.56 minutes. This puts us near the lower left-hand corner of Figure 10.3.

Figure 10.4 also shows the frequencies of the $\ell = 2$ g-modes and the locations of their nodes. Their regions of propagation are confined to the deeper interior because, as is apparent and as we have already discussed, the squares of their eigenfrequencies must be *less* than both of S_ℓ^2 and N^2 to propagate locally. They are ordered in frequency opposite to those of p-modes: the more complicated modes have the lower frequencies. The period of the lowest-order mode, g_1, in this calculation is slightly over an hour, which means observation times must be long to detect it. (You should not take this period too seriously because `PULS.FOR` uses the Cowling approximation which, for solar g-modes, does not give entirely reliable results. But they are good enough for here.)

Detection is also complicated by the structure of the eigenfunctions. Figure 10.5 shows the radial displacements, ξ_r, for $\ell = 2$ p_2 and g_2 modes as a function of radius with the normalization $\xi_r(\mathcal{R}) = 1$. Note that the p-mode is confined to the outer layers and it is only as the surface is approached that it attains significant amplitude; that is, the motions associated with these modes effect the surface layers and leave the interior relatively untouched. The g-modes, on the other hand, have their largest amplitudes in the deep interior. Were one to exist in the sun and have a surface amplitude comparable to a p-mode, the deep interior would have large motions indeed. If they do exist, their surface amplitudes are probably very small. As of this writing, there has been no unambiguous detection of solar g-modes. This is a shame because these modes, in principle, probe deeply into the sun.

Finally, Figure 10.4 shows the location of the "f-mode," which we have not yet discussed. It has a frequency intermediate to those of the p– and g–modes, has no nodes, and is interpreted as a surface gravity wave that causes the whole structure to slosh in unison.

Another way to interpret what is going on is to consider our short-wavelength result $\sigma_p^2 \approx (k_r^2 + k_t^2) v_s^2$ of (10.102). Suppose a wave originates at the solar surface and travels inward. As it does so, it encounters an increase in sound speed because temperature increases inward. But since σ_p^2 is a constant, this implies that $K^2 = k_r^2 + k_t^2$ must decrease. If we assume that k_r is the more rapidly varying term, then K decreases inward until $K = k_t = \sqrt{\ell(\ell+1)}/r_\mathrm{t}$ where r_t is the "turning radius." The reason for this terminology is that the vanishing of k_r means the wave can no longer progress radially inward but must start to move out again. The turning point is then established by the condition

$$\frac{v_s(r_\mathrm{t})}{r_\mathrm{t}} = \frac{2\pi f_{n\ell}}{[\ell(\ell+1)]^{1/2}} \tag{10.107}$$

which is the same as the condition $\sigma_p^2 = S_\ell^2(r_t)$. This picture is similar to

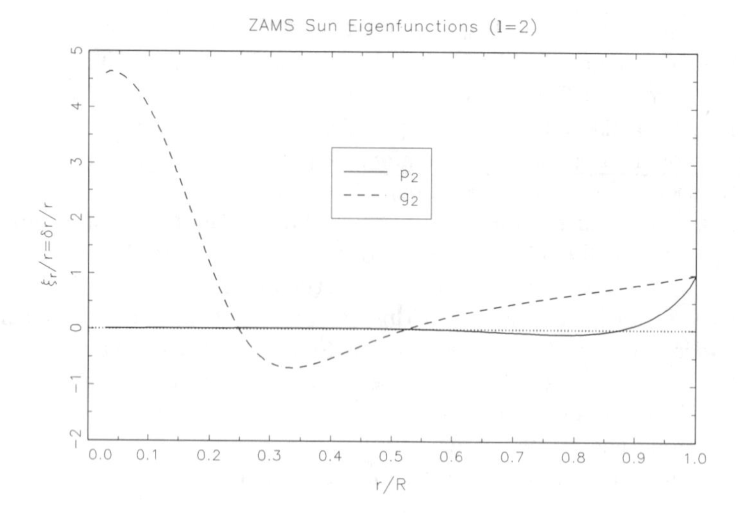

FIGURE 10.5. Shown are the radial displacements, ξ_r, for $\ell = 2$ p_2 and g_2 modes as a function of radius for a ZAMS sun model. These are from the same calculations used to make Figure 10.4.

that of refractive optics where ray paths are determined by the properties of the medium through which the light passes. Figure 10.6 (from Christensen-Dalsgaard et al. 1985) shows "ray paths" for solar oscillations with different values of n/ℓ. The smaller this ratio, the shallower the penetration.

An approximate value for the turning radius for modes that do not penetrate too deeply may be estimated with the following argument. The local sound speed for an ideal gas is $v_s = \sqrt{\Gamma_1 N_{\mathrm{A}} kT/\mu}$ where temperature as a function of radius is given by (7.144) for zero boundary conditions. (We use this despite the observation that the sun is convective just below the photosphere.) If Kramers' is the opacity source (so that $n_{\mathrm{eff}} = 3.25$), $\mu = 0.6$, and Γ_1 is 5/3, then, after little effort, (10.107) becomes

$$\frac{2.56 \times 10^8 f_{n\ell}^2}{\ell(\ell+1)} x_{\mathrm{t}}^3 + x_{\mathrm{t}} - 1 = 0 \tag{10.108}$$

where $x_{\mathrm{t}} = r_{\mathrm{t}}/\mathcal{R}_\odot$.

As an illustration, consider the p_1 mode in the $\ell = 2$ propagation diagram (Fig. 10.4). The frequency of that mode is $f_{1,2} \approx 4 \times 10^{-4}$ Hz ($\Pi \approx 2500$ seconds, $\sigma^2 \approx 6.24 \times 10^{-6}$). The above equation for the turning point then gives $r_{\mathrm{t}} \approx 0.44\mathcal{R}$. If you examine the figure, this is the radius at which $\sigma^2 \approx S_\ell^2$ as should be expected: the wave becomes evanescent interior to that radius.

It would appear from Figure 10.3 that the theoretical results for p-mode frequencies are very good indeed because the results match the observations as well as the eye can detect. This tells us that our model for the sun is

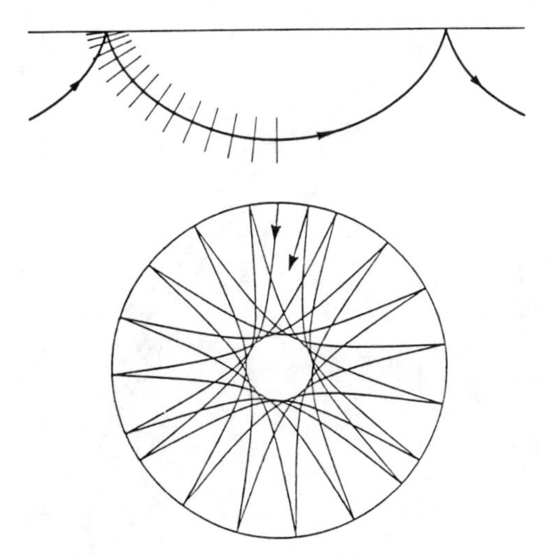

FIGURE 10.6. Shown are ray paths in the sun for two different values of n/ℓ. The wave that penetrates deeply has $n/\ell = 5$ whereas the the shallow wave is for a mode with $n/\ell = 1/20$. Reproduced, with permission, from Toomre (1984).

"accurate" down to appreciable depths into the interior. However, there are discrepancies that cannot be ignored. Shown in Figure 10.7 are the differences between observed and calculated frequencies for ℓs ranging from $\ell = 1$ to 200. Note that the ordinate is in μHz. The theoretical frequencies are for a particular standard solar model (see Christensen-Dalsgaard and Gough 1984) but the result is representative of many investigations.

The most obvious thing of note in Figure 10.7 is the trend that low frequencies seem to be underestimated compared to the observations whereas higher theoretical frequencies are too high (although the overall agreement is still remarkable). The larger errors at high frequency for ℓ greater than about forty imply that there are problems in modeling the very surface layer because that is where those modes primarily probe. This is perhaps not too surprising because the surface of the sun is dynamic with a forming chromosphere, flares, etc., and the formulation of nonradial surface boundary conditions is sensitive to those dynamics. For ℓ less than about 20, errors are independent of ℓ to within the observational errors.

Low-frequency p-modes probe deeply into the sun's interior. This is especially true for modes with low ℓ. (For a tabulation see Duvall et al. 1988.) The most deeply penetrating are the radial modes with $\ell = 0$ that pass right through the center; investigation of these modes bears directly on the solar neutrino problem. Much effort has thus gone into searching for possible errors and effects that make both theoretical oscillation and neutrino calculations match the observations. Bahcall and Ulrich (1988) review such attempts but, sad to say, the results are not all that encouraging. It may be that some elements in our understanding of fundamental physics, such

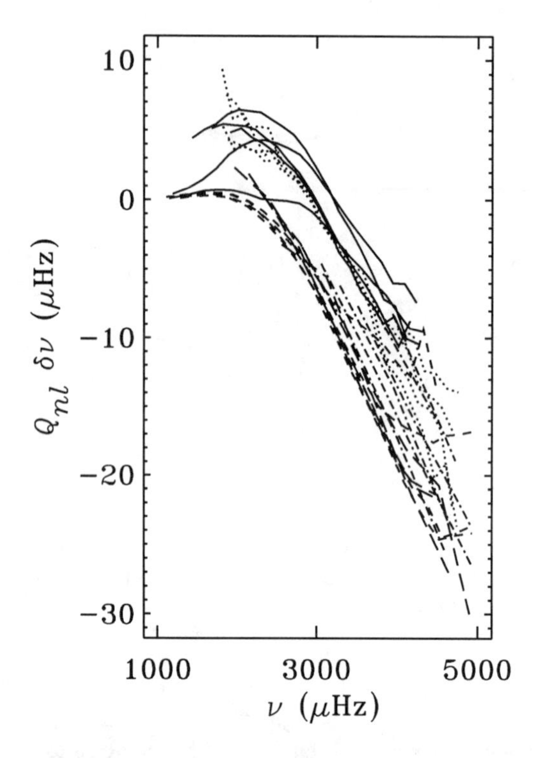

FIGURE 10.7. The ordinate shows the difference, $\Delta\nu = \nu_{\text{obs}} - \nu_{\text{theo}}$, with ν as frequency, between observed solar p-mode frequencies (ν_{obs}) and theoretical frequencies (ν_{theo}). Solid lines refer to modes with $\ell \leq 20$ and dashed lines to $\ell \geq 40$. Reproduced, with permission, from Christensen-Dalsgaard (1991).

as opacities, may be leading us astray but there is as yet no firm evidence for this. Various versions of nonstandard models have been tried with conflicting results. Thus, for example, models with lower amounts of metals in the deep interior ("low–Z models") help the neutrino problem but worsen the match for low ℓ modes. One promising avenue has been to postulate that the primordial helium abundance in the deep interior of the sun was higher than in the surface layers. This tends to bring low ℓ frequencies in line and decreases the calculated neutrino fluxes—but not by enough. Of course, the solar neutrino "problem" may, as indicated earlier, be more a reflection of our ignorance of fundamental particle physics than of more traditional astrophysics.

10.4.2 The Inverse Problem and Rotation

Thus far we have described what is called the "forward problem" in solar seismology, where a particular model is examined with regard to its oscillation properties and these are then compared to observations. If the comparison fails in some respects—as it usually does—then the model is known to be incomplete or somehow wrong. The trouble is that it is not always clear what causes discrepancies or what to do about it. On the other hand, there is a wealth of information contained in the oscillation spectrum with its millions of modes. These modes probe the interior to different depths depending on ℓ and n and whether they are p-modes or perhaps g-modes. The "inverse problem" consists of using these probes to either detect where things have gone astray in the model or to yield other information. We cannot treat this topic with anywhere near the detail it deserves and shall only give a hint to how it works and then present one application. For a fuller introduction from the textbook literature see Unno et al. (1989, §41). The subject was originally developed by terrestrial seismologists, who have also contributed directly to the stellar analogue.

One way to look at the inverse problem is to consider the wave equation for nonradial oscillations. We shall not derive it here (see, for example, Chandrasekhar 1964, Lynden-Bell and Ostriker 1967, Cox 1980 §15.2) but it has the same form as (10.16) for radial oscillations except we are dealing with vector displacements. In symbolic form it is really the force equation

$$\frac{\partial^2 \boldsymbol{\xi}}{\partial t^2} = -\mathcal{O}(\boldsymbol{\xi}) \tag{10.109}$$

where $\mathcal{O}$ is a linear operator acting on $\boldsymbol{\xi}(\mathbf{r}, t)$. Using the temporal factor $e^{i\sigma t}$ this becomes

$$\sigma^2 \boldsymbol{\xi} = \mathcal{O}(\boldsymbol{\xi}). \tag{10.110}$$

We now "dot" multiply each side of the above by $\boldsymbol{\xi}^*$ and integrate over mass and solid angle $[\Omega(\theta, \varphi)]$ to find

$$\sigma^2 = \frac{\int \boldsymbol{\xi}^* \bullet \mathcal{O}(\boldsymbol{\xi})\, d\mathcal{M}_r\, d\Omega}{\int \boldsymbol{\xi}^* \bullet \boldsymbol{\xi}\, d\mathcal{M}_r\, d\Omega}. \tag{10.111}$$

This is a variational expression for σ^2 analogous to (10.60). Thus if the eigenfunctions for the problem are known, then an appropriate integration over those eigenfunctions convolved with variables in the problem (in our case pressures, Lamb frequencies, etc.) yield back the eigenfrequencies. A specific formulation of this integral in terms of the variables we have used is given in Kawaler et al. (1985).

A useful property of this representation and its analogues in quantum mechanics is that the eigenfunctions may be changed by a considerable amount while σ^2, as computed from (10.111), changes only slightly.[12] We can thus imagine the following experiment.

---

[12]The integral is "stationary" with respect to $\boldsymbol{\xi}$. Another way to phrase this is to realize that eigenvalues must be calculated very precisely in order to obtain accurate eigenfunctions. This is why `PULS.FOR` of Appendix C demands strict convergence on eigenvalues.

Suppose some quantity such as density is varied only slightly by an amount $\Delta\rho(r)$ through a solar model for which we already have an eigenfrequency, σ_0^2, and eigenfunction, $\boldsymbol{\xi}_0$. This variation in density will change the integrands in (10.111) by a small amount because the model has changed slightly. At first glance it would appear that to find the new eigenfrequency for the altered model we have to redo the complete pulsation problem. This is unnecessary, however, because of the stationary property of the integrals. All that needs to be done is to recast the integrand of the numerator of (10.111) in the form $K(\mathbf{X}_0, \boldsymbol{\xi}_0, r)\Delta\rho/\rho$ (following the nomenclature of Unno et al. 1989) where the function K is the first-order term in the linearization of the integrand with respect to $\Delta\rho$. The arguments of K are the untouched model parameters, $\mathbf{X}_0$, and the *original* eigenfunction $\boldsymbol{\xi}_0$. The relative change in the eigenvalue is then

$$\frac{\Delta\sigma^2}{\sigma_0^2} = \int_0^{\mathcal{M}} K(\mathbf{X}_0, \boldsymbol{\xi}_0, r)\,\frac{\Delta\rho}{\rho}\,d\mathcal{M}_r. \tag{10.112}$$

This procedure is essentially that followed in first-order perturbation theory in quantum mechanics.

If $\Delta\rho$ is specified in (10.112) then changes in the frequencies of various modes may be easily computed; that is, pick a model, calculate K, and find $\Delta\sigma^2$. But this is just the forward problem all over again. The inverse problem consists of computing σ_0^2 for many modes from a fiducial standard model (thus yielding K), calculating $\Delta\sigma^2 = \sigma_{\rm obs}^2 - \sigma_0^2$ as the difference between model and observed frequency $\sigma_{\rm obs}^2$, and, by some means, finding a best fit for what $\Delta\rho(r)$ must be in order to reproduce $\Delta\sigma^2$ for many modes in, perhaps, a least squares sense. Here is where seismology comes in. If many modes are used and if they probe the sun well, then we expect a good estimate for $\Delta\rho(r)$. The application of this inversion technique is both mathematically and computationally difficult but has yielded useful information about the run of the solar sound speed, for example.

Another application of inverse theory is deducing the internal rotation rate of the sun. The equatorial rotation velocity of the solar surface is 2 km s^{-1}, which corresponds to a rotation period of about 25 days or an angular frequency of $\Omega = 2.9 \times 10^{-6}$ rad s^{-1}. The rotation is differential, however, in that the period increases as we move away from the equator to the poles. Before oscillations gave us a probe into the interior, the only evidence available for deducing the interior rotation properties came from observations of surface oblateness or small deviations in the advance of the perihelion of Mercury as calculated from Einstein's general theory of relativity. These are very indirect methods and depend sensitively on theoretical interpretation of the observations. Solar seismology appears to be more promising in this respect and relies on how pulsation modes are influenced by rotation.

The first thing to note is that rotation implies a preferred axis for the sun (assuming that the axis implied by surface rotation is the same as that for the entire interior). This means that the frequency of an oscillation mode depends not only on the mode order, n, and ℓ, but also on m, which means we must talk about $\sigma_{n\ell m}$. Note also that a typical angular frequency for a five-minute mode is $\sigma \approx 0.02$ rad s^{-1}, a value much larger than Ω for the solar surface. We suspect, therefore, that the effect of rotation on pulsation is small and the $2\ell+1$ frequencies $\sigma_{n\ell m}$ (with $m = -\ell, \cdots, 0, \cdots, \ell$) are close in value to those for no rotation ($\sigma_{n\ell \text{ any } m}$), which are degenerate with respect to m. This effect of removing degeneracy is often called "rotational splitting" and, in the most general case, is difficult to analyze. For the sun, however, the rotation is "slow" and this greatly simplifies matters.

The centrifugal acceleration at the solar equator is $\Omega^2\mathcal{R}_\odot \approx 0.6$ cm s^{-2}. This is negligible compared to the surface gravity $g_\odot \approx 3\times10^4$ cm s^{-2}. Thus, unless the solar interior spins at a very rapid rate (which appears not to be true), centrifugal forces may be neglected as a good first approximation in constructing solar models and we may continue to stick with spherical symmetry. The same holds true for Coriolis forces. Since, however, the sun oscillates and fluid moves, then Coriolis forces, which go as $2\boldsymbol{\Omega}\times\mathbf{v}$ per unit mass where $\mathbf{v}$ is the fluid velocity, cause deviations in the flow that were not accounted for in our earlier pulsation analysis. (The centrifugal force may still be neglected because it goes as the square of Ω, which is small in the first place.) We shall outline some of the steps to be taken to correct that analysis for slow rotation. (A more complete discussion may be found in Unno et al. 1989, §19.1 and §§31–34.)

Suppose, as a simple case, the sun were rotating uniformly. We then place ourselves in a noninertial reference frame corotating with the sun. In that frame, the force equation is amended to take account of Coriolis forces by appending the term $-2\boldsymbol{\Omega}\times\mathbf{v}$ to the right-hand side of the force equation (10.68). The velocity here is that caused by pulsation and, after linearization, becomes $\dot{\boldsymbol{\xi}}$. This, in turn, causes terms of order $|\Omega\sigma\boldsymbol{\xi}|$ to appear in the right-hand sides of the pulsation equations (10.82–10.84) after time has been separated out using $e^{i\sigma t}$. The resulting equations are very difficult to solve because there is no longer any guarantee that spherical harmonics will do the trick. But, as mentioned before, the sun rotates slowly in the sense that $\Omega << \sigma$. Thus the Coriolis term containing $\Omega\sigma$ is much smaller than the acceleration term containing σ^2 on the left-hand sides of (10.82–10.84) and the Coriolis force is only a perturbation to the solution for the nonrotating sun. This sounds like what we discussed for the inverse problem; namely, perturb the Hermitian operator (10.110) to account for small effects due to Coriolis force (rather than small perturbations in density), use the eigenfunctions for the unperturbed problem, and then evaluate a few integrals. For uniform rotation this is straightforward (see Unno *et al.* 1979, §18, or Hansen, Cox and Van Horn 1977) and yields the following solution.

Let $\sigma_{n\ell m}$ be the eigenfrequency of a mode in the fictitious uniformly rotating sun and $\sigma_{n\ell,0}$ be the eigenfrequency for the same mode not influenced by rotation (as computed, independent of m, by the methods outlined previously). The two are connected by the small perturbation in frequency $\Delta\sigma_{n\ell m}$ by

$$\sigma_{n\ell m} = \sigma_{n\ell,0} + \Delta\sigma_{n\ell m} \tag{10.113}$$

where, if we now measure eigenfrequencies in the external inertial frame of an observer,

$$\Delta\sigma_{n\ell m} = -m\Omega\left\{1 - \frac{\int_0^{\mathcal{M}} \left[2\xi_r\xi_t + \xi_t^2\right] d\mathcal{M}_r}{\int_0^{\mathcal{M}} \left[\xi_r^2 + \ell(\ell+1)\xi_t^2\right] d\mathcal{M}_r}\right\}. \tag{10.114}$$

Here the eigenfunctions $\xi_r(r)$ and $\xi_t(r)$ are those obtained from the nonrotating model (and depend only on n and ℓ) and thus the ratio of integrals, denoted by $C_{n\ell}$, depends only on n and ℓ. This solution has two parts: $-m\Omega$ and $m\Omega C_{n\ell}$. The first part comes about only because we are viewing the rotating star from an inertial frame where, for $m \neq 0$, we detect the additional red or blue Doppler shift due to the running wave in the azimuthal direction (Cox 1984). That is, if we were to neglect Coriolis forces then the frequency of a mode as viewed in the rotating frame of the star would be $\sigma_{n\ell,0}$. But moving to the inertial frame means we see running azimuthal wave crests (from $e^{i\sigma t + im\varphi}$) either moving faster or slower with respect to use, depending on whether the waves are prograde ($m < 0$ and moving in the same sense as the rotation) or retrograde ($m > 0$ and moving in the opposite sense). The second term contains the effect of the Coriolis force.

In this example of uniform rotation it is easy to see that the modes are equally split in frequency with $\sigma_{n\ell\, m+1} - \sigma_{n\ell\, m}$ equal to $\sigma_{n\ell\, m} - \sigma_{n\ell\, m-1}$ and the degree of splitting (as given by these differences) is proportional to the rotation frequency Ω. We shall see that such a "picket fence" structure in frequency occurs in the oscillation spectra of some variable white dwarfs but the sun is more complicated. We know from the outset that it rotates differentially and not uniformly. Therefore the frequency splitting of solar p-modes is not necessarily uniform and thus should yield information on the rotation of the interior. This is again an inversion problem where information is used from observations of many modes to, in effect, probe the Coriolis forces within the sun. Unfortunately the rotation frequency is actually a function of all three space variables and many degrees of freedom are implied (unlike inversion problems discussed earlier that assumed spherical symmetry). Yet, by assuming some reasonable constraints, considerable progress has been made.

Representative of these attempts is that shown in Figure 10.8, where $\Omega(r, \theta, \varphi)$ is assumed to have the same θ dependence as at the photosphere. In this investigation the rotation down to $r/\mathcal{R}_\odot \approx 0.7$ is similar to that of the surface but for $r/\mathcal{R}_\odot \lesssim 0.6$ more closely resembles solid body rotation. The important questions these results pose is how this rotation comes about and what consequences it has for the entire evolution of the sun.

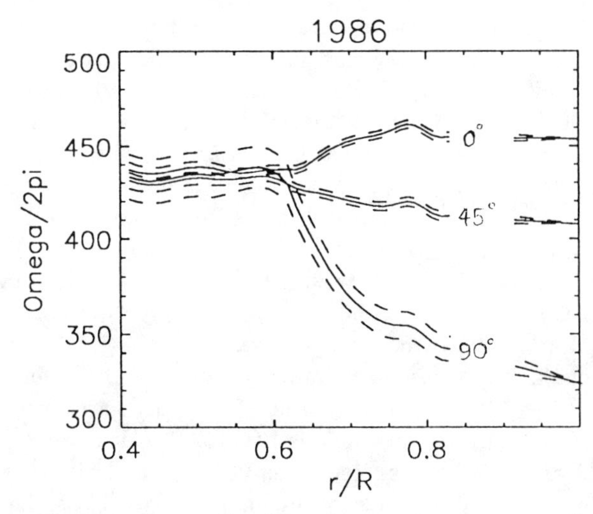

FIGURE 10.8. Shown is the profile of angular rotation frequency for three latitudes of the sun as inferred from rotational splitting using inversion theory. The ordinate is in the units of nanoHertz. The data for deeply penetrating p-modes is inadequate to clearly resolve Ω for $r \lesssim 0.4\mathcal{R}_\odot$. Reproduced, with permission, from Schou (1991).

10.5 Asteroseismology of White Dwarfs

It was once thought that white dwarfs were extremely stable in their light output, so much that they could be used as luminosity standards for faint variable stars. Acting on this assumption, A.U. Landolt observed the white dwarf HL Tau 76 with the intent of using it as a standard star. To his surprise, and as reported in Landolt (1968), he found instead that this star was variable with a period of 12 minutes and with luminosity variations of over a tenth of a magnitude. More than 30 of these variable white dwarfs have been discovered to date and, from considering the statistics of the total white dwarf population, McGraw (1977) (and see Cox 1982) concludes that this class of variable star is the most common in the universe among those that are obviously variable.

The study of variable white dwarfs is still in the stages of active observational and theoretical development and we can only touch on the high points here. Recent reviews include Winget and Fontaine (1982), Cox (1986), Winget (1988), and Kawaler and Hansen (1989).

10.5.1 The Observed Variables

There are three major classes of variable white dwarfs. All are multiperiodic, with periods ranging from roughly 100 to 1000 seconds. The coolest are the hydrogen-surfaced DAVs (with "V" for "variable") or "ZZ Ceti" variables. They lie in the effective temperature range $13,000 \gtrsim T_{\rm eff} \gtrsim 11,000$ K, where the end points may be as uncertain as ± 500 K (Greenstein 1984, Weidemann and Koester 1984, Bergeron et al. 1992). The hot (cool) end of this range is called the "blue (red) edge" because of color. In any case, the interval in temperature and color is narrow and relatively well defined and is called the "instability strip."[13] (The same nomenclature applies to the variable stars of the Cepheid strip discussed in Chapter 2.) Twenty-two of these stars have been discovered.

The next class are the DBVs with helium surfaces lying in the temperature range $24,000 \gtrsim T_{\rm eff} \gtrsim 21,500$ (Koester et al. 1985, Liebert et al. 1986, Theji et al. 1991). There are seven known.

The last class, or subclasses, is not as well defined as the first two. These are very hot ($T_{\rm eff} \gtrsim 10^5$ K) and are either DO white dwarfs or nuclei of planetary nebulae. The prototype DOV is PG1159-035 discovered by McGraw et al. (1979) and is otherwise known as GW Vir and all three names are sometimes used for the eleven known members of the class. The prototype PNN variable is K1–16 that is embedded in a planetary nebulosity, which makes it difficult to observe as a variable star (Grauer and Bond 1984). What is confusing about these variables is that there are other stars that are spectroscopically similar to them but are not variable.

These are the immediate essentials of the observational characteristics of the white dwarf variables. More detailed observational material will be discussed below in considering what the variables tell us about structure and evolution.

10.5.2 White Dwarf Seismology

Because typical periods for well-studied variable white dwarfs are around a few hundred seconds, the observed oscillations cannot be acoustic modes. The estimate for the frequencies of p-modes given by (10.104) sets an upper limit for periods of $\Pi_p \lesssim \pi \int v_s^{-1}\, dr$. But this is essentially the period-mean-density relationship discussed in §1.3.5. A typical average density for white dwarfs is $\langle \rho \rangle \approx 10^6$ g cm^{-3} so that using (1.37) with a coefficient of 0.04 day yields $\Pi_p \lesssim 4$ s. This is too short by at least a factor of 20 to match the observations. What is left are gravity modes. The following outlines the arguments originally developed by Chanmugam (1972) and Warner and Robinson (1972) and put on a firmer numerical footing by Osaki and Hansen (1973).

The periods of gravity modes depend on the run of the Brunt-Väisälä frequency, N^2, as shown by (10.105). There is really no way we can estimate that quantity easily but it does have certain distinctive qualitative features

[13]There is some evidence that not all DAs in the instabilty strip pulsate. If not, then the name is a misnomer. See Kepler and Nelan (1993).

in white dwarfs. For example, it is *very* small in the electron-degenerate interior. This may be seen by comparing (5.35), which gives a definition of N^2, and (3.110), which shows how χ_T varies with temperature for a degenerate gas. The point is that for the (relatively) low temperatures deep inside white dwarfs, N^2 is small. This is not necessarily the case in the envelope and typical values of a few thousand s^{-2} may be encountered. (It will, of course, be negative in convection zones.) On the other hand, the Lamb frequency S_ℓ is large in the interior but becomes very small in the envelope as was the case for the sun.

If we now recall our discussion of the conditions for wave propagation (see 10.101 and following), it becomes clear that g-modes propagate in the envelope regions whereas p-modes (which do not seem to exist in these variables) tend to do so in the deeper interior. Note that this is the opposite from the sun. Thus we have the picture of gravity modes actively waving around in the surface regions but being excluded from the core because of very small values of N^2 deep inside. Detailed numerical calculations, as reviewed in the primary references, yield periods of the length observed.

The cause of the instability has been determined to be the same as that which drives more classical variable stars: it is associated with some combination of the ionization zones of hydrogen or helium and perhaps carbon in the hottest objects (Dziemboski and Koester 1981, Dolez and Vauclair 1981, Winget 1981, Starrfield et al. 1982, and Winget et al. 1982a). Part of the great (and relatively recent) success of this program was the search for, and discovery of, the DB variables (by Winget et al. 1982b). The existence of these variables was predicted by theory. This is the first class of variable star not to have been found by accident.

The calculations that test for stability for g-modes have been remarkably successful for the DAV and DBV stars and the results agree reasonably well with the observed location of their respective blue edges. Although there is some disagreement on the details of precisely how much mass is tied up in surface hydrogen or helium layers, the cause of instability is now understood, and our knowledge of the overall structure of these stars is secure.

The situation for the very hot DOV and PNN variables is not as rosy. Theoretical periods derived from adiabatic pulsation studies have no trouble matching the observed periods for these stars but the exact cause of the instability is still uncertain. The difficulty is that the evidence from spectroscopy is not clear enough to determine the precise composition of their photospheres. If this were known, then some reasonable guesses could be made to model the interior layers close to the surface. It is known, however, that helium, carbon, and oxygen are present in the photospheres and it may well be that ionization of some of these elements is sufficient to drive the star to instability (Starrfield et al. 1985 and references therein).

Another problem that arises with the hot variables is that theoretical studies indicate that these stars should be driven unstable due to shell hydrogen or helium burning left over from the previous evolutionary stages and that oscillations with periods of from 50 to 200 seconds should be seen (Kawaler 1988, Kawaler et al. 1986). These calculations are based on standard evolutionary models in which active burning shells are present. The problem is that the hot variables show no evidence for such short periods (Hine and Nather 1988). This is very disturbing and implies that either some adjustments have to be made in our evolutionary calculations or the pulsation work is incorrect. If it is with the models, then our ideas about how white dwarfs are formed from AGB stars may be flawed. Sounding out such things is one of the roles of asteroseismology.

Another role is detecting evolution in action. This has been accomplished by Winget et al. (1985, 1991). Using data spanning 10 years, they were able to detect a secular change of period in one otherwise very stable oscillation period (at 516 s) in the hot and (presumably) rapidly evolving star PG1159–035. The latest update on the rate of period change for this 516-second g-mode is $\dot{\Pi} = (-2.4 \pm 0.4) \times 10^{-11}$ s s^{-1}, which corresponds to an e-folding time ($\Pi/\dot{\Pi}$) for a decrease in period of about 10^6 years (Winget et al. 1991). This is about the same e-folding time for luminosity decrease derived from evolutionary models of PG1159–035-type stars.

Detecting secular period changes in the cooler white dwarfs is much more difficult because they cool and evolve so slowly. Kepler et al. (1991) have reported a rate of period change for a $\Pi = 215$ second mode of $\dot{\Pi} = (12 \pm 3.5) \times 10^{-15}$ s s^{-1} in the DAV star G117–B15A. This corresponds to an e-folding time of 7×10^8 years and has already ruled out a core consisting of elements much heavier than carbon or oxygen for this star (Bradley et al. 1992).

White Dwarfs and the Whole Earth Telescope

Perhaps the best way to summarize the successes of white dwarf seismology is to review what we know about the DOV star PG1159–035 discussed above and how that information was obtained. This is a case history in asteroseismology.

One of the prime difficulties met in extracting information from a variable star are the constraints placed on observation by the rotation of the earth and the seasonal aspect of the constellations. If we observe from a single telescope, then information is lost during the daylight hours and roughly half a year is lost each year since the star is not in the nighttime sky. This is a serious problem for variable white dwarfs because what is needed is *resolution* of the multiperiodic oscillation structure. As an example, consider a hypothetical variable that is pulsating in two modes whose frequencies are spaced a mere 4 μHz apart. If we were to observe this star for eight hours over only one night then there is no way that we could tell there were actually two periods present. This is because of the relation between length of observation and resolution in frequency, which we can show using the properties of Fourier transforms.

If we observe a sinusoidally periodic signal of frequency f_0 over a finite time span T, then the amplitude of the Fourier transform of that signal is *not* a delta function at f_0. Instead, we find a relatively broad peak around f_0 with a width in frequency of approximately 1/T with "sidelobes" of lower amplitude extending out on either side. This means that we may see a peak at f_0 but the uncertainty in the exact location of the frequency of that peak is perhaps as large as 1/T. (Heisenberg would be amused. This is just another version of the uncertainty principle between time and energy as applied to astronomical observations.) Thus if we observe for eight hours, the uncertainty in frequency is 1/(8 hours) or 3.5×10^{-5} Hz, or 35 μ Hz, which is far larger than the 4-μHz spacing between our two hypothetical modes. The net result is that we would not be able to resolve the two peaks in the Fourier transform and thus would not be able to tell there were two different signals present. The way around this is, obviously, to observe the star long and continuously. And this gets us back to PG1159–035.

Winget et al. (1991) report observations of PG1159–035 taken nearly continuously around the 24-hour clock for a period of two weeks using the "Whole Earth Telescope" (WET). This remarkable instrument consists of up to (depending on the circumstances) 13 individual telescopes, with cooperating observers in attendance, spaced around the world in longitude whose duty it is to observe, during their individual nighttimes, a single variable white dwarf as the earth turns. You might say that even though the earth turns, the telescope doesn't. The data gathered from the high-speed photometers is relayed by electronic mail to a single control site (presently in Austin, Texas) where the information is analyzed in almost real time. The operation of the WET is reported in Nather et al. (1990).

The WET was used for two weeks of "dark time" (no moon) in 1989 to observe PG1159–035 almost continuously. A small sample—only about seven hours out of a total of 264 hours of data—of the light curve from that star is shown in the insert of Figure 10.9, where the ordinate is the relative intensity, in visible light, around the mean. We have pictured this as a continuous curve but it really consists of about 2400 individual points spaced 10 seconds apart, which is the integration time for the photometers.

If you were to take a ruler and measure off a rough spacing between individual swings of data in the insert you would find that, on the whole, they represent a sinusoid-like signal, or signals, with a period of about 500 seconds. This is verified by the main curve in the figure, which shows the Fourier transform of the entire 264-hour data set for a very small interval in frequency. The largest peak (in power or amplitude squared) at $f \approx 1937$ μHz represents a single g-mode oscillation whose period is 516.04 seconds (and is the same one whose rate of change of period has been measured). The peak to the left of this one is lower in frequency by 4.3 μHz. These peaks are well-resolved because the two-week duration of WET implies a frequency uncertainty of only 1/(2 weeks) or about 0.8 μHz. You will note that there are three well-defined and equispaced peaks in the right-hand portion of the figure. (The much smaller peaks around them are due to inevitable noise in the observations.) This triplet is due to the effects of rotation on an $\ell = 1$ g-mode showing the $2\ell + 1$ m modes. If the rotation were uniform, then equation (10.114) gives a rotation period of 1.4 days, which is typical of white dwarfs. There may be no way in the foreseeable future to detect this rotation by spectroscopic means.

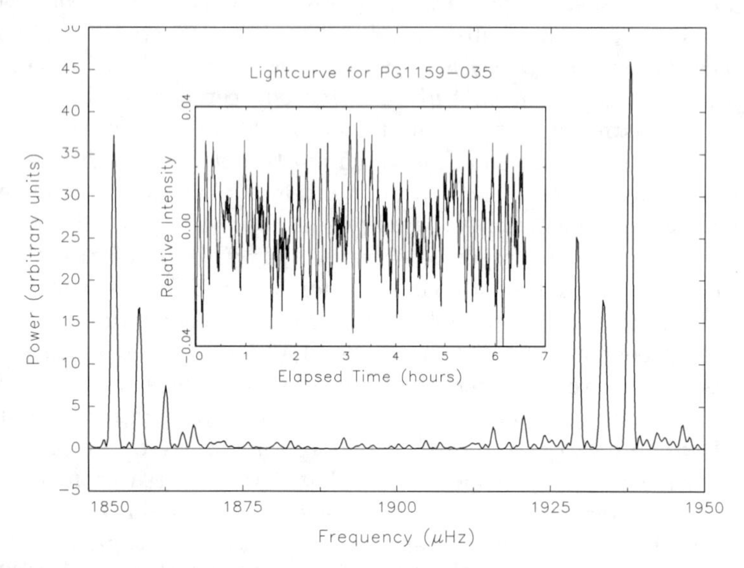

FIGURE 10.9. Shown in the insert is a small segment of the light curve of PG1159–035 from a 1989 WET run. The main frequency peaks, around periods of 500 seconds, are shown in the main figure.

Note that there is another triplet down at about 1850 μHz (periods around 538 seconds) and this is an $\ell = 1$ rotationally split g-mode. From model calculations of DO stars, this mode and its neighbor at 516 s are of harmonic order $n \approx 20$ and the two modes differ by one in n. This complex of strong peaks in the vicinity of 500 seconds is the cause of the curious structure of the light curve, which shows a modulation in intensity with wide swings and nulling superimposed on the main ups and downs of 500 seconds. The situation is analogous to the musical interference beats heard from an orchestra whose members are playing *nearly* the same notes. Here we *see* the beats.

Space prevents us from presenting the entire Fourier spectrum for the DOV star PG–1159, but Winget et al. (1991) have identified 101 modes in this star including many more with $\ell = 1$ and a number of rotationally split quintuplets for which $\ell = 2$. Because even WET observations cannot resolve the disk of the star—unlike the sun—it is unlikely that we can detect modes with $\ell \gtrsim 3$, were they to be present, because of the effects of light cancellation over the disk of the complicated patterns of high ℓ spherical harmonics. Even so, from this wealth of data, it is possible to get a period spacing between modes of given ℓ. Using the $m = 0$ (central component) as the mean for a multiplet, the average spacings between consecutive ns is $\Delta P_{\ell=1} = 21.5$ s and $\Delta P_{\ell=2} = 12.5$ s. Note that the ratio of these two spacings is consistent with (10.105), which contains the factor $[\ell(\ell+1)]^{1/2}$; that is, $\sqrt{21.5/12.7} = 1.72$ is very nearly the same as (within the error bars not quoted here) $[2(2+1)/1(1+1)]^{1/2} = \sqrt{3} = 1.732$. The significance of this result is the following. Kawaler (1986, 1988) has shown that the g-mode period spacing is sensitive to the stellar mass. For DOV models with effective temperatures within the range for PG1159–035 (140,000 K), the mass must be very close to $0.6\,\mathcal{M}_\odot$ to reproduce the period spacings observed in that star. Recall that this figure is the typical mass for single white dwarfs derived from spectroscopy. The WET data and analysis yield the mass using remote seismology.

There are, however, small and systematic deviations from the average spacing between the $\ell = 1$ modes. This is due to the presence of composition discontinuities from helium to carbon-oxygen (and perhaps hydrogen comes in also) due to gravitational settling. Since nature (and eigenfunctions) abhors discontinuities in physical properties, it is not surprising that the periods of some modes are effected in subtle and *predictable* ways. In any case, the observations strongly suggest that we are now able to probe into these stars and sample how their composition varies with depth. The effects of discontinuities on the periods of DA variable white dwarfs is reviewed in Brassard et al. (1992).

We have only touched upon what can presently be accomplished by white dwarf seismology. But it is clear we are probing the insides of these intrinsically dim stars by observing their light variations.

Diamonds in the Sky: An Extended Footnote

We already mentioned in Chapter 9 that in the very late stages of cooling, the carbon-oxygen cores of white dwarfs may crystallize. This makes our earlier analysis incomplete because we must consider how waves travel in a solid. We are not going to go through the analysis but we do point out that the subject now begins to look more like traditional terrestrial seismology where p-waves and shear waves are observed and studied. Very little work has been done for partially crystallized cool white dwarfs (see Hansen and Van Horn 1979) but for those of you who wish a taste of what might be involved we invite you to peruse McDermott et al. (1988) who studied the oscillations of neutron stars with solid crusts and liquid oceans. Instead of just p– and g-modes, that paper discusses *seven* different types of nonradial modes.

10.6 References and Suggested Readings

Introductory Remarks

The most useful textbook references to the theory of variable stars are

- ▹ Cox, J.P. 1980, *Theory of Stellar Pulsation* (Princeton: Princeton University Press)
- ▹ Unno, W., Osaki, Y., Ando, H., and Shibahashi, H. 1979, *Nonradial Oscillations of Stars* (Tokyo: University of Tokyo Press)
- ▹ Unno, W., Osaki, Y., Ando, H., Saio, H., and Shibahashi, H. 1989, *Nonradial Oscillations of Stars*, 2d ed. (Tokyo: University of Tokyo Press).

The last chapter of

- ▹ Cox, J.P. 1968, *Principles of Stellar Structure*, in two volumes (New York: Gordon and Breach)

should also be consulted. A much more difficult but classic reference is

- ▹ Ledoux, P., and Walraven, Th. 1958, in *Handbuch der Physik*, Ed. S. Flügge (Berlin: Springer-Verlag) **51**, 353.

§**10.1**: Adiabatic Radial Pulsations

Stellar pulsation theory can get very mathematical at times. A decent text, among a few of its kind, that will answer many of the questions we raise in this chapter is

- ▹ Arfken, G. 1985, *Mathematical Methods for Physicists*, 3d ed. (New York: Academic Press).

Part VII of

- ▹ Kippenhahn, R., and Weigert, A. 1990, *Stellar Structure and Evolution* (Berlin: Springer-Verlag)

discusses some of the material in our chapter but does not go into any detail about helio- or white dwarf seismology. Their §38.3 gives examples of polytropic oscillations.

- ▹ Tassoul, M., and Tassoul, J.L. 1968, *Ap.J.*, **153**, 127

discuss reduction of the LAWE to an equation that looks more like a wave equation. Their asymptotic results are more complete than ours.

- ▹ Kawaler, S.D. 1993, *Ap.J.*, **404**, 294

discusses the possibility that radial pulsations with periods of around seconds may be present in some white dwarfs.

§**10.2**: Nonadiabatic Radial Pulsations

Although Eddington did not figure out exactly what physical processes were responsible for variable stars he did understand what the thermodynamics had to do. See §134 of

▹ Eddington, A.S. 1926, *The Internal Constitution of the Stars* (Cambridge: Cambridge University Press).

The analysis we quote is also discussed in §5–10 of

▹ Clayton D. D. 1968, *Principles of Stellar Evolution and Nucleosynthesis* (New York: McGraw-Hill).

Numerical investigation of weight functions for radial pulsations was originally reported by

▹ Epstein, I. 1950, *Ap.J.*, **112**, 6.

The tough and important problem of coupling pulsation to convection has not really improved since the review by

▹ Toomre, J. 1982, in *Pulsations in Classical and Cataclysmic Variable Stars*, Eds. J.P. Cox and C.J. Hansen, JILA publication (Boulder, CO), p. 170.

One-dimensional hydrodynamic calculations of radially variable stars have been done since the early 1960s. Recently, however, these have been extended to look into questions of chaos, etc. Reviews and examples may be found in

▹ Buchler, J.R. 1990, *Nonlinear Astrophysical Fluid Dynamics*, Vol. 117 of *Annuals of the New York Academy of Sciences*, p 17

▹ Kovács, G., and Buchler, J.R. 1988a, *Ap.J.*, **334**, 971

▹ *Ibid.* 1988b, *Ap.J.*, **324**, 1026.

§**10.3**: An Introduction to Nonradial Oscillations

There are several good texts on fluid dynamics. One of our favorites is

▹ Landau, L.D., and Lifshitz, E.M. 1959, *Fluid Mechanics* (London: Pergamon).

Our main references for nonradial oscillations use Eulerian perturbations.

▹ Pesnell, W.D. 1990, *Ap.J.*, **363**, 227

discusses the use of a Lagrangian formalism which, under some circumstances, is numerically superior when nonadiabatic calculations are being done.

The Cowling approximation is more than a pedagogical tool. It reduces computational labor (and is used in the pulsation code of Appendix C) without introducing gross errors in many circumstances. See

▹ Cowling, T.G. 1941, *M.N.R.A.S.*, **101**, 367.

The choice of phase for the spherical harmonics is liable to lead to confusion. We choose that of

▹ Jackson, J.D. 1975, *Classical Electrodynamics*, 2d ed. (New York: Wiley & Sons).

You might be able to get a better idea of what nonradial motions look like by consulting

▹ Smeyers, P. 1967, *Bull. Soc. Roy. Sci. Liège*, **36**, 35.

Asymptotic methods for nonradial modes can involve some difficult mathematics. Besides Unno et al., you might also try

▹ Tassoul, M. 1980, *Ap.J. Suppl.*, **43**, 469.

▹ Dziembowski, W. 1971, *Acta Astron.*, **21**, 289

is one of a series of pioneering papers by Dziembowski. Many of us still use the variables he introduced. See, for example,

▹ Osaki, Y., and Hansen, C.J. 1973, *Ap.J.*, **185**, 277.

§10.4: Helioseismology

Some seminal papers in helioseismology include:

▹ Leighton, R.B., Noyes, R.W., and Simon, G.W. 1962, *Ap.J*, **135**, 474

▹ Ulrich, R.K. 1970, *Ap.J.*, **162**, 993

▹ Leibacher, J.W., and Stein, R.F. 1972, *Ap.J.*, **L7**,191.

We recommend the reviews

▹ Christensen-Dalsgaard, J., Gough, D.O., and Toomre, J. 1985, *Science*, **229**, 923

▹ Leibacher, J., Noyes, R.W., Toomre, J., and Ulrich, R.K. 1985, *Sci. Am.*, **253**, 48

▹ Wentzel, D.G. 1991, *Mercury*, **20**, No. 3, 77 (May/June)

▹ Libbrecht, K.G., and Woodard, M.F. 1991, *Science*, **253**, 152

▹ Brown, T.M., Mihalas, B.W., and Rhodes, E.L. Jr. 1986, in *Physics of the Sun*, Vol. I, Ed. P.A. Sturrock (Dordrecht: Reidel), p.177

▹ Toomre, J. 1986, in *Seismology of the Sun*, Ed. D.O. Gough (Boston: Reidel), p.1.

▹ Libbrecht, K.G. 1988, *Sp. Sci. Rev.*, **47**, 275.

Figure 10.3 is from

▹ Libbrecht, K.G. 1988, *Sp. Sci. Rev.*, **47**, 275

who uses data from

▹ Duvall, T.L., Harvey, J.W., Libbrecht, K.G., Popp, B.D., and Pomerantz, M.A. 1988, *Ap.J.*, **324**, 1158.

See also

▹ Libbrecht, K.G., Woodard, M.F., and Kaufman, J.M. 1990, *Ap.J. Suppl.* **74**, 1129.

Figure 10.6 illustrating ray paths is from

▹ Toomre, J. 1984, in JPL Publication 84-84: *Solar Seismology from Space*, Eds. R.K. Ulrich, J.W. Harvey, E.J. Rhodes, Jr., and J. Toomre, (NASA: JPL), p. 7.

This volume, incidentally, has several well–writen papers that discuss the (at the time, potential) uses of helioseismology for other areas in solar and stellar physics.

Figure 10.7 is from

▷ Christensen-Dalsgaard, J. 1991, in *Challenges to Theories of the Structure of Moderate–Mass Stars*, Ed. D. Gough and J. Toomre (Berlin: Springer), p. 11.

There have been several tabulations of solar p-mode frequencies such as Duvall et al. (1988). For lower-order modes see

▷ Duvall, T.L. Jr., Harvey, J.W., Libbrecht, K.G., Popp, B.D., and Pomerantz, M.A. 1988, *Ap.J.*, **324**, 1158.

We have already discussed solar neutrinos in Chapter 8.

▷ Bahcall, J., and Ulrich, R.K. 1988, *Rev. Mod. Phy.*, **60**, 297

tie these in with solar p-mode studies.

Derivations of the nonradial wave equation, among other important items, may be found in

▷ Chandrasekhar, S. 1964, *Ap.J.*, **139**, 644

▷ Lynden-Bell, D., and Ostriker, J.P. 1967, *M.N.R.A.S.*, **136**, 293.

One version of an integral formulation of the equation is given by

▷ Kawaler, S.D., Hansen, C.J., and Winget, D.E. 1985, *Ap.J.*, **295**, 547.

The effects of both uniform and cylindrically symmetic slow rotation on nonradial frequencies is discussed in

▷ Hansen, C.J., Cox, J.P., and Van Horn, H.M. 1977, *Ap.J.*, **217**, 151

and, for an easy introduction to rotational splitting, see

▷ Cox , J.P. 1984, *Publ. Astron. Soc. Pacific*, **96**, 577.

The inferred internal rotation curves for the sun of Figure 10.8 are from

▷ Schou, J. 1991, in *Challenges to Theories of the Structure of Moderate–Mass Stars*, Ed. D. Gough and J. Toomre (Berlin: Springer), p. 84.

§**10.5**: Asteroseismology of White Dwarfs

The initial discovery of the variable white dwarfs was reported in

▷ Landolt, A.U. 1968, *Ap.J.*, **153**, 151.

That these variables are the most common in (at least) our galaxy has been discussed by

▷ McGraw, J.T. 1977, *The ZZ Ceti Stars: A New Class of Pulsating White Dwarfs*, Ph.D. Dissertation, University of Texas, p. 228

▷ Cox, J.P. 1982, *Nature*, **299**, 402.

The dissertation by McGraw reports the first comprehensive study of these stars.

Useful reviews include:

▷ Winget, D.E., and Fontaine, G. 1982, in *Pulsations in Classical and Cataclysmic Variable Stars*, Eds. J.P. Cox and C.J. Hansen (Boulder: Joint Institute for Laboratory Astrophysics)

▷ Cox, A.N. 1986, *Highlights of Astronomy*, Ed. J.P. Swings (Dordrecht; Reidel)

▹ Winget, D.E. 1988, in *Advances in Helio- and Asteroseismolgy*, Eds. J. Christensen-Dalsgaard and S. Frandsen (Reidel: Dordrecht), 305

▹ Kawaler, S.D., and Hansen, C.J. 1989, in IAU Colloqium 114, *White Dwarfs*, Ed. G. Wegner (Berlin: Springer–Verlag), p. 97

▹ Bradley, P.A. 1993, *Theoretical Asteroseismology of White Dwarf Stars: The Encyclopedia Seismologica*, Ph.D. Dissertation, University of Texas at Austin.

The width of the ZZ Ceti instability strip is discussed in

▹ Greenstein, J.L. 1984, *Pub. Astron. Soc. Pac.*, **96**, 62

▹ Weidemann, V., and Koester, D. 1984, *Astron. Ap.*, **132**, 195

▹ Fontaine, G., Bergeron, P., Lacombe, P., Lomontange, R., and Talon, A. 1985, *A.J.*, **90**, 1094

▹ Bergeron, P., Wesemael, F., and Fontaine, G. 1992, *Ap.J.*, **387**, 288.

Temperatures for the DBVs were established by

▹ Koester, D., Vauclair, G., Dolez, N., Oke, J.B., Greenstein, J.L., and Weidemann, V. 1985, *Astron. Ap.*, **149**, 423

▹ Liebert, J., Wesemael, F., Hansen, C.J., Fontaine, G., Shipman, H.L., Sion, E.M., Winget, D.E., and Green, R.F. 1986, *Ap.J.*, **309**, 241

▹ Theji, P., Vennes, S., and Shipman, H.L. 1991, *Ap.J.*, **370**, 355.

But perhaps not all DAs in the strip pulsate. See

▹ Kepler, S.O., and Nelan, E.P. 1993, *A.J.*, **105**, 608.

The prototype of the DOVs was discovered by

▹ McGraw, J.T., Starrfield, S.G., Liebert, J., and Green, R.F. 1979, in *IAU Colloquium 53, White Dwarfs and Variable Degenerate Stars*, Eds. H.M. Van Horn and V. Weidemann (Rochester: University of Rochester), p. 377

and, for K1-16, see

▹ Grauer, A.D., and Bond, H.E. 1984, *Ap.J.*, **277**, 211.

The essentials of the arguments for the g-mode character of the white dwarf variables was laid down by

▹ Chanmugam, G. 1972, *Nature Phys. Sci.*, **236**, 83

▹ Warner, B., and Robinson, E.L. 1972, *Nature Phys. Sci*, **234**, 2.

Shortly after the first numerical experiments were performed by

▹ Osaki, Y., and Hansen, C.J. 1973, *AP.J.*, **185**, 277.

The cause of variablity of the DAV and DBV variables was established by

▹ Dziemboski, W., and Koester, D. 1981, *Astron. Ap.*, **97**, 16

▹ Dolez, N., and Vauclair, G. 1981, *Astron. Ap.*, **102**, 375

▹ Winget, D.E. 1981, *Gravity Mode Instabilities in DA White Dwarfs*, Ph.D. Dissertation, University of Rochester, Rochester, N.Y.

▹ Starrfield, S.G., Cox, A.N., Hodson, S., and Pesnell, W.D. 1982, in *Pulsations in Classical and Cataclysmic Variable Stars*, Eds. J.P. Cox and C.J. Hansen (Boulder: Joint Institute for Laboratory Astrophysics) p. 46

▹ Winget, D.E., Van Horn, H.M., Tassoul, M., Hansen, C.J., Fontaine, G., and Carroll, B.W. 1982a, *Ap.J. Letters*, **252**, L65.

The discovery of the first DBV is reported in

▹ Winget, D.E., Robinson, E.L., Nather, R.E., and Fontaine, G. 1982b, *Ap.J. Letters*, **262**, L11.

Causes of instabilty in the DOV variables is discussed in

▹ Starrfield, S., Cox, A., Kidman, R., and Pesnell, W.D. 1985, *Ap.J. Letters*, **293**, L23.

All the theoretical papers cited above use some version of the nonadiabatic methods of

▹ Saio, H., and Cox, J.P. 1980, *Ap.J.*, **236**, 549

to search for instability.

Possible shell-burning instabilities in the DOV's are reviewed in

▹ Kawaler, S.D., Winget, D.E., Hansen, C.J., and Iben, I. Jr. 1986, *Ap.J. Letters*, **306**, L41

▹ Kawaler, S.D. 1988, *Ap.J.*, **334**, 220

but, for a null search for variables with the requisite periods, see

▹ Hine, B.P., and Nather, R.E. 1988, in IAU Colloquium 95, *The Second Conference on Faint Blue Stars*, Eds. A.G.D. Philip, D.S. Hayes, and J. Liebert (Schenectady: L. Davis Press), p 627.

The very time-consuming measurements of the secular drift of period in one of the modes of PG-1159 are reported in

▹ Winget, D.E., Kepler, S.O., Robinson, E.L., Nather, R.E., and O'Donoghue, D. 1985, *Ap.J.*, **292**, 606

▹ Winget et al. 1991, *Ap. J.*, **378**. 326.

Corresponding measurements for G117-B15A are given by

▹ Kepler, S.O., et al. 1991, *Ap.J.*, **378**, L45

and constraints on core composition for this star based on evolutionary models have been computed by

▹ Bradley, P.A., Winget, D.E., and Wood, M.A. 1992, *Ap.J.*, **391**, L33.

Observations of PG1159 by the Whole Earth Telescope are reported in

▹ Winget et al. 1991, *Ap. J.*, **378**. 326

and the operation of that instrument is described by

▹ Nather, R.E., Winget, D.E., Clemens, J.C., Hansen, C.J., and Hine, B.P. 1990, *Ap.J.*, **361**, 309.

Figure 10.9 is derived from the original 1989 WET data for PG1159.

▹ Kawaler, S.D. 1986, Ph.D. Dissertation, University of Texas

▷ Kawaler, S.D. 1988, in IAU Symposium 123, *Advances in Helio– and Asteroseismology*, Eds. J.Christiansen–Dalsgaard and S. Fransden (Dordrecht: Reidel), p 329

describe how period spacing is related to stellar mass. Layering of elements and the resulting deviations in period spacings for DAVs is discussed in

▷ Brassard, P., Fontaine, G., Wesemael, F., and Hansen, C.J. 1992, *Ap.J.*, **80**, 369.

▷ Hansen, C.J., and Van Horn, H.M. 1979, *Ap.J.*, **233**, 253

▷ McDermott, P.N., Van Horn, H.M., and Hansen, C.J. 1988, *Ap.J.*, **325**, 725

discuss the effects of solid material on nonradial oscillations in white dwarfs and neutron stars.

Appendix A
Glossary

This short glossary of elementary astronomical terms associated with stars is not intended to be complete or very detailed. It is meant mostly for those of you who have no earlier experience in the subject. The only terms we list are those not specifically treated in the main text. An excellent overall reference to this material is Mihalas, D., and Binney, J. 1981, *Galactic Astronomy*, 3d ed. (San Francisco: Freeman & Co.).

1. **Stellar populations**: These are useful shorthand designations for stars sharing common properties of kinematics, location in a galaxy, and composition.
   (a) **Population I** stars have a small scale height (confined to the disk of a spiral galaxy), rotate with the disk, generally have a surface composition not too different from the sun's, and have a large range of masses since the young ones are still on the main sequence.
   (b) **Population II** stars usually have a very large scale height (mostly found in the halo of a spiral galaxy), high space velocities, are poor in metals, and are of low mass since the more massive stars have already evolved. Hence Pop II stars are old stars.

2. **Star clusters**: These are useful since they are generally composed of many stars of roughly the same composition and age. The turn-off point from the main sequence provides an age when compared to models or relative ages when one cluster is compared to another.

(a) **"O–B" associations** consist of a loose cluster dominated in light by bright stars that are still associated with the interstellar gas that begat them. The cluster is not gravitationally bound and the stars are associated only because of their tender youth.

(b) **Open clusters** or **galactic clusters** are disk (Pop I) stars bound by gravitation. They contain both massive and low-mass stars. They are conspicuous because of the bright massive stars. Compared to globular clusters they have fewer stars in total number. The Pleiades is a conspicuous open cluster in the northern night sky.

(c) **Globular clusters** are gravitationally tightly bound and contain many low-mass Pop II stars. They are associated with the halo of a galaxy and were presumably formed early in the history of the galaxy.

3. **Observation of stars—Photometry**: Photometry refers to observing stars over one or more wavelength bands where details of the spectrum are not necessarily important.

(a) The **magnitude scale** is an astronomical scale for brightness constructed along the same lines as the decibel scale for sound. It is logarithmic, as are all such scales, so that the mind can handle the broad dynamic range of real external stimuli. There are two main magnitude scales. (Note that in what immediately follows we assume that the stars in question are observed over the same band of wavelengths or colors.)

i. **Apparent magnitude**, m, is the brightness of a star as observed from earth and is thus a function of the intrinsic brightness of the star and its distance. Unlike the decibel scale it runs backwards: the larger the magnitude, the apparently dimmer the star. An arithmetic *difference* of 5 in magnitude means a *multiplicative* factor of 100 in brightness (say apparent luminosity). If b_i is the apparent brightness in physical units of star i then the rule is

$$\frac{b_1}{b_2} = 2.512^{m_2 - m_1} .$$

A rough guide to apparent magnitudes is that a star with magnitude 6 is just visible to the naked eye while stars of magnitude 0 are among the brightest in the sky.

ii. The **absolute magnitude**, M, of a star is the apparent magnitude the star would have if we placed the star at a standard distance of 10 parsecs (3.086×10^{19} cm). If absorption of light by intervening gas or dust may be ignored, then the relation between apparent and absolute magnitude is

$$M = m + 5 - 5 \log d$$

where d is the actual distance to the star in parsecs. $m - M$ is called the distance modulus.

iii. The **bolometric magnitude** is the magnitude integrated over all wavelengths (i.e., the entire electromagnetic spectrum). Since absolute magnitudes are standardized by the common distance 10 parsecs there must be a relation between luminosity and absolute bolometric magnitude, $M_{\rm bol}$. Using the sun as a normalization this relation for a star ($\star$) is

$$\log(\mathcal{L}_\star/\mathcal{L}_\odot) = [M_{\rm bol}(\odot) - M_{\rm bol}(\star)]/2.5$$

where $M_{\rm bol}(\odot)$ is +4.75.

(b) **Colors** of stars are a reflection of the relative dominance of various wavelengths in their spectra and hence their effective temperatures. Observational magnitudes are always quoted for some range of wavelengths, which is standardized. An example of a standardized system is the Johnson–Morgan UBV system. The U filter lets in a band of light centered around 3650Å in the ultraviolet. The B (blue) and V (visual) filters are centered around 4400Å and 5500Å respectively. The magnitudes in these wavelength ranges are usually denoted by their letters. Thus the absolute magnitude of the sun in the V range of the UBV system is $V = -26.7$. The "color" of a star can be described by the difference in brightness in two filters. Thus, for example, and keeping in mind the backward scale for magnitudes, $(B - V)$ is positive for blue (hot) stars and negative for red (cool) stars.

(c) **Bolometric correction**: To obtain the bolometric magnitude you must assume an energy distribution as a function of wavelength to find the spectrum at unobserved wavelengths (as, e.g., in the far ultraviolet for surface-based telescopes). Since the color, as well as the energy distribution, is related to the temperature, given the color you can estimate the correction to a magnitude to give the bolometric magnitude. Specifically, the bolometric correction, $B.C.$, is defined as

$$M_{\rm bol} = M_V + B.C.$$

where M_V is the absolute visual magnitude (usually V).

(d) **Color-magnitude diagram**: This is the observer's version of the Hertzsprung-Russell diagram (as in Figs. 2.10–2.11 and 9.1). The abscissa is color ($B - V$, for example) and the ordinate is some magnitude (V, M_V, etc.). Color is the effective temperature surrogate and the single magnitude plays the role of luminosity. Needless to say, there is a lot of witchcraft involved in passing back and forth between these variables.

(e) **Time-resolved photometry** is what the name implies. Snapshots are taken of the star to get magnitudes or intensities over intervals of time. The resulting times series is then used to infer dynamic properties such as those seen in variable stars.

4. **Observations of stars—Spectroscopy**: Here the details are important and one or more spectral absorption or emission lines are used for diagnostic purposes.

   (a) **Spectral types or classes**: This refers to the Henry Draper classification scheme, which is based on the appearance of spectral lines of hydrogen, helium, and various metals. An "O" star shows strong HeII (second ionization) lines of helium. Since the ionization potential is high, such stars are the hottest in terms of effective temperature. Spectral type "B" stars show strong HeI plus some HI lines. The Balmer lines reach their peak in the "A" stars and begin to disappear in "F." Calcium H&K lines strengthen in "G" and the "K" stars show other metallic lines. The "M" stars have strong molecular bands (particularly TiO). The effective temperatures decrease in the order O, B, A, F, G, K, M. The historical reasons for the lettering of this sequence are worth looking up in the literature. Subdivisions such as G0, G1, G2, $\cdots$ are in order of decreasing temperature. The sun is a G2 spectral class star.

   (b) **Luminosity classes** were introduced because stars of the same spectral type may have very different luminosities. The Morgan-Keenan (MK) classification uses spectral class and luminosity class as a two-dimensional scheme to phrase the H–R diagram another way. The implementation of the scheme depends on details of spectral lines (such as width) and reflects the density and gravity of the photosphere. Luminosity class I stars are the most intrinsically luminous. Because they are usually large (especially the cooler ones) they are also called **supergiants**. Luminosity class II (**giants**) are intrinsically bright, but not as bright as class I. The numbers increase until luminosity class V is reached. These are the **dwarfs** and primarily refer to main sequence stars. The sun is a G2V star. An M3I star is huge, intrinsically very luminous, and very cool. There are some intermediate classes such as Ia and Ib. White dwarfs are not included in this luminosity scheme.

(c) **Abundances** are derived from spectral lines and are usually compared to the sun or the abundance of hydrogen in the star—or both. Thus the number density of iron, $n(\mathrm{Fe})$, in a star ($\star$) as compared to the sun might be expressed as

$$[\mathrm{Fe/H}] = \log\left[n(\mathrm{Fe})/n(\mathrm{H})\right]_\star - \log\left[n(\mathrm{Fe})/n(\mathrm{H})\right]_\odot .$$

As an example, the metal–poor and old globular cluster M92 has $[\mathrm{Fe/H}] = -2.2$. Since the surface hydrogen abundance of the sun is probably not too different from stars in M92, the conclusion is that the sun's surface abundance of iron is roughly 100 times greater.

Appendix B
Physical and Astronomical Constants

Listed below are some useful physical and astronomical constants derived from various sources. The prime reference for the physical constants is Cohen, E.R., and Taylor, B.N. 1987, *Rev. Mod. Phys.*, **59**, 1121 (who use MKS units). In some cases we have rounded the values.

speed of light in vacuum: $c = 2.99792458 \times 10^{10}$ cm s^{-1}

Newtonian constant of gravitation: $G = 6.6726 \times 10^{-8}$ g^{-1} cm^3 s^{-2}

Planck's constant: $h = 2\pi\hbar = 6.6260755 \times 10^{-27}$ erg s

elementary charge: $e = 4.8032068 \times 10^{-10}$ e.s.u.

electron mass: $m_e = 9.1093898 \times 10^{-28}$ g, 5.4858×10^{-4} amu, 0.5109991 MeV c^{-2}

proton mass: $m_p = 1.6726231 \times 10^{-24}$ g, 1.00727647 amu, 938.27231 MeV c^{-2}

neutron mass: $m_n = 1.6749286 \times 10^{-24}$ g, 1.0086649 amu, 939.56563 MeV c^{-2}

Avogadro's constant: $N_A = 6.0221367 \times 10^{23}$ mole^{-1}

Boltzmann's constant: $k = 1.380658 \times 10^{-16}$ erg K^{-1}, 8.617386×10^{-5} eV K^{-1}

gas constant: $\mathcal{R}_{gas} = N_A k = 8.314511 \times 10^7$ erg K^{-1} mole^{-1}

Stefan-Boltzmann constant: $\sigma = 5.67051 \times 10^{-5}$ erg cm^{-2} K^{-4} s^{-1}

radiation density constant: $a = 4\sigma/c = 7.56591 \times 10^{-15}$ erg cm^{-3} K^{-4}

eV to erg conversion: 1 eV $= 1.60217733 \times 10^{-12}$ erg

amu to gram conversion: 1 amu $= 1.6605402 \times 10^{-24}$ g

solar luminosity: $\mathcal{L}_\odot = (3.847 \pm 0.003) \times 10^{33}$ erg s^{-1}

solar mass: $\mathcal{M}_\odot = (1.9891 \pm 0.0004) \times 10^{33}$ g

solar radius: $\mathcal{R}_\odot = 6.96 \times 10^{10}$ cm

solar effective temperature: $T_{\rm eff}(\odot) = 5780$ K

parsec (pc): 1 pc $= 3.086 \times 10^{18}$ cm

astronomical unit (AU): 1 AU $= 1.496 \times 10^{13}$ cm

Appendix C
Sample Computer Codes

This appendix contains documentation for the two FORTRAN computer codes called ZAMS.FOR and PULS.FOR, which are included as ASCII files on the diskette attached to the back cover of this text. The first program makes ZAMS models which, in turn, may be analyzed for radial or nonradial pulsations in the second code. Also included on the diskette are sample input and output files, executable versions of both codes for use on personal computers (PCs), and a READ.ME file which should be consulted first for further details.

C.1 The ZAMS Model Builder

The FORTRAN program ZAMS.FOR constructs homogeneous zero-age main sequence models using the fitting technique outlined in Chapter 7. It is not terribly sophisticated, but it will make very decent models. For those of you who would prefer to use either C or PASCAL the transition is not too difficult, but you will need a proficiency in FORTRAN to effect the translation. We have tested this program on personal computers (with a Microsoft compiler), UNIX workstations, and VAX VMS mainframes but cannot guarantee that your particular system will happily accept all that follows. We have made an effort to use standard (even old-fashioned) FORTRAN for portability. If you use a PC, we suggest a fast one with a math coprocessor. On some machines you may have to work in extended precision (e.g., G–float in VAX FORTRAN). In addition, we have included what we hope is sufficient documentation within the code. Note that provision is made to write output quantities that may then be used for the radial and nonradial pulsation PULS code, which is discussed in the next section. ZAMS is written in double precision to provide a smooth and accurate model for the PULS code.

The running of the code is fairly straightforward but requires guesses of initial quantities. Some hints are given later. Because we use very simple photospheric conditions and the fit to opacity (c. 1970s) does not cover very cool temperatures, we do not recommend trying to construct models with low effective temperatures. Models with masses much below that of the sun will not be very successful. Also, do not try compositions that differ substantially from Pop I because the opacity fit does not cover a wide range of composition. You are, of course, welcome to substitute your own opacities, etc., including tabulated versions. Because the code assumes models in complete equilibrium, this code cannot be used to construct even warm white dwarfs (and the equation of state has no provisions for degeneracy). You are welcome to try to change things so that it might.

TABLE C.1. Some guesses for the ZAMS model builder

$\mathcal{M}/\mathcal{M}_\odot$	P_c	T_c	$\mathcal{R}$ (cm)	$\mathcal{L}/\mathcal{L}_\odot$
1	1.482×10^{17}	1.442×10^{7}	6.932×10^{10}	0.9083
3	1.141×10^{17}	2.347×10^{7}	1.276×10^{11}	89.35
15	2.769×10^{16}	3.275×10^{7}	3.289×10^{11}	1.960×10^{4}

The program includes public domain routines from LINPACK that are used to solve systems of linear equations (in this case only a four-by-four system). If you are familiar with such routines and can identify the input and output quantities, you may substitute your own favorite programs. We also use a Runge-Kutta integrator called RKF to integrate all four stellar structure equations in this fitting program. (See the documentation before subroutine RKF.) RKF, and its associated routines, is a general-purpose integrator that works well with all but "stiff" systems of differential equations. You are again invited to use your own integrator (such as those in *Numerical Recipes* by Press et al., 1992).

There is quite a bit of output generated by this code and sample output is included on the diskette. As a function of $\mathcal{M}_r$ (actually $1 - \mathcal{M}_r/\mathcal{M}$) it will tabulate r, P, T, ρ, and $\mathcal{L}_r$, ε, κ, $\mathcal{L}_r$(conv), $\mathcal{L}_r(\text{conv})/\mathcal{L}_r(\text{tot})$, ∇, ∇_{ad}, and ∇_{rad} for 201 points in the model. With this information you should be able to figure out what makes the model tick.

Finally, we will list guesses for central pressure, central temperature, total radius, and luminosity for ZAMS models of three different masses (and see Tables 2.1 and 2.2). The composition is $X = 0.74$ and $Y = 0.24$. If the program has compiled properly (among other things), then these models should converge in one iteration of the fitting method. Note that poor guesses can be fatal. You are allowed **NTRY** attempts to converge and it is set to 15. If the code wants to take more iterations than this, then your initial guesses were probably not very good. If you are not satisfied with the composition used in Tables 2.1 and 2.2 then we advise you to creep up on a new composition by slowly adjusting X and Y away from that used in those tables and use graphical interpolation or extrapolation to obtain decent guesses.

C.2 Adiabatic Pulsation Code

The program **PULS.FOR** does an adiabatic pulsation analysis of either nonradial modes in the Cowling approximation or radial modes for the full second order system (see Chapter 10). Most of the input is from **ZAMS** where the first block of data is, in order and usual notation, $x = \ln(r/P)$, r, g, and ρ. The quantity x is the independent variable and replaces r in the pulsation equations. It is used because it "stretches" the mesh at the surface and center to allow for better zoning in integration. (See Osaki and Hansen 1973, *Ap.J.*, **185**, 277.) The second block of data from **ZAMS** consists of x, g/r, $(g/r)\ell(\ell+1)/S_\ell^2$, $(r/g)N^2$, U (of 7.48), and $(1+V)^{-1}$ (also of 7.48). These data are used as is in the nonradial part of the code but is modified (see below) for the radial portion.

The method of solution is by shooting from the center to the surface using the same **RKF** integrator used in **ZAMS**. The solution is started by guessing a value of the period, Π (in seconds), setting one eigenfunction to unity (and this is arbitrary because one normalization is arbitrary), and applying the central boundary condition to yield the value of the second eigenfunction at the center. With these starting values, the pulsation equations are then integrated until the surface is reached. Unless the eigenvalue has been guessed correctly, the outer boundary condition is not satisfied. A Newton's method is used to satisfy that boundary condition by iteration on the eigenvalue.

The non–**ZAMS** input, which you must provide for the code, is a guess for the period and an integer choice for ℓ. An ℓ of zero yields the radial case. The variables for the radial case are $y_1 = \delta r/r$ and $y_2 = \delta P/P$.

The nonradial case is a bit more complicated and uses the "Dziembowski variables" y_1 and y_2 discussed in Unno *et al.* (1989), §18.1 (but in the Cowling approximation). These are more stable for numeric computation then the horizontal and transverse displacements discussed in Chapter 10.

The output of **PULS** consists of information on convergence and, if desired, the eigenfunctions (called y_1 and y_2 in both radial and nonradial problems). For the radial case, y_1 and y_2 are printed out directly. The nonradial output is different. To obtain y_1 and y_2, you must multiply the tabulated output by $(r/\mathcal{R})^{\ell-2}$. This is not done in the code itself because for very high ℓ you get overflow problems. (The computational variables for the code are really

y_1 and y_2 divided by that factor.) In any case, the overall normalization is $y_1(\mathcal{R}) = 1$.

For guesses you might try the following. Both are for the ZAMS sun listed in the model guesses given for the **ZAMS** program. For the fundamental radial mode ($\ell = 0$ and no nodes in the y_1 eigenfunction), try a period guess of 3583 seconds. (This period, of roughly an hour, corresponds to the dynamic or pulsation time scale for the sun as discussed in Chapter 1.) For a high ℓ nonradial mode try $\Pi = 254$ and $\ell = 100$. **PULS** should respond by saying this mode has a "phase diagram mode" of -10. The minus sign means that you have found a p-mode and the 10 implies that the order of the mode is 10; i.e., a p_{10} mode. The phase diagram method of classifying modes is discussed in Unno *et al.* (1989), §17. The results for p-modes you obtain should agree to within a few seconds of those of the present-day sun. You will, however, not get *excellent* agreement using **PULS** because it is a Cowling code and does not take into account gravitational perturbations. Note also that very high overtone modes may give you some difficulty because they have many wiggles in their eigenfunctions. Experiments you might wish to try include using models of different masses and compositions and looking for modes of low frequency (periods longer than the radial fundamental). The latter are gravity modes and tend to be concentrated in the inner portions of the model. The phase diagram mode for a gravity mode is signaled by a + sign. You should, for example, find a g_1, $\ell = 1$, mode at about 35,120 seconds period in the $15\,\mathcal{M}_\odot$ model generated by the guesses given for **ZAMS**. A first overtone radial mode (one node in y_1) is at 6814 seconds for this model. Other periods for upper ZAMS models may be found in Aizenman, Hansen, and Ross (1975), *Ap. J.*, **201**, 387.

Index